Gmelin Handbook of Inorganic Chemistry

8th Edition

Gmelin Handbook of Inorganic Chemistry

8th Edition

Gmelin Handbuch der Anorganischen Chemie

Achte, völlig neu bearbeitete Auflage

Prepared and issued by

Gmelin-Institut für Anorganische Chemie
der Max-Planck-Gesellschaft
zur Förderung der Wissenschaften
Director: Ekkehard Fluck

Founded by — Leopold Gmelin

8th Edition — 8th Edition begun under the auspices of the Deutsche Chemische Gesellschaft by R. J. Meyer

Continued by — E.H.E. Pietsch and A. Kotowski, and by Margot Becke-Goehring

Springer-Verlag Berlin · Heidelberg · New York · Tokyo 1986

STAFF GMELIN HANDBOOK

D. Barthel, Dr. N. Baumann, Dr. W. Behrendt, Dr. L. Berg, Dipl.-Chem. E. Best, Dipl.-Ing. V. A. Chavizon, E. Cloos, Dipl.-Phys. G. Czack, I. Deim, Dipl.-Chem. H. Demmer, R. Dowideit, Dipl.-Chem. M. Drößmar, U. Duggen, M. Engels, Dr. H.-J. Fachmann, Dr. J. Faust, Dr. W.-D. Fleischmann, V. Frick, G. Funk, Dipl.-Ing. N. Gagel, E. Gerhardt, Dr. U. W. Gerwarth, M.-L. Gerwien, C. Gorr, Dipl.-Phys. D. Gras, Dipl.-Bibl. W. Grieser, H. Hartwig, Dipl.-Min. H. Hein, G. Heinrich-Sterzel, H.-P. Hente, H. W. Herold, U. Hettwer, Dr. I. Hinz, Dr. W. Hoffmann, Dipl.-Chem. K. Holzapfel, Dr. S. Jäger, Dr. R. Jotter, Dr. J. von Jouanne, Dipl.-Chem. W. Karl, H.-G. Karrenberg, Dipl.-Phys. H. Keller-Rudek, A. Klusch, Dipl.-Phys. E. Koch, Dipl.-Chem. K. Koeber, Dipl.-Chem. H. Köttelwesch, R. Kolb, E. Kranz, Dipl.-Chem. I. Kreuzbichler, Dr. A. Kubny, Dr. P. Kuhn, Dr. W. Kurtz, M. Langer, Dr. A. Leonard, Dipl.-Chem. H. List, H. Mathis, E. Meinhard, K. Meyer, Dr. M. Mirbach, Dr. U. Neu-Becker, K. Nöring, Dipl.-Chem. R. Nohl, Dipl.-Min. U. Nohl, Dr. W. Petz, I. Rangnow, Dipl.-Phys. H.-J. Richter-Ditten, Dipl.-Chem. H. Rieger, B. Riegert, E. Rieth, A. Rosenberger, E. Rudolph, G. Rudolph, Dipl.-Chem. S. Ruprecht, V. Schlicht, Dipl.-Chem. D. Schneider, Dipl.-Min. P. Schubert, A. Schwärzel, Dipl.-Ing. H.M. Somer, E. Sommer, M. Teichmann, Dr. W. Töpper, Dipl.-Ing. H. Vanecek, Dipl.-Chem. P. Velić, Dipl.-Ing. U. Vetter, H.-M. Wagner, Dipl.-Phys. J. Wagner, R. Wagner, Dr. E. Warkentin, Dr. C. Weber, Dr. B. Wöbke, K. Wolff, U. Ziegler

STAFF GMELIN ONLINE DATA SYSTEM

Dr. L. Kießling — Dr. R. Froböse, Dr. B. Roth
Dipl.-Min. M.-B. Cinarz, Dr. J. Fippinger, Dr. G. Friedrich, Dr. B. Fröhlich, I. Güllich, M. Klöffler, R. Lohmeyer, Dr. M. Pauluth, Dr. H.-J. Schröder, Dr. K. Schücke

Volume published on "Beryllium" (Syst. No. 26)
Main Volume: Element and Compounds — 1930

Gmelin Handbook of Inorganic Chemistry

8th Edition

Be Beryllium

Supplement Volume A1

The Element. Production, Atom, Molecules, Chemical Behavior, Toxicology

With 68 illustrations

AUTHORS Ingeborg Hinz, Karl Koeber, Irmingard Kreuzbichler, Peter Kuhn

Arnulf Seidel
Institut für Genetik und für Toxikologie von Spaltstoffen, Kernforschungszentrum Karlsruhe

EDITORS Ingeborg Hinz, Hans Karl Kugler, Joachim Wagner

CHIEF EDITOR Hans Karl Kugler

System Number 26

Springer-Verlag Berlin · Heidelberg · New York · Tokyo 1986

LITERATURE CLOSING DATE: 1985

Library of Congress Catalog Card Number: Agr 25-1383

ISBN 3-540-93534-7 Springer-Verlag, Berlin · Heidelberg · New York · Tokyo
ISBN 0-387-93534-7 Springer-Verlag, New York · Heidelberg · Berlin · Tokyo

Typesetting, printing, and bookbinding: Universitätsdruckerei H. Stürtz AG, Würzburg

Preface

The present volume is the first in a series of supplement volumes to the beryllium volume which appeared in 1930. This volume "Beryllium" Supplement Volume A 1 is divided into the following chapters:

1. The Production of Beryllium
2. Uses
3. Nuclides
4. Atoms and Ions
5. Molecules
6. Chemical Reactions
7. The Chemical Behavior of Be^{2+} in Solution
8. Toxicology of Beryllium

Chapter 1 describes the steps from ore dressing to obtaining the metal and then further refining and preparing special forms. No differentiation is made between processes performed on an industrial scale and a laboratory scale.

In Chapter 2 are shown various uses, taken from review literature, of Be as a metal, in alloys, and in compounds.

Chapter 6 presents the reactions of Be metal with various elements and compounds. In the section on the reactions with metals is included its behavior in binary metal systems (e.g. diffusion).

In Chapter 7 the behavior of Be^{2+} in solution is limited to hydration, hydrolysis, and a short survey of the analytically most important precipitation reactions. The complex chemical behavior will be described in detail later in a special volume.

The crystallographic and physical properties, and the electrochemical behavior will be treated in a later volume of the series "Beryllium" Supplement A.

Frankfurt/Main, June 1986 — Hans Karl Kugler

Table of Contents

Page

Page

Page

Page

Page

Beryllium

Atomic Number 4 **Atomic Weight 9.012**

Beryllium is a steel-gray, very hard, light metal. It is brittle at ordinary temperatures and ductile at red heat. It has a density of 1.8477, a melting point of 1285 °C, a boiling point of 2477 °C. Its electrical conductivity is 1/12 of that of copper. Beryllium has a metallic luster in dry air at ordinary temperatures and the powder burns on heating to form BeO. In H_2O, Be is coated with a thin skin of hydroxide which has a solubility product of 2.7×10^{-19}. Aqueous solutions of beryllium salts are colorless and have a sweet taste.

Beryllium possesses some special chemical properties due to the "diagonal relation" in the periodic system. It occupies the first position of Group II A and is therefore formally an alkaline earth metal. However, in many respects beryllium resembles aluminium more than magnesium. $BeCl_2$, like $AlCl_3$, is very sensitive to hydrolysis and is sublimable. It forms covalent type bonds, whereas alkaline earth chlorides are salt-like. The high-temperature modification of BeO, like that of Al_2O_3 is an extremely hard, high melting, nonvolatile covalent oxide that is insoluble in acids and bases. Both oxides also exist in low-temperature forms which are soluble in acids and bases. The other alkaline earth oxides are soluble only in acids. Both $Be(OH)_2$ and $Al(OH)_3$ are amphoteric; they are soluble in acids and bases and do not react with CO_2 to form stable carbonates as do the typical alkaline earth hydroxides. Beryllium and aluminium react with acids and bases to produce hydrogen. Beryllan (BeH_2) exists as a nonvolatile polymeric mass, while MgH_2 has ionic lattice. $BeCl_2$ molecules are linear and polymerize to chains. $BeCO_3$ readily decomposes to CO_2 and BeO, and is stable only in a CO_2 atmosphere.

^{9}Be is the only natural isotope of beryllium. The other isotopes are ^{7}Be, ^{10}Be, ^{11}Be, and ^{12}Be, with half-lives of 53 d, 1.6×10^{6} a, 14 s, and 24 ms, respectively.

Beryllium is one of the rarest alkaline earth metals. The earth's crust contains only 5×10^{-4} wt% Be. Beryllium is concentrated in some minerals; the most important ore is beryl, $Be_3Al_2[Si_6O_{18}]$. A well known variety of beryl is the gem stone emerald; a blue variety is known as aquamarine. Another important beryllium mineral is bertrandite $Be_4[(OH)_2Si_2O_7]$.

The industrially interesting minerals are silicates, which require a difficult ore decomposition to recover the Be. Some important separation processes include: chlorination (ore mixed with carbon is heated in chlorine), fluorination (mainly with Na_2SiF_6), the alkaline methods, and the sulfate process (decomposition with H_2SO_4 after heat or alkaline pretreatment). These processes produce compounds (e.g., $BeCl_2$ or BeF_2) which are reduced to beryllium metal by electrolysis or with metals, especially magnesium. The resulting beryllium metal is fabricated by powder metallurgy techniques.

Because of its low atomic number, low atomic weight, low density, and high melting point, beryllium metal is suitable for numerous purposes. Its low X-ray absorption makes beryllium foil an ideal material for windows in X-ray tubes. Its low density and great hardness make it a suitable material for aeronautical and astronautical construction. However, because of its high price, it is normally used only for military purposes. Its once promising future in the nuclear field as a moderator has not developed due to certain difficulties encountered in practice. Recently, beryllium single crystals have also been used as neutron monochromators.

Beryllium has great industrial importance as an alloying metal. The main Be-rich alloy is Lockalloy (Be with 38% Al). Beryllium is most often used as an addition metal in copper alloys. In amounts of 2 to 3% Be, it increases the hardness of Cu fivefold, the strength sevenfold, and the breaking strength and the bending strength threefold without decreasing the high electrical conductivity. The dust and vapor of beryllium and its compounds are very toxic. Therefore, its applications are limited.

1 The Production of Beryllium

General References:

K. A. Walsh in: D. R. Floyd, J. N. Lowe, Beryllium Science and Technology, Vol. 2, Plenum, New York - London 1979, pp. 1/11, 3/7.

D. W. White, J. E. Burke, The Metal Beryllium, Cleveland, Ohio, 1955, pp. 63/123.

G. E. Darwin, J. H. Buddery, Metallurgy of the Rarer Metals-7: Beryllium, Butterworth, London 1960, pp. 1/392, 4/36.

Kirk-Othmer, Encycl. Chem. Technol. 3rd Ed. **3** [1978] 803/23, 806/9.

W. D. Jamrack, Rare Metal Extraction by Chemical Engineering Techniques, Vol. 2, Pergamon, Oxford - London - New York - Paris 1963, pp. 1/360, 30/2, 38, 62/4, 342/6.

W. Schreiter, Seltene Metalle, Band 1: Beryllium, Bor, Cäsium, Gallium, Germanium, Hafnium; 2nd Ed., V.E.B. Deutscher Verlag für Grundstoffindustrie, Leipzig 1963, pp. 1/343, 107/29.

P. S. Bryant, Extraction and Refining of the Rarer Metals, The Institution of Mining and Metallurgy, London 1957, pp. 310/22.

S. I. Pol'kin, Flotatsiya Rud Redkikh Metallov i Olova, Flotation of Ores of Rare Metals and Tin, Pt. II, Chapter IV, Moscow 1960, pp. 386/407.

D. A. Everest, The Chemistry of Beryllium, Elsevier Publishing Company, Amsterdam - London - New York 1964, Chapter 8, pp. 102/16.

1.1 Mineral Dressing

The main source of beryllium in commercial deposits is beryl, $Be_3Al_2[Si_6O_{18}]$. Beryl is generally found in pegmatites or micaceous greisens and is therefore mainly associated with feldspar, quartz, and micas. Other sources of beryllium are bertrandite, $Be_4(OH)_2Si_2O_7$, and minor minerals such as chrysoberyl, Al_2BeO_4, phenakite, Be_2SiO_4, barylite, $BaBe_2Si_2O_7$, and helvite, $((Mn, Fe, Zn)_4Be_3Si_3O_{12}S)$. These minor minerals, however, rarely occur in workable deposits. The most important method of concentration is flotation, but other methods such as picking, electrostatic and magnetic separation, and separation with heavy liquids have also been employed.

1.1.1 Picking

All commercial supplies of beryl are at present obtained by hand picking because no satisfactory mechanical or flotation process has been developed. The minimum size of crystals suitable for hand picking is 3/8 inch ($\approx$1 cm). Thus, beryl mining is largely confined to coarse-grained zoned pegmatites containing at least 1% beryl, and the amount of beryl recovered is estimated at not more than 30% of that in the pegmatite [1].

Automatic picking utilizes the nuclear reaction $Be^9(\gamma,n)Be^8$ which has the lowest γ-threshold (1.63 MeV) of all (γ,n) reactions. Because the next threshold is $^2D(\gamma,n)^1H$ at 2.2 MeV, all γ-energy between 1.63 and 2.2 MeV is considered to be specific for beryllium. Thus, beryl bearing ore is passed under a suitable γ-source and sorted by a sorting arm which is operated by a neutron counter at a sufficiently high rate of neutron emission. The arm is able to pick up pieces weighing $\geqq$1 g at a rate of 5 per second and allows the concentration of beryl:quartz mixtures from 1:80 to 1:10 [2], see also [3, 4]. The method is limited by the fact that there is a lower limit of feed size and that milling and flotation are required for large deposits containing low-grade beryl [1].

References:

[1] G. E. Darwin, J. H. Buddery (Beryllium, Butterworths, London 1960, pp. 4/5).
[2] A. M. Gaudin, J. Dasher, J. H. Pannell, W. L. Freyberger (Mining Eng. **2** [1950] 495/8).
[3] J. H. Pannel, W. L. Freyberger (MITG-224 [1949] 1/18; N.S.A. **4** [1950] No. 576).
[4] A. M. Gaudin, U.S. Atomic Energy Commission (U.S. 2707555 [1950/55] 1/4; N.S.A. **9** [1955] No. 700).

1.1.2 Electrostatic Separation

A beryl containing pegmatite with 0.46 wt% Be was crushed, pulped, and conditioned with an aqueous solution of NaOH (≈0.45 kg/t ore). Water was added to provide a pulp density of 25 to 75% solids. After 5 to 30 min of agitation, the mass was drained, washed with water, and drained again. This procedure was repeated four times after which aqueous HF (≈0.45 kg/t ore) was added. After draining, the ore was washed with water containing a soap (226 g/t) with a high content of stearic, palmitic, and napthenoic acid (pH 6 to 7) until the wash water attained the same pH range. After the final draining, the ore was dried at a temperature below 100 °C and subjected to electrostatic separation at 50 °C in an atmosphere of 30% relative humidity. The ore was then passed through a four roll collector with negative electrodes to yield a rough concentrate which was then repassed through a six roll collector with positive electrodes. The process yielded 3.5 wt% of a recleaned concentrate containing 11.4% BeO resulting in the recovery of 88% of the beryl content of the initial ore. Similar treatment of a pegmatite containing 0.86% BeO yielded 7.7 wt% of a recleaned concentrate containing 10.0% BeO. When a twofold excess of the conditioning reagents was used, the BeO content of the concentrate increased only slightly to 10.7%, F. Fraas (U.S. 2769536 [1953/56] 1/6; C.A. **1957** 2514).

1.1.3 Separation by Heavy Liquids

The beryl content of a crushed (<9.5 mm) Australian (Wodgina) pegmatite (4.35 wt% BeO) was 80 to 90% recovered in a concentrate containing >8% BeO by sink-float separation with a tetrabromoethane dibutylphthalate mixture of a suitable composition. The rejection of minerals with densities equal to or greater than that of beryl (2.63 to 2.8 g/cm^3), such as muscovite and lepidolite (>2.75 g/cm^3), was not readily accomplished, and therefore, the BeO content in the concentrate could be increased only at the expense of beryl recovery, Heyes, Trahar [1].

Spodumene (D = 3.1 g/cm^3) was separated from two bulk pilot plant flotation concentrates containing about 43 or 23% beryl (≈6 or 4.5% BeO), 37 or 61% spodumene, along with 19 or 14% mica, quartz, and feldspar (particle size between 74 and 500 μm), using tetrabromoethane (D = 2.9 g/cm^3) in a cyclone separation. The best separations resulted when a cyclone pulp containing 10 to 20% solids was used at a pressure between p_e = 4 and 5.4 atm. Under these conditions, a yield of about 8 to 10% BeO in the overflow pulp was realized with the recovery of about 90% of the BeO. An effective separation occurs using one cyclone stage, but exceptionally high-grade products result when four cyclones are used in a series, Tipper, Browning [2].

A short, anonymous report describes the separation of beryl (D ≈ 2.7 g/cm^3) from lighter and heavier pegmatitic gangue minerals (after electrostatic removal of mica) by the heavy liquid method using various mixtures of $C_2H_2Br_4$ and naphtha in the density range from 2.63 to 2.915 g/cm^3 [3]. The mica was removed by electrostatic separation from the coarse (>0.5 mm) fraction of a pegmatite ore with 0.375 wt% BeO containing beryl, quartz, feldspars, muscovite, garnets, and tourmalin. The beryl was separated by treatment with tetrabromoethane-naphtha mixtures of D = 2.672 and 2.875 g/cm^3 and was ground to pass a

0.208 mm (65 mesh) sieve. After treatment with a long-chain alkylamine in an acidic circuit to float off some residual quartz and mica, the resulting final concentrate assayed 8.0% BeO with an overall recovery of about 80% of the original BeO [4], see also [5].

References:

[1] G. W. Heyes, W. J. Trahar (Rept. Australia C.S.I.R.O. Mining Dept. Univ. Melbourne Ore Dressing Invest. No. 641 [1963] 1/6; C.A. **60** [1964] 1363).
[2] R. B. Tippin, J. S. Browning (U.S. Bur. Mines Rept. Invest. No. 7134 [1968] 41/7, 52/3; C.A. **69** [1968] No. 45356).
[3] Anonymous (Mining Mag. **105** [1961] 118/9).
[4] A. M. Baniel, A. Mitzmager, J. Mizrahi, S. Star (Trans. AIME **226** [1963] 146/54, 152).
[5] Anonymous (Eng. Mining J. **162** No. 9 [1961] 91/3).

1.1.4 Enrichment of Beryllium Ores by Fuming and Magnetic Concentration

A beryllium bearing magnetite ore (1.44% BeO) from Iron Mountain in New Mexico was mixed with coke (both 2 mm particle size) in a 4:1 weight ratio. The mixture was heated at 1325 to 1380 °C in a carbon crucible for 6 h to produce a fume concentrate containing 10.41% BeO (and 1.90% Fe) which recovered 85.2% of the BeO from the ore. The iron oxide was reduced to the metal, and the ZnO was volatilized together with the BeO in the fumed product. For magnetic concentration, the ore was ground to a grain size <0.074 mm (−200 mesh) and separated in a Frantz ferro filter at a current of 2.0 A. The resulting magnetic fraction was separated in a Davis tube at a current of 1.0 A. The nonmagnetic fraction of the latter treatment assayed 2.80% BeO and recovered 52.8% of the BeO from the ore. The enrichment of low-grade beryl ores (0.24% BeO) by magnetic concentration in a Davis tube (1.0 A) was not efficient, W. R. Storms (U.S. Bur. Mines Rept. Invest. No. 4024 [1947] 1/13, 5/8; C.A. **1947** 4067).

1.1.5 Flotation

1.1.5.1 General

Flotation is a common method for beneficiation of BeO (mostly as beryl) containing ores. Suitable collectors for beryl flotation are 8 to 18 carbon fatty acids (especially oleic acid) or their salts, mixtures of fatty acids (e.g., in coconut oil), and petroleum sulfonate. Pretreatment (conditioning) with activating agents, mainly fluoric acid or NaOH, improves the action of the collector and the flotation behavior. A frothing agent, such as Emulsol X-1, is also often used. The addition of selective depressing agents is necessary to separate the minerals in an ore by selective flotation.

1.1.5.2 Flotability of Beryl

Pure surfaces of beryl react feebly with fatty acids, but the treatment of beryl with aqueous NaOH reduces the contact time necessary for adherence of the oleic acid collectors and increases beryl recovery by flotation [1].

The pretreatment of beryl (74 μm size) with alkali hydroxides (NaOH, KOH) induces a highly negative electrokinetic zeta potential on the beryl which improves the flotability by oleic acid increasing from pH = 8.5 to 10.4. It is assumed that the OH^- ions (causing the potential) are adsorbed to break off silica bonds and, thus, participate in the construction of the inner part of a binary electrical layer, Plaksin et al. [2], see also the IR studies of Plaksin, Solnyshkin [3]. Treatment with $Ca(OH)_2$ gives positive zeta potentials, due to Ca^{2+} adsorption on the beryl surface [2].

Flotation studies of beryl with oleic acid after preconditioning with NaOH, KOH, and $Ca(OH)_2$ reveal that flotation of beryl is best promoted by KOH. With NaOH, best results were obtained after 5 min of alkali pretreatment, washing twice, and flotation in a solution of Na_2CO_3 at pH=7.5. No linear dependence of the flotation of beryl on the amount of collector adsorbed (on the mineral) could be detected. The best recovery of beryl was attained by adding small amounts (60 to 120 g per ton) of kerosene and floating at pH=9. Addition of more kerosene increased the adsorption of the collector on beryl, whereas addition of more collector or prolonged contact with beryl decreased the adsorption of kerosene [4].

IR studies indicate that oleic acid and sodium oleate are adsorbed as ions on the surface of pretreated (NaOH, HF) beryl, since flotation occurs only when a minimum number of these collectors are present in the ionic state. Upon treatment with H_2SO_4, the fixation of the collector becomes molecular and causes depression of the mineral [5].

Pretreatment (activation) of beryl by HF (1 M) prior to flotation with oleic acid is suggested because IR studies have shown that oleic acid is not adsorbed at the surface of pure beryl. It is assumed that adsorption of oleic acid on activated beryl occurs through hydrogen bonding of the oleic acid monomers to the F^- surface bridging sites produced by the reaction of HF with surface hydroxyl groups or chemisorbed water. Adsorption of oleic acid on beryl of <0.25 mm particle size, activated by HF, attains a maximum at pH=7 [6]. Hydrofluoric acid activates beryl by selective dissociation of surface silicate groups, which produces favorable conditions for chemisorption, Bogdanov et al. [7]. HCl does not activate beryl [6].

Beryl treated with 5% hydrochloric acid and carefully washed is not floated by oleic acid in amounts up to 1 kg/t [1]. The zeta potential of an acid (HCl) washed beryl, measured by the streaming-current technique, is shown in **Fig. 1-1**. The zero point of charge falls in the range from pH 3.7 to 4.6 depending on the type of beryl. At higher pH values, the zeta potential is negative, while at lower values, it is positive, Moir et al. [8]. Acid (5% HCl or H_2SO_4) pretreatment of powdered beryl followed by washing with twice distilled water produced a negative charge on the mineral surface in the pH=2.6 to 11.1 range. Addition of $FeCl_3$ converted this negative charge into a positive charge. Such activation yields a maximum flotability of beryl by sodium oleate at pH≈7.0 (86.0%) at a small negative electrokinetic potential of −2.5 mV [9].

Fig. 1-1

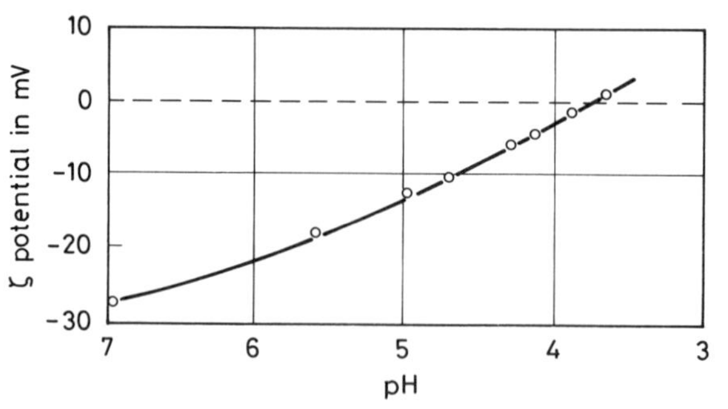

ζ potential of acid-washed beryl.

Treatment of beryl with H_2SO_4 (50 mg/L) at pH≈6 for 30 min strongly depresses the flotability of beryl by sodium oleate (from 65.5% to 12.4%) because a surface film of oleic acid forms which prevents flotation [10], also see [11]. Generally, the flotability of beryl from different sources and of varying grain size depends on its nature and may require between 0.5 and 1.5 kg of oleic acid per 1000 kg of ore at pH values ranging between 5.5 and 12. The flotability of beryl may be increased by up to 18% by using sodium oleate instead of oleic acid and by applying higher pH values (pH 11 to 12 for some). However, the extent of improvement strongly depends on the nature of the beryl ores and their content of other metal oxides [12].

The iron was removed from a sample of beryl by leaching with strong mineral acid (aqua regia and HCl). The flotation behavior of this iron free beryl by alkyl-aryl sulfonate (molecular weight 450 to 470, alkyl chain length 25 to 30) was studied in an aqueous solution (38.2 mg/L, 8.46×10^{-5} M) made up with conductivity water in the presence of several metal salts (15.2 mg/L) at pH≈3.2. Except for $FeCl_3$ (100% recovery at pH=3.16), all of the salts studied depressed the flotability of the beryl. In solutions containing 2.12×10^{-5} M sulfonate and 1.37×10^{-5} M Fe^{3+}, complete flotation of beryl was observed at pH=3.37 to 3.70, whereas no flotation occurred at pH<3.0 and >4.10. The use of tap water containing 30 ppm $Ca^{2+}+Mg^{2+}$ instead of conductivity water reduced the flotability of beryl strongly. However, the depressing effect of Ca^{2+} could be compensated by the enhancing effect of Fe^{3+} which apparently forms a less soluble sulfonate than Ca^{2+}. Thus, with 1.89 ppm Fe^{3+} and 6.31×10^{-5} M sulfonate, 81% of the leached beryl could be recovered at pH=2.84 compared with 2.4% without Fe^{3+}. If unleached beryl ground to a 0.32×0.1 mm size in a mild steel rod mill and tap water were used for flotation, a 95% recovery of beryl was achieved with a concentration of only 1.28×10^{-5} M sulfonate in the pH=3.13 to 3.25 range. Since leached beryl could not have floated under these same conditions, it is apparent that a considerable amount of iron was imparted by the grinding step. The experimental data show that the H^+ ion concentration (pH) assumes a significant role in flotation. It establishes the electrical charge at the mineral-liquid interface and may determine the extent of hydrolysis of certain ions in the system. If H_2SO_4 is used instead of HCl to adjust the pH value, the anion (SO_4^{2-}) can significantly affect the results of flotation by surface effects [13].

The flotability of beryl and chrysoberyl in solutions of various anionic and cationic surfactants (collectors) was compared to that of the main gangue minerals. Studies employing microflotation and contact-angle methods reveal that selective flotation (5 min) of beryl (and chrysoberyl) from fluorite, topaz, or feldspar is possible with sodium isooctyl phosphate (50 mg/L) at pH<2. Sodium oleate (20 mg/L) at pH 8.5 (with and without modifiers such as sodium silicate, dextrin, starch, etc.) is not a selective collector for beryl and chrysoberyl because recoveries of calcite, fluorite, and topaz are equal or higher. Other collectors (with and without modifiers) such as sodium dioctyl sulfosuccinate at pH=4.0, a (C_{12}) α-amine acetate at pH 11.0, a (C_{15}) β-amine chloride at pH=8.5, a (C_{16} to C_{18}) diamine acetate at pH=9.0, a tertiary (C_{18} to C_{22}) amine acetate at pH=8.0, and a (C_{12} to C_{18}) quaternary ammonium chloride at pH=9 were found to be unselective [14]. Flotation of beryl with cationic collectors such as octyl thioisouronium bromide, octadecyl (+stearyl) ammonium acetate (Armac 18D), a primary tallow fat amine acetate (Armac NT), octadecyl trimethyl ammonium iodide or a dialkyl-dimethylammoniumchloride derived from soya fats (Arquad 2S) was possible in the range from pH=3 to 8 with an optimum effect between pH=5.5 and 7.0 (except for the thioisouronium salt). At pH<2 to 3 or >8 the collector activity decreased rapidly. Unfortunately, minerals such as quartz and feldspars also float and thus the flotation is not specific to beryl. However, the float selectivity may possibly be improved by the addition of suitable modifiers [15].

References:

[1] M. A. Eigeles, I. T. Leviush (Proc. 2nd Intern. Conf. Peaceful Uses At. Energy, Geneva 1958, Vol. 3, pp. 162/6 (paper 2065); C.A. **1959** 6554).

[2] I. N. Plaksin, G. N. Khazhinskaya, E. A. Shrader (Izv. Vysshikh Uchebn. Zavedenii Gorn. Zh. **5** No. 7 [1962] 132/6; C.A. **58** [1963] 6472/3).

[3] I. N. Plaksin, V. I. Solnyshkin (Dokl. Akad. Nauk SSSR **139** [1961] 936/7; Proc. Acad. Sci. USSR Phys. Chem. Sect. **136/141** [1961] 597/8).

[4] G. N. Khazhinskaya, D. V. Maksimov (Data on the 2nd Leningrad Conference on the Use of Radioactive Isotopes in the Coal and Mining Industry, Leningrad-Moscow 1961, pp. 60/73 from Ref. Zh. Met. **1963** No. 10 G56; Abstr. J. Met. B **1963** No. 10 Ref. 311).

[5] V. I. Bogdanov, Wei-Chung Hsing (Obogashch. Rud. **6** No. 6 [1961] 35/8; C.A. **57** [1962] 3113).

[6] A. S. Peck, M. E. Wadsworth (Trans. AIME **238** [1967] 264/8).

[7] O. S. Bogdanov, Wei-Chung Hsing, N. A. Yanis (Tr. Vses. Nauchn. Issled. Proekt. Inst. Mekh. Obrab. Polez. Iskop. No. 128 [1961] 26/36, 29; C.A. **57** [1962] 4393).

[8] D. N. Moir, D. N. Collins, H. C. Curwen, R. M. Manser (Proc 6th Intern. Congr. Mineral. Process., Cannes 1963 [1965], pp. 651/67, 653, 657; Met. Abstr. [3] **1** [1966] 449).

[9] V. V. Troitskii (Tsvetn. Met. **33** No. 8 [1960] 73/4; C.A. **1961** 9195/6).

[10] K. A. Razumov, Wei-Chung Hsing (Izv. Vysshikh Uchebn. Zavedenii Gorn. Zh. **5** No. 6 [1962] 178/81; C.A. **57** [1962] 16218).

[11] M. A. Eigeles, E. P. Sakharova (Tr. Vses. Nauchn. Issled. Inst. Mineral'n. Syr'ya **1967** No. 18, pp. 82/101, 94, 99; C.A. **70** [1969] No. 13575).

[12] I. N. Plaksin, G. N. Kazhinskaya (Dokl. Akad. Nauk SSSR **135** [1960] 389/90; Proc. Acad. Sci. USSR Chem. Technol. Sect. **130/135** [1960] 187/8).

[13] M. C. Fuerstenau, R. B. Bhappu (Trans. AIME **226** [1963] 164/74, 166/8).

[14] A. J. Fergus, G. V. Sullivan, G. F. Workentine (U.S. Bur. Mines Rept. Invest. No. 7188 [1958] 1/22, 11/2, 15/21; C.A. **70** [1969] No. 13573).

[15] I. N. Plaksin, D. V. Maximow (Izv. Vysshikh Uchebn. Zavedenii Gorn. Zh. **5** No. 5 [1962] 187/90; C.A. **57** [1962] 12099).

1.1.5.3 Flotation of Synthetic Beryl-Mineral Mixtures

Complete separation of beryl from a 50% quartz-beryl mixture ground to <100 μm (−150 mesh) was attained by flotation. A mixture of sodium oleate (100 g/t) and pine oil (150 g/t) at pH ≈ 6.9 served to separate the beryl from a pulp of 20% solids (after the addition of $Pb(NO_3)_2$, 100 g/t) [1]. After pretreatment with dilute aqueous HF or HCl and desliming, the flotation of beryl from admixtures of quartz was done with a mixture of oleic acid (≈0.45 kg/t) and terpineol (0.23 kg/t). The pretreated beryl-quartz mixture was conditioned for 20 min at pH = 6 to give a concentrate containing about 11 wt% BeO and a recovery of 84% of the beryl from the original mixture [2]. Beryl can be separated from ground (50 to 150 μm) mixtures of 80 to 90% feldspar by flotation. The mixture was conditioned for 1/2 min with sodium lauryl sulfate (45 g/t, twice) and floated for 1 min in two stages at 25 °C. From a mixture containing 80% feldspar (2.63% BeO), the process yielded an optimum recovery of beryl (69.44%) at pH = 2.8 in a concentrate containing 8.11% BeO. A concentrate with 9.13% BeO, which had been floated at pH = 2.6, recovered only 48.77% of the beryl. Mixtures containing less beryl (e.g., 10%) gave concentrates with a similar recovery (59.4%), but of far lower gradation (4.48% BeO). Poorer results were obtained at higher and lower pH and on prolonged (30 min) flotation. Removal of the sodium laurylsulfate coating by washing with water prevented flotation of beryl in the pH = 2 to 3 region [3]. Attempts were made to separate beryl from other minerals in binary (1:1) mixtures by flotation with laurylamine hydrochloride and terpineol in acid (H_2SO_4, HF) or alkaline (NaOH) circuits, followed by several cleaning steps. Admixtures studied include biotite at pH = 2.5 and 12.0, muscovite at pH = 2.6, topaz at pH = 2.0, quartz at pH = 2.5 (HF) and microcline in HF. In the mixtures with biotite, topaz, and muscovite, beryl was left in the tailing which retained about 85% of the BeO. From quartz mixtures, the beryl was floated to give an almost quantitative recovery (99.1% BeO) in the concentrate. Microcline and beryl were floated in the HF circuit, and after an aqueous Na_2CO_3 wash, the beryl was floated by sodium oleate which recovered around 95% of the BeO in the concentrate [4]. Selective flotation of topaz from beryl and microcline by fatty acids (oleic, tridecylic) was achieved by pretreatment with H_2SO_4 (5%) [7]. Treatment of the oleic acid collector pulp with γ-irradiation (2.27 MeV from $^{90}Sr + ^{90}Y$ or 3.15 MeV from $^{144}Ce + ^{144}Pr$ as the sources) for 2 to 20 h prior to flotation improved the separation of beryl from spodumene. After irradiation, the froth contained more spodumene (+15%) and less beryl (−15 to −20%). The irradiation possibly converts the oleic acid into the stereoisomeric elaidic acid or stearic acid which have other collector properties [5, 6].

References:

[1] L. Usoni (Giorn. Chim. Ind. Appl. **15** [1933] 13/5; C. **1933** I 2742).

[2] N. R. Srinivasan, H. S. Aswath (J. Indian Inst. Sci. B **38** [1956] 135/42, 139/41).

[3] D. N. Moir, J. R. Stevens (Bull. Inst. Mining Met. No. 688 [1964] 373/91, 385; C.A. **60** [1964] 15480).

[4] N. A. Yanis (Tr. Nauchn. Tekhn. Konf. Inst. MEKHAOBR **1** [1965/67] 551/73, 554, 559/61, 564/6; C.A. **69** [1968] No. 98551).

[5] I. N. Plaksin, N. G. Malysheva, L. P. Starchik (Dokl. Akad. Nauk SSSR **175** [1967] 663/4; Dokl. Chem. Technol. **172/177** [1967] 126/7).

[6] I. N. Plaksin, N. G. Malysheva, L. P. Starchik, V. I. Solnyshkin (Tsvetn. Met. **41** No. 9 [1968] 26/7; Soviet J. Non Ferrous Metals **9** No. 9 [1968] 35/6).

[7] M. A. Eigeles, E. P. Sakharova (Tr. Vses. Nauchn. Issled. Inst. Mineral'n. Syr'ya **1967** No. 18, pp. 82/101, 94, 99; C.A. **70** [1969] No. 13575).

1.1.5.4 Flotation of Beryl Bearing Ores

Selective flotation of beryl in the presence of mica, quartz, and feldspar can be attained by using Na-cationated soft (demineralized) water to prevent and depress gangue activation. Removal of slimes from the ground ore is necessary because they sink slowly and contaminate the beryl concentrate. After conditioning with Na_2S (500 g/t) for 5 min to depress the gangue minerals, the collector, generally oleic acid (200 g/t of solids), is added to the pulp (50% solids) at 80 to 85 °C. Thus, concentrates with 8.1 to 11.5 wt% BeO were obtained from ores with even $<0.1\%$ BeO in a 75 to 90% BeO recovery. Varying the BeO content of the ore, grain size, hardness of water, pretreatment with NaOH, temperature, pH of pulp, amount of Na_2S and oleic acid, and time of conditioning and floating confirmed the results [1]. For pilot plant tests using this method, see [2]. Thus, the selectivity of beryl flotation with respect to the other components of the ores was enhanced by the use of soft water containing Na^+ ions (as Na_2S) and heating the pulp at 85 to 90 °C prior to the addition of oleic acid [3], also see [4, 5].

Finely ground (<0.15 mm $\triangleq$ −100 mesh) beryl ore was conditioned with NaF ($\approx$1.35 to 2.70 kg/t ore) and H_2SO_4 ($\approx$3.2 to 6.8 kg/t ore), instead of HF, at pH = 2.0 to 2.5. After washing or repeated decantation, the repulped ore was floated at pH $\approx$ 6 to 7 with NaF (90 g/t), oleic acid (45 to 115 g/t), and a small amount of Emulsol X-1 [6].

A preconditioned (HF, ≈1.8 kg/t), ground, and deslimed pegmatite containing 0.25 wt% BeO was washed thoroughly with water and floated directly with cold oleic acid and fuel oil at pH 6.5 to 7.5. The flotation yielded 3.4 wt% of a concentrate containing 6.15% BeO and recovered 87.7% of the original beryl content [2].

For detailed operation conditional involving washing with sodium silicate solution and removal of the slime with grain size <24 μm, conditioning with HF at a pulp density of 50% solids at pH=2.2 for 5 to 10 min, washing with soft water, addition of oleic acid, and adjusting the pH to 6.5 to 7.5 to prevent flotation of mica, see [7]. After mechanical removal of coarse mica by screening, the deslimed beryl ore containing 5.6 wt% beryl (0.75 wt% BeO) was conditioned with HF (47%, ≈2.1 kg/t ore) in a pulp of 50% solids for 30 min. The mixture was washed with water and floated with oleic acid (≈160 g/t) and pine oil (≈80 g/t) at pH 7. The resulting rough concentrate froth was cleaned twice without additional reagents to give 8.1 wt% of a final concentrate containing 7.62% BeO recovering 81.0% of the original beryl [8]. This procedure depressed the gangue excepting tourmaline [8, 9] (see p. 12 for the separation from tourmaline). The mica from low grade pegmatite (1.4% BeO) was removed by cocoamine acetate in an $H_2SO_4+Al_2(SO_4)_3$ circuit. After conditioning with HF (900 g/t), flotation with oleic acid in two steps (454+90 g/t ore) yielded a final concentrate containing 8.05% BeO recovering 69.3% of the beryl [10]. Beneficiation of pegmatitic beryl ores containing 3.9 to 12% beryl (0.55 to 1.7% BeO) was achieved by stage grinding to <0.208 mm (−65 Tyler mesh), desliming with H_2O (10 kg/1 kg ore), and conditioning with NaOH (≈0.45 to 2.25 kg/1000 kg ore) in an agitated pulp with 50% solids for 10 to 30 min. After washing with water until a pH=6.5 to 8 was attained, the pulp was diluted with water to 20% solids. The pulp was then conditioned with coconut fatty acid (≈360 g/t ore) and Emulsol X-1 (180 g/t ore) for 5 min and floated for 5 min. The beryl collected in the froth was cleaned to yield a final concentrate assaying 87 or 94% beryl (12.2 or 13.2% BeO) recovering 74.7 or 71.4% from ores with ≈10 or 3.9% beryl, respectively. The greatest loss of beryl was in the slimes. The concentrates contained beryl as well as minor minerals, mainly feldspar and mica [11]. In laboratory tests, mixtures (≈2.7 kg/t ore) of 19% peltogen (sulfonated sperm whale oil), 31% linseed fatty acids, 12% cresylic acid, and 38% fuel oil were used to condition ground (0.21 mm, −72 mesh) pegmatitic beryl ores containing 1.8% BeO. Pulps of 55% solids were conditioned for 10 min and floated with the addition of Na_2CO_3 (≈2 kg/t ore), at pH=9.8. Cleaning and recleaning of the beryl bearing froth yielded final concentrates containing 10.0 to 12.45% BeO recovering up to 61.8% of the original beryl. Other grind sizings, reagent additions, conditioning densities and times, pH, flotation times, and cleaning steps gave, in most cases, poorer concentrate gradation and beryl recovery [12].

"Heavy minerals" such as apatite, tourmaline, spodumene, amblygonite etc. are floated with beryl, but can be separated from pegmatite ore using a mixture of NaF and dextrin as the depressing agent for beryl (and spodumene). An ore containing 0.21% BeO was ground to a 12 to 24 μm grain size and deslimed. The heavy minerals were floated off using a 1:1 mixture of NaF and dextrin (45 g/t ore) and oleic acid (≈115 g/t) at pH=6.5. The residue (tailing) was conditioned with HF (≈1.35 kg/t) or with $NaF+H_2SO_4$ at pH=2.5 to 3.2, washed thoroughly with water, and reconditioned with NaF (≈90 g/t) and oleic acid (≈115 g/t). Beryl was floated off at pH=6.5 to 7.3 to give a concentrate which, after cleaning, contained 10.87% BeO and recovered 82.2% of the original beryl [2].

After previous conditioning with $Ca(OCl)_2$ or NaOCl and washing, beryl was floated directly from a mica-free pegmatite. Water soluble petroleum sulfonates were used as the collector in a sulfuric acid circuit in the presence of a frothing agent (methyl amyl alcohol). Beryl contents of the concentrates and recovery rates of beryl from the ore were generally somewhat low [13, 14].

Mica, feldspar, beryl, and quartz were successively separated and recovered by a single integrated process. The pegmatite material (≈0.25% beryl) was ground to free the beryl and other minerals from one another and deslimed. Muscovite was selectively removed, and a bulk beryl feldspar concentrate was isolated from the remaining material, which was mainly quartz. Finally, the beryl was selectively floated from the feldspar. The muscovite was removed by conditioning the deslimed ore with H_2SO_4 (≈0.45 to 2.25 kg/t) for 2 to 3 min and flotation with a long-chain alkylamine salt such as tallow amine acetate (≈22.5 to 67.5 g/t) and a frother (methyl amyl alcohol, 4.5 g/t) at a pulp density of about 25% solids. The resulting tailing was conditioned with HF (≈0.45 to 1.8 kg/t) for 2 to 4 min, an amine collector (same as above, ≈45 to 225 g/t), and a frother (≈4.5 to 22.5 g/t) to isolate the beryl feldspar concentrate. The beryl was separated from the feldspar by conditioning the bulk concentrate with $Ca(OCl)_2$ (≈0.23 to 0.9 kg/t) at a pulp density of about 50% solids for at most 5 min. All reagents were removed by washing with water and the bulk product was conditioned with H_2SO_4 (≈0.45 to 1.8 kg/t) and petroleum sulfonate (molecular weight 400±30, ≈0.22 to 1.35 kg/t). After 2 to 3 min of conditioning at pH=2 to 3 and a pulp density of at least 50% solids, the pulp was diluted with water to 20 to 25% solids. Frother was added if required and then the beryl was floated off in the froth. This froth was cleaned stepwise until a beryl concentrate of 66.5% (8.3% BeO) was attained (74% BeO recovery). Similar treatment of a pegmatite with 10.1% beryl (~1.25% BeO) yielded a concentrate of 97.4% beryl (~12.0% BeO) in an 82.0% recovery [14], see also [2, 15, 16]. A similar separation procedure was applied by Ashton and Weir [17]. For further flotation tests using different collectors, see Weir, Moskovits [18].

The rough beryl concentrate resulting from treatment of pegmatite ores containing 0.1 to 2.5% beryl, as described above, was enriched by treatment with HCl, to depress the beryl, and with an amine salt collector, to float off the admixed minerals. Since the finer particles of beryl tended to float with the admixtures, separation was not satisfactory. Commercially suitable concentrates, with a beryl content above 85% (~10% BeO) were produced only from ores containing ≧2.5% beryl (≈0.35% BeO) with recovery rates around 90% [19]. Beryl from a pegmatite containing 1.01% BeO (Roebling mine), which had been ground to a <0.5 mm (−35 mesh) grain size, was separated by direct flotation. The ore was conditioned with H_2SO_4 (≈0.54 kg/t), NaF (≈0.45 kg/t) (to depress the gangue), and petroleum sulfonate (≈1.575 kg/t) at pH=5.8. The resulting rough concentrate, which was scrubbed with NaOH (≈0.9 kg/t), deslimed, and conditioned with lauryl amine hydrochloride (to float off the mica), yielded a final concentrate (7.0 wt%) with 9.0% BeO recovering 62.3% of the beryl. Similar treatment of a pegmatite containing 0.43% BeO (Colony mine) yielded a final concentrate assaying 7.8% BeO and recovering 63.5% of the beryl. Alternatively, the Roebling ore was conditioned with H_2SiF_6 (0.9 kg/t) to depress the beryl and lauryl amine hydrochloride (0.45 kg/t) to float off the mica at pH=5.0. The beryl bearing tailings were conditioned with H_2SiF_6 and petroleum sulfonate (0.9 kg/t each) and floated at pH=4.9. The final concentrate, obtained by twofold cleaning, assayed 11.9% BeO and recovered 80.2% of the beryl. Similar treatment of the colony ore produced a final concentrate with 6.3% BeO recovering 70.0% of the beryl. The BeO content could be increased to 7.5% with a lower (61.1%) recovery of beryl [20].

Spodumene was separated from beryl and mica in a laboratory test. A finely ground (<0.32 mm ≙ −48 mesh) pegmatite, containing spodumene (18%), mica (5.0%), and beryl (0.5%), was deslimed and pulped to about 40 to 45% solids. The pulp was conditioned with NaF and alkali-, ammonium-, or magnesium lignin sulfonate (≈0.9 kg/t each) for 5 min, and with oleic acid (≈0.42 kg/t) at pH=9.6 to float off the spodumene. The spodumene tailing was refloated to remove additional spodumene, deslimed, and conditioned with H_2SO_4 (0.9 kg/t) and cocoamine acetate (≈0.07 kg/t) to separate the mica by floating (at pH=3.6).

The mica tailing was thickened and conditioned with HF (≈0.9 kg/t of dry feed) for 5 min (pH=3.6) and washed to remove the acid. The washed pulp was conditioned with NaOH (≈0.09 kg/t) and oleic acid (≈0.37 kg/t) and floated at pH=7.6. The resulting rough concentrate was triply cleaned to give a final concentrate with 1.57% BeO and 3.0% Li_2O recovering 75% of the original beryl [21]. In a pilot plant test, ground pegmatite was similarly treated to remove 83% of the spodumene content by floating. The resulting tailings were pulped and conditioned with HF (0.9 kg/t of dry feed) at pH=3.8 for 5 min, washed to remove the acid, and repulped to 30% solids. The slurry was conditioned with ≈0.42 kg sodium silicate, ≈0.14 kg NaOH, and ≈0.42 kg of oleic acid per ton of dry feed at pH=7.3. The pulp was floated to give a rough spodumene-beryl concentrate, which after a threefold cleaning, assayed 1.25% BeO and 4.45% Li_2O and recovered about 87% of the beryllium and 66% of the lithium. This concentrate, which also contained mica (12.5%), feldspar (10.5%), and quartz (5.5%), was conditioned with H_2SO_4 (≈45 kg/t of dry feed) to remove the fatty acid collector. The pulp was then reconditioned with H_2SO_4 (≈0.92 kg/t) and "Armac CD" (cocoamine acetate, ≈0.09 kg/t) and the mica floated off. The mica tailings were conditioned with magnesium lignin sulfonate (≈0.68 kg/t), NaF (0.69 kg/t), and oleic acid (0.37 kg/t of feed). The spodumene was floated and cleaned twice. The combined spodumene tailings and middlings, which contained about 80% of the beryl, were conditioned with HF (≈0.9 kg/t of bulk concentrate). The acid was washed off and the pulp reconditioned with NaOH (≈0.14 kg/t of feed) and oleic acid (0.22 kg/t of feed). Floating at pH≈7 yielded a rough concentrate which, after twofold cleaning, contained 6.37% BeO and 2.57% Li_2O and recovered 76% of the beryl [39]. For other related procedures, see [22 to 25]. If the mica tailings were conditioned with Na_2S (≈0.46 kg/t) instead of NaF and lignin sulfonate, the spodumene flotation left 94% of the beryl content in the tailings which, upon floating at pH=7 (as above), yielded a concentrate assaying 7.48% BeO and recovering 87.4% of the beryl [22, 23].

A deslimed beryl (0.71% BeO) bearing pegmatite, composed mainly of muscovite (a mica), albite (a feldspar), and tourmaline, was conditioned with HF (47%, ≈4.2 kg/t) at pH=2.0. Flotation with oleic acid (≈0.33 kg/t) and pine oil (≈165 g/t) produced a rough concentrate consisting mainly of beryl and tourmaline. The mixture was separated by conditioning with H_2SO_4 (0.92 kg/t) for 5 min and flotation with a long-chain amine salt (hydrochloride) at pH=2.0 for 1 min. The flotation of tourmaline in the H_2SO_4 circuit produced a high grade (10.0% BeO) concentrate recovering 72.4% of the beryl from the clean tailing, whereas the flotation of beryl in the HF circuit produced a somewhat lower grade concentrate (8.7% BeO) from the froth [8], see also [9].

Beryl was separated from tourmaline, garnet, and mica by flotation. The mixture was treated first with oleic acid, then with Na_2CO_3 (≧15 g/L) for 1 h at 80 to 90 °C, and finally floated at 22 to 25 °C for 7 to 8 min to remove the tourmaline and garnet. The remaining concentrate contained beryl (5% BeO) and muscovite in the tailing. Another method involved treatment of the mixture with a cationic collector in an acid medium (H_2SO_4 or HCl, pH=2) to float off mica and tourmaline. Garnet was separated by flotation with oleic acid and Na_2CO_3 from the remaining beryl and garnet containing residue [26]. Separation of beryl from fluorite and mica was achieved by floating off the fluorite with oleic acid and pine oil (ratio 1:2) in an aqueous emulsion containing sodium silicate (to depress the beryl) at pH≈8. The mica was then removed by a cationic collector in an alkaline circuit after depressing the beryl with Na_2S, Na_2CO_3, or sodium silicate. The beryl in the tailing was ground to a −0.1 to −0.15 mm size, deslimed, activated with NaOH, and floated with oleic acid at 25 °C and pH=8. After several cleaning steps, a final concentrate containing 6.35% BeO was isolated in a 72.35% recovery from the ore [27]. For separation of beryl from tungsten and molybdenum ores (quartz-topaz greisen), see [28].

Sodium alkyl-aryl sulfonate (see p. 7) was used to achieve an optimum selective separation of beryl from other components of a natural ore which assayed 0.55% BeO and contained quartz, feldspar, apatite, mica schist, and iron bearing silicates. The ore was ground to <0.32 mm size, deslimed by decantation at 20 μ, repulped with tap water, and the pH adjusted to 2.40 by H_2SO_4. The pulp was conditioned for 3 min with primary tallow amine acetate (0.33 kg/t) and two drops (54 g/t) of methyl isobutyl carbinol (frother). The mica schist was then floated off over 3 min. The pulp was thickened, repulped, and the pH was adjusted to 2.60 with H_2SO_4. Stage conditioning for 3 min with sodium alkyl aryl sulfonate (0.441 kg/t) and one drop of frother produced a rough beryl concentrate after 3 min of floating at a final pH = 2.65. One drop of frother was added and the concentrate was cleaned twice at pH = 2.7. The clean concentrate was dried and passed through a high intensity magnetic separator to remove apatite and iron bearing gangue minerals. The final concentrate assayed 8.0% BeO and recovered about 68% of the beryl content [29].

References:

[1] M. A. Eigeles, I. T. Leviush (Proc. 2nd Intern. Conf. Peaceful Uses At. Energy, Geneva 1958, Vol. 3, pp. 162/6 (paper 2065); C.A. **1959** 6554).
[2] D. N. Moir, D. N. Collins, H. C. Curwen, R. M. Manser (Proc. 6th Intern. Congr. Mineral. Process., Cannes 1963 [1965], pp. 651/67, 653, 657; Met. Abstr. [3] **1** [1966] 449).
[3] I. T. Leviush, M. A. Eigeles (Tr. Vses. Nauchn. Issled. Inst. Mineral'n. Syr'ya **1961** No. 6, pp. 106/22, 113/4; C.A. **57** [1962] 14771).
[4] M. A. Eigeles, I. V. Fuki (Tr. Vses. Nauchn. Issled. Inst. Mineral'n. Syr'ya **1967** No. 18, pp. 46/62, 62/82, 79; C.A. **70** [1969] No. 13570, No. 13572).
[5] I. T. Leviush, P. N. Fedorov (Mineral'n. Syr'e **1960** No. 1, pp. 141/5; C.A. **1961** 26915/6).
[6] H. D. Sneddon, H. L. Gibbs (U.S. Bur. Mines Rept. Invest. No. 4071 [1947] 1/18, 8; C.A. **1947** 4067).
[7] D. N. Moir, H. C. Curwen (Brit. 952246 [1962/64] 1/3; C. **1967** No. 30-2427; N.S.A. **18** [1964] No. 16154).
[8] J. S. Kennedy, R. G. O'Meara (U.S. Bur. Mines Rept. Invest. No. 4166 [1948] 1/18, 5, 8, 11/4; C.A. **1948** 2212/3).
[9] J. S. Kennedy, R. G. O'Meara (U.S. 2414815 [1943/47] 1/8; C.A. **1947** 1593).
[10] F. D. Lamb (U.S. Bur. Mines Rept. Invest. No. 4040 [1947] 1/9, 5/7; C.A. **1947** 4066/7).
[11] F. D. Lamb, L. Banning (U.S. 2385819 [1943/45] 1/3; C.A. **1946** 47).
[12] L. J. Weir (Mining Rev. [Adelaide]) No. 1965 [1967] 22/32, 25/8; C.A. **69** [1968] No. 98533).
[13] S. M. Runke (U.S. 2666587 [1952/54] 1/10, 5/7; C.A. **1954** 6086).
[14] S. M. Runke (U.S. Bur. Mines Rept. Invest. No. 5067 [1954] 1/19, 3/4, 9, 12/5; C.A. **1954** 11999).
[15] S. M. Runke, E. O. Binyon, J. B. Cunningham (U.S. Bur. Mines Rept. Invest. No. 5061 [1954] 1/21, 12; C.A. **1954** 9285).
[16] Anonymous (Eng. Mining J. **161** No. 9 [1960] 93/102, 99/100).
[17] B. E. Ashton, L. J. Weir (Min. Rev. [Adelaide] No. 112 [1960/61] 47/61, 52, 56/9; C.A. **57** [1962] 9499/500).
[18] L. J. Weir, E. E. Moskovits (Australasian Inst. Mining Met. Proc. [2] No. 206 [1963] 143/52, 149/51; C.A. **60** [1964] 2569).
[19] S. M. Runke, J. M. Riley (U.S. Bur. Mines Rept. Invest. No. 5339 [1957] 1/18, 7/11; C.A. **1957** 11190/1).
[20] J. E. Shelton (U.S. Bur. Mines Rept. Invest. No. 5767 [1961] 1/10, 5, 9; C.A. **1961** 14211).
[21] J. S. Browning (Trans AIME **220** [1961] 420/3).
[22] J. S. Browning, B. H. Clemmons (U.S. 3028008 [1960/62] 1/10; C.A. **57** [1962] 1913/4).

[23] J. S. Browning, B. H. Clemmons, T. L. McVay (U.S. Bur. Mines Rept. Invest. No. 5750 [1961] 1/20, 14/8; C.A. **1961** 16924).
[24] A. Hegarty (Mining Mag. [London] **108** [1963] 202/7, 207).
[25] O. S. Bogdanov, Wei-Chung Hsing, N. A. Yanis (Tr. Vses. Nauchn. Issled. Proekt. Inst. Mekh. Obrab. Polez. Iskop. No. 128 [1961] 26/36, 29; C.A. **57** [1962] 4393).
[26] I. V. Chipanin, A. N. Kozhukhovskaya (Nauchn. Tr. Irkutsk. Gos. Nauchn. Issled. Inst. Redkikh Tsvetn. Metal No. 16 [1967] 186/90; C.A. **70** [1969] No. 21961).
[27] M. A. Eigeles, I. T. Leviush, E. P. Sakharova, P. N. Fedorov, V. P. Kuznetsov (Tr. Vses. Nauchn. Issled. Inst. Mineral'n. Syr'ya **1967** No. 18, pp. 284/99, 291; C.A. **70** [1969] No. 21971).
[28] I. V. Fuki, P. N. Fedorov (Tr. Vses. Nauchn. Issled. Inst. Mineral'n. Syr'ya **1967** No. 18, pp. 299/311, 303, 308; C.A. **69** [1968] No. 108815).
[29] M. C. Fuerstenau, R. B. Bhappu (Trans. AIME **226** [1963] 164/74, 170).

1.1.5.5 Flotation of Bertrandite, Phenacite, and Other Beryllium Minerals

A pegmatitic ore containing 0.24% BeO as bertrandite and phenacite (crushed to <150 μm size) was heated for 30 min at 788 °C in natural (hydrocarbon) gas and allowed to cool out of contact with air. The heating changed the BeO components from moderately hydrophilic to highly hydrophobic ones. Thus, BeO separation from heat treated ore was achieved by flotation at pH = 12 without a collector and with a minimum amount of a frother. The process yielded a concentrate containing 16.2% BeO and tailings essentially free from beryllium, Marvin [1].

A bertrandite, phenacite, and fluorite containing ore (0.75% BeO), crushed to a ≦0.1 mm size, was pretreated with NaOH and BP-103 (a frother). Flotation was accomplished using a small amount of oleic acid to collect the bertrandite, phenacite, and fluorite in the concentrate. After being recleaned several times, large amounts of sodium silicate were added to the concentrate. The mixture was heated to 80 °C and the fluorite selectively floated using small amounts of oleic acid as collector. After three cleaning steps and washing with HCl (to remove soluble admixtures), the initial BeO content of the concentrate tailing was increased from 5.34 to 7.8% with the recovery of 73.6% of the beryllium [2]. Selective flotation of the fluorspar and the majority of the calcite from bertrandite can be accomplished using alkali silicates and fluorosilicates as depressing agents: A mixture of the ore with Na_2CO_3 (0.92 to 2.76 kg/t), an alkali silicate and/or fluorosilicate (0.45 to 0.90 kg/t), and H_2O (0.5 to 0.7 t H_2O/t ore) was ground to ≈60 μm average particle size. The mixture was conditioned for 4 to 12 min with slight stirring at a temperature between 30 and 100 °C and then a carboxylic acid (C_8 to C_{24}), such as oleic acid or its alkali salts (0.23 to 2.3 kg/t ore) and H_2O (1.15 to 1.38 t/t ore) were added. The fluorspar and calcite were froth floated (at 40 to 60 °C) to the surface of the pulp by passing air through the pulp. The bertrandite containing rough tail could be refined by refroth or by known techniques. If additional tannin (≈0.35 kg/t ore) was added after grinding during the conditioning step, it was possible to remove most of the fluorspar in a previous flotation step and thereby to improve the final BeO content. The residual tannin and carboxylic acid were scrubbed from the resulting rough tailing, and all of the calcite and any remaining fluorspar were removed as described above, Becker, Foster [3].

The beneficiation of a bertrandite containing feldspatitic-quartzitic ore with 0.23% BeO, was accomplished using sodium silicate as depressing agent and distilled tall oil as collector, Chaikova et al. [4].

The main gangue was removed in ores containing bertrandite, beryl, mica, quartz, and other silicates. The ore was crushed to a <0.6 mm (−30 mesh) particle size, deslimed and conditioned with H_2SO_4 (≈1.84 kg/t) for 5 min in a pulp of 60 to 70% solids. The mica was flotationally removed by the addition of a long-chain, fatty amine salt such as tallow amine acetate (Armac T, ≈0.45 kg/t), kerosene (≈0.23 kg/t), and small amounts of a frother (pine oil or methyl-amyl alcohol). The tailing was refloated by conditioning with mineral oil (0.23 kg/t pulp) for 2 min, leaving a cleaner tailing containing about 8 to 10% BeO which recovered 70 to 90% of the BeO. Bertrandite can be separated from the beryl present in the tailing by a magnetic separator [5]. Studies were made to confirm the efficiency of a batch scale procedure for the concentration and separation of beryllium minerals from gangues composed mainly of calcite, fluorite, quartz, and mica. The continuous selective froth flotation of bertrandite and phenacite (from ores with 0.47, 0.78, and 4.7% BeO, samples I to III) at a feeding rate of ≈23 kg/h was investigated. In batch tests, the wet-ground ores (<0.23 mm particle size) were preconditioned during grinding by NaF. Next, the ore was conditioned with $(NaPO_3)_6$ in the flotation cell in quantities depending on the amount of calcite and fluorite to be depressed. After 5 min of conditioning, bath flotation was done from pulps of 25 to 35% solids with a mixture of oleic acid and kerosene (1), fuel oil (2), or turpentine (3) as the collector. The following results were obtained:

sample	concentrate (wt%)			modifiers in kg/t		collectors [a)]	
(% BeO)	total	BeO assay	BeO recovery	NaF	$(NaPO_3)_6$	oleic acid	others
I (0.47)	2.9	14.1	82	1.8	1.13	0.68	0.34 (1)
II (0.78)	3.5	17.3	78	1.91	0.8	0.6	0.45 (3)
III (4.7)	11.7	33.7	85	1.8	0.34	1.5	0.6 (2)

[a)] in kg/t, dissolved in ethyl alcohol

Since continuous operation-type flotation tests (on sample I) using the same reagent quantities and conditioning factors as in the batch test gave poor results, the optimum continuous flotation conditions for the three samples were studied. Thus, the amount of reagent used differed and the required conditioning times were longer. Conditioning with $(NaPO_3)_6$ was done in 6 tanks in a series, and the total conditioning times were 83, 71, and 90 min for the three samples. The conditioning time for NaF was about 31 min. Flotation occurred in 5 to 7 min at pH=9.8 or 9.5 with oleic acid (samples I and III, respectively) and turpentine as collector (with addition of a frother for samples I and III). Under these conditions, the following results were obtained (in % BeO): Assay 12.2 and recovery 75.3% for sample I, assay 21.0% and recovery 78.3% for sample II, and assay 25.0% and recovery 88.5% for sample III, Havens, Nissen [6]. For earlier batch tests under similar conditions, see Havens et al. [7]. For batch tests employing NaF, $(NaPO_3)_6$, $Na_4P_2O_7$, and borax as depressing agents, see Havens [8]. Fine particles (<5 and <44 μm) of phenacite are much less readily floated by oleic acid than larger ones because they require a longer treatment by the collector as a consequence of their large absorption capacity [9]. IR studies of colloidal suspensions of a phenacite mineral (<250 μm) conditioned with oleic acid show that phenacite reacts with oleic acid to form a chemisorbed oleate monolayer on its surface which is associated with a displacement of the OH^- and silicate anions from the mineral surface. Flotation recovery is directly related to the monolayer surface coverage of the chemisorbed oleate, Peck, Wadsworth [10].

A phenacite and bavenite containing beryllium ore, ground to a ≦0.2 mm grain size, was pulped and conditioned with Na_2S, Na_2CO_3, and tall oil (without preheating). The main phenacite was floated, and after three cleaning flotations, the bulk concentrate was treated with HCl at 70 °C and washed. The fluorite admixture was floated off with an (long-chain) alkyl sulfate at pH=3, and the bulk phenacite concentration was concluded by magnetic separation in a 12 kOe magnetic field to yield a final concentrate containing 6.1% BeO in a 49.5% recovery [11].

Beryllium bearing margarite was floated with other mica minerals by collective flotation using a long-chain alkylamine as the collector at pH=3 or 10. Selective flotation of the beryllium bearing margarite was done at pH=3 after addition of $AlCl_3$ as the depressing agent for other mica minerals contained in the concentrate. Thus, a margarite concentrate assaying 20% BeO was obtained in an 85% recovery from the ore [12]. A helvite bearing magnetite ore containing 1.44% BeO (from the Iron Mount in New Mexico) was ground to a <0.15 mm (−100 mesh) particle size and scrubbed with HF (≈8.28 kg/t ore) for 15 min. Flotation with oleic acid (≈243 g/t ore) and pine oil (≈81 g/t) at pH=7.5 yielded a concentrate which, after twofold cleaning contained 4.73% BeO and recovered 84.3% of the BeO content in the ore [13].

References:

[1] O. F. Marvin (U.S. 3377159 [1964/68]; C.A. **69** [1968] No. 4490).
[2] I. T. Leviush, N. I. Barashnev, V. A. Podkosova (Tr. Vses. Nauchn. Issled. Inst. Mineral'n. Syr'ya **1967** No. 18, pp. 311/26, 323/5; C.A. **70** [1969] No. 21986).
[3] C. W. Becker, R. S. Foster (U.S. 3295767 [1963/67] 1/5; C.A. **67** [1967] No. 110763).
[4] N. A. Chainikova, K. V. Kacher, G. A. Fedorishcheva (Khim. Issled. Mineral. Rast. Zhivotn. Syr'ya Dal'nevost. **1966** 4/6 from C.A. **68** [1968] No. 5081).
[5] C. C. Cook (U.S. 3112260 [1961/63] 1/2; C.A. **60** [1964] 3771).
[6] R. Havens, W. I. Nissen (U.S. Bur. Mines Rept. Invest. No. 6386 [1964] 1/18; C.A. **60** [1964] 15481/2).
[7] R. Havens, W. I. Nissen, J. B. Rosenbaum (U.S. Bur. Mines Rept. Invest. No. 5875 [1961] 1/14, 6/9, 11/3; C.A. **56** [1962] 5694; N.S.A. **16** [1962] No. 1840).
[8] R. Havens (U.S. 3078997 [1961/63] 1/5, 4; C.A. **58** [1963] 13496/7).
[9] M. Ya. Yampol'skaya, V. I. Kashtaeva, S. I. Pol'kin (Izv. Vysshikh. Uchebn. Zavedenii Tsvetn. Met. **11** No. 2 [1968] 15/8; C.A. **70** [1969] No. 39872).
[10] A. S. Peck, M. E. Wadsworth (Trans. AIME **238** [1967] 245/8; C.A. **67** [1967] No. 119310).
[11] P. N. Fedorov, Z. G. Farukshina (Tr. Vses. Nauchn. Issled. Inst. Mineral'n. Syr'ya **1967** No. 18, pp. 326/36, 331, 335; C.A. **70** [1969] No. 21996).
[12] A. I. Kakorin (Obogashch. Rud **12** No. 6 [1967] 5/8; C.A. **69** [1968] No. 45350; U.S.S.R. 188926 [1965/66]; C.A. **67** [1967] No. 56305).
[13] W. R. Storms (U.S. Bur. Mines Rept. Invest. No. 4024 [1947] 1/13, 10; C.A. **1947** 4067).

1.2 Chemical Decomposition of Beryllium Minerals

1.2.1 With Clorinating Agents

Coarsely ground beryl can be chlorinated with dry gaseous Cl_2 at 900 °C in an Ni reaction vessel lined with gold. The gaseous reaction products containing $BeCl_2$, $AlCl_3$, and other volatile chlorides can be separated from each other by fractional condensation, Fink [1].

A ceramic reaction vessel and temperatures above 1000 °C were employed by Kangro, Lindner [2]; see also Spitzin [3]. The reaction between BeO and Cl_2 is very slow at ≦1000 °C, but a mixture of BeO and carbon black can be chlorinated at 800 °C [3]. The chlorination of an intimate mixture of beryl and carbon at 800 °C with a mixture of Cl_2 and CCl_4 gives better yields than with Cl_2 alone. Condensation of the resulting chloride vapors at 375 °C gives $BeCl_2$ free of Al, Si, and Fe chlorides, which condense at lower temperatures, Winters,

Yntema [4]. HCl, CCl_4, $COCl_2$, SCl_2, and other chlorinating agents can be also used instead of Cl_2, Jaeger [5]. The volatile chlorides can be separated either by fractional condensation or by extraction with suitable solvents such as the liquid chlorides or oxychlorides of S, P, C, or B, which dissolve $AlCl_3$ and $SiCl_4$, but not $BeCl_2$, Strauss [6].

Arc melting of beryl-carbon mixtures leads to carbides which can be chlorinated by Cl_2 at temperatures of a few hundred degrees [5]. The two stages (formation of carbide and chloride) are combined by using electrodes composed of mixtures of beryl and 30% soft coal and by using electric arc ignition in a Cl_2 atmosphere, Sheer, Korman, et al. [7].

A briquetted mixture of beryl and coke was chlorinated by only 38% at 925 °C and by 85% at 1300 °C, Zverev, Barsukova [8]. A mixture of beryl and charcoal briquetted with coal tar gave yields $\leqq$76% after 3 h at 1250 to 1300 °C, Firsanova, Belyaev [9], see also McTaggart [10]. For the chlorination of beryl-carbon mixtures under various parameters of grinding, briquetting, temperature, and addition of 2% NaCl, see [10]. Almost complete chlorination was obtained by treating a mixture of powdered (0.046 mm) beryl (12 parts) and coke (0.058 mm, 6 to 8 parts) in a melt of NaCl+KCl (36 to 40 parts) at 950 °C with Cl_2 at a flow rate of 130 mL/min for 3 h [8].

The chlorination of a briquetted mixture of charcoal with a BeO bearing margarite (1.55 wt% BeO) at 1000 °C yielded a volatile fraction containing about 95% of the BeO as $BeCl_2$ and a total decomposition of 85%, Pisitsina [11]. Chlorination at 700 to 900 °C of beryl and phenacite containing ores bearing 0.1 to 2.8% BeO which had been pelleted with carbon in a 10:4 or 10:1 weight ratio and dried at 110 °C for at least 12 h yielded practicable results only for the helvite ore (1.56% BeO) from which (at 700 °C) 62.2% of its BeO content could be recovered as $BeCl_2$ in the volatile product. Beryl ores with 2.80% BeO gave rather poor results when treated with Cl_2, CCl_4, or dry HCl at 850 °C (<30%), however, chlorination with $SOCl_2$ and S_2Cl_2 at 700 °C (45.7 and 54.2%) was more successful. Addition of Na_3AlF_6 (but not of NaCl) to the mineral-carbon mixture (4:10:2 and 10:10:2 ratios) and chlorination with Cl_2 at 700 °C gave improved results (70.2 and 85.4% recovery) for a beryl ore containing 2.80% BeO and also for beryl-spodumene concentrates with 1.15% BeO and 1.96% BeO at 800 to 900 °C, May, Hoatson [12]. For chlorination of BeO containing ores in mixture with carbon (as pellets) by Cl_2 up to 1200 °C, see [13]; in mixtures with C or carbides at $\geqq$1800 °C, see [14].

Chlorination in the presence of Si was accomplished using pea size briquettes of finely ground (<74 µm) beryl and Si which were compacted with silicone grease (ratio 3:1:0.3), and sintered at about 500 °C. After being heated in a Vycor tube in a Cl_2 stream between 1285 and 1390 °C, the total yield of $BeCl_2$, from the briquettes, in terms of original BeO, was 30%. Addition of finely ground $CaCl_2$ to beryl in a 1:4 ratio slightly improved the yield of $BeCl_2$ [15, 16].

Better yields were obtained from mixtures with $CaCl_2$ at higher temperatures using $SiCl_4$ as the chlorinating agent. The best yield (93.2%) resulted from an ore with 12.4% BeO ground to a 5 to 15 µm size after addition of $CaCl_2$ (3/8 of the ore weight) and heating the mixture containing 476 mg BeO under Ar at 1140 °C for 1 h. Then the Ar stream (24 mL/min) was saturated with $SiCl_4$ vapor (at first at 27 °C, then at 54 °C, corresponding to a partial vapor pressure of 710 Torr), and it was heated to 1400 °C in 85 min and up to 1525 °C in 150 min followed by slow cooling. The total time of heating above 1400 °C was about 4.5 h and 92.5% of the beryllium was recovered in the chloride condensate (mixed with $AlCl_3$), while 0.7% was found as a precipitate in the $SiCl_4$ condensate. If no $CaCl_2$ was added to the ore, only 6% of the beryllium could be recovered as $BeCl_2$ [17, 18].

Chlorination of BeO was also done in the presence of carbon and sulfur using a crushed beryl ore mixed with coke or coal in a 1:(1 to 3) weight ratio. The mixture was heated at 900 to 1200 °C in the presence of sulfur vapor, CS_2, or H_2S to convert the BeO to BeC and BeS, which form $BeCl_2$ when treated with dry Cl_2 or HCl at 800 to 900 °C. The $BeCl_2$ was separated from the other volatile chlorides ($AlCl_3$, $SiCl_4$) by fractional condensation, Gardner [19].

References:

[1] C. G. Fink (U.S. 2104741 [1935/38] 1/3; C. **1938** II 916).
[2] W. Kangro, A. Lindner (Brit. 356380 [1929/31] 1/4; C. **1931** II 3242).
[3] V. Spitzin (Z. Anorg. Allgem. Chem. **189** [1930] 337/66, 347, 361).
[4] R. W. Winters, L. F. Yntema (Trans. Am. Electrochem. Soc. **55** [1929] 205/7).
[5] G. Jaeger (Metall **4** [1950] 183/91, 186).
[6] R. Strauss (Angew. Chem. **48** [1935] 745/50, 747).
[7] C. Sheer, S. Korman, P. H. Sellew, W. R. Rice (NYO-1040 [1950] 1/127, 101/3; N.S.A. **9** [1955] No. 6597).
[8] L. V. Zverev, Z. S. Barsukova (Dokl. Akad. Nauk SSSR **130** [1960] 593/5; Proc Acad. Sci. USSR Chem. Technol. Sect. **130/135** [1960] 15/7).
[9] L. A. Firsanova, A. I. Belyaev (Sb. Nauchn. Tr. Inst. Tsvetn. Met. M. I. Kalinina **1957** No. 26, pp. 184/92 from C.A. **1961** 285).
[10] F. K. McTaggart (J. Council Sci. Ind. Res. [Australia] **20** [1947] 564/84, 573/4, 583; C.A. **1948** 5623).

[11] T. I. Pisitsina (Tsvetn. Metal. **33** No. 4 [1960] 56/8; Soviet J. Non-Ferrous Metals **1** No. 4 [1960] 70/2).
[12] J. T. May, C. L. Hoatson (U.S. Bur. Mines Rept. Invest. No. 6037 [1962] 1/19, 7/9; C.A. **57** [1962] 14775).
[13] Deutsche Gold- und Silber-Scheideanstalt vorm. Roessler (Fr. 858443 [1938/40] 1/5; C. **1941** II 807).
[14] H. von Zeppelin, K. Hoffmann, H. Kalhammer (Ger. 1200269 [1960/65] 1/4; C.A. **64** [1966] 4674).
[15] R. O. Bach, Beryllium Metals & Chemicals Corp. (U.S. 3250592 [1964/66] 1/2; C.A. **65** [1966] 373).
[16] Beryllium Metals & Chemicals Corp. (Fr. 1469062 [1965/67] 1/3; C.A. **67** [1967] No. 75525).
[17] R. O. Bach, Beryllium Metals & Chemicals Corp. (Brit. 1023357 [1963/66] 1/4; C.A. **64** [1966] 17114).
[18] Beryllium Metals & Chemicals Corp. (Fr. 1372644 [1963/64] 1/3; C.A. **62** [1965] 11466).
[19] D. Gardner (Brit. 482531 [1936/38] 1/3; C. **1938** II 764).

1.2.2 With Fluorinating Agents

1.2.2.1 HF and SiF_4

Beryl concentrates (2.80% BeO) or beryl spodumene concentrates (1.96% BeO) were pelleted with carbon in a 10:4 weight ratio and dried at 110 °C. Treatment with anhydrous gaseous HF at 700 °C for 2 h converted 99.2 and 96.5%, respectively, of the BeO content into soluble BeF_2 which was extracted by hot water. Precipitation with ammonium hydroxide and ignition to the oxides gave a product which contained 62 to 70% BeO with Fe, Al, and Si as the major impurities, May, Hoatson [1]. Treatment of the crude beryl ore with gaseous HF at 400 to 600 °C produced soluble BeF_2 and volatile SiF_4 leaving insoluble CaF_2, Al_2O_3, and Fe_2O_3. The oxides apparently originate from hydrolyzed fluorides, Vogel

[2], also see [3]. Treatment of a beryl containing 12% BeO with HF at 600 °C until the evolution of SiF_4 had ceased or with equal volumes of HF and SiF_4 at 500 °C in a rotatory furnace, leaching the powdered product with hot water, and precipitation of $Be(OH)_2$ from the extract by ammonia was proposed by [4]. For the removal of F^- and Fe^{3+} ions by precipitation with Ca acetate, see [5]. The reaction of powdered beryl with gaseous SiF_4 in a tubular rotating furnace at 600 to 650 °C yielded water-soluble BeF_2 and insoluble AlF_3 and SiO_2 [7]; see also [3]. Complete disintegration of a beryl ore containing 11.90% BeO occurred with a mixture of HF (80 mol%) and NO_2 (20 mol%) at 80 °C in 10 h to yield $(NO)_2BeF_4$ (along with $NOAlF_4$ and $(NO)_2SiF_6$) which could be converted into BeF_2 at 400 °C. The $(NO)_2SiF_6$ is volatilized and the $NOAlF_4$ is decomposed leaving insoluble AlF_3 [6].

References:

[1] J. T. May, C. L. Hoatson (U.S. Bur. Mines Rept. Invest. No. 6037 [1962] 1/19, 10/1; C.A. **57** [1962] 14775).

[2] F. Vogel (Z. Prakt. Geol. **42** [1934] 120/4, 122; C. **1936** I 162).

[3] G. Jaeger (Metall **4** [1950] 183/91, 185).

[4] I. G. Farbenindustrie A.-G. (Brit. 345902 [1930/31] 1/2; C. **1931** I 3596).

[5] M. Zimmermann, I.G. Farbenindustrie A.-G. (Ger. 550758 [1931/32] 1/2; C. **1932** II 586).

[6] A. Kigoshi, K. Okada, A. Ohkawa, M. Ohmi (Sci. Rept. Res. Inst. Tohoku Univ. Ser. A **22** [1971] 192/201, 194/6; C.A. **75** [1971] No. 120337).

[7] B. Wempe (Ger. 651471 [1933/37]; C. **1938** I 3373).

1.2.2.2 With Alkali Fluorides, CaF_2, FeF_3, and ZnF_2

Complete and rapid conversion of the BeO component of beryl into BeF_2 was accomplished by treating the finely ground (<74 mm) ore in a fluidized state with vaporized NH_4HF_2 at 450 to 600 °C under atmospheric pressure. A highly pure product resulted when about 95% of the BeO in the beryl had reacted [1]. The BeO in beryl or other beryllium minerals was converted into $(NH_4)_2BeF_4$ by heating with about half or slightly more than half of the stoichiometric amount of NH_4F or NH_4HF_2 at 250 to 660 °C in a closed vessel. The resulting fluoroberyllate was converted into an oxide fluoride enriched with BeF_2 [2], see also [3]. After separation (by water), the BeF_2 could be converted into BeO by treatment with steam at 400 to 600 °C under fluidizing conditions [1].

Beryl bearing ore was also completely decomposed when heated with a stoichiometric amount of $NaHF_2$ at 600 to 650 °C in a muffle or rotating furnace. Reduction of the amount of $NaHF_2$ to that required for the formation of the water soluble Na_2BeF_4 (not regarding Al_2O_3 and SiO_2) recovered at least 90% of the Be content in the ore [4]. The reaction with $NaHF_2$ was the basis for an earlier technical process, applied at SAPPI, Italy. A briquetted mixture of powdered beryl and $NaHF_2$ (weight ratio 2:1) was heated at 680 °C for 8 h to produce insoluble Na_3AlF_6 along with Na_2BeF_4. The mixture was extracted with hot (80 to 95 °C) water containing $KMnO_4$ to oxidize and render insoluble admixtures of iron. Beryllium hydroxide was precipitated from the filtrate by NaOH, washed, filtered, repulped, and pumped into a hydraulic press. A description and flow sheet of the process is reported in [5], also see [6]. In another technical process, a finely ground (<43 µm) beryllium bearing ore was mixed with NaF and Na_2CO_3, pelleted, and fused in a sintering furnace at 955 to 980 °C. The sinter was crushed, slurried with water containing H_2O_2, and filtered several times to remove all solids. NaOH was added to the filtrate containing the beryllium (15 to 20 g BeO/L) and some impurities (Si, Fe, Al, B, Mg, Li), in order to precipitate $Be(OH)_2$ which was further purified in several steps [7].

Complete decomposition of the ore (for the analytical determination of Be) was possible at lower temperatures by fusion with low-melting KF (41 °C) of KHF_2 in a 1:6 weight ratio. Gentle heating of the mixture for about 10 to 15 min followed by melting for 2 to 3 min in a strong flame gave a resolidified product that was readily decomposed by water and H_2SO_4 [8].

Fusion of a phenacite concentrate with CaF_2 (fluorite) and Na_2CO_3 in a 1:0.5:0.75 weight ratio at 750 °C for 3 h (or with NaCl instead of Na_2CO_3 at 900 °C) yielded a product from which >95% of the BeO content could be recovered by hot aqueous 10% H_2SO_4 [9].

The beryllium content of finely ground mixtures containing beryl or bertrandite and fluorite (CaF_2) in a 1:(100 to 500) weight ratio was converted into the soluble sulfate with a yield of 92%. Mixtures were treated with a 1.3- to 1.5-fold excess (by weight) of H_2SO_4 (96%) at 220 to 300 °C and atmospheric pressure for 2 h. The HF formed during the reaction promoted the conversion of the BeO component in the ore into $BeSO_4$ [10]. An earlier method involved the treatment of a pegmatitic beryl ore with CaF_2 (470 kg per 1000 kg) and a solution of $NaNO_3$ and HNO_3 for some hours at 85 °C. The mixture was filtered with pressure and the resulting filter cake was heated for 2 h at 650 °C followed by leaching and precipitation of $Be(OH)_2$ [11].

Good recovery of beryllium was obtained by roasting (in an open system) briquetted mixtures of finely ground (<149 μm) beryl ore with $FeF_3 \cdot xH_2O$ (F^- to beryl mole ratio of 6:1) at 550 °C for 4 h or with ZnF_2 in a 5:1 mole ratio at 750 °C for 2 h. The process produced yields of 82.2 and 96.6%, respectively, of solubilized BeO which could be extracted by an aqueous 5% solution of NH_4F as $(NH_4)_2BeF_4$. Other simple or complex fluorides gave far lower results [12].

References:

[1] M. H. Furland (U.S. 2399178 [1945/46] 1/5; C.A. **1946** 6764).
[2] Deutsche Gold- und Silber-Scheideanstalt vorm. Roessler (Brit. 457315 [1936] from C.A. **1937** 3220).
[3] A. Wille, G. Jaeger, Deutsche Gold- und Silber-Scheideanstalt vorm. Roessler (U.S. 2162323 [1939] from C.A. **1939** 7501).
[4] C. Adamoli, G. Panebianco, S. Opatowski (Ger. 599101 [1933/34] 1/5, 3; C. **1934** II 3420).
[5] H. W. West. W. F. Randall, G. L. Miller, R. Turner, E. M. Foster, D. W. Crossley (BIOS-FR-550 [1945] 1/8, 22/4, 81).
[6] H. A. Sloman, C. B. Sawyer (FIAT-FR-522 [1946] 1/102, 63/8; N.S.A. **1** [1948] No. 1845).
[7] Anonymous (Mining Eng. **13** [1961] 1144/5).
[8] E. P. Ozhigov, V. I. Shevchenko (Soobshch. Dal'nevost. Filiala Sibirsk. Otd. Akad. Nauk SSSR **1961** No. 14, pp. 95/7; C.A. **60** [1964] 17).
[9] Yu. M. Putilin, A. D. Romanova, L. V. Favorskaya (Tekhnol. Mineral. Syr'ya **1972** 33/47, 37/9; C.A. **85** [1976] No. 97309).
[10] R. S. Olson, E. C. Tveter, J. P. Surls, R. G. Shaw, Dow Chemical Co. (U.S. 3375060 [1964/68] 1/3; C.A. **68** [1968] No. 97719).
[11] C. Adamoli, Seri Holding Soc. An. Luxemburg (Ger. 745085 [1938/44] 1/2; C.A. **1946** 20; Fr. 845666 [1938/39]; C.A. **1941** 1195).
[12] A. J. Stonehouse (NYO-1201 [1954/57] 1/29, 4/6; N.S.A. **11** [1957] No. 13095).

1.2.2.3 With Alkali Hexafluorosilicates and/or Na_3FeF_6

Copaux [1] first described the decomposition of beryl by heating with Na_2SiF_6. Extended studies of sintered mixtures of beryl and Na_2SiF_6 [2, 3] have shown that the reaction roughly follows the equation: $3BeO \cdot Al_2O_3 \cdot 6SiO_2 + 3Na_2SiF_6 \rightarrow 6NaF + 3BeF_2 + 2AlF_3 + 9SiO_2$ followed by the competing reactions: $2NaF + BeF_2 \rightarrow Na_2BeF_4$ (soluble in water) and $3NaF + AlF_3 \rightarrow Na_3AlF_6$ (insoluble). The best results were obtained when a briquetted mixture of finely ground (40 to 75 µm) beryl and Na_2SiF_6, mole ratio 1:3, was heated at 700 °C for 4 h in a static atmosphere. The beryllium was converted (96.2%) into a mixture of Na_2BeF_4 and BeF_2 which were readily extracted by water. A finer fraction (5 to 15 µm) of beryl, heated with Na_2SiF_6 (mole ratio 1:3, static atmosphere) at 750 °C, reacted far more rapidly and yielded 97% of extractable Be compounds within 0.5 h. The reaction was accompanied by a sharp rise in pressure which was due to the formation of SiF_4 from the thermal decomposition of Na_2SiF_6. Since the pressure decreased to the starting value within 30 min, the SiF_4 must be consumed in the reaction steps leading to the complete conversion of BeO to Na_2BeF_4. Lower beryl to Na_2SiF_6 ratios (if not compensated by NaF), heating below 700 or above 750 °C (melting), as well as prolonged (>4 h) roasting reduce the amount of Be compounds that are leachable by water at 20 °C. This is evidently due to the conversion of the initially formed cryolite and cristobalite into albite which incorporates Be^{2+} ions as shown by X-ray studies [2]. Thus, a temperature of 725 °C and a time not exceeding 3 h represent the optimum conditions for a beryl (<74 µm) mixed with Na_2SiF_6 in a 1:3 mole ratio [2, 3]. Replacement of Na_2SiF_6 by K_2SiF_6, Rb_2SiF_6, or Cs_2SiF_6 gave good yields ($\geqq \sim 90\%$) of extracted Be after 4 h of sintering at 700 or 750 °C. Rb_2SiF_6 and Cs_2SiF_6 gave good yields even at 600 °C, whereas the efficiency of Li_2SiF_6 was considerably lower (~70%) even at 750 °C. At 700 °C, the efficiency of Na_2SiF_6 and Li_2SiF_6 in a streaming argon atmosphere is lower than that of the other fluorosilicates because their temperature of decomposition is lower [3]. A crushed beryl ore or concentrate containing at least 10% BeO was ball-milled to a 0.075 mm grain size, briquetted with the calculated amount of Na_2SiF_6, and sintered at 750 °C for 2 h. After cooling, the briquettes were crushed, ground, and leached three times with water at room temperature [4]. Briquetted mixtures of low grade beryllium ores (0.6 and 1.3% BeO) were crushed to <0.15 mm and sintered with Na_2SiF_6 (30 or 20 kg per kg BeO in the low or high grade clayey ore) at 750 °C for 3 h. After grinding the sinter to <0.15 mm, at least 80% of the Be could be leached as Na_2BeF_4 by water. Addition of NaOH precipitated $Be(OH)_2$ at pH = 10.8 which was redissolved in aqueous 1.5 N NaOH (ratio Na:Be = 5.6:1). The resulting solution containing 7.1 g BeO per L was filtered and boiled for 1 h to precipitate granular $Be(OH)_2$ in a 61% yield (calculated from the Be content in the ore) [5]. For earlier work based mainly on the method of Copaux [1], see [6 to 9]; patents, see [10 to 12]. For information concerning the precipitation of $Be(OH)_2$ from fluoroberyllate solutions by $Ba(OH)_2$ or $Ca(OH)_2$ instead of NaOH, see Fischer [10]. Sodium fluoride and gaseous SiF_4 can replace Na_2SiF_6. They react according to the equation: $2(3BeO \cdot Al_2O_3 \cdot 6SiO_2) + 12NaF + 3SiF_4 \rightarrow 6Na_2BeF_4 + 2Al_2O_3 + 15SiO_2$. In practice, a mixture of beryl and NaF is simply heated at 600 to 700 °C under SiF_4 [13 to 15].

Part of the Na_2SiF_6 (up to half) can be replaced by an equivalent amount of NaF according to the reaction: $2(3BeO \cdot Al_2O_3 \cdot 6SiO_2) + 3Na_2SiF_6 + 6NaF \rightarrow 6Na_2BeF_4 + 2Al_2O_3 + 15SiO_2$. Heating at 600 to 900 °C for about 3 h gave results comparable to those obtained with pure Na_2SiF_6 (see above) [15, 16], see also the review of Higbie, Farmer [17]. Claflin has suggested that Na_2SiF_6 and NaF can be replaced by $CaSiF_6$ and CaF_2 and that the resulting sinter cake containing $CaBeF_4$ and SiO_2 (from phenacite, Be_2SiO_4) can be leached with H_2SO_4 [16].

To reduce the cost of production, part of the Na_2SiF_6 (up to a third) was replaced by Na_2CO_3 which reacts according to: $3BeO \cdot Al_2O_3 \cdot 6SiO_2 + 2Na_2SiF_6 + Na_2CO_3 \rightarrow 3Na_2BeF_4 + Al_2O_3 + 8SiO_2 + CO_2$. Pellets of the finely ground beryl (<74 µm) were heated in a mixture containing BeO, Na_2SiF_6, and Na_2CO_3 in a 1.0:5.0:1.4 weight ratio at 750 °C [18], also see [19]. A 7% excess of fluoride served to compensate for the loss of SiF_4 during heating

[19]. A typical batch, which was the prototype of a technical procedure (applied by Murex, England), is described by Bryant [20]. A crushed and ball-milled beryl ore containing 12.0% BeO (95% <0.15 mm size) was mixed with Na_2SiF_6, Na_2CO_3, and water in an approximately 10:8:1:1 weight ratio. The mixture was briquetted and processed following the flow sheet shown in **Fig. 1-2** (taken from Darwin, Buddery [24]). After firing at 780 °C for 6 h, the briquettes were crushed, ground in a wet ball mill, and the resulting slurry treated with $KMnO_4$ to oxidize divalent iron. The slurry was filtered to give a filtrate containing about 13 g BeF_2 per L. Sodium hydroxide was added with stirring to the filtrate until all the Be was precipitated (pH = 11.3 to 11.6) as the crude hydroxide. The gelatinous hydroxide, contaminated with $Fe(OH)_3$, NaF, and silica as the main impurities, was filtered and redissolved

Fig. 1-2

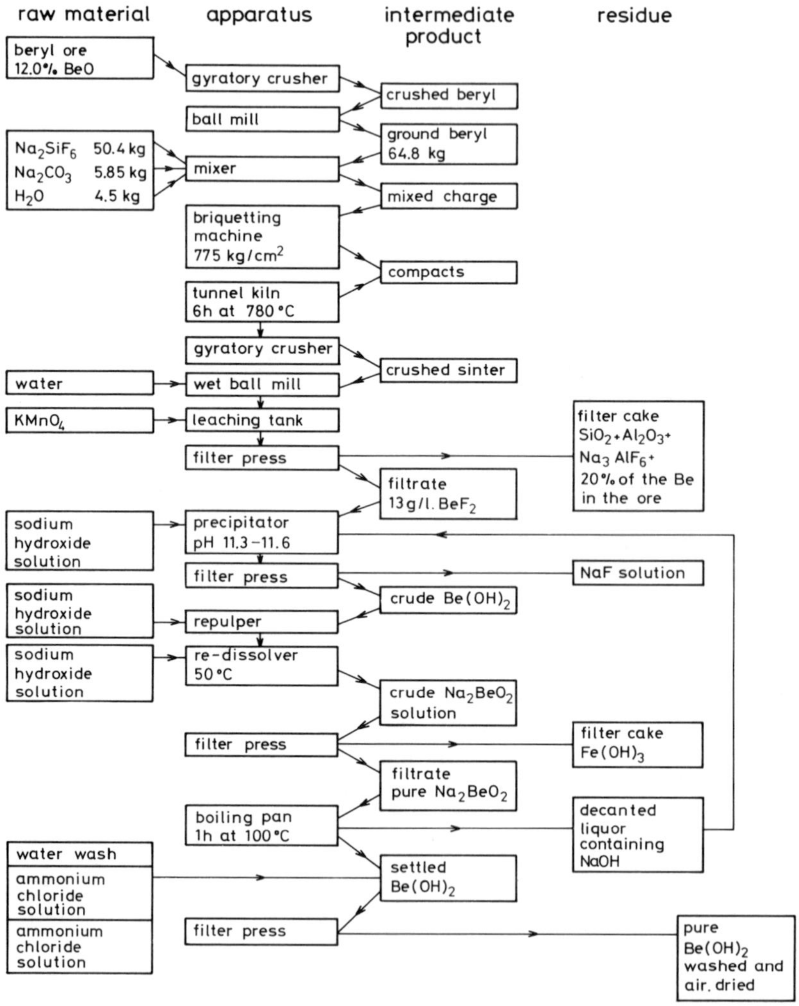

Flow sheet for the production of beryllium hydroxide from beryl at Murex.

in an excess of NaOH solution at 50 °C. After the coagulation of $Fe(OH)_3$ was complete, the solution was filtered and the filtrate containing the beryllate kept near boiling for 1 h to precipitate granular $Be(OH)_2$. The precipitate was decanted and washed with demineralized water and NH_4Cl solution until free from NaOH. The air-dried product contained 30% moisture as well as 0.3% SiO_2 and 0.5 to 1.0% NaF impurities (% based on BeO), Bryant [20]. A briquetted mixture containing beryl (11% BeO, <74 µm grain), Na_2SiF_6, and Na_2CO_3 in a 11.15:8.1:0.95 weight ratio produced the best results. The wet ground sinter (<0.74 µm) was leached three times with water to give a combined extract which recovered up to 96% of the beryllium from the ore [21], also see Morana, Simons [19]. Heating of a mixture of beryl, Na_2SiF_6, and Na_2CO_3, Na_2O or NaOH in a 11:6.1:0.93 weight ratio (referred to Na_2O and 92% Na_2SiF_6) at 660 to 690 °C for 2 h was proposed by Opatowski [22]. Sintering of a beryl spodumene concentrate crushed to <0.15 mm which contained 30% beryl (=3.7% BeO) with 10 kg Na_2SiF_6 and 3 kg Na_2CO_3 per kg of BeO at 750 °C for 3 h and leaching the powdered sinter with hot (75 °C) water in four steps recovered 84% of the BeO in the concentrate which, after suitable precipitation with NaOH, yielded a rather pure $Be(OH)_2$. Leaching with 0.1 N H_2SO_4 recovered 95% of the BeO, but gave a $Be(OH)_2$ product of much lower purity (10.2% SiO_2, 3.2% P_2O_5). Extraction with cooler water (25 °C) yielded an inferior recovery of BeO [23]. For low grade beryl ores (0.6 to 1.3% BeO), partial replacement of Na_2SiF_6 by Na_2CO_3 was ineffective in small amounts and detrimental in large amounts [5]. An easily filterable granular precipitate of $Be(OH)_2$ was produced by adding the required amount of NaOH to 20% of the aqueous Na_2BeF_4 extract followed by heating to 95 °C with steam coils. The remaining extract was then slowly added with stirring and the solution kept at 95 °C. After all of the extract had been added, the pH was adjusted to pH 12 by further addition of NaOH or extract. The precipitate was filtered, washed with hot water, and pressed to give a press cake containing 75% moisture which was dried and calcined to the oxide, Lundin [18], Paul et al. [21], see also Morana, Simons [19].

Replacement of Na_2SiF_6 by Na_3FeF_6 results in the following reaction at 750 °C: $3BeO \cdot Al_2O_3 \cdot 6SiO_2 + 2Na_3FeF_6 \rightarrow 3Na_2BeF_4 + Fe_2O_3 + Al_2O_3 + 6SiO_2$, Kawecki [25, 26]. Finely ground (<74 µm) beryl ore (10 to 12% BeO) was mixed with 90% of the stoichiometric amount of Na_3FeF_6, briquetted and sintered at 750 °C for 1 h to convert at least 95% of the beryllium content into soluble Na_2BeF_4. The sinter cake was ground and extracted with water to give, after filtration of the unreacted oxides of Al, Si, and Fe, a solution containing 3 g BeO per L, Kawecki [27], also see [25, 28]. A low grade clayey beryl ore (0.5% BeO), which had been sintered with 30 kg Na_3FeF_6 per 1 kg BeO at 750 °C for 3 h, allowed to leach 80% of its beryllium content with water [5, p. 6]. As in the case of Na_2SiF_6 (see above), part of the Na_3FeF_6 can be replaced by Na_2CO_3 (up to 20%) under the same general conditions [25]. In a technical procedure applied by the Beryllium Corporation of America, beryllium ore was crushed, ground to <74 µm by 70% and mixed with a slight stoichiometric excess of moist Na_3FeF_6, Na_2SiF_6, and Na_2CO_3. The mixture was then briquetted and fired at 750 °C for 2 h as outlined in the flow sheet given in **Fig. 1-3**, p. 24 (taken from [24] according to [26]). Fusion of the briquettes must be avoided because of the formation of insoluble Be compounds, whereas lower temperatures would require too much heating time. The sintered briquettes were crushed, ball-milled wet, and leached by water with repeated agitating, settling, and decanting through a filter (the contact between water and sinter was kept at a minimum) [26]. Granular $Be(OH)_2$ was precipitated by NaOH at pH=12 as described above [18]. The Na_2BeF_4 formed in the reaction was converted into the more soluble $(NH_4)_2BeF_4$ by treating the finely ground sinter with a small stoichiometric excess of $(NH_4)_2SiF_6$ in the presence of water at room temperature. The resulting solution of $(NH_4)_2BeF_4$ was separated from the very slightly soluble Na_2SiF_6 (and the oxides) by filtration and evaporated to give the crystallized salt [28].

Fig. 1-3

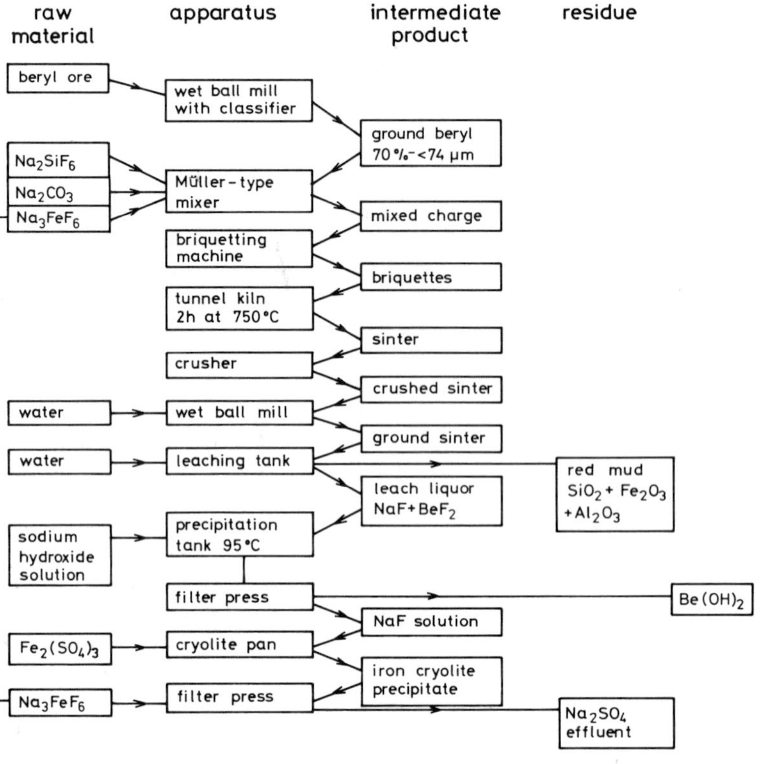

Flow sheet for the production of beryllium hydroxide from beryl at the Beryllium Corporation of America.

It is desirable to remove phosphate from beryl concentrates because it seriously interferes with the fluorosilicate process. This is done by heating the ore concentrate at 700 °C for 1 h followed by treatment with cold 2N HCl. The P_2O_5 content of the concentrate was reduced from 8.1 to 0.22 wt% and the CaO-content from 7.2 to 1.4 wt%. Treatment with H_2SO_4 instead of HCl was less efficient, Everest, Napier [29].

References:

[1] H. Copaux (Compt. Rend. **168** [1919] 610/2).
[2] K. R. Hyde, P. L. Robinson, M. J. Waterman, J. M. Waters (Bull. Inst. Mining Met. No. 653 [1961] 397/406, 400/2, 605/13, 605/6; C.A. **1961** 14218).
[3] B. F. Greenfield, K. R. Hyde, F. H. Moore, J. M. Fletcher, P. I. Robinson (AERE-R-4182 [1963] 1/21, 4/11; C.A. **59** [1963] 7186).
[4] F. Hochstetter (Chemiker-Ztg. **86** [1962] 108/9).
[5] R. O. Dannenberg, D. W. Bridges, J. B. Rosenbaum (U.S. Bur. Mines Rept. Invest. No. 6156 [1963] 1/12; C.A. **58** [1963] 8690).
[6] I. E. Newnham (Res. Appl. Ind. **11** [1958] 185/91, 186/8).
[7] D. C. McLaren (Mining Mag. [London] **69** [1943] 273/82, 280).

[8] W. Hessenbruch (Metall Erz **32** [1935] 234/7).
[9] R. Gadeau (Rev. Met. [Paris] **32**[1936] 627/37, 628).
[10] H. Fischer, Metal & Thermit Corp. (U.S. 1815056 [1927/31] 1/2; C. **1931** II 3139).
[11] H. Fischer, Siemens & Halske A.-G. (Ger. 604575 [1927/34] 1/2; C. **1935** I 943).
[12] I.G. Farbenindustrie A.-G. (Neth. 50629 [1938/41] 1/2; C. **1941** II 2480; Brit. 524171 [1939/40] 1/2; C. **1940** I 614).
[13] H. C. Claflin, D. O. Hubbard, Beryllium Development Corp. (Brit. 391369 [1932/33] 1/2; C. **1933** II 1581).
[14] H. C. Claflin, D. O. Hubbard, Beryllium Development Corp. (U.S. 1991272 [1931/35] 1/2; C. **1935** I 1958).
[15] G. Jaeger (Metall **4** [1950] 183/91, 185).
[16] H. C. Claflin, Beryllium Development Corp. (Brit. 374705 [1931/32] 1/2; C. **1932** II 2105; Ger. 577629 [1931/33] 1/2; C. **1933** II 762; U.S. 1929014 [1930/33] 1/2).
[17] K. B. Higbie, M. C. Farmer (Chem. Eng. Progr. **54** [1958] 51/4).
[18] H. Lundin (Trans. Am. Inst. Chem. Eng. **41** [1945] 671/91, 672/4).
[19] S. J. Morana, G. F. Simons (J. Metals **14** [1962] 571/4, 572).
[20] P. S. Bryant (Extr. Refining Rarer Metals Proc. Symp., London 1956 [1957], pp. 310/22, 311/6; C.A. **1958** 19783).
[21] C. M. Paul, B. P. Sharma, K. S. Subba Rao, M. G. Rajadhyaksha, C. V. Sundaram (Trans. Indian Inst. Metals **28** [1975] 373/9, 374; C.A. **84** [1976] No. 138814; BARC-794 [1975] 1/18, 6; C.A. **84** [1976] No. 153549).
[22] R. S. Opatowski (U.S. 2209131 [1938/40] 1/3; C.A. **1941** 285).
[23] R. O. Dannenberg, D. W. Bridges, J. W. Rosenbaum (U.S. Bur. Mines Rept. Invest. No. 6153 [1963] 1/18, 6/12; C.A. **58** [1963] 8692).
[24] G. E. Darwin, J. H. Buddery (Metallurgy of the Rarer Metals, Vol. 7: Beryllium, London 1960, pp. 27, 29).
[25] H. Kawecki, Reconstruction Finance Corp. (U.S. 2312297 [1941/43] 1/3; C.A. **1943** 4864).
[26] H. Kawecki (in: D. W. White, J. E. Burke, The Metal Beryllium, Cleveland, Ohio, 1955, pp. 63/70, 66).
[27] H. Kawecki (Trans. Electrochem. Soc. **89** [1946] 229/36, 233).
[28] H. Kawecki, G. F. Simons, Beryllium Corp. (U.S. 2532102 [1948/50] 1/4, 2; C.A. **1951** 1489).
[29] D. A. Everest, E. Napier (Proc. 6th Intern. Congr. Mineral. Process., Cannes 1963 [1964], pp. 231/43, 231/2; C.A. **62** [1965] 15811).

1.2.3 With Alkaline Agents

Beryl or Be concentrates can be decomposed by sintering or melting with alkali hydroxides (300 to 400 °C), alkali carbonates (700 to 1000 °C), CaO or $CaCO_3$ (1200 to 1300 °C), $Ca(OH)_2$ at $p_e=30$ atm, or by heating with concentrated aqueous NaOH (or KOH) under $p_e=3$ to 6 atm at temperatures below 200 °C. In most cases, the alkaline treatment is followed by an acid leaching (generally H_2SO_4) to give solutions containing Be, Al, and Fe sulfates as well as alkali sulfates [1]. Melting (sintering) with lime (CaO) found commercial use during Second World War in the Degussa sulfate process. The process involves (1) fusion with lime, (2) H_2SO_4 acid leaching, (3) alum crystallization to remove Al, (4) neutralization with limestone to pH 4 ($+H_2O_2$) to remove Fe, (5) ammonia precipitation of $Be(OH)_2$, and (6) calcination of $Be(OH)_2$ to form BeO, see p. 34 for the flow sheet.

A special case is the soda-lime-acid process in which beryl is heated with a mixture of Na_2CO_3 and CaO at about 1000 °C. Treatment of the sinter with dilute acid (HCl or HNO_3) leaves insoluble BeO and SiO_2 [1].

NaOH. A mixture of 20 g beryl, 30 g NaOH, and 30 g H_2O was heated in an Ni crucible for 4 h at 250 °C. The cool roast cake was washed with H_2O and leached with hot H_2SO_4 (50 to about 96% concentration) or with H_2SO_3 (generated by bubbling SO_2 into the aqueous slurry). The recovery of Be in the solution was >90%, Surls [2]. A mixture of 3 kg finely powdered beryl and 3.6 kg NaOH (+10 to 12% H_2O) was slowly heated in an iron crucible to about 400 °C, and the temperature was maintained for 10 to 12 h. Alternatively, a mixture of 2 kg beryl, 6 kg NaOH, and 1 L H_2O was heated with stirring in an autoclave at 160 to 170 °C for about 6 h. The cooled product, soluble in dilute HCl, showed a total decomposition of the beryl. When leached with H_2O Be is left in the residue and about 60% of the alkali and ~20% of the silicate is dissolved. Treatment of the residue with concentrated H_2SO_4 separated SiO_2 from the water soluble sulfates of Be, Al, and Fe. Addition of ammonia precipitated the hydroxides from which Be was extracted with a solution of $(NH_4)_2CO_3$ [3]. Beryl or a Be bearing mineral was heated with NaOH (or KOH) in a mole ratio 1:16 (related to BeO) in a closed vessel at 150 to 170 °C in the presence of 5 to 25% H_2O (based on alkali) and then extracted with water [4]. Powdered beryl was treated with aqueous NaOH (40%) at a pressure of 40 kg/cm^2 for 4 h (or with saturated Na_2CO_3 solution), and then the cake was washed and leached with hydrochloric acid [5].

The finely ground beryllium-bearing mineral was stirred with a 50% aqueous NaOH solution at 225 °C for 12 h. Water was then added, and the slurry was filtered to remove the residual alkali. The filter cake containing the $Be(OH)_2$ was treated at 50 °C with an approximately 14% solution of $NaHCO_3$ for 7 h to dissolve the $Be(OH)_2$. The solution was filtered and boiled to precipitate granular $Be(OH)_2$ which was washed with water, dried, and converted into BeO (99% purity) by heating at about 500 °C [29].

Finely ground (<0.149 mm) beryllium ore (bertrandite) was leached with aqueous 10 N NaOH, boiled at 110 °C for about 20 min, and centrifuged to separate the leach liquor containing the beryllium. The leaching was repeated and the pulp washed with water. The liquids were combined to recover almost 90% of the BeO content of the ore in a >8 N NaOH solution. This solution was treated with excess CaO or $Ca(OH)_2$ and heated, under pressure, for 150 min at about 150 °C in an autoclave. The SiO_2 was almost completely precipitated leaving the beryllium in solution. If, prior to the addition of excess (80%) $Ca(OH)_2$, the solution was diluted to 1.5 N NaOH and then autoclaved at 150 °C, the beryllium was precipitated together with the silicate and essentially all the other contaminants were left in solution [6].

Na_2CO_3 and $Na_2Be_4O_7$. The powdered beryl (or concentrate) was heated with Na_2CO_3 in a 1:2 weight ratio for 2 to 3 h at about 700 °C [7, 8], with Na_2CO_3 at 820 °C for 2 to 4 h [9], at about 850 °C in a 1:1.8 weight ratio [10], at 900 °C for 4 h in a 1:1 weight ratio [11], or at 950 to 1000 °C for 25 min [12]. A mixture of beryl and borax (1:1 weight ratio) was heated for 10 min at 1100 °C [13] and the molten or sintered cake leached with H_2SO_4 [9 to 11, 13]. The crushed sinter was treated with HCl to form a thick pulp which was added to a warmed (50 °C) solution of $NaHCO_3$. Up to 90% of the Be content in the solution was recovered with complete separation of beryllium from the other components (Si, Al, Fe) of the ore which remained in the insoluble residue [7, 8]. A beryl concentrate, made soluble by heating with Na_2CO_3 (mole ratio 1:10), was treated with aqueous HNO_3 (150 g/L) at a solid-to-liquid ratio of 1:14. A 20% stoichiometric excess of HNO_3 (32 mol/mol beryl) was sufficient to afford almost complete (99%) extraction of beryllium, which at 25 °C required 15 min and less time at higher (up to 60 °C) temperatures [12].

CaO, $Ca(OH)_2$, or $CaCO_3$. In an earlier technical process (applied at Degussa, Germany) finely ground beryl was mixed with CaO in a 2:1 weight ratio and sintered (or melted) in a rotary furnace at 1400 to 1500 °C for 1.5 to 2 h. The hot melt was added dropwise to a pan containing H_2O. The (glassy) drops broke into small particles which were dried and ground. To 300 kg of this powder, 480 kg H_2SO_4 and 40 L H_2O were added. A description of the process and the flow sheet, see Fig. 1-5, p. 34, are reported in [15, 16], see also [14]. For melting with CaO (or $CaCO_3$) and leaching with H_2SO_4, see also [17 to 21]. For the separation of Be, Al, and Fe in sulfate solutions, refer to the sulfate process, pp. 32/5. Beryl is decomposed by CaO at 1200 to 1300 °C or by anhydrous $Ca(OH)_2$ at 30 atm [1], see also [22]. Decomposition of beryl by $Ca(OH)_2$ occurs after 2 h at 1000 °C and requires 5 kg $Ca(OH)_2$ per kg of beryl [23]. A far lower decomposition temperature was needed when beryl was heated with a mixture of $Ca(OH)_2$ (60%) and anhydrous $Ca(NO_3)_2$ (40%) in a 1 BeO:1 CaO mole ratio. Heating the mixture at 575 °C for 2 h made about 80% of the BeO content soluble [24]. In a modified Degussa process, a ground beryl concentrate with 2.7% BeO was mixed with powdered limestone ($CaCO_3$), 2% bentonite for binder, and 1% Na_2SiF_6. The mixture was briquetted and heated at 1050 to 1100 °C (without melting) for 45 min followed by quenching in water. The addition of $CaCO_3$ was computed to yield a sinter having a $CaO:SiO_2$ molar ratio of 0.8 [25]. A beryl containing concentrate (2 to 7% BeO or more) was mixed with $CaCO_3$ in a 1:1 weight ratio and with 2 wt% Na_2SiF_6 as a mineralizer. The mixture was sintered, in powder form or as pellets, for 4 h at 1100 to 1190 °C (without melting). The ground sinter was treated with a 15% excess of strong H_2SO_4 at about 100 °C and finally at ca. 300 °C (cured) to render the silica insoluble [26]. A briquetted mixture of 100 kg beryl (12.5% BeO) and 150 kg $CaCO_3$ was heated for 4 to 6 h at 1200 °C, cooled, and crushed in the presence of water. The pulp was treated with 12 L of boiling aqueous 30% NaOH solution per kg of ore until the reaction was complete. Then CO_2 was bubbled through the boiling solution to precipitate $CaCO_3$, silica, and basic Al carbonate which were filtered off. The filtrate, which contained a complex sodium beryllium carbonate, was boiled to precipitate $BeCO_3$. Calcination of the precipitate left BeO in a 92% yield [27]. When a mixture of 11 to 12 mol $CaCO_3$ per mol beryl (e.g., 1100 g $CaCO_3$ (=11 mol) and 540 g beryl (=1 mol)) was heated for 2 h at 1200 °C, BeO and SiO_2 were left in the residue after leaching with dilute (7%) hydrochloric acid (the powdered reaction product was introduced in an excess of the acid). Treatment of the filtered residue (91.5 g with 61% BeO) with hot concentrated H_2SO_4 produced soluble $BeSO_4$ and insoluble SiO_2. After leaching with H_2O, precipitation of $Be(OH)_2$, and calcining, 64 g BeO (85% yield) were obtained. Similarly, a mixture of 1080 g beryl (11% BeO), 1060 g Na_2CO_3, and 1800 g $CaCO_3$ was heated to about 1000 °C and maintained at this temperature for 2 to 3 h. Then 2.04 kg of the reaction product were stirred for 1 h with 40 L of dilute HCl (~6.7%). The filtered residue of BeO and SiO_2 (223 g) was stirred with 240 L of H_2SO_4 and heated until fuming. The yield after precipitation with ammonia was 96% [28]. When a sintered mixture of 10 g beryl (10.8% BeO), 5 g Na_2CO_3, and 5 or 10 g $CaCO_3$ (4 h at 900 °C) was treated with 5 N H_2SO_4 instead of dilute HCl, 99 to 100% of the BeO was leached. In mixtures without Na_2CO_3 (10 g beryl + 10 g $CaCO_3$ sintered 8 h at 1000 °C, or 10 g beryl + 7 g $CaCO_3$ sintered 4 h at 1100 °C) only 63 or 60% of the BeO could be leached by 10 or 2 N H_2SO_4, respectively, although no undecomposed beryl was detectable in the reaction product by X-ray diffraction. Similarly, when beryl was sintered with gypsum (1:0.7 weight ratio) for 2 h at 1200 °C, only about 50% of BeO could be leached with 1 or 10 N H_2SO_4 [11].

References:

[1] G. Jaeger (Metall **4** [1950] 183/91, 184).

[2] J. P. Surls, Dow Chemical Co. (U.S. 3369860 [1963/68] 1/3; C.A. **68** [1968] No. 80662).

[3] Deutsche Gold- und Silber-Scheideanstalt, vorm. Roessler (Fr. 788243 [1934/35] 1/3; Brit. 435092 [1935] 1/3; C. **1936** I 1081).

[4] G. Jaeger, Deutsche Gold- und Silber-Scheideanstalt, vorm. Roessler (Ger. 635047 [1936]; C.A. **1937** 224; U.S. 2063811 [1936]; C.A. **1937** 822).
[5] G. A. Blanc, F. Jourdan (Fr. 839301 [1938/39] 1/9, 7; C. **1939** II 2264).
[6] C. K. Hanson, M. E. Wadsworth (U.S. 3615260 [1966/69] 1/6; C.A. **76** [1972] No. 27322).
[7] H. Fischer, Metal & Thermit Corp. (U.S. 1820655 [1928/31] 1/3; C. **1931** II 3139).
[8] H. Fischer, Siemens & Halske A.-G. (Ger. 519622 [1927/31] 1/2; C. **1931** I 3039; Ger. 557228 [1927/32] 1/2; C. **1932** II 2221).
[9] M. Lazarević (Extract. Nucl. Technol. Yugoslaw-Polish Colloquium, Kowary, Poland, [1961], pp. 1/305, 267/72; N.S.A. **16** [1962] No. 13142).
[10] V. Turovschi, O. Popa, Ministery of Petroleum and Chemical Industry Roumania (Brit. 992803 [1961/65] 1/4; C.A. **63** [1965] 3928).

[11] R. W. Cattrall (Australia Dept. Mines Mining Rev. No. 110 [1959/60] 12/21).
[12] T. A. Adilov, M. M. Madzhidova (Khim. Khim. Tekhnol. Redk. Tsvetn. Metal. **1974** 94/6; C.A. **83** [1975] No. 196683).
[13] G. H. Osborn, W. Stross, International Alloys Ltd. (Brit. 587483 [1944/47] 1/3; C.A. **1947** 6520; N.S.A. **1** [1948] No. 144).
[14] G. T. Motock (U.S. Bur. Mines Inform. Circ. No. 7357 [1946] 1/12, 4/6; C.A. **1946** 7105).
[15] H. W. West, W. F. Randall, G. L. Miller et al. (BIOS-FR-550 [1945] 1/81, 7/8, 80).
[16] H. A. Sloman, C. B. Sawyer (FIAT-FR-522 [1946] 1/102, 43/9, 101).
[17] J. E. Bucher, Antioch Industrial Institute Inc. (U.S. 2010844 [1932/35] 1/3; C. **1936** I 1284).
[18] C. James, H. C. Fogg, E. D. Coughlin (Ind. Eng. Chem. **23** [1931] 318/20).
[19] G. A. Meerson, D. D. Sokolov, N. F. Mironov et al. (At. Energiya SSSR **5** [1958] 624/30; Soviet J. At. Energy **5** [1958] 1555/62, 1555; C.A. **1961** 7218).
[20] H. M. Finniston (Res. Appl. Ind. **15** [1962] 109/18, 110).

[21] J. B. Rosenbaum, R. O. Dannenberg, D'Arcy R. George (Solvent Extract. Chem. Metals Proc. Intern. Conf., Harwell, Engl., 1965 [1967], pp. 315/26, 319; C.A. **69** [1968] No. 79358).
[22] P. Mouret, A. Rigaud (CEA-1090 [1958/59] 1/22, 13/4; N.S.A. **14** [1960] No. 8660).
[23] K. Vetejška, J. Mazáček (Sb. Praci. Ustavu Vyzk. Rud **4** [1962] 153/6; C.A. **57** [1962] 14775).
[24] B. S. Brčić, A. Šmalc (Vestn. Sloven. Kem. Drustva **7** [1960] 91/6; C.A. **56** [1962] 6870).
[25] J. M. Riley (U.S. Bur. Mines Rept. Invest. No. 5963 [1962] 1/10, 4; C.A. **56** [1962] 15216).
[26] D. A. Everest, E. Napier (Proc. 6th Intern. Congr. Mineral. Process., Cannes 1963 [1964], pp. 231/43, 235/7).
[27] C. Adamoli, Perosa Corp. Wilmington (U.S. 2344480 [1941/44] 1/3; C.A. **1944** 3428).
[28] Deutsche Gold- und Silber-Scheideanstalt, vorm. Roessler (Brit. 384926 [1932/33] 1/3; C. **1933** I 2295).
[29] R. H. McKee (U.S. 2298800 [1939/42] 1/6, 3; C.A. **1943** 1838).

1.2.4 With Aqueous CO_2

A beryllium bearing pegmatite (Val Musul, Bozen) was heated at 850 to 900 °C for 10 h, ground with water until it passed a fine sieve, decanted, and dried in the air. A vigorously stirred (>100 r/min) suspension of 5 kg of the dry powder in very pure water containing about 50 g $(NH_4)_2CO_3$ per 10 kg H_2O was treated with CO_2 for several days. After saturation, the solution was neutralized with stirring by slow and cautious addition of HCl. The residue was centrifuged off and evaporated. The acid evaporation was repeated to precipitate all of the silica. The filtered solution was neutralized with ammonia, and excess $NaHCO_3$ added. After digesting for about 2 h, the solution was filtered and the filtrate acidified with HCl. The solution was heated to expel the CO_2 and ammonia was added to precipitate the $Be(OH)_2$ which was filtered off, washed, dried, and calcined. Although the first extraction recovered only about 20% of the BeO from the mineral, almost complete recovery of all of the BeO was achieved by repeated extraction. The extraction was possible with or without alkaline or acid catalysts, C. Adamoli (Ger. 541544 [1931/32] 1/2; C. **1932** I 3332).

1.2.5 With Organic (Carboxylic) Acids

A finely ground (<0.074 mm) beryllium ore could be extracted by treating the aqueous slurry with an aqueous solution of oxalic acid under the action of an alternating current of 60 Hz and a current density of about 1.1 W/cm^2 at 25 °C and pH=4 to 5 until all of the Be was recovered. The oxalate solution was filtered and the oxalate was converted by HCl into $BeCl_2$ [1]. Extraction of beryllium ores by a mixture of oxalic and citric acids in 0.1 M aqueous solutions was found to be six times more efficient than with 0.01 M HNO_3 [2].

References:

[1] W. A. Rhodes (U.S. 3441405 [1966/69] 1/4; C.A. **71** [1969] No. 23873).
[2] J. C. Goni (U.S. 3511645 [1967/70] 1/3; C.A. **73** [1970] No. 17637).

1.2.6 With Mineral Acids and SO_3

HCl and HNO_3. The beryllium minerals beryl ($3BeO \cdot Al_2O_3 \cdot 6SiO_2$), chrysoberyl ($BeO \cdot Al_2O_3$), phenacite ($Be_2SiO_4$), and bertrandite ($4BeO \cdot 2SiO_2 \cdot H_2O$) are almost insoluble in hot concentrated hydrochloric acid, whereas Be can be completely leached from helvine ($3(Fe,Mn)O \cdot 3BeO \cdot 3SiO_2 \cdot MnS$), and danalite ($3(Fe,Zn,Mn)O \cdot 3BeO \cdot 3SiO_2 \cdot (Fe,Zn)S$) even by dilute HCl at 95 °C. Complete decomposition of the vesuvianite ($2(Mg,Mn,Zn)O \cdot 6CaO \cdot 4BeO \cdot Al_2O_3 \cdot 6SiO$) required roasting at 700 to 800 °C for 1 h and treatment with 6 M HCl [1]. Also, the powdered rare earth mineral gadolinite (2 kg), containing 9% BeO, was decomposed completely in about 36 h by heating under stirring in an open die. Thickening with 7 L concentrated HCl at 120 to 130 °C left a residue of SiO_2. After repeated HCl treatment, the rare earths were precipitated as oxalates and the Be^{2+} was precipitated from the filtrate as $Be(OH)_2$ [2]. A phenacite containing concentrate (2.3% BeO) released only about 7% of its Be content even after refluxing with concentrated HCl (30%) for 6 h at 92±3 °C. Preheating the concentrate at 800 °C for 4 h increased the HCl recovery of BeO to 11.5% [3]. A fluorspar containing 0.09 or 0.9 wt% Be (as the bertrandite) was heated with an excess of 20% HCl for 3 h at 108 °C to recover 95 or 92% of the beryllium as the soluble chloride. In contrast, beryl containing fluorspars released only 6% of the Be content with the same treatment. Treatment of a bertrandite containing fluorspar (0.22% Be) with an excess of aqueous HNO_3 (20%) at 105 °C for 3 h released 88% of the beryllium content into the acidic extract [4].

H_2SO_4. Treatment of a phenacite concentrate (2.3% BeO) with H_2SO_4 in a 1:2 (90% acid) or 1:3 (60% acid) weight ratio at 220 °C for 2 h recovered about 80% of the BeO content [3]. A wet, crushed ore (<1.4 mm) containing <1 wt% BeO (as the mineral bertrandite) was leached with 225 kg H_2SO_4 per 1000 kg ore at 82 to 93 °C and pH≈1.4 for 6 to 20 h (the shorter time was in the presence of a soluble fluoride) and released about 95% of its Be content into the solution. NH_3 gas was slowly introduced into the leach solution at 82 °C first to pH=3.8, to precipitate Al and Fe, and then to pH=8.5 (at ~77 °C), to precipitate $Be(OH)_2$. The $Be(OH)_2$ was further purified by dissolution in aqueous NaOH and separation of impurities such as the insoluble sulfides [5]. About 89% of the Be was leached from a fluorspar containing 0.22 wt% Be (as bertrandite) by heating with an excess of 9 M H_2SO_4 at 105 °C for 1 h [4]. A calcareous ore (from the topaz mountain, Utah) containing 0.46% BeO and finely ground pyrite (FeS_2, 43% S) in a 17.7:1 weight ratio was heated with water (30%) and H_2SO_4 (146 kg per 1000 kg ore) at 200 °C for 30 min in an autoclave

(34 atm). The process made essentially 100% of the BeO soluble in a solution assaying 1.31 g BeO (and 0.05 g Fe) per liter [6].

Treatment of berylite bearing ores (crushed to <0.1 mm) containing 0.4 to 0.8% BeO with strong H_2SO_4 (75 vol%) at 325 °C for 3.5 to 7.5 h recovered 90% of the BeO content as the soluble Be sulfate. In contrast, treatment with concentrated H_2SO_4 at 445 °C for 7 h yielded a recovery of only 75% BeO (maximum). Prolonged heating decreased the yield even further [7]. A herderite ore ($CaBePO_4(F,OH)$) ground to a grain size of <100 μm and treated with a three- to fourfold stoichiometric excess of concentrated H_2SO_4 at 180 °C for about 2 h allowed to recover 96 to 97% of the BeO content of the ore as the soluble sulfate. Prolonged heating at higher temperatures reduced the yield of extraction [8].

Four BeO containing ores, montmorillonite (1.4% BeO), saponite (0.8% BeO), a calcareous ore (0.6% BeO), and an unaltered tuff (0.5% BeO), were leached by stirring with H_2SO_4 (270 to 360 kg per ton of ore) at 95 °C in a pulp of 35% solids (pH=0.3 to 0.9) for 3 h (or at 65 °C for 24 h). About 95% of the BeO content was recovered as the soluble sulfate. The addition of ~0.27 to 0.36 kg of polyacryl flocculant and 4 t of water per ton of ore was required to settle the pulp for filtration. About 95% of the BeO was recovered by a five-step extraction. Less polyacrylic flocculant (0.135 kg) and less water (2.6 t) were needed when 360 kg H_2SO_4 per ton of ore were used for leaching or when the ores were prerosted at 500 °C. Depending on the ore, the extracts contained 2 to 7 g BeO per liter [9].

Sealed glass ampules containing 15 g of a Seal Lake ore concentrate (<0.074 mm, 0.35 to 0.40% Be), 30 mL of 2.03 N H_2SO_4, and sufficient $KClO_4$ to oxidize all Fe^{II} to Fe^{III} were heated at 230 °C for 1 h. At the end of the heating, 95% of the BeO was recovered in solution (pH≈0.5), and the majority of the contaminant metals (Fe, Al, Th, rare earths) were left in the residue. The use of HNO_3 or HCl instead of H_2SO_4 is possible, but the results are poorer than with H_2SO_4 [10].

When 50 g of a finely ground Be ore were treated with 87 g H_2SO_4 (81%) at 300 °C and then slowly heated to 630 °C for about 3 h, 99.7% of the BeO content was recovered in the sulfate solution [11].

Beryl reacts very slowly with concentrated H_2SO_4 even under extreme conditions, e.g., when beryl was heated for several hours in an autoclave at 400 °C, only 50 to 60% of the Be was converted to the soluble sulfate. However, the reaction is catalyzed by fluoride addition, Jaeger [12]. Because of the slow rates of reaction and serious corrosion problems, in commercial use, beryl must be pretreated before leaching with H_2SO_4 (for previous alkaline attack, see pp. 25/7, for premelting without flux, see p. 31).

Instead of leaching Be from beryl, treatment with strong H_2SO_4 can enrich the Be content of the solid according to [13]. Thus, a beryl flotation concentrate containing 3.4 wt% BeO (along with 54.7% Al_2O_3, 13.2% CaF_2, 12.9% Fe_2O_3, 3.4% SiO_2) was enriched up to 5 wt% BeO by treatment with 94 and 60% H_2SO_4 in a 1:3.5 solid to liquid ratio at 220 to 300 °C for 2 h. The separated residue was leached with water and then with 10% HCl (solid to liquid ratio 1:10) at 92 to 95 °C. The loss of BeO during aqueous leaching was between 5 and 8 wt% when concentrated H_2SO_4 was used, but became 11 to 13% when 60% H_2SO_4 was used. The enrichment of the concentrate was increased to 6 wt% BeO when it was mixed with half its weight in Na_2SiF_6 prior to the acid treatment at 220 °C. The loss of BeO during leaching was only 5 wt% as compared with around 15 wt% when NaF was added instead of Na_2SiF_6. Lower degrees of enrichment were obtained when the concentrate contained less CaF_2 (<13%) or more SiO_2 (>55%). Quartz could not be removed by this treatment, but fluorite, heavy metal ores, and part of the feldspars were eliminated [13].

H_2SO_4 with Premelted Beryl Ores. Beryl reacts very slowly with hot concentrated H_2SO_4 [12], but after melting (~1650 °C) and quick quenching, the beryl structure is broken down and the product is much more readily attacked by hot concentrated H_2SO_4, see, e.g. [14 to 26]. Only 50 to 60% of the Be can be extracted, but yield can be increased to 90 to 95% by a subsequent heat treatment of the quenched glassy product to segregate BeO from the glassy mass [14, 17, 21]. Various heat treatments have been employed such as heating at a temperature between 900 and 950 °C [15], heating for 2 h at 920 °C [22], and heating for at least 30 min between 900 and 1000 °C [23]. This "fuse-quench process" is used by the Brush Beryllium Co. to break down high-grade beryl containing >10% BeO. However, the process is unsatisfactory for the treatment of concentrates containing only 2 to 7% BeO [15]. In the Brush process, fusion of the beryl is carried out in a carbon-lined three-phase electric furnace. The rapidly quenched melt is screened to collect particles <1.27 cm which are heat treated in a gas-fired rotary kiln and ground to <74 μm. They are then mixed with 93% H_2SO_4 in an amount sufficient to convert all Be and Al into the sulfates. The slurry is pumped as a small jet into a preheated (250 to 300 °C) gas-fired plain steel mill which assures an almost instantaneous reaction. The sulfated ore is then leached with H_2O and the resulting liquor is filtered. The liquor contains aproximately (g/L): 13 Be, 15 Al, 2 Fe, 0.1 Si and is about 1 N in H_2SO_4 [14]. The high Be:Al ratio is not in accordance with the beryl composition (Be:Al weight ratio of 1:2) unless the sulfating reaction selectively favors the Be. For the separation of Be from Al and Fe in sulfate solutions, refer to the sulfate process, p. 35. When a siliceous beryllium ore (such as beryl) was melted with portions of carbon and iron, an iron-silicon alloy was formed which sank to the bottom of the melt. The Be-rich molten slag was poured into H_2O and sulfated with H_2SO_4. Alternatively, a treatment with a hot caustic solution is possible [27].

SO_3. A finely ground (<149 μm) montmorillonite ore (10 g) containing 0.2% Be was treated with a stream of SO_3 and air at 580 to 610 °C for 15 min (under fluidizing conditions). Leaching with water at 60 to 80 °C for 2 h (slurry with pH=2.2) recovered about 85% of the Be in the aqueous phase [28]. Using 2 g of a very finely ground (<37 μm) clay mineral containing 0.87% BeO (and ~6.6% CaF_2) allowed the solubilization of 85% of the Be content by SO_3 at lower temperatures [29, 30]. The samples were dried at 200 °C for 1 h and treated (in a thin bed) with SO_3 at 200 to 300 °C for 1 h [30]. When coarser material was heated for 15 min between 300 and 500 °C, only 65% of the Be became soluble [29]. See pp. 29/30 for leaching of ores with H_2SO_4.

References:

[1] G. N. Samorokova (Zavodsk. Lab. **33** [1967] 270/4; Ind. Lab. [USSR] **33** [1967] 320/4).
[2] W. Fischer, P. Herbach, H. Plempe, G. Wirths (Z. Anorg. Allgem. Chem. **250** [1942] 72/81).
[3] Yu. M. Putilin, A. D. Romanova, L. V. Favorskaya (Tekhnol. Mineral. Syr'ya **1972** 74/82; C.A. **85** [1976] No. 180611).
[4] W. A. Mod, C. W. Becker, Dow Chemical Co. (U.S. 3177068 [1959/65] 1/3; C.A. **62** [1965] 15818).
[5] J. K. Grunig, W. B. Davis, W. C. Aitkenhead (U.S. 3685961 [1963/72] 1/9, 6; C.A. **77** [1972] No. 166770; Ger. 1911141 [1968/69] 1/10, 6; C.A. **72** [1970] No. 23740).
[6] P. H. Johnson, H. E. Johnson & Associates (U.S. 3264099 [1962/66] 1/10, 7; C.A. **65** [1966] 11856).
[7] W. J. Biermann, Beryloy, Ltd. (U.S. 3295962 [1962/67] 1/3; C.A. **66** [1967] No. 58128; Can. 767045 [1962/67] 1/8, 7; C.A. **68** [1968] No. 23836).
[8] M. Delcorte, A. Lecocq, O. Stulzaft (Fr. 1407265 [1964/65] 1/3; C.A. **65** [1966] 10200).
[9] L. Crocker, R. O. Dannenberg, D. W. Bridges (U.S. Bur. Mines Rept. Invest. No. 6322 [1963] 1/16, 15; C.A. **60** [1964] 2571; N.S.A. **18** [1964] No. 341).

[10] R. S. Olson, J. P. Surls, Dow Chemical Co. (U.S. 3511597 [1965/70] 1/5, 3/4; C.A. **73** [1970] No. 27831; U.S. 3669649 [1965/72] 1/5, 3; C.A. **77** [1972] No. 104246).

[11] S. J. Anderson, Chemsep. Corp. (U.S. 3958985 [1975/76] 1/12, 7, 11; C.A. **85** [1976] No. 49758).

[12] G. Jaeger (Metall **4** [1950] 183/91, 184/6).

[13] Yu. M. Putilin, A. D. Romanova, L. V. Favorskaya (Tekhnol. Mineral. Syr'ya **1972** 48/57, 51; C.A. **85** [1976] No. 97310).

[14] C. W. Schwenzfeier (in: D. W. White, J. E. Burke, The Metal Beryllium, Cleveland, Ohio, 1955, pp. 71/101, 73/84).

[15] D. A. Everest, E. Napier (Proc. 6th Intern. Congr. Mineral. Process., Cannes 1963 [1965], pp. 231/43, 232, 235/9; C.A. **62** [1965] 15811).

[16] C. R. Hayward (An Outline of Metallurgical Practice, New York - Toronto - London 1952, pp. 419/24, 421).

[17] W. W. Beaver (TID-2503 [1952] 21/51, 25/7; N.S.A. **12** [1958] No. 17340).

[18] A. G. Thomson (Mining J. [London] **241** [1953] 661/2).

[19] G. C. Ellis (Metallurgia **58** [1958] 173/6, 175).

[20] J. Schubert (McGraw-Hill Encyclopedia of Science and Technology, New York 1960, pp. 176/8; N.S.A. **15** [1961] No. 9493).

[21] C. W. Schwenzfeier (J. Metals **12** [1960] 793/7).

[22] R. K. Bayliss, R. Derry (J. Appl. Chem. **16** [1966] 114/21, 115/7).

[23] R. Derry, R. K. Bayliss, United Kingdom Atomic Energy Authority and Council for Scientific and Industrial Research [India] (Brit. 1027461 [1961/66] 1/4; C.A. **65** [1966] 372).

[24] B. R. F. Kjellgren (Trans. Electrochem. Soc. **89** [1946] 247/62, 248/9).

[25] C. B. Sawyer, B. R. F. Kjellgren, Brush Beryllium Co. (U.S. 2018473 [1935]; C.A. **1936** 253; U.S. 1823864 [1930/31] 1/9, 7; C.A. **1932** 72).

[26] I. A. Meerson, D. D. Sokolov, N. F. Mironov et al. (At. Energiya SSSR **5** [1958] 624/30; Soviet J. At. Energy **5** [1958] 1555/62, 1555).

[27] B. R. Kjellgren, C. R. Sawyers, Brush Beryllium Co. (U.S. 2092621 [1932/37] 1/2; C. **1938** I 725).

[28] I. M. Kruse, E. I. Du Pont de Nemours & Co. (U.S. 3148022 [1961/64] 1/2; C.A. **61** [1964] 12991).

[29] F. Habashi, R. Dugdale (Trans. AIME **252** [1972] 423/5; C.A. **78** [1973] No. 87411).

[30] F. Habashi, R. Dugdale, Anaconda Co. (Ger. 2100921 [1970/71] 1/4, 3; C.A. **75** [1971] No. 132029).

1.2.7 The Sulfate Process

The sulfate process is based on the H_2SO_4 extraction of Be from beryl concentrates which are pretreated (solubilized) either by fuse-quenching without flux (see p. 31) or by alkaline attack (see pp. 25/7). The first method, applied by the Brush Co., is shown in the flow sheet, **Fig.** 1-**4** (taken from [2]). The steps leading up to the leaching with H_2SO_4 are described on p. 31. After removal of the silica gel by continuous centrifugation, the leach liquor is blended with an amount of ammonium hydroxide equivalent to its Al content and held in a crystallizer at 20 °C until about 75% of the Al crystallizes as ammonium alum, $(NH_4)Al(SO_4)_2 \cdot 12H_2O$. The alum crystals are separated and washed in a centrifuge. The filtrate passes into a water cooled stainless steel beryllating reactor where a chelating agent (EDTA) and dilute aqueous NaOH are added to a basic normality of about 1.5. The sodium beryllate solution passes to the hydrolyzer where $Be(OH)_2$ is precipitated by boiling and subsequently separated by continuous centrifugation. The chelated impurities, mainly Fe and Al, remain in solution. Approximately 7% of the $Be(OH)_2$ is too fine and escapes

Fig. 1-4

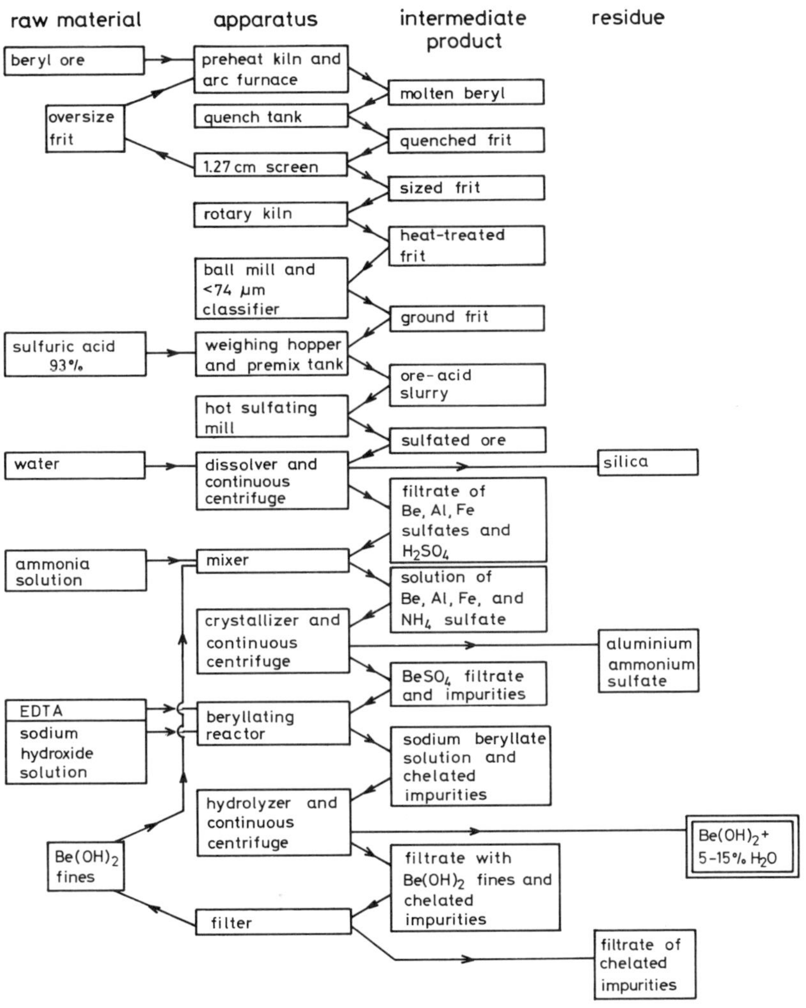

Flow sheet for the production of beryllium hydroxide at the Brush Beryllium Company.

the centrifuge; it is separated by filtration and recycled to the alum crystallization step [1], see also [3, 4]. Prior to the use of the EDTA process, the $BeSO_4$ present in the leaching solution was recovered by a fractional crystallization process. The bulk of the ammonium alum was separated in the first crystallization step by adding an amount of $(NH_4)_2SO_4$ equivalent to the Al content. The mother liquor was then evaporated and a crude beryllium sulfate, containing some ammonium and iron alums, was crystallized. This salt was purified by dissolving in an ammonium sulfate solution (giving a saturated solution of $BeSO_4$), separation of crystallized alum, and crystallization of an Al free beryllium sulfate from the evaporated filtrate [5], see also [6].

The flow sheet of the Degussa sulfate process, in which beryl concentrate (10 to 12% BeO) is melted with CaO before leaching with H_2SO_4 (see pp. 26/7), is shown in

Fig. 1-5

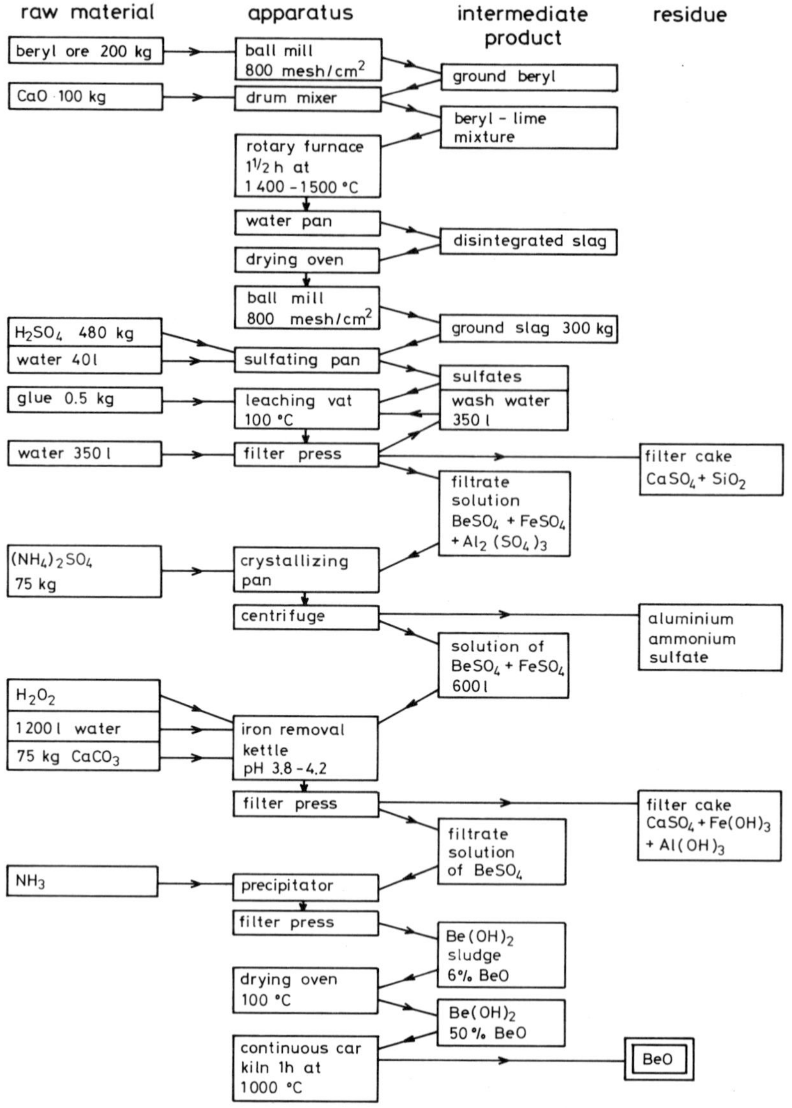

Flow sheet for the production of beryllium oxide from beryl at Degussa.

Fig. 1-**5** (taken from [2], also shown in [8] and first in [7, 9, 10]). The sulfate leaching mass (prior step, see pp. 26/7) is mixed with 350 L of wash water and 0.5 kg of glue and heated with steam to 105 °C to separate the silica and the $CaSO_4$. The filtered sulfate solution of Be, Al, and Fe is mixed with 75 kg of ammonium sulfate. Ammonium alum crystallizes at about 10 °C and is centrifuged. Iron is precipitated by mixing 600 L of the solution with 1200 L of H_2O and 3 L H_2O_2 and then with 75 kg $CaCO_3$ (pH=3.8 to 4.2). In the filtrate,

$Be(OH)_2$ is precipitated by gaseous NH_3. The recovery of Be from the ore is 78 to 80% [9, 10]. For the separation of Al (as alum) from Be in the leach liquor using four leaching stages, see [11]. In a modified Degussa process using a beryl concentrate with only 2.7% BeO, the resulting leach liquor contained (in g/L): 2.4 BeO, 20 Al_2O_3, and 6 Fe_2O_3 and recovered 80% of the Be. About 10% of the dissolved Be was lost in the alum and lime purification steps which remove 2/3 of the Al and Fe. Thus, 70% of the Be was recovered in a "purified solution" which still contained three times as much Al and twice as much Fe than Be, Riley [8].

Beryllium sulfate was also separated from the aluminium sulfate by leaching the sulfatized beryl-lime sinter with aqueous ethanol (85 wt%). Repeated refluxing recovered up to 93% of the BeO in the solvent phase from a concentrate with 23% BeO and increased the Be:Al weight ratio from 0.15 in the concentrate to about 11 to 12 in the extract [11], also see [12]. A very effective method for the separation of Be from Al and Fe is the ammonium carbonate method. An aqueous saturated solution of ammonium carbonate dissolves $Be(OH)_2$, but not the hydroxides of Al, Fe, and other heavy metals. Boiling the filtered solution precipitates a pure basic beryllium carbonate in an easily filterable form. Traces of heavy metals can be removed as the sulfides before boiling the solution [13]. BeO of very high purity was obtained by dissolving the basic beryllium carbonate in acetic acid and purifying the resulting beryllium acetate by distillation. The purified beryllium acetate was then decomposed by pyrolysis [14].

References:

[1] C. W. Schwenzfeier (in: D. W. White, E. Burke, The Metal Beryllium, Cleveland, Ohio, 1955, pp. 71/101).
[2] G. E. Darwin, J. H. Buddery (Beryllium, London 1960, pp. 17, 20).
[3] C. W. Schwenzfeier (J. Metals **12** [1960] 793/7).
[4] W. W. Beaver (TID-2503 [1952] 21/51, 25/7; N.S.A. **12** [1958] No. 17340).
[5] B. R. F. Kjellgren (Trans. Electrochem. Soc. **89** [1946] 247/62, 249).
[6] A. G. Thomson (Mining J. [London] **241** [1953] 661/2).
[7] G. T. Motock (U.S. Bur. Mines Inform. Circ. No. 7357 [1946] 1/12, 4/6; C.A. **1946** 7105).
[8] J. M. Riley (U.S. Bur. Mines Rept. Invest. No. 5963 [1962] 1/10; C.A. **56** [1962] 15216).
[9] H. W. West, W. F. Randall, G. L. Miller, et al. (BIOS-FR-550 [1945] 1/81, 7/8, 80).
[10] H. A. Sloman, C. B. Sawyer (FIAT-FR-522 [1946] 1/102, 43/9, 101; N.S.A. **1** [1948] No. 1845).
[11] D. A. Everest, E. Napier (Proc. 6th Intern. Congr. Mineral. Process., Cannes 1963 [1964], pp. 231/43, 238/9).
[12] H. Barber, D. Ali (Mikrochemie **35** [1950] 542/52, 551).
[13] G. Jaeger (Metall **4** [1950] 183/91, 184).
[14] G. A. Meerson, D. D. Sokolov, N. F. Mironov, et al. (At. Energiya SSSR **5** [1958] 624/30; Soviet J. At. Energy **5** [1958] 1555/62).

1.3 Separation of Beryllium from Solutions

1.3.1 Extraction of Beryllium from Aqueous Solutions

1.3.1.1 With Fatty Acids

Beryllium can be extracted from aqueous solutions of $BeCl_2$ (obtained by dissolving the $Be(OH)_2$ precipitated from a solution of $BeSO_4$ in aqueous HCl) by solutions of butyric acid (330 to 500 mL per 1 g BeO) in organic solvents after adjusting the pH value of the aqueous phase at 9.3 to 9.5 with ammonia. Of the solvents tested ($CHCl_3$, CCl_4, ether, benzene, ethyl- and amylacetates) best results were obtained with $CHCl_3$ (99% after four extractions) followed by ethyl acetate as the next [1]. The extraction of beryllium from an aqueous 0.2 M beryllium sulfate solution containing butyric acid by $CHCl_3$ depends somewhat on the concentration of the acid and increases from 87.5% for a 2.0 M acid solution to 94.7% for a 5.0 M acid solution. But the extraction is not selective since Al is also completely extracted. Using hexanoic or octanoic acid (fatty acids insoluble in H_2O), a 100% extraction of Be (and Al) occurs without $CHCl_3$ [2]. Selective extraction of Be from sulfate solutions by ethylhexanoic acid in kerosene solution (33%) could be achieved by complexing the other metal ions, mainly Al^{3+}, with EDTA at pH=4 to 5. The extraction equilibrium was attained within 30 to 60 min, after which the organic phase was scrubbed with 4 M H_2SO_4 for further purification. The purified Be compound can be removed either by strong aqueous HCl or HF. The stripping by these acids proceeds very slowly at room temperature but the rate increased noticeably upon heating to 60 or 70 °C when all Be could be stripped by a twofold excess of HF (48%) or HCl (12 M) within about 15 to 90 min [3], also see [4]. By extraction of a mixed fluoride solution of Be, Fe^{III}, and Al with fatty acids (C_7 to C_9) at pH=9 to 10 only around 6% of the Be was extracted, but about 89% of the Fe and 67% of the Al could be extracted [5].

In further tests it was shown that beryllium could be extracted at pH=9 by fatty acids (C_7 to C_9) from fluoride solutions containing 1 g Be^{2+} per L to a noticeable extent (70 to 80%) only in the presence of a 1.7-fold molar amount of Al^{3+} which apparently prevents the formation of the less readily extracted $Be(OH)^+$ ions [6], also see [7].

References:

[1] A. K. Sundaram, S. Banerjee (Anal. Chim. Acta **8** [1953] 526/9).
[2] W. J. Biermann, R. McCorkell (Can. J. Chem. **40** [1962] 1368/73, 1371).
[3] R. R. Grinstead, J. P. Surls (Nucl. Sci. Eng. **28** [1967] 346/52).
[4] R. R. Grinstead, Dow Chemical Co. (U.S. 3328116 [1963/67] 1/5; C.A. **67** [1967] No. 66741).
[5] Yu. M. Putilin, A. D. Romanova, L. V. Favorskaya (Tekhnol. Mineral. Syr'ya **1972** 196/202, 201; C.A. **85** [1976] No. 127614).
[6] Yu. M. Putilin, A. D. Romanova, L. V. Favorskaya (Tekhnol. Mineral. Syr'ya **1972** 203/7; C.A. **85** [1976] No. 127615).
[7] L. V. Favorskaya, Yu. M. Putilin, A. D. Romanova, O. P. Reznyakova (Geol. Razved. Metody Izuch. Mestorozhd. Polezn. Iskop. **1969** 240; Ref. Zh. Met. **1970** No. 12G214; C.A. **75** [1971] No. 91799).

1.3.1.2 With Long-Chain Aliphatic Amines

Extraction of Be from an aqueous 1 M sulfate leach solution containing 0.5 g Be/L by a 0.3 M solution of 6-amino-3,9-diethyl-tridecane in Solvesso 100 (a kerosene) at pH=2.9 (1.7 to 3.0) recovered 92% of the beryllium after six extractions. After 4 scrubbing steps with 0.02 M H_2SO_4, Be was stripped from the organic phase with 1.0 M H_2SO_4. If aluminium (5 g/L) was present in the aqueous phase, the ratio Al:Be changed from 10:1 to 1:22 [1]. The distribution coefficient (D_{Be}=10 resulting at pH=2.6) was higher than with other primary amines ($R_1R_2CHNH_2$) such as 1-nonyldecyl-, 1-heptyloctyl- or 1-methyloctadecylamine in the same solvent [1, 2], also see patents [3, 4]. In the presence of fluoride (0.01 M F^-) the extraction of Be is severely reduced but the separation from Al in process liquors is strongly improved. Be is readily stripped from the organic phase by aqueous 0.5 to 1.0 M H_2SO_4 or dilute solutions of NH_4F, which forms $(NH_4)_2\ BeF_4$ [2]. For extraction of Be from acidified (0.2 N H^+) solutions of $LiNO_3$ or Li_2SO_4 (total NO_3^-=0.5 to 10 N, SO_4^{2-}=0.3 to 5 N) [5] and

LiCl (total $Cl^-=0.5$ to 10 N) [6] by solutions of long-chain primary, secondary, or tertiary amines in diethylbenzene with a distribution coefficient $D_{Be}<1$, see [5, 6].

Extraction of Be from sulfate solutions (5 g Be/L) by solutions of tri-isooctylamine and tricaprylamine in kerosene recovered >50% of the metal in the organic phase [7]. Beryllium could be extracted from aqueous solutions (0.01 M) by solutions of tri-isooctylamine in $CHCl_3$ as the complex salt $[(i\text{-}C_8H_{17})_3NH]_2[Be(C_2O_4)_2]$. Equilibrium studies revealed that at 25 °C (and pH = 4) 87% of the Be was complexed in $[Be(C_2O_4)_2]^{2-}$ and up to 85% could be extracted; similar data (80 and 88%) were obtained at pH = 3 and considerably lower results (49 and 78%) at pH = 2. If malonic acid was used as the complexant instead of oxalic acid the extraction at pH = 4 gave somewhat lower (88 and 60%) and at pH = 3 considerably poorer (48.5 and 36%) results. Extraction of Be with malic, succinic, and phthalic acids as the complexants was very poor (<5%) even at pH = 4 [8]. The observed initial increase of Be extraction with rising pH is apparently due to the increased dissociation of oxalic acid with a consequent increase of complex formation [9]. Beryllium was extracted from an aqueous, 0.2 to 1.0 M alkali or ammonium carbonate solution at pH = 10 by a 0.3 to 0.5 N solution of a quaternary ammonium carbonate $(R_1R_2R_3NCH_3)_2CO_3$ (with R_1 = alkyl or aryl, R_2 and R_3 = alkyl of ≧6 carbon atoms if R_1 = alkyl or R_2 = methyl, and R_3 = alkyl of ≧17 C atoms if R_1 = aryl) in a solvent immiscible with water such as aliphatic, aromatic, or chlorinated hydrocarbons containing 3.5% tridecanol. Up to 98% of the Be in the organic phase could be recovered, most effectively with $[(C_{12}H_{25})_2(CH_3)_2N)]_2CO_3$. Be can be stripped from the organic phase by aqueous 1.5 to 2.0 M alkali carbonate solutions containing CO_3^{2-} and HCO_3^- in a 1:1 mole ratio [10].

References:

[1] D. J. Crouse, F. G. Seeley (U.S. 3359064 [1964/67] 1/3; C.A. **68** [1968] No. 52222).

[2] D. J. Crouse, K. B. Brown, F. G. Seeley (ORNL-P-1602 [1965] 1/18, 7, 11/5; C.A. **64** [1966] 1706).

[3] G. Bourat, Commissariat a l'Energie Atomique (Fr. 375862 [1963/64] 1/3; C.A. **62** [1965] 7415).

[4] M. Delcorte, A. Lecocq, O. Stulzaft (Fr. 1407265 [1964/65] 1/3; C.A. **65** [1966] 10200).

[5] F. G. Seeley, D. J. Crouse (J. Chem. Eng. Data **16** [1971] 393/7, 394).

[6] F. G. Seeley, D. J. Crouse (J. Chem. Eng. Data **11** [1966] 424/9, 425).

[7] P. Boutin, F. W. Melvanin (U.S. 3295932 [1963/67] 1/5, 3; C.A. **66** [1967] No. 68035).

[8] M. J. DeBruin, D. Kairaitis, R. B. Temple (Australian J. Chem. **15** [1962] 457/66, 465; AAEC-E-77 [1961] 1/14, 4; N.S.A. **16** [1962] No. 1851).

[9] H. J. DeBruin, R. B. Temple (AAEC-E-68 [1961] 1/15, 6; N.S.A. **16** [1962] No. 1850).

[10] United States Atomic Energy Commission (Brit. 1188759 [1967/70] 1/7, 3/5; C.A. **73** [1970] No. 89629).

1.3.1.3 With Acetylacetone and Derivatives

Extraction of beryllium from aqueous solutions by 1% solutions of acetylacetone in $CHCl_3$ was studied in the pH = 2 to 12 range and it was shown that extraction of Be is complete (100%) between pH = 6 and 9 [2, 3]. Selective extraction is possible with EDTA as a masking agent for other metals; addition of NaCl, $NaNO_3$, or Na_2SO_4 improves the separation. 97% of the beryllium could be recovered at pH = 7 to 8 by a three-step extraction with $CHCl_3$. From the organic phase Be could be stripped by an aqueous 0.1 N solution of NaOH at pH > 10 [3]. Extraction of Be by a solution of acetylacetone in CCl_4 at pH > 5 up to 8 (about 90%) and stripping of the organic phase with HNO_3 followed by precipitation of $Be(OH)_2$ with gaseous ammonia at 75 °C was recommended by [4]. Extraction of Be by acetylacetone saturated with water (without diluents) was reported to be almost complete (97.5%) at pH = 2 by [5]. (Addition of ammonia to a mixed solution of aqueous $BeSO_4$ and acetylacetone precipitates white Be acetylacetonate [5], see also [1] for the precipitation and purification of the precipitate by sublimation).

Extraction of beryllium from an aqueous solution by solutions of trifluoroacetylacetone in n-butanol (equal volumes) at pH = 5.5 to 6.5 recovered 99.9% of the Be in the organic phase [6]. Extraction of Be by a saturated solution of copper trifluoroacetylacetonate in benzene (equal volumes) at pH ≈5 was reported to collect about 80% of the Be in the benzene phase within 2 h, probably through a metal exchange reaction [7]. Extraction of Be by solutions of thenoyl trifluoroacetone in benzene was reported to recover about 97% of the Be (maximum) in the organic phase at pH = 4.0 or 95% at pH = 3.25 [8].

References:

[1] E. W. Berg, A. D. Shendrikar (Anal. Chim. Acta **44** [1969] 159/64, 161).
[2] T. Shigematsu, M. Tabushi (Bull. Inst. Chem. Res. Kyoto Univ. **39** [1961] 35/42, 37; C.A. **56** [1962] 3172).
[3] T. Shigematsu (Bull. Inst. Chem. Res. Kyoto Univ. **36** [1958/59] 156/62, 160; C.A. **1959** 8928/9).
[4] C. E. L. Bamberger, C. F. Baes Jr., H. F. McDuffie (ORNL-3417 [1963] 297/300; N.S.A. **17** [1963] No. 25092).
[5] J. F. Steinbach, H. Freiser (Anal. Chem. **25** [1953] 881/4; NYO-3070 [1952] 1/18; N.S.A. **7** [1953] No. 109).
[6] W. G. Scribner, G. K. Schweitzer, Monsanto Research Corp. (U.S. 3451807 [1966/69] 1/6, 2; C.A. **71** [1969] No. 52462).
[7] W. G. Scribner (AD-666459 [1967] 1/90, 31/2; C.A. **69** [1968] No. 62005).
[8] E. Sheperd, W. W. Meinke (AECU-3879 [1958] 1/5; N.S.A. **13** [1959] No. 1201).

1.3.1.4 With Acidic Esters of Phosphoric Acid

Di-n-butylphosphate. Beryllium could be extracted from a 0.05 M solution of a Be^{2+} salt in aqueous 1.0 M HNO_3 by an equal volume of a 1 M suspension of di-n-butylphosphate $(C_4H_9O)_2POOH$ in kerosene to an extent of 91.1% within 5 min whereas a 0.2 M suspension recovered only 50.8% of the metal. The extraction is not selective and a clean separation of the liquid phases is difficult [1]. The extraction of Be from a 0.1 M sulfate solution in aqueous 0.25 N HNO_3 by a 0.50 M solution of di-n-butylphosphate in toluene yielded a distribution coefficient $D_{Be} = 34.5$; after 15 min equilibration is attained. Higher concentrations of acid in the aqueous phase decrease the extraction of Be very noticeably ($D_{Be} = 0.74$ in 2 M HNO_3). The value of D_{Be} depends also on the solvent and decreases in the order: kerosene > CCl_4 > toluene > benzene > $CHCl_3$. It is assumed that Be is extracted from dilute aqueous acid solutions (< 2 M) mainly as the 1:4 complex $[BeH_2(PO_2(OC_4H_9)_2)_4]$, [2].

Di(-2-ethylhexyl)phosphoric Acid (abbreviated HDEHP or, sometimes, DEHPA, in the literature). The extraction of Be from weakly acid solutions (< 2 M HNO_3 or $HClO_4$ [2], or < 1 M HCl [3]) by a dialkylester of phosphoric acid such as HDEHP (in a nonpolar diluent like kerosene) occurs predominantly as the complex $Be(DEHP)_2(HDEHP)_2$. At higher acid concentration, complexes containing the anion of the aqueous acid are probably also extracted [2]. The distribution coefficient D_{Be} (ratio of Be content in organic and aqueous phase) studied with the radioactive tracer method for the extraction of a (^7Be labelled) 0.1 M Be solution in HCl at pH = 3 by 0.18 M HDEHP in various diluents depends on the diluent and decreases in the order: hexane > kerosene > CCl_4 > benzene > toluene > xylene > $CHCl_3$ [3].

Extraction studies at 25 °C in 10^{-3} to 5 M HCl with 0.1 M HDEHP in kerosene show that D_{Be} is highest at pH=3 (10^{-3} M HCl), then slowly decreases up to pH≈1 and sharply decreases at pH<1. A sharp decrease of D_{Be} was found with increasing Be concentration between 0.1 and 0.5 M (studied at pH=3 and extractant concentrations of 0.2 and 0.4 M) as well as with decreasing extractant concentration between 0.6 and 0.05 M (0.1 M Be). D_{Be} is directly proportional to the second power of the extractant concentration at pH=3 and 1. Shaking of equal parts (10 mL) of aqueous and organic phase in the separatory funnels for 2 h was sufficient to attain equilibrium [3].

Extraction studies of 0.1 M Be in 0.25 and 2 M HNO_3 with 0.5 M HDEHP in toluene showed in comparison to other dialkyl esters (see p. 40) a very slow extraction rate at room temperature. The time required to reach equilibrium was about 10 and 20 h in 0.25 and 2 M HNO_3, respectively, and D_{Be} decreased from 19 to 0.35 as the acid concentration increased. A similar decrease of D_{Be} with acid concentration was observed also in other acids, such as $HClO_4$ and HCl [2].

In H_2SO_4, D_{Be} increases by a factor of 20 between pH=0 and 2 with decreasing acidity (0.5 M HDEHP in kerosene). The rate of extraction increases with increasing temperature. The equilibrium at pH=1.5 is reached after contact for 300 min at 25 °C, or after only 100 min at 45 °C [4]. The effect of temperature (25 to 60 °C) at pH=1 and the effect of pH (0 to 2) at 45 °C were also studied by [5]. The rate of extraction, studied at pH=3 and 1.1 at 30 °C (0.5 M HDEHP in kerosene+4 wt% octan-2-ol) is also dependent on the speed of the turbine run: at 1250, 1500, 1750, and 2500 rpm the efficiency of Be extraction within 5 min at pH=3 was 59.5, 64.5, 69.8, and 71.4% of the amount extracted at equilibrium (92.6%). At pH=1.1 the corresponding values within 30 min are 68.4, 77.4, 83.5, and 85.9% of the equilibrium value (82.0%) [6]. For similar studies, see [5]. The distribution coefficient D for the extraction of 10^{-4} M Be with 0.125 M HDEHP (in benzene) in the presence of 4 M $NaNO_3$ increases with rising pH from about D_{Be}=0.1 at pH=0 to a maximum of about 5×10^{4} at pH=4. (D_{Be} is higher than that for the other alkaline earth ions by several orders of magnitude) [7]. The extraction of Be with a concentrated solution (80 vol%) of HDEHP in toluene (regarded as a cation exchanger) at contact times of 1 min (20 mL of aqueous and organic phase each) is <5% at pH<1 (1 N HCl or HNO_3), 59% at pH=1, and 72% at pH=4 to 5 [8].

The extraction of Be from sulfate leach solutions (from ores) which commonly contain Fe and Al along with Be is not selective at pH=3 (HDEHP in kerosene); the order of preference is Fe^{III}>Be>Al [9]. The coextraction of Fe is depressed by a precedent reduction to Fe^{II} with SO_2 or NaHS or a removal of Fe by amine solvent extraction. The coextraction of Al increases with increasing pH between 0 and 3 [10]. The extraction of Be in the presence of Al was studied using synthetic salt solutions at 30 °C and pH=0.5 to 4 with 0.1 M HDEHP in kerosene containing 4 vol% nonanol. The 1:1 aqueous-organic phases were stirred for 2 h. At pH>2.5 Be and Al were extracted. At pH=2.2 the amounts of Be and Al extracted were 70 and 12%, respectively, giving a separation factor D_{Be}/D_{Al}=17 [11]. Upon extracting with 0.1 and 0.2 M HDEHP (in kerosene) at pH=2.2 or 1.8 to 2.0, respectively, the coextracted Al amounts to $^1/_7$ to $^1/_{10}$ of the Be. In the presence of large amounts of Al, the optimum pH value is 1.6 to 1.8 [12]. A separation factor Be/Al of about 7 and 12.7 at pH=1.5 and 1, respectively, was reported by [13] using 0.5 M HDEHP in kerosene at 40 °C.

Extracting sulfate leach solutions with 0.5 M HDEHP in kerosene at pH=0 to 2 gives the best Be-Al separation at pH=0, although the loading capacity of the organic phase then was severely reduced. The separation factor found is 160 at pH=0, 52 at pH=0.5, and about 70 at pH=1.5 to 2 [4]. For the recovery of Be from sulfate leach solutions (containing in g/L: 7.9 BeO, 23.2 Al_2O_3, 0.9 Fe_2O_3, 0.6 FeO, et al.) by consecutive countercurrent extraction at pH=0.7 to 1 without reduction of Fe^{III} (extractant 0.52 M HDEHP in kerosene containing 4 wt% octan-2-ol), see [6]. After reduction of Fe^{III} with NaHS, the best conditions for the Be extraction from a sulfate leach solution containing 1.6 g BeO/L are: 0.25 N HDEHP in kerosene containing 2 g isodecanol/100 mL, 45 °C, and pH = 2. A commercial plant would achieve 95% overall extraction operating with a strong agitator with a power input of 393 W/1000 L (20 hp/1000 gal) dispersion in each of the seven mixer-settlers with 16 min residence time per mixer [5]; for similar tests see [14]. Removal of Fe from the sulfate leach liquor by precedent amine solvent extraction or reduction of Fe^{III} by SO_2 prior to the countercurrent multistage extraction at 45 °C with 0.5 M HDEHP in kerosene (containing 0.5 mol = 80 g/L of primary decyl alcohol) at pH = 0.7 and 0.55 was studied by [15]. The addition of decanol to the organic phase prevents the separation of sodium di-2-ethylhexyl phosphate from the kerosene during caustic stripping [15].

Be was recovered from the organic extract by stripping with aqueous 1.5 N NaOH (applied mole ratio Na_2O:BeO = 2:1) according to the equation: $BeH_2(DEHP)_4 + 6NaOH = Na_2BeO_2(aq) + 4NaDEHP(org) + 4H_2O$ [5, 15]; 3 N NaOH [6] or 4.2 N NaOH [14] are also applied. Upon boiling the strip liquor deposited 96 to 98% of the beryllium as $Be(OH)_2$ which after calcination yielded BeO of 98 to 99% purity, suitable for most non-nuclear uses [5], also see [15]. Acid stripping of the loaded kerosene phase by aqueous 6 N HCl or 6 N H_2SO_4 and precipitation of $Be(OH)_2$ from the strip liquor at pH = 9 in the presence of a chelating agent was recommended by [4]. Stripping with 9 M H_2SO_4 at 60 to 65 °C afforded the crystallization of rather pure $BeSO_4 \cdot 2H_2O$ [13]. Stripping with aqueous 1 M NH_4HF_2 solution with an aqueous to organic flow ratio of 1:3 in three steps recovered 99.7% of the beryllium as the crystallizable fluoro complex $(NH_4)_2BeF_4$ [6]. Almost complete recovery of beryllium (98%) by stripping with 1 M aqueous solutions of HF, NaF (formation of BeF_4^{2-}), and $(NH_4)_2CO_3$ was reported by [9].

Other Di- and Monoalkylphosphoric Acids. The extraction of Be from solutions in 0.25 M HNO_3 was studied using various dialkylphosphoric acids $HPO_2(OR)_2$ (I to VII) and one monoalkylphosphoric acid $H_2PO_3(OR')$ (VIII) with R = n-butyl (I), R = 2-ethylhexyl (II), R = n-octyl (III), R = iso-octyl (IV), R = n-decyl (V), R = iso-decyl (VI), R = R′ = 1-isobutyl-3,5 dimethylhexyl (VII) and (VIII). The following table includes the distribution coefficients D_{Be}, the times t_e to reach equilibrium (t_e in min), and the equilibrium constants K of the overall reactions in 0.25 M HNO_3,

$$Be^{2+}_{aq} + 2(HA)_{2,\,org} \rightleftharpoons BeA_2(HA)_{2,\,org} + 2H^+_{aq}$$

where $(HA)_2$ is the hydrogen-bonded dimeric form of the dialkyl esters (contained in the 0.5 M solutions in toluene):

extractant . .	I	II	III	IV	V	VI	VII	VIII
D_{Be}	34.5	19.0	62	28	54.0	30.2	60	31.4
t_e	15	≈600	250	30	45	90	250	30
K	1.55	1.28	1.71	1.45	1.73	1.48	1.78	—

The D_{Be} values for the extraction from 2 and 10 M HNO_3 with the same extractants are <1, or 1 to 1.2, in two cases [2]. Further extraction studies with 0.5 M solutions of several mono- and dialkyl phosphoric acids in kerosene (with addition of octan-2-ol) revealed that only the di-nonyl and two di-octyl phosphoric acids with alkyl chains branching at the 2-positions were promising as selective extractants for Be from Al-containing sulfate solutions at pH = 3.0 and 1.0. Comparative screening tests showed that di(2-propyl-4-methylpentyl)phosphoric acid had the lowest extraction coefficient and that di(2-ethyl-4-methylpentyl)phosphoric acid and di(2-ethylhexyl)phosphoric acid (HDEHP, see above) differed only in marginal respects. The latter product is preferred because of its larger commercial availability [6].

As for HDEHP, with the extractants monododecylphosphoric acid (H_2DoDP) and monoheptadecylphosphoric acid (H_2HDP) the rate of extraction increases with increasing temperature (25 to 45 °C) and pH (0 to 2), but D_{Be} and D_{Be}/D_{Al} at pH = 1.5 are smaller than with HDEHP [4]. For short time extractions with H_2DoDP (and HDEHP) in kerosene studied at various pH values with and without the presence of Al and Fe, see [9]. For some tests for the extraction of ^7Be labelled Be by 0.55 M solutions of H_2HDP (in mineral spirits) followed by stripping with 8.2 N H_2SO_4, see [16]. Extraction of sulfate leach solutions with H_2HxDP (monohexadecylphosphoric acid) in kerosene at 40 °C gives at pH = 1 a higher value of D_{Be} than with HDEHP and a lower selectivity (D_{Be}/D_{Al}), but at pH = 1.5 the selectivity is higher. A process of two-stage countercurrent extraction of the leach solution (2.18 g Be and 4.3 g Al per liter) with 0.5 M H_2HxDP (in kerosene) at 60 to 65 °C and pH = 1.5 with a contact time of 1 h in each step was described. The organic phase was subsequently stripped with 9 M H_2SO_4. The phase ratios for leach:organic:strip are 2.8:5:1. Testing of the proposed process through several cycles with recycling of both organic and strip solutions produced products containing a Be:Al weight ratio of about 1:0.1. Further reduction of the Al content of the solid can be accomplished by repulping with H_2SO_4 of slightly lower concentration [13]. A similar procedure is described in [17]; furthermore the extraction with H_2HxDP, H_2DoDP and dioctylphosphoric acid (in kerosene) was studied at pH = 1 to 3.5, or at pH = 7 after addition of a complexing agent (EDTA) [17].

Alkylaryl Phosphoric Acids. For the extraction of Be from aqueous sulfate solutions (1 g BeO/L) by a 0.1 M solution of di-(p-tertiary-octylphenyl)phosphoric acid (HDOPP) at pH = 5 and an organic to aqueous ratio of 5:1 (33% in 5 min), see [9]. Extraction of Be from a sulfate solution containing 3.0 g BeO/L by a 0.5 M solution of methyl-p-tert-octylphenylphosphoric acid (HMOPP) in kerosene containing 10% capryl alcohol at pH = 1 to 3 recovered 96 to 99.4% of the beryllium in the organic phase within 2 to 10 min and was much faster than the extraction by HDEHP (see p. 39) at comparable conditions. The selectivity with respect to Al was also studied [17]. For the extraction of Be from sulfate solutions containing 2.6 g BeO/L by (alkyl substituted) diphenylphosphoric acids in a mixture of kerosene (or a similar solvent) and hexane at pH = 2.5, see [18].

References:

[1] W. J. Biermann, R. McCorkell (Can. J. Chem. **40** [1962] 1368/73, 1371).
[2] C. J. Hardy, B. F. Greenfield, D. Scargill (J. Chem. Soc. **1961** 174/82, 176, 180).
[3] J. S. El-Yamani, M. Y. Farah, E. N. Abd El-Messieh (J. Radioanal. Chem. **45** [1978] 147/53, 148/51).
[4] R. O. Dannenberg, D. W. Bridges, J. B. Rosenbaum (U.S. Bur. Mines Rept. Invest. No. 5941 [1962] 1/16, 4/9; C.A. **56** [1962] 13870; N.S.A. **16** [1962] No. 10096).
[5] R. O. Dannenberg, L. Crocker, D. W. Bridges (U.S. Bur. Mines Rept. Invest. No. 6469 [1963] 1/31, 27/9; N.S.A. **18** [1964] No. 27613).
[6] R. A. Wells, D. A. Everest, A. A. North (Nucl. Sci. Eng. **17** [1963] 259/67, 263/6).
[7] W. J. McDowell, C. F. Coleman (J. Inorg. Nucl. Chem. **28** [1966] 1083/9, 1085).
[8] F. E. Butler, A. R. Boulogne, E. A. Whitley (Health Phys. **12** [1966] 927/33, 928; C.A. **65** [1966] 6678).
[9] A. A. Büggs, J. V. Martin, G. L. Milward, G. M. E. Sims (NCL-AE-172 [1958] 1/15; N.S.A. **16** [1962] No. 25475).
[10] J. B. Rosenbaum, R. O. Dannenberg, D'Arcy R. George (Solvent Extract. Chem. Metals Proc. Intern. Conf., Harwell, Engl., 1965 [1967], pp. 315/26, 320/2; C.A. **69** [1968] No. 79358; N.S.A. **22** [1968] No. 16690).

[11] R. W. Cattrall (Australian J. Chem. **14** [1961] 163/6).

[12] V. Jiřele (Chem. Prumysl **17** [1967] 175/9, 177; C.A. **67** [1967] No. 66699).
[13] J. P. Surls, R. R. Grinstead (Nucl. Sci. Eng. **28** [1967] 338/45, 339/41).
[14] R. O. Dannenberg, J. M. Maurice (U.S. Bur. Mines Rept. Invest. No. 6841 [1966] 1/12, 5/8; C.A. **66** [1967] No. 5002).
[15] L. Crocker, R. O. Dannenberg, D. W. Bridges, J. B. Rosenbaum (U.S. Bur. Mines Rept. Invest. No. 6173 [1963] 1/27, 7/12, 25; C.A. **58** [1963] 9900; N.S.A. **17** [1963] No. 12357).
[16] A. M. Poston Jr., J. V. Batty, H. L. Gibbs (U.S. Bur. Mines Rept. Invest. No. 5980 [1961] 1/10, 6/8; N.S.A. **16** [1962] No. 20950).
[17] Dow Chemical Co. (Fr. 1369803 [1963/64] 1/5; C.A. **62** [1964] 4959).
[18] G. L. Milward, J. C. Reeve (Brit. 960457 [1961/64] 1/4, 2; C.A. **62** [1965] 2538).
[19] J. K. Grunig, R. J. Anderson, B. L. Vance, Anaconda Co. (U.S. 3729541 [1971/73] 1/3; C.A. **79** [1973] No. 33981).

1.3.1.5 With Various Extractants from Thiocyanate Solutions

Beryllium could be extracted from a 0.1 M aqueous sulfate solution containing 3 M NH_4SCN at pH=3.0 by 1-pentanol to an optimum extent of 81%. Substantially higher or lower pH values [1] and lower concentrations of NH_4SCN [2] will decrease the extraction of Be [1, 2] either because of a lowered concentration of the NCS^- ions (pH<3) or formation of species like $[BeOH]^+$ (pH>3). The most efficient solvents seem to be alcohols with six to eight carbon atoms or solvents containing double-bonded oxygen associated with five or six carbon atoms. If the separation from Al is also considered, 1-hexanol, 2-octanol, methyl isobutyl ketone, and isoamyl acetate appear to be most promising ones [1]. However, far better results were obtained with tri-n-butylphosphate, which at pH=3.1 extracted 80% of the beryllium even from a solution containing only 0.5 M NH_4SCN, 98% at 3.0 M, and 99.5% at 5.0 M NH_4SCN [3]. The extraction of Be by methyl isobutyl ketone could be increased from 71.0 to 86.6% upon raising the concentration of NH_4SCN from 3 M to 7 M [2], also see [3], and by 1-hexanol from 72.1 to 94.5% [2], for isoamyl alcohol (similar values), see [3]. With additional 0.1 M Al^{3+} in solution, a considerable part of aluminium was coextracted, especially by 1-hexanol (41.2% at 3.0 M NH_4SCN). Improved separation from Al could be achieved only by lowering the concentration of NH_4SCN to 2.0 M, which in the case of methyl isobutyl ketone, decreased the extraction of Be^{2+} from 71.0 to 42.3% and that of Al from 12.7 to 3.7%. Thus, for a good recovery of Be with effective separation from Al a two solvent countercurrent extraction of Be from a suitable thiocyanate solution (2 M NH_4SCN at pH=3.0) was proposed [2], also see [4]. With methyl isobutyl ketone as the solvent, the distribution coefficient D_{Be} depends only on the concentration of thiocyanate, as given by the equation $D_{Be}=0.3\,[SCN^-]^2$, and is independent of the pH value [5].

Iron can be removed from $BeSO_4$ solutions as the thiocyanato complex by adding a solution of NH_4SCN and kerosene containing 10% tributylphosphate. Complete removal of iron was attained at pH=1.8 to 2.3 and a mole ratio $NH_4SCN:Fe^{3+}=6:1$ (3 to 6 g NH_4SCN/L) upon adding an equal volume of the organic solvent which extracts the iron (III) thiocyanato complex [6].

Beryllium could be extracted from aqueous 4 M NH_4SCN solutions containing 0.04 M Be^{2+} by alkyl amines (0.3 M) with polar organic diluents such as alcohols, ketones, or esters. The best extraction (>60%) at pH=6 with amyl acetate as the solvent was obtained by trihexylamine (80.5%) and trioctylamine (72.3%). If diamylamine was used as the amine component the following solvents were most efficient: cyclohexanol (86.1%), methyl isobutyl ketone (83.3%), acetophenone (75.0%) 1-hexanol, isoamyl alcohol (69.4% each), 2-octanol (66.6%), diethyl ether (63.8%), and ethyl acetate (61.1%). It is assumed that Be is extracted as the organoammonium salt of an anionic thiocyanato complex $[amineH]_n[Be(SCN)_{2+n}]$

where the ammonium ion is N-hydrogen bonded to the solvent oxygen. This is supported by the fact that the Be complex is not extracted into nonpolar solvents like $CHCl_3$ or CCl_4 [7]. Using tricapryl methyl ammonium thiocyanate (Aliquat 336), dissolved in 4-methyl-2-pentanon, n-hexane, benzene, or chloroform, the optimal extraction occurs at 4 M KCNS and pH 3, El-Yamani, El-Messieh [8].

References:

[1] W. J. Biermann, R. McCorkell (Can. J. Chem. **40** [1962] 1368/73, 1372).
[2] W. J. Biermann, R. McCorkell (Can. J. Chem. **41** [1963] 112/6, 114).
[3] C. Różycki, W. Suszczewski (Chem. Anal. [Warsaw] **17** [1972] 1209/17, 1214; C.A. **79** [1973] No. 21991).
[4] W. J. Biermann, Beryloy Ltd. (Can. 751298 [1962/67] 1/8, 5/6; C.A. **66** [1967] No. 118010).
[5] W. J. Biermann, R. H. McCorkell (Can. J. Chem. **45** [1967] 2846/9).
[6] R. B. Byersmith, C. S. Pomelee, J. Birnbaum (NYO-1116 [1953] 1/61, 54/6; N.S.A. **10** [1956] No. 1045).
[7] A. V. Novoselova, T. I. Pochkaeva, N. S. Tamm, G. A. Trubacheva (Vestn. Mosk. Univ. Khim. **24** No. 3 [1969] 44/8; C.A. **71** [1969] No. 85108).
[8] I. S. El-Yamani, E. N. Abd El-Messieh (J. Radioanal. Nucl. Chem. Letters **86** [1984] 327/36).

1.3.2 Ionic Flotation

Beryllium can be separated from Al and Fe by treating 100 mL of an aqueous solution containing 760 mg Be, 5 mg Al, 5 mg Fe with 0.07 g potassium stearate (1/50 of the theoretical value) at 60 °C and pH = 4.5 which recovered 90% of the Al and 97.4% of the Fe, but only 1.8% of the Be in the stearate froth. However, considerably more Be (7.2%) will be found in the sublate (froth) at higher contents of Al (11 g/L) and Fe (1.1 g/L) in the solution. The increased flotation of the metals at higher pH values indicates hydrolysis of the metal ions to form hydroxy cations [1], also see [2, 3]. Separation of Be from Al and Fe could be achieved also by masking these metals with EDTA (8.0% solution) and floating Be as $Be(C_{17}H_{35}COO)_2$ with stoichiometric amounts of potassium stearate (50 g/L) at 60 °C and pH = 3.50 to 4.75. Maximum, almost complete recovery of Be (99.74%) with only little admixtures of Al (~1.1%) and Fe (~1.2%) was attained at an initial pH = 4.25 and a final pH = 5.25. A purer product (0.7% Al, 0.2% Fe), recovering somewhat less Be (96.64%), resulted at pH = 4.0 to 4.5 [4]. Studies on the collection of Be from waste waters (25 to 30 mg Be/L) by aqueous solutions of the potassium salts of long-chain (C_9 to C_{16}) fatty acids (0.015 to 1.260%) showed the best results with potassium myristate (C_{14}), which at 20 °C and pH > 6 recovered about 85% of the Be within 8 min. The extraction of Be could be increased by 5 to 10% upon adding ethanol or propanol in small amounts (25 mmol/L) and raising the concentration of the collector. The same amount of butanol, NaCl, or Na_2SO_4 decreased the extraction of Be by 10 to 20% [5].

References:

[1] P. Tadzhibaev, T. A. Adilov (Khim. Redk. Tsvetn. Metal. **1975** 33/6; C.A. **84** [1976] No. 33979).
[2] G. A. Tsyganov, P. Tadzhibaev (Gidromet. Tsvetn. Redk. Metal. **1971** 53/7; C.A. **76** [1972] No. 88847).
[3] P. Tadzhibaev, G. A. Tsyganov, T. A. Adylov, T. V. Azizov (Flotatsionnye Metody Izvlecheniya Tsennykh Kompon. Rastvorov Ochistki Stochnykh Vod. Mater. 1st Vses. Semin., Moscow 1972 [1972/73], pp. 52/62, 54/6; C.A. **83** [1975] No. 13826).

[4] P. Tadzhibaev, G. A. Tsyganov (Uzb. Khim. Zh. **15** No. 5 [1971] 15/6; C.A. **76** [1972] No. 132202).
[5] L. D. Skrylev, L. A. Dashuk, Yu. F. Sin'kov (Izv. Vysshikh Uchebn. Zavedenii Tsvetn. Met. **1976** No. 6, pp. 8/11; C.A. **86** [1977] No. 143431).

1.3.3 Extraction of Beryllium by Cation or Anion Exchange Resins

Cation Exchange. Information concerning the extraction of beryllium from acetate buffered (pH=4.07) aqueous solutions of $BeSO_4$ (~10^{-6} M) with the cation exchange resin Dowex 50 and determination of the distribution of Be in the resin and the aqueous phase as a cationic acetato complex is found in [1]. Absorption of beryllium from an aqueous chloride solution (0.16 N NaCl) by the cation exchanger Dowex 50 was complete at pH<2.5 and dropped sharply to zero at pH>3.2. Selective and rapid elution of the absorbed Be was achieved by a >0.02 M aqueous solution of sulfosalicylic or gentisic acid (2.5-dihydroxy-benzoic acid) at pH=3.5 to 4.5. A neutral 1:1 complex was formed with the eluant [2, 3]. Complete separation of Be^{2+} from Ca^{2+}, Cu^{2+}, and UO_2^{2+} was possible at pH=3.5 to 3.8 since no elution of these metal ions occurred in that pH range [3]. Studies on the distribution of Be between the cationic KU-2 (a styrene-divinylbenzene copolymer with $-SO_3H$ groups) and the aqueous phase in perchloric acid solutions (Be^{2+} 10^{-2} to 10^{-4} M, $HClO_4$ 1 to 10^{-4} M) revealed that the extraction of Be by the resin increased at lower acid concentrations but was independent of the Be concentration for Be solutions of 10^{-3} M or less. The extraction rate of Be is ~78% for 10^{-4} M Be^{2+} in 0.1 M $HClO_4$. A far lower extraction of Be was observed in H_2SO_4 (~15% in 10^{-3} M H_2SO_4), probably due to the formation of sulfato complexes [4]. Quantitative recovery of Be by cationic SBS from solutions in HCl was attained at pH>2, and almost complete elution of Be (99.8%) from the resin was achieved with 2 N HCl [5]. Complete elution of Be from the exchanger resins KU-1 and KU-2 by aqueous 1 M NH_4Cl solutions was attained at 8 °C and pH=5.3. Elution was much faster with KU-1 (a strongly acidic exchanger prepared from 4-phenolsulfonic acid and formaldehyde) than with KU-2 and required less (1/3) eluant. Far more rapid elution occurred with 1 M HCl which converted the dimeric hydroxo species $[Be(OH)_2Be]^{2+}$, assumed at pH=5.3, into the more versatile Be^{2+} ions [6]. Elution of Be and other metals from the Dowex 50 W-X8 cation exchanger by a 0.05 M solution of diethylenetriamine pentaacetic acid at 70 °C and pH=6 yielded the elution sequence Cu>Zn>Al>Mn>Be>Mg>Ca>Sr>Ba [7].

Anion Exchange. Amberlite IRA 410, a strongly basic (quaternary amine) anion exchanger absorbed beryllium as an anionic sulfosalicylate complex from aqueous solutions containing sulfosalicylic acid (0.02 M) at ≧6 [3]. Absorption of Be from solutions in dilute HCl (0.05 M) by the anion exchanger Dowex-1, a strongly basic quaternary amine resin, was negligible but was significant in the presence of 1 M HF. However, the absorption of Be decreased in more concentrated HCl until at concentrations >1 M HCl it was essentially insignificant [8]. For absorption of Be from solutions in aqueous 0.1 M NH_4Cl by Dowex-1 in the presence of 0.25 M Na-ethylenediamine tetraacetate at 25 °C, see [9], and as the anionic oxalato complex $[Be(C_2O_4)_2]^{2-}$ by the anion exchange resin IRA-400 from 10^{-4} M to 10^{-3} M solutions of Be^{2+} at pH=4, see [10].

References:

[1] C. Bamberger, A. Suner (CNEA-191 [1966] 1/20, 12/3; C.A. **67** [1967] No. 6233).
[2] J. Schubert, A. Lindenbaum, W. Westfall (Chimia [Aarau] **11** [1957] 50/1).
[3] J. Schubert, A. Lindenbaum, W. Westfall (J. Phys. Chem. **62** [1958] 390/4).
[4] T. A. Belyavskaya, I. F. Kolosova (Vestn. Mosk. Univ. Ser. II Khim. **17** No. 5 [1962] 55/9; C.A. **58** [1963] 6427).

[5] T. A. Belyavskaya, V. I. Fadeeva (Vestn. Mosk. Univ. Ser. VI **11** [1956] No. 4, pp. 73/9, 74; C.A. **1957** 11162).
[6] A. I. Zhukov, G. P. Baranov, P. V. Plyasunov (Zh. Neorgan. Khim. **7** [1962] 1452/7; Russ. J. Inorg. Chem. **7** [1962] 745/8, 746).
[7] W. H. Hale, C. A. Hammer (Ion Exchange Membr. **1** [1972] 81/5, 82; C.A. **78** [1973] No. 105171).
[8] F. Nelson, R. M. Rush, K. A. Kraus (J. Am. Chem. Soc. **82** [1960] 339/48, 340).
[9] F. Nelson, R. A. Day, K. A. Kraus (J. Inorg. Nucl. Chem. **15** [1960] 140/50, 146).
[10] A. Lusher, F. Selba (J. Appl. Chem. [London] **15** [1965] 577/80, 579).

1.3.4 Separation of Beryllium by Sublimation of the Acetylacetonate

Aqueous solutions containing Be^{2+} together with other metals (such as Mg^{2+}, Al^{3+}, Cu^{2+}, Mn^{2+}, or Zn^{2+}) were made 0.5 M with NH_4SCN (to mask Zn^{2+}) and the pH was adjusted to about 7 by adding sodium acetate. Then an alcoholic solution of acetylacetone was added to precipitate the acetylacetonates. The precipitate was filtered off, washed, dried, and fractionally sublimed at 170 °C under a pressure of 1 Torr for 2 h with air as the carrier gas. The resulting sublimate of beryllium acetylacetonate was recovered mechanically or by dissolving in a suitable solvent such as acetone; it contained $<0.002\%$ of Zn (or other metals) after a single sublimation. However, the average yield was only 62% due to incomplete precipitation.

Reference:

E. W. Berg, A. D. Shendrikar (Anal. Chim. Acta **44** [1969] 159/64, 161).

1.4 Production of the Metal

General References:

K. A. Walsh, Extraction of Beryllium, Beryllium Sci. Technol. **2** [1979] 1/11; C.A. **93** [1980] No. 29663.
J. Ballance, A. J. Stonehouse, R. Sweeney, K. Walsh, Beryllium and Beryllium Alloys, Kirk-Othmer Encycl. Chem. Technol. 3rd Ed. **3** [1978] 803/23.
K. S. Subbarao, B. P. Sharma, C. M. Paul, C. V. Sundaram, Beryllium Development Program in India, Beryllium 1977 Conf. Preprint 4th Intern. Conf. Beryllium, London 1977, pp. 1/10; C.A. **92** [1980] No. 151299.
G. Petzow, F. Aldinger, Beryllium und Beryllium-Verbindungen, Ullmanns Enzykl. Tech. Chem. 4th Ed. **8** [1974] 442/61.
G. Dressler, M. Rühle, Beryllium in: K. Winnacker, L. Küchler, Chemische Technologie, 3rd Ed., Vol. 6, München 1973, pp. 549/601, 590/4.
E. H. Hohmann, Beryllium, Chem. Age India **20** [1969] 169/74.
H. H. Hausner, Beryllium Its Metallurgy and Properties, Univ. Calif. Press, Berkeley, Calif., 1965, pp. 1/30, 279/303.
D. A. Everest, The Chemistry of Beryllium, Topics in Inorganic and General Chemistry, Vol. 1, Elsevier, New York 1964, pp. 102/16, 133/40.
R. G. Bellamy, N. A. Hill, Extraction and Metallurgy of Uranium, Thorium, and Beryllium, Macmillan, New York 1963, pp. 11/6, 45/52, 72/84, 187/92.
W. Schreiter, Seltene Metalle, 2nd Ed., Vol. 1, Leipzig 1963, pp. 91/162.
W. D. Jamrack, Rare Metal Extraction by Chemical Engineering Techniques, Vol. 2, Pergamon, New York 1963, 30/2, 38, 62/5, 233/7, 280/1, 342/6.
B. R. F. Kjellgren, Beryllium in: C. A. Hampel, Rare Metals Handbook, 2nd Ed., New York 1961, pp. 32/57.

G. E. Darwin, J. H. Buddery, Beryllium, Butterworth, London 1960, pp. 1/392.
L. R. Williams, P. B. Eyre, Beryllium in: A. B. McIntosh, T. J. Heal, Materials for Nuclear Engineers, Interscience, New York 1960, pp. 269/318; N.S.A. **15** [1961] No. 11485.
P. Silber, Glucinium in: P. Pascal, Nouveau Traité de Chimie Minérale, Vol. 4, Masson, Paris 1958, pp. 7/30.
G. A. Meerson, D. D. Sokolov, N. F. Mironov, N. M. Bogorad, I. D. Pokhomov, D. S. Lvovsky, E. S. Ivanov, V. M. Shmelev, Beryllium, At. Energiya SSSR **5** [1958] 624/30; Soviet At. Energy **5** [1958] 1555/62; C.A. **1961** 7218.
W. W. Beaver, Technology of Beryllium and Beryllium Oxide, Progr. Nucl. Energy V **1** [1956] 277/99; Met. Abstr. [2] **24** [1956/57] 963.
D. W. White Jr., J. E. Burke, The Metal Beryllium, ASM, Cleveland, Ohio, 1955, pp. 1/703.
K. Illig, H. Fischer, W. Birett, Beryllium in: G. Eger, Die Technische Elektrolyse im Schmelzfluß, 2nd Ed., Leipzig 1955, pp. 571/641.
G. Jaeger, Über die Gewinnung des Berylliums, Metall **4** [1950] 183/91.

Review

The preparation and production of beryllium metal is one of the more difficult tasks in metallurgy because of the specific chemical and physical properties of the metal and its compounds, and because of its high toxicity [1, 2].

Beryllium metal was first obtained in 1828 by Wöhler [3] and independently by Bussy, see [4]. Both reduced beryllium chloride with potassium in a platinum crucible. Since that time hundreds of methods for its preparation and production have been reported, and more than one hundred have been patented. However, relatively few have a sound experimental basis. In many patents, methods suitable for the preparation or production of other metals are extended to beryllium without further investigation or consideration of its special properties. Most of the really viable methods can only be carried out on a laboratory scale or small scale production. Since the 1940's, only two methods have been successfully applied on a commercial scale. These methods are the reduction of beryllium fluoride with magnesium metal, and the "low-temperature" electrolysis of beryllium chloride/sodium chloride mixtures. Today, the demand for beryllium metal has decreased, and only the magnesium metal reduction process is still in operation [1, 2, 5 to 10], cf. also [12].

In preparation, production, and handling, strict safety provisions (completely automatic, closed equipment; mist, dust, and fume filtering; working clothes and medical supervision for the operators) must be observed [2, 9, 11].

Methods for the preparation and production of beryllium metal published before 1930 are described in "Beryllium" 1930, pp. 39/45.

References:

[1] W. J. Kroll (U.S. Bur. Mines Inform. Circ. No. 7326 [1945] 1/15).
[2] R. G. Bellamy, N. A. Hill (Extraction and Metallurgy of Uranium, Thorium and Beryllium, Macmillan, New York 1963, pp. 11/6, 12, 184, 187/9).
[3] F. Wöhler (Ann. Physik Chem. [2] **13** [1828] 577/82).
[4] J. Berzelius (Jahresber. Fortschr. Phys. Wiss. **9** [1830] 96/8).
[5] G. Jaeger (Metall **4** [1950] 183/91, 187, 190).
[6] S. B. Roboff (in: H. H. Hausner, Beryllium, Its Metallurgy and Properties, Univ. California Press, Berkeley 1965, pp. 17/30, 18/22; C.A. **64** [1966] 10879).
[7] G. E. Darwin, J. H. Buddery (Beryllium, Butterworth, London 1960, pp. 74/5).

[8] T. T. Magel (in: D. W. White Jr., J. E. Burke, The Metal Beryllium, ASM, Cleveland 1955, pp. 124/35).
[9] J. Ballance, A. J. Stonehouse, R. Sweeney, K. Walsh (in: Kirk-Othmer Encycl. Chem. Technol. 3rd Ed. **3** [1978] 803/23, 809, 814).
[10] K. A. Walsh (in: D. R. Floyd, J. N. Lowe, Beryllium Science and Technology, Vol. 2, Plenum, New York - London 1979, pp. 1/11).
[11] A. R. Kaufmann, B. R. F. Kjellgren (Proc. Intern. Conf. Peaceful Uses At. Energy, Geneva 1955, Vol. 8, pp. 590/9, 598).
[12] F. E. Block, R. E. Mussler, T. T. Campbell (Advan. Extr. Met. Proc. Symp., London 1967 [1968], pp. 551/71; C.A. **69** [1968] No. 69070).

Comparison of the Advantages of Commercial and Alternative Production Processes

In comparing the major processes developed over the years for commercial Be metal production, namely (I) the reduction of BeF_2 with Mg metal, and (II) the low-temperature electrolysis of $BeCl_2$, certain advantages can be stated for each of them.

Advantages of Process I: 1) Process (I) is more economical than electrolysis (II), especially at low production rates [1 to 3, 5, 12]. 2) There exists more operational experience than with electrolysis (II) [2, 4]. 3) The Be metal is produced in relatively large amounts, in a convenient, handleable form (pebbles or beads) [2, 3]. 4) The starting material BeF_2 is more easily prepared than $BeCl_2$, the starting material for process (II) [1, 5, 6, 9]. 5) Higher yields are obtained than by process (II) [1, 4, 5].

Advantages of Process II: 1) Large-scale production with inclusion of certain processing improvements promises lower production costs [1, 2, 4]. 2) Easily grindable Be metal flakes in relatively high purity are produced with lower levels of metallic impurities than in the product of process (I), [1, 2, 5 to 8, 11, 12]. 3) The flakes are suitable for repurification by re-electrolysis [2].

For comparative purity data for the Be metal products of both processes (I and II), see [2], cf. [7, 8].

An existing alternative production process involving a Kroll-type reduction of $BeCl_2$ vapor with molten sodium metal reductant has the following advantages over the commercial processes: 1) Be sponge is obtained in higher purity and higher yields. 2) The sponge is readily comminutable into fine powder, eliminating the need for vacuum casting. 3) It promises to offer economic advantages over current commercial practices [10, 11].

References:

[1] G. E. Darwin, J. H. Buddery (Beryllium, Butterworth, London 1960, pp. 78/9, 98).
[2] S. B. Roboff (in: H. H. Hausner, Beryllium, Its Metallurgy and Properties, Univ. California Press, Berkeley 1965, pp. 17/30, 29/30; C.A. **64** [1966] 10879).
[3] J. Ballance, A. J. Stonehouse, R. Sweeney, K. Walsh (in: Kirk-Othmer Encycl. Chem. Technol. 3rd Ed. **3** [1978] 803/23, 809/10).
[4] C. W. Schwenzfeier Jr. (in: D. W. White Jr., J. E. Burke, The Metal Beryllium, ASM, Cleveland 1955, pp. 71/101, 85, 95).
[5] G. A. Meerson, D. D. Sokolov, N. F. Mironov, N. M. Bogorad, I. D. Pokhomov, D. S. Lvovsky, E. S. Ivanov, V. M. Shmelev (At. Energiya SSSR **5** [1958] 624/30; Soviet At. Energy **5** [1958] 1555/62, 1556; C.A. **1961** 7218).
[6] W. W. Beaver (Progr. Nucl. Energy V **1** [1956] 277/97, 283, 285/6; Met. Abstr. [2] **24** [1956/57] 963).

[7] L. R. Williams, P. B. Eyre (Nucl. Eng. **3** [1958] 9/18, 10/11).
[8] L. R. Williams, P. B. Eyre (in: A. B. McIntosh, T. J. Heal, Materials for Nuclear Engineers, Interscience, New York 1960, pp. 269/318, 272/5; N.S.A. **15** [1961] No. 11485).
[9] A. I. Evstyukhin (Met. Metalloved. Chist. Metal. No. 1 [1959] 91/105, 94; C.A. **1961** 4292).
[10] T. T. Campbell, R. E. Mussler, F. E. Block (Met. Trans. **1** [1970] 2881/7, 2885; C.A. **73** [1970] No. 122649).
[11] F. E. Block, R. E. Mussler, T. T. Campbell (Advan. Extr. Met. Proc. Symp., London 1967 [1968], pp. 551/71, 552, 569; C.A. **69** [1968] No. 69070).
[12] S. J. Morana, G. F. Simons (J. Metals **14** [1962] 571/4).

1.4.1 Reduction of Beryllium Oxide with Metals or Nonmetals

Only limited work has been done on the direct reduction of beryllium oxide to beryllium metal with metals or nonmetals [1, 2]. The successful reduction methods are usually expensive and not very suitable for commercial processes [1, 3, 4]. One method, however, the small scale production of beryllium metal from beryllium oxide with zirconium metal, has found some value for special applications [3, 5].

References:

[1] T. T. Campbell, R. E. Mussler, F. E. Block (Trans. AIME **236** [1966] 1456/61).
[2] K. Illig, H. Fischer, W. Birett (in: G. Eger, Die Technische Elektrolyse im Schmelzfluß, 2nd Ed., Leipzig 1955, pp. 571/641, 593/5).
[3] G. E. Darwin, J. H. Buddery (Beryllium, Butterworth, London 1960, pp. 75/6, 93/4).
[4] H. H. Kellogg (in: D. W. White Jr., J. E. Burke, The Metal Beryllium, ASM, Cleveland 1955, pp. 49/62, 52/6, 53).
[5] T. T. Magel (in: D. W. White Jr., J. E. Burke, The Metal Beryllium, ASM, Cleveland 1955, pp. 124/35, 129/32).

1.4.1.1 Reduction with Metals

With Calcium

From a study of the standard free enthalpies of formation of beryllium oxide and a series of other metal oxides as a function of temperature (see [2, 3]), it is evident that beryllium oxide is extremely stable. Under standard state conditions only calcium can reduce BeO to the metal [1 to 4], cf. [7].

The reduction of beryllium oxide with calcium, however, yields a mixture of calcium oxide and a relatively stable intermetallic Ca-Be phase [4, 7], thought to be $CaBe_{13}$[1)] [4]. The intermetallic phase always forms, since a large excess of calcium is necessary to achieve nearly complete reduction of the beryllium oxide [1, 4, 7]. A successful removal of calcium from the intermetallic phase has been reported by [7].

The reduction was carried out under 1 atm argon gas using a mixture of chemically pure BeO and 99.99% Ca (produced by redistillation of electrolytic calcium metal at 850 °C and 10^{-4} Torr) [7]. The weight ratio BeO:Ca was 1:6, the reaction temperature was 1200 °C, and the reaction time was 5 h. Under these reaction conditions calcium vaporizes quite rapidly and the calcium vapor reacts upon contact with the BeO. The reaction products obtained (intermetallic phase and CaO), are then treated with a cold aqueous solution of NH_4Cl (200 g/L) to remove Ca and CaO. Beryllium metal is obtained in the form of a highly

---

1) $CaBe_{13}$ has been identified by chemical analysis [4] and by X-ray study [5]; cf. also [6].

pure powder, containing only spectrographical traces of Fe, Ca, Mg, and Al. By this process 100% utilization of BeO is attained. The Be metal powder produced is used for Be powder metallurgy [7].

Another method for the removal of calcium from the intermetallic Ca-Be phase containing CaO inclusions involves high vacuum distillation at high temperatures. Here, however, the purity of the resulting Be metal is only 85% [7].

The apparatus used for the reduction of BeO with Ca metal consists of a stainless steel retort, containing a reaction crucible made of magnesia or mild steel [7].

In a previous study [4], the reduction was performed under an Ar atmosphere on compacts of BeO (completely waterfree material of maximum activity, obtained by heating Be hydroxide very slowly to 700 °C) containing a 100% excess of calcium (double distilled, particle size −12+20 mesh) at ~1000 °C. The intermetallic phase, $CaBe_{13}$, was readily separated from the CaO and excess Ca by leaching in cold dilute acetic acid.

Attempts to remove the calcium from the $CaBe_{13}$ by vacuum distillation and by reaction with hydrogen, magnesium, bismuth, or lead were all unsuccessful. The expected compounds, Ca_3Mg_4, Ca_3Bi_2, and Ca_2Pb were not detected by X-ray methods [4].

For earlier, less successful studies on the reduction of Be oxide with calcium metal, see [14].

With Zirconium and Other Metals

The reduction of beryllium oxide with zirconium metal is theoretically possible if the normally unfavorable reduction equilibrium is shifted by continuously removing the Be metal that is formed. In order to do this, the reaction must be carried out at a high temperature in a high vacuum. Beryllium is much more volatile than the other three components, and is distilled off [2, 3, 8, 9].

As early as 1939, Kroll [10] proposed that a reduction of beryllium oxide with zirconium metal might be possible. Later, in 1952, Magel [8, 11] studied the beryllium oxide reduction by zirconium (or zirconium hydride) on "small scale" (milligram yields) and "larger scale" (gram yields) experiments. In 1966, an exploratory study of this reduction reaction was undertaken by Campbell et al. [9]. The study was not completed [9].

Magel [8, 11, 12] developed a technique to produce beryllium metal of relatively low oxygen content. An intimate stoichiometric mixture of finely ground (−300 mesh) zirconium (or its hydride $ZrH_{1.3}$) and beryllium oxide powder (−300 mesh) is compacted by pressing at 12000 psi, and the compacts are broken into pieces of about 1.27 cm in diameter. Heating in vacuum at 1525 °C for 2 to 4 h yields about 50% of the total available metal; additional heating (2 to 4 h) at 1725 °C increases the yield to 93%. In one seven-hour experiment at 1725 °C, a yield of 87% Be metal was attained. The Be metal contained 0.15% BeO and 0.15% Fe [3, 8, 11, 12].

The apparatus (see the paper) basically consists of a high-frequency heated, tantalum-lined graphite crucible, above which is mounted a tantalum-lined graphite condenser [8, 12].

The zirconium reduction of BeO can be adapted to very small scale operations (milligram amounts) and easily gives small quantities of high quality beryllium [8]:

Zirconium converts radioactive BeO (e.g., 287 mg) to radioactive Be metal in a 75% yield of consolidated metal. The conversion requires one hour of heating under vacuum at 1600 °C [12].

Yields as high as 95.4% were obtained in other small scale runs with inactive BeO [8, 12].

The reduction apparatus for the very small scale operations is described in [8, 12]. The zirconium reduction of BeO has definite advantages when only a very small amount of BeO is available. It is doubtful that any other small scale method of reducing beryllium compounds gives comparable yields of metal in a coherent usable form [8].

The studies of Campbell et al. [9] on the reduction reaction were carried out with charges of 150 to 200 g BeO (−200 mesh powder) and a 25% excess of zirconium metal compacted into pellets and carefully degassed.

The dehydriding of zirconium hydride usually starts at about 500 °C and is complete at 1000 °C. The BeO reduction is best performed at 1600 to 1750 °C for periods of ~30 h, giving an 85.8% yield of Be metal. The Be vapors formed in the reaction are condensed as a coherent deposit on a heated surface. Zirconium metal contamination and minor or trace amounts of intermetallic compounds $ZrBe_{13}$, $ZrBe_5$, and Zr_2Be_{17} were identified in the Be deposit. Oxygen contamination is significantly lower than that usually found in commercial beryllium. For analytical data for Zr, O, and other impurities, see the table in the paper. The highest purity Be metal (~99.6%) is obtained at a reduction temperature of 1600 °C. For a description of the apparatus, see the figure in [9].

Analogous criteria as for the selection of zirconium (see p. 49) as a suitable reductant for Be oxide are also found for yttrium, lanthanum, thorium [9], and titanium [10, 13].

The same type of charge, procedure, and apparatus as used for the reduction with zirconium [9], are used for the reductions with yttrium, lanthanum, and thorium.

With yttrium metal (25% excess), the best yield (~97%) of Be metal is attained at 1600 °C after a reaction period of 30 h [9].

Lower yields (~70%) are obtained with lanthanum metal (stoichiometric amount) at 1450 °C after 40 h [9].

The use of thorium metal (10% excess) results in a low yield (~32%) of Be metal after 60 h at 1500 °C. Thorium oxide probably forms on the surface of the beryllium oxide particles, and the subsequent diffusion of thorium through this barrier inhibits the reaction [9].

When yttrium or lanthanum are used as reductants, contaminations of 2 to 2.8% YBe_{13} and minor amounts of $LaBe_{13}$, respectively, are found in the Be deposits. For further analytical data on the impurities in the resulting Be metal, see [9].

The reduction of Be oxide (briquettes) with powdered titanium metal is performed on a small scale under high vacuum at 1400 °C. The resulting coherent Be metal deposit is quite pure. Operation at higher temperatures, however, results in a Be metal product contaminated with TiO [10, 13].

The apparatus for the reduction (see figure in the paper) consists of a high frequency induction heated quartz tube containing a reaction crucible made of Be oxide which is sealed with a Be oxide condenser hood [10].

Further investigations on the reduction of Be oxide with Li, Ce, Mg, Al, and Si reducing agents all show negative results [13 to 15, 25]. Important criteria for this deficiency are the normally unfavorable reduction equilibria, the appreciable vapor pressures of reductants like Mg, and Al at the reduction temperatures (>1350 °C), and the formation of stable beryllides [3, 9, 26].

Beryllium metal may be obtained by treating beryl ore ($3BeO \cdot Al_2O_3 \cdot 6SiO_2$) with fused Al metal. The Be metal is extracted from the initially formed Be-Al-Si alloy by fused Cd metal. The resulting Be-Cd alloy is then decomposed into its components by distillation [19].

1.4.1.2 Reduction with Nonmetals

With Carbon

The thermal reduction of beryllium oxide with carbon or carbonaceous material as the reducing agent, is reported in the patent literature between 1934 and 1943, see, e.g. [15 to 18]. This method, however, has not attained commercial importance [8, 20].

The reduction of beryllium oxide with carbon is made possible by shifting the normally unfavorable equilibrium $BeO + C \rightleftharpoons Be + CO$ to the right. This is accomplished by withdrawing both reaction products as vapors from the system. The principle of the method is to protect the vaporized Be metal at the reduction temperature (>1900 °C) from contact with carbon and the simultaneously evolved carbon monoxide by means of an inert gas stream, so that formation of beryllium carbide and reformation of beryllium oxide are avoided. The Be vapor is recovered by condensation and the carbon monoxide is drawn off [16, 17, 19]. A similar process carried out under vacuum instead of with inert gas flushing seems to be of minor importance [17].

Instead of a direct reduction of beryllium oxide with stoichiometric amounts of carbon, the carbide, Be_2C, may be formed as an intermediate by using excess carbon reductant at ~1900 °C. This intermediate is subsequently decomposed into the elements at temperatures above 2100 °C [18], cf. [19, 22].

In a process developed by Kruh [16], a stoichiometric mixture of BeO and carbon or carbonaceous material (e.g., C_2H_2 or other hydrides of carbon) is heated to at least 1900 °C in an electric furnace in a stream of protective gas that is inert to beryllium (hydrogen or argon). The amount of hydrogen is ~3 to 4 times the amount of the carbon monoxide formed. The volatile mixture of Be metal vapor, inert gas, and gaseous reaction products (mainly carbon monoxide) formed during the reduction is continuously carried over to the cooled surfaces of a chilling device which transforms the vaporized Be metal into a solid, finely divided deposit. The gaseous byproducts are drawn off with some of the beryllium metal. The latter can be recovered by mechanical or electrical filtration. The yield of Be metal is 98 to 99.5% (remainder is BeO) [16].

The apparatus for carrying out the reduction process is shown and described in [16].

In principle, the same process is used by Kjellgren and Sawyer [17] who proposed the use of high purity BeO with an apparent density of less than 0.5 to prevent the sintering of BeO which causes a loss in the amount of the exposed surface. Powdered graphite, coke, or resistor carbon is used as the reductant. A flow sheet for the process is given in [17].

Another modification of the process uses an extremely intimate reduction mixture of BeO and carbon. This mixture is obtained by subjecting a stream of gaseous hydrocarbon (usually methane in the form of natural gas) in contact with porous BeO at the carbonizing temperature. The hydrocarbon decomposes into hydrogen and finely divided carbon; the latter deposits in the mass of BeO [21].

In an improved form of Kruh's process, BeO is reduced by means of carbon or carbonaceous material at high temperatures. A sufficiently high pressure is maintained to retain the reduced metal within the reduction zone until the reaction substantially goes to completion. The carbon monoxide and other gaseous reduction products are removed from the system by means of an inert gas stream. The pressure is then lowered, and while maintaining the inert gas stream, the Be metal is distilled off and condensed. No pure raw materials are required for this process [18].

In another method, the BeO is heated in the presence of an excess of carbon or carbonaceous material to form beryllium carbide. The charge is maintained at a predetermined pressure and temperature to assure the formation of the carbide and its retention in the reaction zone. After carbon monoxide is removed from the system by means of an inert gas stream, the temperature is raised until the Be carbide is decomposed (2150 to 2300 °C). The evolved Be metal vapor is passed in the presence of inert gas to a condenser zone [18, 19]. For the apparatus used in this process, see [18].

With Hydrogen

The reduction of BeO with atomic hydrogen has been proposed by [23]. The H_2 gas, purified of O_2 and H_2O, is fed into a reducing furnace where it passes through an electric arc for dissociation (~1700 °C at atmospheric pressure) and immediately thereafter onto BeO in the form of briquettes compressed from light, fluffy powder. Simultaneously, in order to avoid recombination, the H_2O vapor formed in the reduction reaction is quickly and completely removed from the formed beryllium, e.g., by means of a stream of molecular hydrogen. The Be vapors may be condensed after a short distance. A flow sheet and apparatus for the process are given in [23], cf. [19].

The reduction of BeO is also said to be possible by the use of CaH_2 and Ca in an H_2 atmosphere [19].

A specially designed vacuum reduction furnace for the reduction of BeO with a reducing agent (not specified) and condensation of the Be vapors at reduced pressure is described by [24]. In this furnace, the deposition of fine pyrophoric crystals from the Be vapors is avoided by arranging a metallic hood above the BeO-containing heating crucible [24], cf. [15].

References:

[1] S. B. Roboff (in: H. H. Hausner, Beryllium, Its Metallurgy and Properties, Univ. California Press, Berkeley 1965, pp. 17/30, 18/9; C.A. **64** [1966] 10879).
[2] H. H. Kellogg (in: D. W. White Jr., J. E. Burke, The Metal Beryllium, ASM, Cleveland 1955, pp. 49/62, 52/6, 53).
[3] G. E. Darwin, J. H. Buddery (Beryllium, Butterworth, London 1960, pp. 75/6, 93/4).
[4] J. H. Buddery, R. W. Thackray (J. Inorg. Nucl. Chem. **3** [1956] 190/3).
[5] T. W. Baker, J. Williams (Acta Cryst. **8** [1955] 519).
[6] S. J. Morana, G. F. Simons (J. Metals **14** [1962] 571/4).
[7] Hachie Sawamoto, Takeo Oki, Akira Nishina (Mem. Fac. Eng. Nagoya Univ. **12** [1960] 130/5; C.A. **1961** 24458).
[8] T. T. Magel (in: D. W. White Jr., J. E. Burke, The Metal Beryllium, ASM, Cleveland 1955, pp. 124/35, 129/32).
[9] T. T. Campbell, R. E. Mussler, F. E. Block (Trans. AIME **236** [1966] 1456/61).
[10] W. Kroll (Z. Anorg. Allgem. Chem. **240** [1939] 331/6, 332/3).

[11] T. T. Magel (AECD-3321 [1952] 1/15 from [9]).
[12] T. T. Magel, Massachusetts Institute of Technology (unpublished information from [8]).
[13] W. J. Kroll (U.S. Bur. Mines Inform. Circ. No. 7326 [1945] 1/15, 2, 6).
[14] W. Kroll (Z. Anorg. Allgem. Chem. **219** [1934] 301/4).

[15] K. Illig, H. Fischer, W. Birett (in: G. Eger, Die Technische Elektrolyse im Schmelzfluß, 2nd Ed., Leipzig 1955, pp. 571/641, 593/5).
[16] O. Kruh (U.S. 2121084 [1934/38]; C.A. **1938** 6223).
[17] B. R. F. Kjellgren, C. B. Sawyer, Brush Beryllium Co. (U.S. 2204221 [1939/40]; C.A. **1940** 6920).
[18] O. Kruh (U.S. 2255549 [1938/41]; C.A. **1942** 76).
[19] G. Jaeger (Metall **4** [1950] 183/91, 188).
[20] G. Petzow, F. Aldinger (Ullmanns Encykl. Tech. Chem. 4th Ed. **8** [1974] 451).

[21] J. D. Hanawalt, J. S. Peake, The Dow Chemical Co. (U.S. 2256161 [1940/41]; C.A. **1942** 72).
[22] F. S. Muratov, A. V. Novoselova (Dokl. Akad. Nauk SSSR **129** [1959] 334/6; Proc. Acad. Sci. USSR Chem. Sect. **124/129** [1959] 979/81; C.A. **1961** 23234).
[23] P. M. Dolan (U.S. 2226525 [1937/40]; C.A. **1941** 2463).
[24] N. V. Philips' Gloeilampenfabrieken (Fr. 862427 [1940/41]; C. **1941** II 1324).
[25] W. Kroll (Wiss. Veröffentl. Siemens-Konzern **11** No. 2 [1932] 88/92).
[26] G. Dressler, M. Rühle (in: K. Winnacker, L. Küchler, Chemische Technologie, 3rd Ed., Vol. 6, München 1973, pp. 549/601, 591).

1.4.2 Reduction of Beryllium Halides with Metals or Hydrogen

Review

The temperature dependence of the standard free enthalpies of formation of beryllium fluoride and beryllium chloride, and of the fluorides and chlorides of several other metals (cf. figures in [1, p. 77], and [2, pp. 59/60]), indicates that beryllium fluoride and beryllium chloride are much less stable than beryllium oxide, and that they may be reduced by many of the more reactive metals. Thus, it should be possible to reduce beryllium chloride and fluoride to beryllium metal with lithium, sodium, potassium, magnesium, or calcium according to the reaction

$$BeX_2(l) + 2M^+(l) = Be(c, l) + 2MX(c, l) \qquad (I)$$

or

$$BeX_2(l) + M^{2+}(l) = Be(c, l) + MX_2(c, l) \qquad (II)$$

where M^+ is an alkali metal, and M^{2+} an alkaline earth metal. These reactions are all exothermic, and the equilibrium position is far to the right under most experimental conditions [1 to 3].

Of the usual reductants used in extraction metallurgy, calcium, magnesium, and sodium, the use of calcium is not desirable because it tends to form a stable $CaBe_{13}$ phase [1, 3]. Sodium and magnesium offer no such problem and hence, usually yield beryllium metal [3].

The reduction of BeF_2 with magnesium (according to equation II) has developed into the process (see p. 56) that is currently producing most of the beryllium sold today. The preferential use of BeF_2 rather than $BeCl_2$ in this process is probably due to the low volatility and easy handling of BeF_2 in comparison with $BeCl_2$. Magnesium metal is chosen because of its relatively easy handling and its low price [1 to 3].

The sodium reduction of BeF_2 or $BeCl_2$ (according to equation I) necessitates a modified version of the magnesium reduction process. This is because sodium is more difficult to handle due to its lower boiling point (880 °C [3]) in comparison with magnesium (1120 °C [1]), and because of the tendency of sodium to attack graphite, the usual crucible material. Furthermore, the increased handling difficulties (boiling point 520 °C [4]) of $BeCl_2$ must be taken into consideration [1, 3, 4].

All methods for the reduction of BeF_2 and $BeCl_2$ by sodium and of $BeCl_2$ by magnesium, respectively, according to equations I and II (see p. 53), have been performed on a laboratory scale only [1, 4].

However, a reduction (not covered by the above equation I) of $BeCl_2$ vapor with molten sodium metal by a "Kroll-type" process has evolved into a larger laboratory process and is considered suitable for an economical commercial operation. In general, the purity of the Kroll-process beryllium product is superior to that of commercial grade beryllium [7, 8], cf. also [1, 4 to 6].

The hydrogen reduction of beryllium halides has only been attempted with $BeCl_2$ on a laboratory scale. The reported results are often contradictory [9, 10]. Whereas $BeCl_2$ vapor reduction on hot tungsten-wire is successful [9], reduction in glow discharge fails [10]. Available thermodynamic data point to a weak hydrogen reduction activity [2].

References:

[1] G. E. Darwin, J. H. Buddery (Beryllium, Butterworth, London 1960, pp. 75/9, 94/6).
[2] H. H. Kellogg (in: D. W. White Jr., J. E. Burke, The Metal Beryllium, ASM, Cleveland 1955, pp. 49/62, 56/60).
[3] S. B. Roboff (in: H. H. Hausner, Beryllium, Its Metallurgy and Properties, Univ. California Press, Berkeley 1965, pp. 17/30, 19/21; C.A. **64** [1966] 10879).
[4] G. Petzow, F. Aldinger (Ullmanns Encykl. Tech. Chem. 4th Ed. **8** [1974] 451).
[5] J. M. Tien (Trans. Electrochem. Soc. **89** [1946] 237/45, 242/3).
[6] Brush Beryllium Co. (Brush Beryllium Co. Progr. Rept. No. 12/17 [1947] from [1]).
[7] T. T. Campbell, R. E. Mussler, F. E. Block (Met. Trans. **1** [1970] 2881/7; C.A. **73** [1970] No. 122649).
[8] F. E. Block, R. E. Mussler, T. T. Campbell (Advan. Extr. Met. Proc. Symp., London 1967 [1968], pp. 551/71, 553, 562/9; C.A. **69** [1968] No. 69070).
[9] L. Hackspill, J. Besson (Bull. Soc. Chim. France **1949** 113/6; C.A. **1949** 4967).
[10] B. Kopelman, Sylvania Electric Products (unpublished information from [11]).

[11] T. T. Magel (in: D. W. White Jr., J. E. Burke, The Metal Beryllium, ASM, Cleveland 1955, pp. 124/35, 133).

1.4.2.1 Reduction of Beryllium Fluoride and Double Fluoride with Metals

1.4.2.1.1 Introduction

Beryllium fluoride can be reduced by lithium, sodium, magnesium, and calcium [1 to 3]. Only magnesium is used for the commercial reduction of beryllium fluoride (see pp. 53, 56) [1, 2]. Reductions with the other metals are restricted to laboratory or small scale work [2, 3]. A process involving the reduction of beryllium-alkali metal double fluorides has also attained no technical significance [16].

Difficulties encountered in the reduction of BeF_2 with metals include: When using lithium or magnesium as reductant, the low boiling points of those metals (1347 and 1105 °C [4], respectively) have to be considered. Because the melting point of Be metal is 1283 °C [4], the boiling point of the reducing agent is readily reached in this reaction, due to a noticeable evolution of heat. Calcium, boiling at 1484 °C [5], develops more heat, thus offering no special advantage [3].

These difficulties may be overcome by the use of closed bombs able to withstand the vapor pressure of the reducing metal. When using magnesium, the vapor pressure in such a closed system is about 4.78 kp/cm^2 (68 lb/in^2) at 1300 °C, a temperature just above the melting point of Be metal [3]. These conditions are attained only by adding external heat to the system or by the use of boosters (e.g., $ZnCl_2$, $PbCl_2$) [12], cf. also [22]. They allow molten beryllium to be obtained [3], cf. [12]. Alternatively, reduction, especially with Mg metal, is carried out under external heating at atmospheric pressure in the presence of damping additives such as zinc [6], cf. [7], of Mg-Ca mixtures or alloys [12, 13], or, preferably, of fluxes of the single fluoride type (e.g., BeF_2, CaF_2, BaF_2 [8 to 11]), [1 to 3, 12]. Sintered beryllium metal, which has to be remelted, is obtained [3].

Of the two possibilities for overcoming the difficulties encountered in the reduction of BeF_2, only the atmospheric pressure reduction has gained industrial importance, see [1].

Reduction of BeF_2 with sodium is reported to be only partial because of the formation of stable $NaF \cdot BeF_2$, which cannot be reduced by an excess of sodium [2, 3, 15, 17]. The low boiling point of sodium, 883 °C [4], is another disadvantage if reduction is to be conducted at atmospheric pressure [2].

Early attempts for preparation of Be metal by the reduction of double fluorides of beryllium and alkali metal (Na, K) with alkali or alkaline earth metals (mainly magnesium) are reported in "Beryllium" 1930, pp. 39/40, cf. also [3, 16, 17]. Nearly theoretical yields of Be metal are attained with the alkali double fluoride reduction at atmospheric pressure with alkaline earth metals or Li, Kroll [14]. The formation of sodium or potassium vapor is minimized by using an excess of BeF_2 more than one mol per mol MF (M = Na, K) [3, 14], cf. [7, 18, 19]. Double fluoride reductions of the type mentioned above, e.g., by means of Mg metal in the presence of MgF_2, CaF_2, SrF_2, BaF_2, or LiF are more highly recommended because of their inertness to the reductant [20], cf. [3, 19]. Another reduction process involving $BeF_2 \cdot NaF$ with Mg metal is carried out under vacuum without any further addition of fluoride salt and is also reported to proceed quantitatively [21].

References:

[1] G. E. Darwin, J. H. Buddery (Beryllium, Butterworth, London 1960, pp. 76/8).
[2] L. J. Derham, D. A. Temple (Extr. Refining Rarer Metals Proc. Symp., London 1956 [1957], pp. 323/36, 328; C.A. **1958** 19783).
[3] W. J. Kroll (U.S. Bur. Mines Inform. Circ. No. 7326 [1945] 1/15, 3).
[4] D. R. Stull (JANAF Thermochemical Tables, PB-168370 [1965]; NSRDS-NBS-37 [1971]).
[5] C. J. Kunesh (in: Kirk-Othmer, Encycl. Chem. Technol. 3rd Ed. **4** [1978] 412/21, 413).
[6] L. Losana (Alluminio **8** [1939] 67/75; C.A. **1939** 7630).
[7] Compagnie de Produits Chimiques et Électrométallurgiques Alais, Froges et Camargue (Brit. 435747 [1935]; C.A. **1936** 1353 from [3]).
[8] B. R. F. Kjellgren, The Brush Beryllium Co. (U.S. 2381291 [1941/45]).
[9] C. W. Schwenzfeier Jr. (in: D. W. White Jr., J. E. Burke, The Metal Beryllium, ASM, Cleveland 1955, pp. 71/101, 93/4).
[10] H. C. Kawecki (U.S. 2486475 [1945/49]).
[11] G. A. Meyerson (Proc. 1st Intern. Conf. Peaceful Uses At. Energy, Geneva 1955, Vol. 8, pp. 587/9).
[12] F. H. Spedding, H. A. Wilhelm, W. H. Keller, C. Neher (TID-5212 [1955] 43/8; C.A. **1956** 7691).
[13] T. H. Schofield (Brit. 709552 [1947/54]).
[14] W. Kroll (Ger. 480128 [1926]).
[15] K. Illig, M. Hosenfeld (Wiss. Veröffentl. Siemens-Konzern **8** No. 1 [1929/30] 26/9).
[16] G. Jaeger (Metall **4** [1950] 183/91, 188).

[17] T. T. Magel (in: D. W. White Jr., J. E. Burke, The Metal Beryllium, ASM, Cleveland 1955, pp. 124/35, 127).
[18] Compagnie de Produits Chimiques et Électrométallurgiques Alais, Froges et Camargue (Ger. 675526 [1939]; C.A. **1939** 6786 from [3]).
[19] K. Illig, H. Fischer, W. Birett (in: G. Eger, Die Technische Elektrolyse im Schmelzfluß, 2nd Ed., Leipzig 1955, pp. 571/641, 592/3).
[20] C. Adamoli, Perosa Corp. (U.S. 2193363 [1937/40]; U.S. 2193364 [1939/40]; C.A. **1940** 4721).
[21] R. Rohmer (Compt. Rend. **214** [1942] 744/6; C.A. **1944** 6142).
[22] Ames Laboratory (CT-1985 [1944/55] 1/27; N.S.A. **10** [1956] No. 5278).

1.4.2.1.2 Reduction of Beryllium Fluoride with Magnesium

The reduction of beryllium fluoride with magnesium metal is still today the established process for the large scale commercial production of Be metal. This process is economically more favorable than all of the electrowinning processes developed so far [1 to 3, 14], cf. also [46].

The first investigations of the reduction of beryllium fluoride with magnesium metal were carried out in 1927 by Kroll [4], cf. [5]. Other researchers [6 to 8, 38] followed, and several patents were issued [6, 8, 38]. The first process for industrial scale production was developed around 1940 by Kjellgren [9] and taken over by the Brush Beryllium Company, USA [10, 11]. With increasing demand for pure Be metal for atomic energy reactors, this firm became, and remains today, the world's main supplier of Be metal (outside communist countries) [1 to 3, 12, 13]. Up until the 1960's, the Beryllium Corporation of America [2, 15], and up until the early 1970's the Kawecki Berylco Ind. Inc., USA [3] also produced Be metal in large quantities, both using various modifications of the Brush Beryllium Company process [16, 17]. Furthermore, a pilot plant version of the Brush Beryllium Company process has been in operation since 1956. It is operated for the United Kingdom Atomic Energy Authority by the Imperial Smelting Corporation in Great Britain [1, 18]. Related production technology has been applied in Canada, India, and the USSR [20 to 22]. With decreasing demand for Be metal, most plants have ceased to produce. Today only one plant (Brush Wellman Inc., USA) continues to operate in the noncommunist world [14, 23].

1.4.2.1.2.1 Basic Process Design.

The preparation of Be metal by the Mg reduction process is normally carried out in two major steps:

1) Conversion of BeO to anhydrous BeF_2 according to the reactions

$$BeO \cdot H_2O + 2NH_4 \cdot HF_2 \rightarrow (NH_4)_2BeF_4 + 2H_2O$$

$$(NH_4)_2BeF_4 \xrightarrow{\text{heating}} BeF_2 + 2NH_4F$$

2) Conversion of BeF_2 to Be metal according to the reaction

$$BeF_2 + Mg \rightarrow Be + MgF_2 \quad [2, 26, 27, 30, 36]$$

High purity BeF_2 must be produced in step 1) because there is little chance of removing impurities during the reduction with Mg metal in step 2) [27].

The raw material for step 1) is normally technical beryllium hydroxide ($Be(OH)_2$ or $BeO \cdot H_2O$) [2, 11, 14, 15, 20, 22, 27]. Basic beryllium carbonate [22], Be metal, beryllia scrap [27], beryllium oxide, and Be scrap or recycle products [2, 12, 13, 28] are also used.

$(NH_4)_2BeF_4$ is produced by dissolving the raw material in aqueous NH_4HF_2. It is a highly soluble salt which crystallizes by evaporation of the water without hydrolysis or adding water. Its dissociation is very slight in solution. Consequently, its pH can be varied over a wide range without precipitating beryllium hydroxide. Impurities can therefore be precipitated and removed with comparative ease from virtually any raw material, thus allowing a high degree of purification [1, 2, 12, 24].

The thermal dissociation of $(NH_4)_2BeF_4$ into anhydrous BeF_2 and volatile NH_4F begins at about 125 °C. Since any decomposition below the melting point (803 °C) of BeF_2 leaves a porous friable mass which hydrolyzes on contact with air, decomposition is carried out above the melting point of BeF_2. After solidification, the resulting glassy, dense fluoride has a small surface area and can be handled and stored easily [1, 2, 13, 16, 18, 27].

Reaction step 2) is performed by mixing the reaction components in a suitable crucible and heating to ~1300 °C. If an intermetallic compound $MgBe_{13}$ should form during the reduction, as was suggested by [16], it would decompose at 950 °C (under vacuum), and it would be absent at the completion of the reaction [16, 26]. The reaction products, Be metal and MgF_2 slag, separate from each other just above the melting point of Be metal when the reaction is complete [1, 27]. For the physical properties see Table 1/1.

Since the reduction reaction is moderately exothermic ($\Delta H^{\circ}_{298} = -39$ kcal/mol Be [29]), the rapid increase in temperature must be restrained to avoid sudden volatilization of unreacted Mg metal. Use of large pieces of BeF_2 and Mg metal results in a fairly moderate reaction, so that external heating becomes necessary. The large BeF_2 pieces are further advantageous due to toxicity and hygroscopicity considerations. The only crucible material that can be used under the process conditions is impervious graphite [1, 3, 14, 27 to 29].

Table 1/1
Physical Properties of BeF_2, MgF_2, Be, and Mg [1].

property	BeF_2	MgF_2	Be	Mg
melting point, °C	803 [43]	1263	1283	650 [a]
boiling point, °C	1159 [43]	2260 [43]	2477 [44] [a]	1117 [44] [a]
density, g/cm^3	2.0	3.1	1.85	1.74
solubility, g/100 g H_2O at room temperature	50	0.012	—	—

[a] More recent data are given in Stull [45] for the boiling point of Be: 2484 °C; melting and boiling point, respectively, of Mg: 649 and 1105 °C.

Be metal and MgF_2 slag do not separate cleanly into two fused layers probably because of the small density difference between the two. Consequently, in order to lower the melting point of the MgF_2 slag and to improve Be metal separation, it is current practice to use an excess of BeF_2 as a flux-forming component[1] in the reduction reaction [1, 3, 9, 14, 27, 28, 30, 37]. The requirement for an excess of BeF_2 has stimulated considerable work on finding a satisfactory substitute through the use of other fluxes such as CaF_2 [17], $CaCl_2$ [18, 19, 31], or BaF_2 [32, 37] as replacements for the BeF_2 excess. However, no completely satisfactory solution to this problem has been found [1, 2].

[1] Some investigations on the BeF_2-MgF_2 system are reported by [37, 39].

The following sequence of events occurs during reduction: A mixture of Mg lumps and an excess of BeF_2 pellets is heated in a graphite crucible. At 650 °C the Mg metal melts and starts reacting with the BeF_2. The reaction is slow because both reaction products are solid. At 803 °C the BeF_2 also melts. The reaction rate then increases, but the reaction products remain solid. Both starting materials remain unevaporated until all of the Mg metal is consumed on reaching a temperature of 1120 °C. The excess BeF_2 fluxes with the MgF_2, thereby reducing its activity, and consequently its vapor pressure [1, 11, 26, 33].

At 1283 °C, the Be metal melts and separates from the slag into globules, some of which coalesce into fairly large pieces. In subsequent leaching with water to reclaim the Be metal, the network of BeF_2 which surrounds each insoluble MgF_2 crystal readily dissolves in water causing the slag to disintegrate rapidly and to release the Be metal particles [1, 2, 26].

Because maintenance costs of the equipment for all operations from dissolving beryllium hydroxide to beryllium fluoride production, constitute a major portion of the production costs of Be metal, special attention must be paid to the careful choice of construction materials to minimize corrosion [2].

1.4.2.1.2.2 Commercial Scale Production Processes

Brush Beryllium Company Practice

This practice consists of four basic stages. The first two stages concern the production of the starting material (BeF_2) for the reduction stage. (1) Production of high purity $(NH_4)_2BeF_4$. (2) Thermal decomposition of $(NH_4)_2BeF_4$ to BeF_2. (3) Reduction of BeF_2 to Be metal with Mg metal. (4) Vacuum casting. Stage (4) is covered under the Chapter "Separation of Metal from Slag and Refining of Metal" (see p. 91) [12].

A flow sheet of the Brush practice is given in **Fig. 1-6**.

Production of High Purity $(NH_4)_2BeF_4$. Beryllium hydroxide (produced by the sulfate or the bertrandite extraction process [14, 25], see pp. 29, 32), and beryllium oxide or beryllium metal scrap is pulped with aqueous ammonium bifluoride solution. It dissolves rapidly to give a solution of $(NH_4)_2BeF_4$. The pH of the solution is adjusted to approximately 5.5. The ammonium fluoride content is adjusted to approximately 90% of the stoichiometric requirement (see equation under (1), p. 56). The balanced solution is heated to boiling, and then purified according to the Schwenzfeier-Pomelee process, by adding $CaCO_3$ flour to precipitate the aluminium as the hydroxide. Thereafter, PbO_2 is added to oxidize the manganese and chromium and to precipitate them as MnO_2 and $PbCrO_4$ [1, 2, 12, 14, 24]. The PbO_2 treatment is omitted when beryllium hydroxide obtained from bertrandite is the input material [25]. Ammonium polysulfide is added to the filtrate, which at this point has a pH of ~8.3. The copper, lead, and nickel impurities precipitate as sulfides and are filtered off. The ammonium fluoride-beryllium fluoride ratio is again adjusted as near as possible to the stoichiometric ratio in order to favor $(NH_4)_2BeF_4$ crystallization. The crystals contain only trace amounts of impurities. The high purity mother liquor can be returned to the evaporator without recycling [1, 2, 11, 12, 14, 24].

Thermal Decomposition of $(NH_4)_2BeF_4$ to BeF_2. The $(NH_4)_2BeF_4$ crystals are decomposed between 900 and 1100 °C. At this process stage, the potential toxicity is very high, particularly on tapping the fused BeF_2. In order to reduce this toxicity to acceptable limits, a process has been developed which operates continuously and is completely automatic. The $(NH_4)_2BeF_4$ crystals are fed continuously to a high-frequency heated CS graphite retort. The BeF_2 formed in the retort in the fused state flows continuously onto a casting wheel

Fig. 1-6

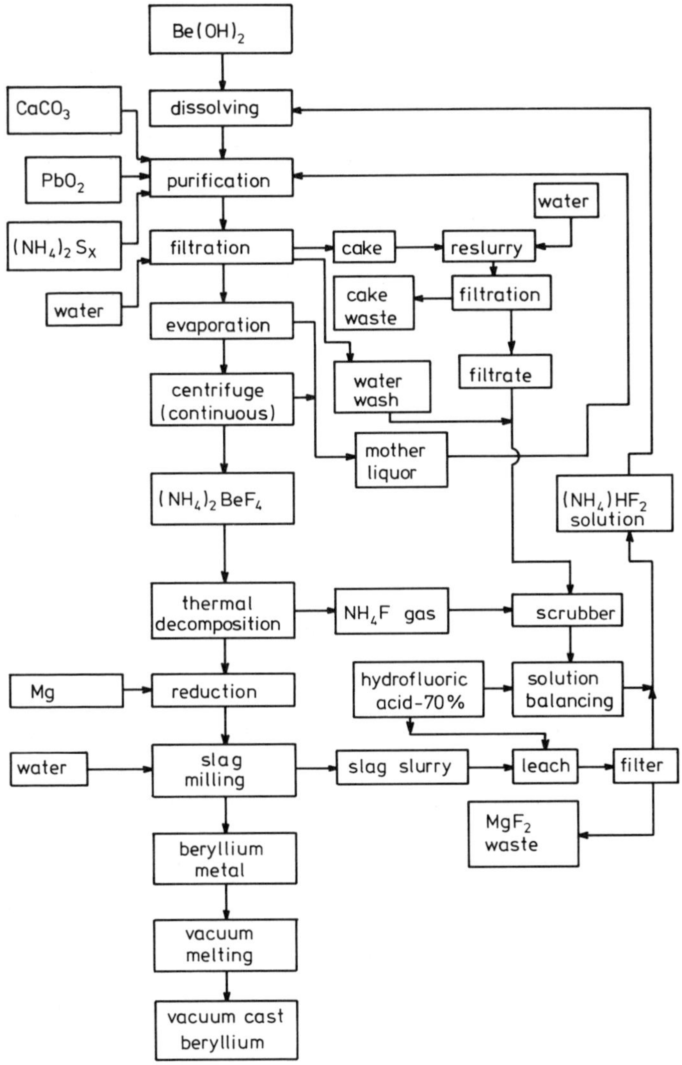

Flow sheet for the production of Be metal from $Be(OH)_2$ at Brush Beryllium Company [2].

where it cools rapidly and forms glassy pellets. The NH_4F gas is removed and absorbed in a scrubbing tower; any NH_4F not collected in the tower is caught in an electrostatic precipitator. The BeF_2 pellets are stored for the reduction process. It is claimed that this processing gives practically 100% yields of BeF_2 and NH_4F [1, 2, 9, 11, 12, 14].

The equipment for this continuous automatic decomposition stage consists of a 100-kW, 3000-cycle induction furnace with provisions for charging $(NH_4)_2BeF_4$ at the top [2, p. 92].

Reduction of BeF_2 with Mg Metal. The reduction is carried out as a batch operation in a high frequency electric furnace provided with a graphite crucible [2, 11, 14].

A furnace charge consists of 117.9 kg (260 lb) of BeF_2 pellets and 43.54 kg (96 lb) of magnesium lumps of approximately 1 in (2.54 cm) diameter; this is about 70% of the stoichiometric magnesium requirement. If more than 75% is used, the encapsulation of the MgF_2 crystals by the BeF_2 network is incomplete and the disintegration of the slag on leaching is more difficult. By using more than 85% of the stoichiometric magnesium requirement it is virtually impossible to separate the Be metal from slag to any degree [2, 11, 12, 14].

After completing the reaction of the furnace charge at about 900 °C, the temperature is raised to ~1300 °C. The fused charge is then poured into a graphite receiving crucible attached to the furnace. The filled crucible is covered and allowed to cool. The furnace cycle lasts approximately 3.5 h with 75 to 100 kW power input. Thereafter, the entire solid block of metal and slag is put into a ball mill for wet milling to release the Be metal pebbles. Water is pumped continuously through the mill at a sufficiently high rate to carry out the MgF_2 crystals as they are liberated by milling. The solution leaving the mill passes to a tank for settling out the MgF_2 crystals and then returns to the mill until it becomes saturated with BeF_2. The MgF_2 in the settling tank is washed with dilute hydrofluoric acid to dissolve any fine metallic beryllium carried over from the milling step; it is then filtered off, washed, and discarded. The washings and the mill effluent are returned to the wet chemical process stage (p. 58) [2, 11, 12, 14].

The pebbles are washed with water and pickled in acetic acid [14]. After drying, they are passed over a magnetic separator which removes the steel stars used in the ball mill. Pieces of slag, occasionally included in the mixture of pebbles and steel stars, are removed in a simple sink-float process utilizing a mixture of ethylene dibromide and light mineral oil as the heavy medium. However, a high degree of separation is not attempted, because it is much more economical to completely rid the metal of slag in a subsequent vacuum melting and casting stage (see p. 95). The pebbles are washed with isopropyl alcohol, then with water, and dried [2], cf. [11, 14].

The purity of the pebbles is approximately 97%; impurity contents (in %) are: Mg 1.5, other metals 0.2 to 0.5, C 0.02, BeO 0.1 [2, 12].

Approximately 14 kg (31 lb) of beryllium pebbles can be recovered from the milling operation from each furnace charge; this corresponds to a yield of 62%. The overall beryllium recovery in the reduction and milling operations is 96 to 97% [1, 2, 12, 14].

The furnace is a 100-kW, 3000-cycle induction furnace. The crucible used is 24 inches (60.9 cm) in diameter and made of CS grade graphite [2, 12].

Again, due to the high toxic hazard, the reduction furnace is provided with a close fitting primary ventilation system, which very effectively removes any fumes released during the reduction [2], cf. [11].

Modifications of the Brush Practice

Beryllium Corporation Practice. This practice differs in detail from that used by the Brush Beryllium Company. It is based on the use of highly purified beryllium hydroxide (obtained from the fluoride decomposition process with subsequent purification, see p. 19) as raw material. In the Brush practice, the BeF_2 production step is performed with a 100% excess of NH_4HF_2 (see equations under (1), p. 56). The Beryllium Corp. practice uses a 30 to 35% excess of NH_4HF_2 over the amount theoretically required according to: $Be(OH)_2 + NH_4HF_2 \rightarrow BeF_2 + NH_3 + 2H_2O$. If BeF_2 is produced without an excess of NH_4HF_2, a sirupy mass, rather than crystals forms upon evaporation. A certain minimum of $(NH_4)_2BeF_4$ (cf. equations under (1), p. 56) crystal nuclei is required to prevent the dried product from becoming tacky. Since $(NH_4)_2BeF_4$ is more insoluble in water than BeF_2, it crystallizes first. Upon evaporation of the slurry, BeF_2 remains deposited onto each $(NH_4)_2BeF_4$ nucleus. The residual BeF_2 product is obtained without tackiness as long as the excess NH_4HF_2 added is in the range of 30 to 35% [15, 16].

For the production of BeF_2, purified beryllium hydroxide in the form of a filter cake containing 40 to 50% free water is reacted with the required excess of NH_4HF_2 flake. This results in an aqueous slurry containing BeF_2, NH_3, and an amount of $(NH_4)_2BeF_4$ equivalent to the excess of NH_4HF_2 used. The slurry is partially dehydrated to coarse flakes by means of a single drum steel drum dryer. It thus becomes a suitable feed material for induction-heated graphite decomposition furnaces, where the moisture and excess fluorine are driven off.

The BeF_2 product from the decomposition furnaces is in a fused glassy state and is collected on a water-cooled surface, where it solidifies in the form of thin wafers [15, 16].

The reduction of BeF_2 with Mg metal is performed according to the Brush practice. The resulting Be metal beads generally contain Mg impurity contents from 0.5 to 1%. For further purification, see p. 95 [15, 16].

Imperial Smelting Corporation Practice. This practice varies only in detail from the Brush practice. The beryllium hydroxide used (produced from the Murex plant at Milford Haven, U.K., according to a fluorosilicate method [34]) is highly pure, and the purification process of the $(NH_4)_2BeF_4$ (cf. in [18]) is simplified [18].

The decomposition stage of the dried, free-flowing $(NH_4)_2BeF_4$ crystals is carried out at 800 to 850 °C. During the early stages of decomposition the evolution of gaseous NH_4F from the viscous mass causes frothing. A specially shaped graphite retort that is heated in a furnace mainly from above, speeds up the decomposition rate, and overcomes this difficulty. The retort is operated continuously. The rate of decomposition is approximately 15.8 kg/h (35 lb/h). Before periodically tapping off the fused BeF_2 in 27.2 kg to 36.2 kg (60 to 80 lb) batches into graphite moulds at 150 °C, the retort temperature is increased to 900 °C in order to give quicker flow. The NH_4F and decomposition products (NH_3 and HF) are led away from the furnace to condensers and scrubbers for recovery. The glassy solid BeF_2 obtained after cooling to room temperature, contains $<0.001\%$ NH_3 and ~0.2 to 0.6% BeO [18].

The reduction stage employs stoichiometric amounts of crushed BeF_2 and small high purity Mg metal raspings. Calcium chloride is used as flux and as a substitute for the excess BeF_2, used in the Brush practice. Calcium chloride has the potential advantage that BeF_2 can be completely reduced to Be metal and recycling is eliminated [18, 19].

The reduction reaction is carried out in a furnace provided with a graphite crucible. The compacted BeF_2-Mg mixture is added to the crucible containing fused $CaCl_2$ at 1000 to 1100 °C. The reaction takes place rapidly, since its initiation temperature is about 650 °C. The products are heated above the melting point of Be (1283 °C) for about 2 h. The Be metal floats to the surface and aggregates into small ingots or beads of about 2.54 cm (1 in). After cooling the reaction products in the crucible to 1050 to 1100 °C, the Be beads are skimmed from immediately below the surface of the fused flux. The MgF_2 settles as a crystalline deposit on the crucible bottom. The beads are cooled to room temperature in an argon-filled vessel. The MgF_2 slag can then be run from the crucible through a graphite valve at the bottom before the next charge is added to the $CaCl_2$ melt. The almost spherical beads (~2.54 cm (~1 in) in diameter) are washed with hot H_2O in a rubber-lined tumbler vessel to leach out $CaCl_2$ and to remove any adhering MgF_2 crystals. They are next rinsed

Fig. 1-7

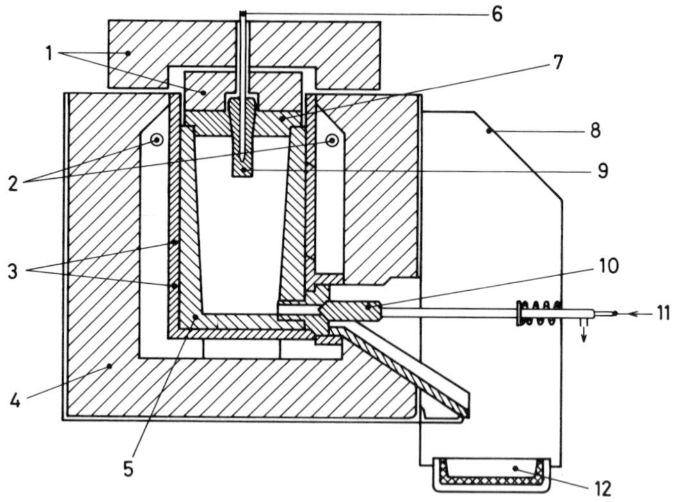

Gas heated furnace for the magnesium reduction of BeF_2: 1) Loose refractory covers; 2) gas burners; 3) sheathing rings (silicon carbide); 4) aluminous firebrick setting; 5) graphite crucible; 6) thermocouple; 7) crucible lid; 8) ventilation hood; 9) stalk (graphite); 10) tapping valve; 11) water cooling; 12) slag mould.

in cold H_2O and dried in a steam-heated oven [18, 19]. A gas heated furnace for Mg reduction of BeF_2 is shown in **Fig.** 1-**7**, taken from [18]. Further purification see p. 95.

In comparison with the Brush practice, improved yields of Be metal are obtained and the atmospheric concentration of beryllium near the reduction furnace is greatly reduced. However, substantial amounts of beryllium remain in the MgF_2 slag, and most of the impurities in the $CaCl_2$ are transferred to the beryllium [1]. For a description of the equipment involved, see [18].

Another practice for reducing BeF_2 with Mg metal that has been used in the USA is described by H. C. Kawecki [17]. The practice varies only in detail from that used at the Brush Beryllium Company [17].

The reduction reaction takes place at relatively low temperatures. The process uses approximately equal amounts of BeF_2 and CaF_2, both substantially pure, which are first melted together in an induction-heated graphite crucible that is open to the atmosphere. The melting point of the mixture is about 700 °C, but the temperature is actually raised to about 800 °C. Massive Mg metal in form of ingots or pigs is then added in about 10% stoichiometric excess. The Mg metal is held below the surface of the fused fluoride until it melts and rises to the surface of the melt. The Be metal forms as a powder. The fusion mixture is stirred sufficiently to prevent it from settling out in the upper portion of the fluoride melt and forming an interface layer between it and the fused Mg metal, thus preventing the completion of the reduction reaction. The reaction is relatively rapid, but is moderated somewhat due to the dilution effect of the fused CaF_2. The heat of reaction raises the temperature to ~1000 °C, which is sufficient to prevent solidification of the melt when its final composition becomes MgF_2-CaF_2 [17]. When the reduction is complete, the melt is left unstirred at a temperature just below 1100 °C for an extended time to allow the Be powder to agglomerate near the surface and sinter together [17].

The major portions of the fused fluorides are decanted from the semi-sintered, porous, metallic mass. The beryllium mass is then quickly melted at about 1300 °C, and then rapidly cooled to prevent the reverse reaction from occurring. Both the residual Mg vapor and the fused fluorides prevent atmospheric oxidation of the Be metal during melting. The resulting beryllium ingot can be cleanly separated from the small amount of residual slag. The ingot is remelted and kept molten for a longer period to completely remove the residual magnesium. Be metal of about 99.8% purity is obtained [17].

Development of a flow sheet for Be metal production on a large scale has been undertaken in India. The processing is based on the results of laboratory scale investigations. It differs in detail from the Brush practice [21].

The Be hydroxide used is produced by the fluoride extraction process. The decomposition of $(NH_4)_2BeF_4$ crystals is carried out at about 900 °C and the BeF_2 recovery is 97%. A BeF_2 excess of ~40% is used for the reduction of BeF_2 with Mg metal. After completing the reduction at about 1000 °C, the temperature of the charge is raised to about 1400 °C. The purity of the Be pebbles is 95 to 97%, major impurities being residual Mg and MgF_2. The beryllium yield is ~80% (based on Mg). The overall beryllium recovery is ~95% (based on BeF_2) [21].

The practice in the USSR varies from the Brush practice in several details only [13].

Technical grade beryllium oxide [13, 32, 37], beryllium hydroxide, or basic beryllium carbonate [22] is dissolved in hydrofluoric acid containing NH_4F to form $(NH_4)_2BeF_4$. The solution is purified to remove admixtures of Fe, Mn, and other metals by what is described as "a sorption" method followed by an additional treatment with organic compounds containing sulfur. The thermal decomposition of the $(NH_4)_2BeF_4$ crystals to BeF_2 is carried out at 900 to 1000 °C in graphite crucibles. Fused BeF_2 in graphite crucibles is reduced with distilled Mg metal in induction furnaces. Towards the end of the operation the temperature in the furnace is raised to 1300 °C. The excess BeF_2 used in the Brush practice to maintain a fluid slag in the BeF_2-Mg reduction is replaced by BaF_2. This results in an increase in the Be yield as well as a reduction of the recycle load. The fused Be beads resulting from this process are extracted by washing the slag with water. After washing, the beads are heated under vacuum at ~1100 °C to purify the metal from Mg and other volatile impurities [13, 20, 22, 32, 37]. For further purification, see p. 95.

1.4.2.1.2.3 Small Scale Investigations on the Reduction Process

Methods used to produce Be metal from BeF_2 with a stoichiometric amount of Mg or an Mg-Ca mixture as the reductant have been studied on a small scale (0.25 to 0.5 kg Be metal). In the reduction with Mg metal (alone), $CaCl_2$ is added as a flux. Reduction with an Mg-Ca mixture is performed without the addition of a flux. Good results are obtained with a reductant composition of (in mol%) 58 Mg and 42 Ca, since an MgF_2/CaF_2 slag of this ratio constitutes a eutectic mixture melting at 950 °C. The charge of the components (weight in kg (lb), particle size in mesh: BeF_2, 2.2(5), 5 to 10; Mg, 0.61(1.35), 10 to 40; Ca, 0.8(1.8), 10 to 40), intimately mixed and tightly packed, reacts smoothly when heated to ~1000 °C by induction in a graphite crucible. On continued heating to a final temperature of 1350 to 1400 °C, the metal and slag separate fairly well. Very clean, massive Be metal is obtained [31].

In another method, BeF_2 is reduced with a stoichiometric amount of a 63 Mg-36 Ca alloy. The slag shows the approximate composition of the CaF_2/MgF_2 eutectic, melting at 940 °C [35].

However, these methods have not been further developed, probably because of the high cost of calcium and the possibility of $CaBe_{13}$ formation [1].

An earlier study [7], (see also [40]) reports the small scale production of Be metal in which the reduction of very pure BeF_2 with Mg metal is performed in an Ar atmosphere in the presence of a reaction damping volatile metal such as zinc [7], see also [40], cf. also [41, 42].

Additional studies to produce Be metal by reducing BeF_2 with a stoichiometric amount of Mg metal were carried out on a small scale (~1 kg BeF_2) using closed, steel lined bombs as reaction chambers [31]. The application of Mg metal alone fails to produce sufficient heat to fuse the products and bring about the separation of metal and slag; apparently, additional heat must be supplied. However, the use of $ZnCl_2$ and $PbCl_2$ as boosters fuses both the metal and slag and results in the separation of the metals from each other and from the slag. The best yields of Be metal (50 to 80%) are obtained with 0.05 mol $ZnCl_2$ or $PbCl_2$ per mol BeF_2. However, the resulting beryllium in each case is not as free of slag inclusions as is desirable [31].

When sulfur is used as a booster with mixtures of Mg and Ca metal reductants, relatively clean, massive Be metal is produced. The particle size of the components must be within a certain range. Good results are obtained with particle size (in mesh): BeF_2 ~100, Mg and Ca ~10 to 40, S ~40. The highest yields (85 to 90%) and best separation of Be metal are obtained using 0.75 mol S per mol BeF_2 with 80 mol% Mg and 20 mol% Ca. A pre-heating temperature for the bomb of about 565.6 °C is used. However, excessive pressure frequently causes burn-outs at the gasket of the bomb. The process is therefore dangerous and wasteful [31].

References:

[1] G. E. Darwin, J. H. Buddery (Beryllium, Butterworth, London 1960, pp. 76/8, 87/91, 98).
[2] C. W. Schwenzfeier Jr. (in: D. W. White Jr., J. E. Burke, The Metal Beryllium, ASM, Cleveland 1955, pp. 71/101, 84/97).
[3] G. Petzow, F. Aldinger (Ullmanns Encykl. Tech. Chem. 4th Ed. **8** [1974] 451).
[4] W. Kroll (Ger. 480128 [1926]; U.S. 1740857 [1927/29]).
[5] W. J. Kroll (J. Franklin Inst. **260** [1955] 169/92, 177).
[6] W. B. Donahue (U.S. 2072067 [1934/37]).
[7] L. Losana (Alluminio **8** [1939] 67/75).
[8] C. Adamoli (U.S. 2193363 [1936/40]; U.S. 2193364 [1936/40]).
[9] B. R. F. Kjellgren, The Brush Beryllium Co. (U.S. 2381291 [1941/45]).
[10] B. R. F. Kjellgren (NML Tech. J. **3** No. 1 [1961] 37/45, 39; C.A. **56** [1962] 3225).

[11] B. R. F. Kjellgren (Trans. Electrochem. Soc. **93** [1948] 122/8, 124).
[12] A. R. Kaufmann, B. R. F. Kjellgren (Proc. Intern. Conf. Peaceful Uses At. Energy, Geneva 1955, Vol. 8, pp. 590/9, 592).
[13] W. W. Beaver (Progr. Nucl. Energy V **1** [1956] 277/97, 282/3, 285/6; Met. Abstr. **24** [1956/57] 963).
[14] J. Ballance, A. J. Stonehouse, Ray Sweeney, K. Walsh (in: Kirk-Othmer Encycl. Chem. Technol. 3rd Ed. **3** [1978] 803/23, 809).
[15] C. W. Schwenzfeier Jr. (in: Kirk-Othmer Encycl. Chem. Technol. 2nd Ed. **3** [1964] 450/74, 460).

[16] S. J. Morana, G. F. Simons (J. Metals **14** [1962] 571/4).
[17] H. C. Kawecki (U.S. 2486475 [1945/49]).
[18] L. J. Derham, D. A. Temple (Extr. Refining Rarer Metals Proc. Symp., London 1956 [1957], pp. 323/36, 325/31; C.A. **1958** 19783/4).
[19] L. J. Derham, National Smelting Co., Ltd. (Brit. 781100 [1956/57]; N.S.A. **12** [1958] No. 721).
[20] G. F. Silina, Yu. I. Sarembo, G. J. Kaplan (Tsvetn. Metal. **30** No. 1 [1957] 66/71, 69).

[21] K. S. Subbarao, B. P. Sharma, C. M. Paul, C. V. Sundaram (Beryllium 1977 Conf. Preprint 4th Intern. Conf. Beryllium, London 1977, pp. 1/10, 3/4, 5; C.A. **92** [1980] No. 151299).
[22] G. A. Meerson, D. D. Sokolov, N. F. Mironov, N. M. Bogorad, Ya. D. Pokhomov, D. S. Lvovsky, E. S. Ivanov, V. M. Shmelev (At. Energiya SSSR **5** [1958] 624/30; Soviet At. Energy **5** [1958] 1555/62, 1556/7; C.A. **1961** 7218).
[23] F. Aldinger, W. C. Heraeus GmbH (private information 1984).
[24] C. W. Schwenzfeier Jr., C. S. Pomelee (U.S. 2660515 [1948/53]).
[25] K. A. Walsh (in: D. R. Floyd, J. N. Lowe, Beryllium Science and Technology, Vol. 2, Plenum, New York-London 1979, pp. 1/10, 7/9).
[26] S. B. Roboff (in: H. H. Hausner, Beryllium Its Metallurgy and Properties, Univ. Calif. Press, Berkeley 1965, pp. 17/30, 22/24; C.A. **64** [1966] 10879).
[27] D. A. Everest (Topics in Inorganic and General Chemistry, Vol. 1: The Chemistry of Beryllium, Elsevier, New York 1964, pp. 1/158, 114/5).
[28] W. Schreiter (Freiberger Forschungsh. B No. 17 [1956] 50/77, 63/4).
[29] G. Dressler, M. Rühle (in: K. Winnacker, L. Küchler, Chemische Technologie, 3rd Ed., Vol. 6, München 1973, pp. 549/601, 590/4).
[30] W. D. Jamrack (Rare Metal Extraction by Chemical Engineering Techniques, Vol. 2, Pergamon, New York 1963, pp. 233/7; N.S.A. **18** [1964] No. 13952).

[31] F. H. Spedding, H. A. Wilhelm, W. H. Keller, C. Neher (TID-5212 [1955] 43/8; C.A. **1956** 7691).
[32] G. A. Meyerson (Proc. 1st Intern. Conf. Peaceful Uses At. Energy, Geneva 1955, Vol. 8, pp. 587/9).
[33] L. M. Pidgeon (Can. Mining Met. Bull. No. 461 [1950] 349/58, 355).
[34] P. S. Bryant (Extr. Refining Rarer Metals Proc. Symp., London 1956 [1957], pp. 310/22, 311).
[35] T. H. Schofield (Brit. 709552 [1947/54]).
[36] L. R. Williams, P. B. Eyre (Nucl. Eng. **3** [1958] 9/18, 10).
[37] A. I. Evstyukhin (Met. Metalloved. Chist. Metal. No. 1 [1959] 91/105, 94/9, 103; C.A. **1961** 4292).
[38] B. Wempe (U.S. 2091087 [1932/37]; C.A. **1937** 7389).
[39] G. Venturello (Atti Reale Accad. Sci. Torino **76** I [1941] 556/63; C. **1942** I 1114).
[40] H. A. Sloman, C. B. Sawyer (FIAT Final Rept. No. 522 [1946] 1/102, 36/7).

[41] W. J. Kroll (U.S. Bur. Mines Inform. Circ. No. 7326 [1945] 1/15, 3).
[42] T. T. Magel (in: D. W. White Jr., J. E. Burke, The Metal Beryllium, ASM, Cleveland 1955, pp. 124/35, 127).
[43] Anonymous (NBS-C-500 [1952]).
[44] D. R. Stull, G. C. Sinke (Thermodynamic Properties of the Elements, ACS, No. 18, Washington 1956, pp. 53, 124).
[45] D. R. Stull (JANAF Thermochemical Tables PB-168370 [1965]; NSRDS-NBS-37 [1971]).
[46] F. E. Block, R. E. Mussler, T. T. Campbell (Advan. Extr. Met. Proc. Symp., London 1967 [1968], pp. 551/71; C.A. **69** [1968] No. 69070).

1.4.2.1.3 Reduction of Beryllium Double Fluoride with Magnesium or Calcium

In a reduction of an equimolar mixture in pastil form of BeF_2 and NaF with Mg (or Ca) metal at 10^{-2} Torr and 600 to 700 °C, Be metal is quantitatively liberated according to: $BeF_2 \cdot NaF + 1.5\ Mg \rightarrow Be + Na + 1.5\ MgF_2$. From the Be metal, intimately mixed with MgF_2, Na and residual Mg are distilled off. Then, on heating to 1350 °C Be metal and MgF_2 separate into two distinct layers. The upper layer consists of beryllium. After solidification the powdery Be product contains only traces of fluoride, Mg and Ca.

Reference:

R. Rohmer (Compt. Rend. **214** [1942] 744/6; C.A. **1944** 6142).

1.4.2.2 Reduction of Beryllium Chloride

1.4.2.2.1 With Metals

Beryllium chloride can be readily reduced by lithium, sodium, potassium, magnesium, and calcium [1 to 4, 10]. All of these methods are as of yet restricted to laboratory scale [1, 2, 4].

1.4.2.2.1.1 Basic Process Design

The reduction of beryllium chloride has two major disadvantages in comparison with the reduction of beryllium fluoride: 1) The preparation of pure anhydrous beryllium chloride (cf. p. 69) is more difficult than that of beryllium fluoride (see p. 58), and almost inevitably, the product is contaminated with beryllium oxide and hydroxide [2, 3]. 2) The boiling point of beryllium chloride is so low (520 °C [16]) that the reductions ordinarily carried out in the liquid or solid phase are restricted to temperatures at which the resulting beryllium metal is obtained in a very fine-powdery form or as a collection of beads, but it is not permitted to fuse. This type of product, however, is susceptible to oxidation during the removal of the other reaction product and is difficult to separate from the beryllium oxide introduced with the beryllium chloride [1 to 4].

Wöhler first prepared metallic beryllium in 1828 by heating a mixture of $BeCl_2$ with potassium metal. After removal of the byproduct KCl with H_2O, finely divided, glistening particles of Be metal contaminated with BeO were obtained [5], cf. [19].

The bulk of the work up to 1945 on the metallothermic reduction of $BeCl_2$ with alkali or alkaline-earth metals has been reviewed by Kroll [3] who stresses the above mentioned disadvantages of the reduction process; also see the review of Tien [11].

In 1933, Kangro proposed the use of aluminium as a reducing agent and gave some supporting thermodynamic calculations. The reaction is based on the volatility of the aluminium chloride reaction product [6]. Also in 1933, Fischer and Peters [7] attempted to produce beryllium metal on the basis of Kangro's suggestions. They carried out the reaction at 260 to 350 °C in a hydrogen atmosphere, but the metal powder produced was so fine that it oxidized almost completely on transfer from the reaction vessel [7], also see [8]. On repeating this reaction, Kroll [3] reported that aluminium partially reduced $BeCl_2$ and that "low-grade Be-Al alloys" were produced [3].

Magel [9] (presumably in the 1950's) devised a laboratory method which solved some of the major difficulties by consolidating the Be metal in a "bomb"-type reduction. The reduction was achieved by placing a hollow charge of anhydrous $BeCl_2$ in a well-sintered magnesia crucible, the hollow being filled with a mixture of Ca metal and iodine. The crucible was covered with a magnesia lid and placed in a sealed steel bomb. The reaction was initiated by inductively heating the bomb to about 600 °C, whereupon the wall temperature of the bomb suddenly increased several hundred degrees indicating that the reaction had taken place. The iodine "booster" reacted with the excess calcium to add extra heat to the reacting charge of $BeCl_2+Ca$. The so formed CaI_2 increased the fluidity of the slag, thus helping the initially formed Be metal particles melt and agglomerate into a single billet [2, 9].

Be metal yields of up to 90 to 95% were claimed on an 18 g scale [9]. The expected $CaBe_{13}$ contamination may have been prevented by the presence of iodine [2].

Sulfur may also be used instead of iodine as "booster" [9].

Additional reductions carried out in bomb reaction vessels using sodium or magnesium metal as the reductant have been reviewed by Jaeger [10]. These reductions proceeded to about total completion under anhydrous, oxygen-free conditions in sealed iron vessels [10].

When Na metal is used as a reductant, the reaction components are ground at −90 °C, mixed, and brought to reaction like a thermite mass. During the violent reaction, the Be metal particles melt and agglomerate into a regulus. Other methods are carried out in a high frequency furnace. The Be metal sponge so obtained is fused into a regulus at the end of the reaction by heating above 1350 °C. In both cases, Be metal is separated from the byproduct NaCl by aqueous leaching [10].

Similar conditions have been applied in the Mg metal reduction of $BeCl_2$. The Be metal particles are separated from the byproduct by leaching with aqueous NH_4Cl solution [10].

The reduction of $BeCl_2$ with Mg metal in the presence of NaCl carried out in an open crucible is said to produce Be metal directly in a compact and ductile form [10].

While most of the $BeCl_2$ reduction experiments are conducted at atmospheric or higher pressure, in 1946 Tien [11] reported a small scale sodium reduction of $BeCl_2$ under reduced pressure. Sodium vapor (b.p. of sodium at 1.08×10^{-3} Torr: ~490 °C) was passed over $BeCl_2$ held at 250 °C. When the sodium vapors contacted the $BeCl_2$, the mass glowed until the reduction was complete. Any excess sodium was then distilled under vacuum. After removal from the reaction tube at room temperature under a CO_2 atmosphere, the reaction products were leached in absolute methanol and then in saturated aqueous ammonium carbonate solution. A ~91% yield of ~96% pure Be metal was reported [11].

In 1947 at the Brush Beryllium Co. [12], a vapor phase reduction of $BeCl_2$ with sodium was tried on a large scale. $BeCl_2$ and Na metal were vaporized in separate boilers at an initial operating pressure of 1 to 2 Torr (distillation temperatures: $BeCl_2$, 330 °C; Na, 500 °C). The vapors were led to a reaction vessel at controlled rates by means of an argon flow introduced via the Na containing boiler. The Be metal was obtained as a very fine powder which oxidized rapidly in the separation process. An appreciable increase in particle size and reduction in oxide content of the Be powder was claimed by applying an argon pressure of 1 to 2 atm. As in Tien's process, the reaction product was freed from excess sodium by distillation at 1 Torr and leaching in a methanol-aqueous ammonium carbonate solution. A yield of 65% Be powder of 5 μm average particle size and 3.6% BeO content was obtained from the reaction of 65 g $BeCl_2$ with 65 g Na metal [12].

A new method has been developed by the Brush Beryllium Co. as a result of further investigations into this vapor phase reduction. This method is said to be adaptable to large scale operation. Be metal yields of up to 90% are reported in [13].

In the late 1960's, Block et al. [4] and Campbell et al. [14] worked at devising a more effective process for reducing $BeCl_2$. The techniques studied were similar to the "Kroll-process" which was the first successful method for producing high purity titanium, zirconium, and hafnium and is still the only commercial method for processing these highly reactive metals [4, 14]. Kroll [3] first suggested this approach in 1945, but no record of experimental work has been found [2, 4, 14].

In the "Kroll-process" an oxygen-free, anhydrous, purified metal chloride is fed into a crucible containing the molten reductant. Because the reaction can be controlled by the addition rate of the chloride and because the reaction is carried out at near atmospheric pressure, temperature and pressure fluctuations are easily controlled, and the process can be carried out in very large reactors. The metal product is formed in a bath of the molten chloride of the reduction metal. Metals formed in this way usually agglomerate into a compact metal sponge which is easy to recover free of surface impurities. Since a "Kroll-type" reaction can easily be controlled, corrosion of the reactor materials caused by localized overheating can be minimized. In the final step of this process the excess metal reductant and the chloride salt of the reductant are separated from the metal sponge by vacuum distillation. The contamination problems associated with leaching fine metal powders are thus avoided [4, 14].

Lithium, sodium, and magnesium have commercial possibilities as the metallic reductants in "Kroll-type" processes between $BeCl_2$ vapor and molten metal reductants.

A general consideration of the physical properties shows that magnesium should be the most favorable reductant. The melting points of Mg metal and $MgCl_2$ are close. The vapor pressures are low at the melting point of $MgCl_2$ (714 °C), which is the minimum temperature necessary to carry out the reaction. In addition, $BeCl_2$ has an appreciable vapor pressure at this temperature which allows its vapor phase transport to the reductant. The $MgCl_2$ vapor pressure is sufficient to allow separation from the Be metal sponge by vacuum distillation at 950 °C, the upper temperature limit for performing a salt separation in alloy steel equipment. The densities of molten Mg metal and molten $MgCl_2$ indicate that Mg metal floats above $MgCl_2$ where it remains available to react with the $BeCl_2$ vapors throughout the process [4].

Lithium, disregarding its higher cost, appears to be an even more suitable reductant than magnesium. Because of the low melting point of LiCl (610 °C), the reaction should be operable at temperatures low enough to overcome corrosion of the most common construction materials [4].

Sodium seems to be the least suitable reductant for furnace operation because its vapor pressure is very high at the melting point of NaCl. Therefore, its vaporization must be strongly controlled in order to avoid a vapor phase reaction with deposition of finely divided beryllium powder throughout the reactor. However, the easy handling of molten sodium, its purity, and relatively low price are important factors in considering this metal as a reductant in an industrial process [4].

1.4.2.2.1.2 Industrial Procedure

"Kroll-type" reduction processes of $BeCl_2$ using Li, Na, or Mg metal as reductant have been carried out on larger laboratory scales (kilogram amounts). The complete process involves three stages; two concern the preparation of the starting material ($BeCl_2$): 1) Preparation of crude $BeCl_2$; 2) Purification of crude $BeCl_2$; 3) Reduction of $BeCl_2$ [4, 14].

The proposed daily material flow sheets (in lb) for conversion of BeO to Be sponge by either lithium or sodium reduction of $BeCl_2$ on a commercial scale (100 tons/a Be) are given in [4].

Preparation of Crude $BeCl_2$

Several methods for the preparation of anhydrous crude $BeCl_2$ have been used by [4]. They include reaction of elemental chlorine with either beryllium scrap or a beryllium-aluminium carbide mixture and reaction of Be oxide with carbon tetrachloride or phosgene. The most frequently used method, the direct chlorination of a beryllium oxide-carbon mixture, is described in the following paragraphs [4].

Beryllium oxide is mixed with a ~10% excess of "gas carbon" (gas black?) or ground charcoal. This charge is then blended with a 15% sucrose solution to a plastic consistency. The mixture is air dried and calcined at 800 °C under a protective atmosphere to form strong, but porous beryllium oxide/carbon spheres [4].

The chlorination of the beryllium oxide/carbon spheres is conducted in a vertical quartz reactor. The porous charge is supported by a perforated graphite pedestal and heated (see below) by electrical resistors radiating directly on the vitreous quartz reactor. Chlorine that is fed near the bottom reacts with the charge to form $BeCl_2$ vapor which is prevented from condensing between the reactor and the condenser by heating the connecting conduit to 500 °C. On heating the condenser to 320 °C, virtually all the $BeCl_2$ is collected as a crystalline condensate, which allows some of the more volatile impurities to be separated [4].

If the beryllium oxide/carbon charge is kept at 850 to 900 °C, the chlorination proceeds at a rate of up to 3 L/min, corresponding to a production rate of 0.26 kg $BeCl_2$/h/dm^2 (2.4 kg/h/ft^2) of reactor cross section. The chlorine utilization in forming $BeCl_2$ is 75 to 90%, and the recovery of BeO as condensed $BeCl_2$ is 90 to 94% [4].

A schematic drawing of the chlorination apparatus is shown in the paper. Except for the vitreous quartz reactor, it is constructed entirely of nickel [4].

Purification of Crude $BeCl_2$

The purification of crude $BeCl_2$ is the most critical stage in the process for preparing high purity beryllium metal because most of the impurities in the $BeCl_2$ are transferred quantitatively during "Kroll reduction" to the beryllium sponge product [4, 14]. The purification operation consists of a two step treatment involving (1) bulk sublimation of crude $BeCl_2$ under vacuum followed by (2) fused-salt scrubbing [4, 14].

Vacuum Sublimation. The vacuum sublimation process of crude $BeCl_2$ or of commercial material is carried out in a stainless steel retort 30.48 cm in diameter (12 in) equipped with an Ni charge container and an Ni condenser, see the paper. The sublimation is effected by heating the crude $BeCl_2$ from 350 °C (initial temperature) to 430 °C (final temperature) under reduced pressure. The absolute pressure in the retort ranges from 10 to 40 Torr during the early purification stages and to $\sim 20 \times 10^{-3}$ Torr at the end of a run. The recovery of $BeCl_2$ is ~90 to 97%, depending largely on the hydrate moisture of the crude material which causes BeO formation during the sublimation. The BeO is readily recovered from the charge container and can be recycled to a primary chlorination step. After completion of the sublimation, the condenser is rapidly transferred at room temperature under argon to an inert atmosphere handling chamber. The $BeCl_2$ is removed and stored in Mason jars awaiting further purification. The purification process for a charge of about 12 kg $BeCl_2$ takes approximately 90 h [4, 14].

Fused-Salt Scrubbing. While bulk sublimation is effective in removing O_2, BeO, carbon, and aluminium from impure crude $BeCl_2$, other metallic impurities are not removed to suitable levels. Adequate purification of $BeCl_2$, however, can be effected by passing the $BeCl_2$ vapor from step (1) through a fused-salt bath. A bath composition (in mol%) of 55 LiCl, 36 KCl, and 9 NaCl proves most effective, and its low melting point (346 °C [15], from [14]) allows working temperatures low enough for minimal corrosion of equipment [4, 14].

The fused-salt scrubbing operation is carried out in a stainless steel retort vacuum vessel provided with a single bubble cap tray, an Ni charge container, and an Ni condenser. For the purification procedure, $BeCl_2$ (2 to 3 kg) from step (1) is added to the retort and evacuated to approximately 1 to 10×10^{-4} Torr. After being brought up to operating temperature, the fused-salt scrubbing is initiated by heating the lower section of the retort to 435 °C. The $BeCl_2$ vaporizes and bubbles through the 3.81 to 5.08 cm (1.5 to 2 in) deep molten salt bath (370 g of eutectic salt mixture) and condenses in the upper section of the retort which is maintained at 135 °C. A condenser temperature of 135 °C favors the formation of coarse, dense crystalline $BeCl_2$ which tends to minimize moisture pickup from the atmosphere during handling [4, 14].

A 2 to 3 kg $BeCl_2$ charge requires 20 to 24 h for purification. This is equivalent to a rate of approximately 0.11 to 0.13 kg/h/dm^2 (1.0 to 1.2 kg/h/ft^2) of retort cross section. The yield of purified $BeCl_2$ ranges from 92 to 98.6%. Two major sources of loss are the solution of $BeCl_2$ in the scrubbing salt mixture and minor handling losses [4, 14].

For the effectiveness of the purification operations (1) and (2) in obtaining high purity $BeCl_2$ from crude feed materials of significantly different purity, see the analytical data in [4, 14].

Reduction of $BeCl_2$

Because of the similar physical properties of $BeCl_2$ and $ZrCl_4$, the apparatus and techniques employed for $BeCl_2$ reduction are similar to those used for Zr metal production (cf. [17]) in the Kroll process. The $BeCl_2$ reductions with Li, Na, or Mg metal as reductant are all made in the same apparatus, though the mode of operation is varied to conform with the physical properties of each metal reductant [4], cf. [14].

For a schematic drawing of the reduction apparatus (20.32 cm (8 in) in diameter), see **Fig. 1-8**, p. 71. Purified $BeCl_2$ is loaded into the upper Ni retort of the apparatus, and a metal crucible in the lower retort contains the reductant. An Ni conduit connecting the two retorts serves as a port for adding the reductant to the lower retort. Mg metal is added as solid blocks, Na or Li metal as a liquid. Molybdenum (10 or 20 mm thick sheet) proves to be the least corroding material for crucible construction. The use of stainless steel crucibles coated with molybdenum on the inside by a plasma spraying technique is also possible [4, 14].

The reduction of purified $BeCl_2$ (~5 kg) with a reductant of the highest purity technical grade available (lithium is not obtainable in the same high purity as the other reductants) is carried out at a controlled rate under helium at atmospheric pressure. The lower retort of the apparatus is heated to a temperature sufficient to melt the reductant, and the upper retort is heated to 500 to 550 °C in order to vaporize the $BeCl_2$. The pressure during this heating period is maintained between 0 and 703×10^{-3} kp/cm^2 (0 and 10 lb/in^2) above atmos-

Fig. 1-8

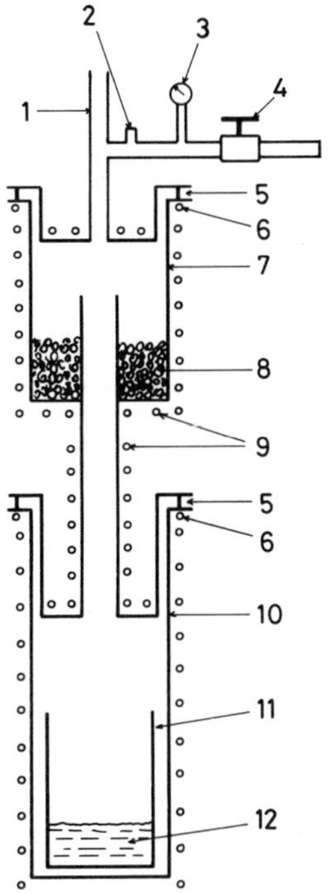

Reduction apparatus, taken from [4, 14]. 1) Reductant inlet; 2) He inlet; 3) pressure gauge; 4) valve; 5) O ring; 6) water cooling; 7) Ni retort; 8) $BeCl_2$; 9) resistance heaters; 10) stainless steel retort; 11) crucible; 12) reductant.

pheric pressure by periodically bleeding helium from the retort. The internal pressure rises again as the reaction progresses indicating decreasing reaction rate due to formation of a byproduct salt which covers the surface of the molten metal reductant. At this time, increased heating of the reduction crucible is necessary to melt this salt and to expose fresh reductant to the $BeCl_2$ vapors. Completion of the reduction is indicated by an abrupt decrease in pressure [4].

The minimum retort temperatures necessary for sustaining the reductions with magnesium, lithium, or sodium are 800, 675, and 810 °C, respectively. The reaction rate of sodium with $BeCl_2$ is equivalent to 0.139 kg/h/dm^2 (1.3 kg/h/ft^2) of reactor cross section. Lithium reductions are conducted at twice this rate because the LiCl formed remains molten, thereby enabling a clean separation of Be metal from the salt and exposing a larger surface of molten Li metal to the reacting $BeCl_2$ vapors [4].

The separation of excess metal reductant and byproduct chloride salt from the Be sponge is carried out in a high vacuum retort. The crucible is usually supported in an inverted position by a stainless steel pedestal. The upper half of the retort is heated to 950 °C, and the retort pressure is maintained below 10^{-5} Torr. After heating the crucible for 20 h, unreacted metal reductant and chloride salt are separated from the Be sponge either as a liquid, dripping from the sponge, or as a vapor, condensing on the lower water-cooled section of the retort wall. The Be sponge is scraped from the crucible after vacuum treatment. It generally appears as very friable dendrites when broken from the crucible walls and as a denser, stronger material when recovered from the lower center of the crucible [4].

Contrary to expectations (cf. p. 68), the yields of Be sponge from reductions with Mg metal are very low, ranging from 20 to 50%. This is attributed to the partial formation of fine metallic powders which are recovered with the byproduct salt. X-ray diffraction analysis indicates that these fine powders are the intermetallic phase $MgBe_{13}$. The use of an excess of Mg metal does not affect the Be yield [4].

The yields of Be sponge from reductions with Li metal, however, are high, ranging between 94 and 99%. The optimum yields are obtained with a stoichiometric amount of Li metal, although a 5% excess is generally employed. Since the reductions with Li metal are performed at lower temperatures than those with Mg or Na metal, there is no evidence of vapor phase reaction products on the inner surfaces of the retort [4].

Be sponge yields of 90 to 93% are attained from Na metal reductions using a 15 to 20% excess of sodium, and by careful control (see p. 68) of the reduction cycle. Increasing the retort pressure to between 703×10^{-3} and 1054×10^{-3} kp/cm^2 (10 to 15 lb/in^2) gauge throughout the cycle reduces the sodium partial pressure in the vapor. The sodium vaporization is also suppressed by prolonging the operation cycle during the earlier stage. The bulk of the sodium reacts, though at a somewhat retarded rate, by holding the crucible temperature at 700 to 750 °C (cf. p. 71). A final stage at 810 to 850 °C completes the reaction and promotes consolidation of the products in the crucible [4].

Lithium- or sodium-reduced Be metal contains considerably fewer impurities than the Be metal currently produced industrially [4].

In a more recent laboratory study by the same authors [14], the above reduction procedure was optimized in several aspects with regard to sodium reduction: 1) Lowering of chlorine contamination in the Be product. 2) Production of Be sponge material that was readily comminutable to a fine powder. 3) Lowering of reduction temperatures. 4) Increase of reduction rate [14].

These higher reduction rates and lower operation temperatures [4] were achieved by reducing the amount of helium in the retort at the beginning of the reaction. This lower level of He causes a correspondingly higher $BeCl_2$ partial pressure in the retort, a minimized vapor-phase reaction, and significantly increased reaction rates. By operating below 750 °C and at higher $BeCl_2$ partial pressures, corrosion of reactor components is minimized with a corresponding decrease in metallic impurities, chlorine, and oxygen in the beryllium product [14].

The apparatus shown previously (Fig. 1-8, p. 71) is used in the following reduction procedure. The purified $BeCl_2$ is added to the upper retort of the reduction apparatus at room temperature and under dry argon atmosphere. Liquid sodium (reactor grade) in 15% excess is introduced into the evacuated, heated (400 °C) reactor through the reductant inlet. The reactor is then backfilled with helium to 500 Torr absolute pressure. The lower (reduction) retort is heated from 650 to 750 °C, and the upper is heated ($BeCl_2$ charge) to 500 to 550 °C. Best results are obtained by maintaining the temperatures in the reduction zone at 650 to 700 °C and the retort pressures at or slightly above atmospheric pressure. The reduction of 3 to 5 kg $BeCl_2$ requires 2.0 to 3.5 h. Reaction rates are about 0.08 to 0.30 kg/h/dm^2 (0.7 to 2.8 kg/h/ft^2) of crucible cross section. After completion of the reduction step, the excess reductant and NaCl are removed from the fine dendritic Be sponge by vacuum distillation at 950 °C and 1×10^{-2} to 2×10^{-5} Torr. After approximately 20 to 24 h, all but minor amounts of sodium and NaCl are removed from the Be sponge. The yields of Be sponge metal (irregular mass of interlocking, lightly sintered dendritic crystals) average from about 90 to 93% [14].

The ease of comminution of Be sponge material increases with rising reduction rate during preparation, due to the more dendritic and less dense constitution of the Be sponge product. Accordingly, grinding of fine, low density, dendritic sponge proceeds more readily and also wet grinding (with H_2O) for removal of chlorine becomes more effective. For experimental data, see [14].

The purity of the Be metal obtained ranges from 99 to 99.6% [14]. The purity of commercial products generally ranges from 98.5 to 99.5% [2].

The analytical data from Be powders prepared at relatively low reduction rates are given in [14].

A comparable reduction of Be halide vapor (presumably also of $BeCl_2$ vapor) with molten Mg metal in a similar apparatus and processing is referred to in [18].

References:

[1] T. T. Magel (in: D. W. White Jr., J. E. Burke, The Metal Beryllium, ASM, Cleveland 1955, pp. 124/35, 125/7).
[2] G. E. Darwin, J. H. Buddery (Beryllium, Butterworth, London 1960, pp. 94/6).
[3] W. J. Kroll (U.S. Bur. Mines Inform. Circ. No. 7326 [1945] 1/15, 3/4).
[4] F. E. Block, R. E. Mussler, T. T. Campbell (Advan. Extr. Met. Proc. Symp., London 1967 [1968], pp. 551/71; C.A. **69** [1968] No. 69070).
[5] F. Wöhler (Ann. Chim. Phys. [2] **39** [1828] 77/84).
[6] W. Kangro (Metall Erz **30** [1933] 389/90).
[7] H. Fischer, N. Peters (Metall Erz **30** [1933] 390/1).
[8] H. E. Martens (Metallbörse **23** [1933] 1438/9).
[9] T. T. Magel, Massachusetts Institute of Technology (unpublished information from [1]).
[10] G. Jaeger (Metall **4** [1950] 183/91, 188).

[11] J. M. Tien (Trans. Electrochem. Soc. **89** [1946] 237/45, 241/3).
[12] Brush Beryllium Co. (Brush Beryllium Co. Progr. Rept. No. 12/17 [1947] from [2]).
[13] P. R. Kalischer, Brush Beryllium Co. (BBC-6 [1948] 1/27; BBC-8 [1948] 1/22; BBC-17 [1948] 1/24; BBC-23 [1948] 1/22; BBC-26 [1948] 1/43; BBC-29 [1948] 1/26 from [4]).
[14] T. T. Campbell, R. E. Mussler, F. E. Block (Met. Trans. **1** [1970] 2881/7, 2882/5).
[15] P. V. Clark (SC-R-65-930 [1965] 1/294 from [14]).
[16] G. Petzow, F. Aldinger (Ullmanns Encykl. Tech. Chem. 4th Ed. **8** [1974] 451, 455).
[17] E. D. Dilling (U.S. Bur. Mines Bull. No. 561 [1956] 16/51 from [4]).
[18] H. Zeppelin, H. Schlageter (Ger. 1274800 [1957/68]; C.A. **69** [1968] No. 69326).
[19] R. G. Bellamy, N. A. Hill (Extraction and Metallurgy of Uranium, Thorium and Beryllium, Macmillan, New York 1963, pp. 75/6).

1.4.2.2.2 With Hydrogen

Hydrogen reduction would be an attractive method because both the reducing agent and the resulting halide are noncontaminating gases. However, hydrogen seems to be a weak reducing agent for $BeCl_2$ as is shown by the free enthalpy ΔG and equilibrium constant K of the reaction $BeCl_2(g) + H_2(g) \rightleftharpoons Be(c) + 2\,HCl(g)$ at 1000 and 1250 °C. These data are, however, considered to be somewhat uncertain [1].

In a laboratory scale attempt by [2], the reduction of $BeCl_2$ vapor with hydrogen on the surface of a hot tungsten wire in a narrow temperature range below 1283 °C (m.p. of Be metal) deposited Be metal. At higher temperatures, a beryllium-tungsten alloy formed instead. The results are based on X-ray (Debye-Scherrer) studies [2].

Experiments carried out at Sylvania Electric Products [3], cf. also [5] on a hydrogen glow discharge reduction of $BeCl_2$ at reduced pressure (temperature not specified) did not show any formation of Be metal [4].

References:

[1] H. H. Kellog (in: D. W. White Jr., J. E. Burke, The Metal Beryllium, ASM, Cleveland 1955, pp. 49/62, 57/8).
[2] L. Hackspill, J. Besson (Bull. Soc. Chim. France **1949** 113/6; C.A. **1949** 4967).
[3] B. Kopelman, Sylvania Electric Products (unpublished information from [4]).
[4] T. T. Magel (in: D. W. White Jr., J. E. Burke, The Metal Beryllium, ASM, Cleveland 1955, pp. 124/35, 133).
[5] C. I. Whitman, R. B. Holden (SEP-123 [1953] 1/29 from N.S.A. **10** [1956] No. 1816).

1.4.3 Reduction of Beryllium Sulfide with Metals or Hydrogen

Be metal can be obtained by heating beryllium sulfide (prepared by reducing beryllium sulfate with charcoal at about 700 °C under a dry, air-free gaseous hydrocarbon or argon atmosphere) with Sn metal in inert atmosphere, preferably O_2, H_2O, and N_2 free argon. The resulting Sn sulfide is distilled off at about 1350 °C [1]. Further, because of its low enthalpy of formation, beryllium sulfide is also said to be easily reduced by Mg, Ca, Sr, Ba, and Al [2].

Be metal can also be obtained by heating beryllium sulfide (prepared by heating $BeCl_2$ with carbonaceous material, e.g., low temperature carbonization coke, in sulfur vapors and (or) a gaseous or vaporized nonmetallic sulfur compound, e.g., CS_2 or H_2S at about 900 to 1200 °C) in a stream of hydrogen at about 1350 °C or above; H_2S is evolved. An admixture of beryllium sulfide with powdered Al metal may also be used for this hydrogen reduction; Al_2S_3 and Be metal are thus obtained. The Be metal formed is distilled off. Both of the beryllium sulfide reduction reactions can be considerably facilitated by using a flux, e.g., BeF_2 and (or) BaF_2 [3].

References:

[1] H. Lavoisier (Fr. 802579 [1936]; C.A. **1937** 1753).
[2] G. Jaeger (Metall **4** [1950] 183/91, 188).
[3] D. Gardner (Brit. 482467 [1936/38]; C.A. **1938** 7008).

1.4.4 Miscellaneous Reductions of Beryllium Compounds

The following methods are currently only performed on a laboratory scale.

In a recently developed method, powdered Be metal was obtained by reduction of $Be[Be(CN)_4]$ crystals (prepared by reaction of Be hydroxide with hydrocyanic acid) with hydrogen. The crystalline $Be[Be(CN)_4]$ is first dried for several hours at 200 °C in a dry, pure hydrogen stream. Then, while maintaining the hydrogen stream, the temperature is increased at a rate of 600 °C/h to the reaction temperature of 1100 °C, giving the Be powder. The hydrocyanic acid formed is removed with the hydrogen stream and recycled for production of the starting material [1].

Finely divided Be metal, free from metallic impurities, can be prepared at moderate temperatures and atmospheric pressure by reduction of dialkyl beryllium compounds with an approximately stoichiometric amount of alkali metal, preferably sodium, under an inert atmosphere, and with or without an inert solvent. In the organo beryllium compound R_2Be, R is a straight- or branched-chain aliphatic hydrocarbon radical containing from 1 to 6 C atoms. The protective atmosphere may be, e.g., N_2, CO, Ar, or a suitable hydrocarbon. The equipment required for the reduction is simple and inexpensive due to the low reaction temperature [2].

The preferred process uses di-tert-butyl beryllium with a suspension of finely divided Na metal in, e.g., xylene, under dry nitrogen at $\leqq$145 °C. The reaction follows the equation: $3R_2Be + 2Na \rightarrow Be + 2NaBeR_3$. The $NaBeR_3$ is reacted with $BeCl_2$ to regenerate R_2Be ($2NaBeR_3 + BeCl_2 \rightarrow 3R_2Be + 2NaCl$) [2].

Examples given in the patent are the reduction of di-tert-butyl beryllium diethyl etherate with Na metal and of di-n-hexyl beryllium with a liquid Na-K alloy [2].

The formation of Be metal from $BeCl_2$ by a "cold" plasma reaction was studied on a very small scale (milligram amounts) by McTaggart [3, 4], cf. [5]. When $BeCl_2$ vapor (sublimation rate 1.2 mmol/h) in a silica vacuum tube (about 0.1 Torr upwards) is passed through an electric discharge of a 28 Mc/s field at 70 W, an unstable Be monochloride species and chlorine atoms or ions are produced. X-ray and other analyses indicate that the Be monochloride species disproportionates on contact with the vessel walls to an equimolar mixture of Be metal (mirror) and $BeCl_2$. The Be metal may be extracted from this mixture by vacuum distillation or solution of the dichloride. Chlorine may be recovered when inert carrier gases such as nitrogen or helium are employed, but hydrogen chloride is formed if hydrogen gas is used. The gases are introduced by a needle valve to a pressure of 0.5 to 10 Torr [3, 4].

Electrons rather than active hydrogen atoms (if used) seem to be the chemically active species in these reactions [3, 4], cf. also [6].

When $BeCl_2$ is reduced in a "thermic" hydrogen plasma jet at 400 to 6000 K and 1 to 10 bars, beryllium is obtained in the gaseous state only. A thermodynamic analysis of this reduction process is given in [7].

References:

[1] R. A. Paris, P. A. Amblard, A. C. Rousset (Ger. 2222173 [1971/72]; C.A. **78** [1973] No. 46952).
[2] P. Kobetz, R. J. Laran, Ethyl Corp. (U.S. 3492 112 [1966/70]; C.A. **72** [1970] No. 92480).
[3] F. K. McTaggart (Advan. Extr. Met. Proc. Symp., London 1967 [1968], pp. 541/50, 545, 549; C.A. **69** [1968] No. 79352).
[4] F. K. McTaggart (Australian J. Chem. **18** [1965] 937/48, 938, 940, 945).
[5] F. K. McTaggart (Nature **206** [1965] 616).
[6] F. K. McTaggart (Australian J. Chem. **18** [1965] 949/57, 956).
[7] A. P. Karpov, A. L. Suris, S. N. Shorin (Tezisy Dokl. 2nd Vses. Simp. Plazmokhim., Riga 1975, Vol. 1, pp. 178/81 from C.A. **88** [1978] No. 26116).

1.4.5 Reduction of Beryllium Compounds by Electrolysis

Review

To our knowledge, the fusion electrolysis process is no longer anywhere currently in operation due to economic reasons.

Previously, the only way to electrolytically obtain beryllium metal from its compounds on a commercial scale was fusion electrolysis.

Attempts to electrolytically obtain beryllium metal from solutions of its compounds in an aqueous milieu failed; those in liquid ammonia as well as in a series of organic solvents gave questionable results. Positive results were only reported from solutions of $BeCl_2$ and $Be(CH_3)_2$ in ethyl ether, and from solutions of $Be(ClO_4)_2$ in acetonitrile.

The literature before 1931 is reviewed in "Beryllium" 1930, pp. 41/5.

Data on the cathodic behavior of beryllium during electrolysis will be presented in a future volume; electrodeposition of beryllium in refining is described in Section 1.5.9, p. 110.

1.4.5.1 Fusion Electrolysis

All technically applicable processes of fusion electrolysis are based on the decomposition of a mixture of beryllium oxide, beryllium fluoride, beryllium oxide fluoride, or beryllium chloride with alkaline halides or alkaline earth halides. This second component is necessary to form a stable, conducting fused electrolyte, since fused beryllium halides or beryllium oxide are of a covalent nature and therefore bad conductors of electricity, and since under atmospheric pressure only beryllium fluoride shows a sufficiently wide melting and boiling point range [1 to 3].

Be metal is produced in the solid state from beryllium oxide and beryllium chloride [3, 4]. It is deposited in fused form from beryllium fluoride and from beryllium oxide fluoride (approximate formula $5\,BeF_2 \cdot 2\,BeO$) [1].

Beryllium chloride is most often used as the starting material for large scale Be production. The other compounds have only been applied in smaller scale Be production [1 to 4].

References:

[1] K. Illig, H. Fischer, W. Birett (in: G. Eger, Die Technische Elektrolyse im Schmelzfluß, 2nd Ed., Leipzig 1955, pp. 571/641, 596/618).
[2] G. Jaeger (Metall **4** [1950] 183/91, 188).
[3] G. Petzow, F. Aldinger (Ullmanns Encykl. Tech. Chem. 4th Ed. **8** [1974] 451).
[4] D. A. O'Keefe, D. E. Couch (U.S. Bur. Mines Rept. Invest. No. 7347 [1970] 1/11; C.A. **73** [1970] No. 101193).

1.4.5.1.1 From Beryllium Oxide

Conditions for a direct electrolysis of fused BeO are very unfavorable because the electrical conductivity of BeO is very low and the melting point is extremely high (2525 °C) [1].

Early attempts (before the 1930's) involving the direct electrolytic production of Be metal from BeO using solutions or suspensions of BeO in fused metal compounds failed, see [1, 2]. Because of the potential economic advantages over the existing commercial method (reduction of BeF_2 by Mg metal, see p. 56), renewed endeavours were undertaken by [3 to 5] in the years 1968 to 1972. The efforts produced a process that was seeming suitable for commercial production.

Basic Process Design

The electrowinning process for Be metal comprises two steps:

1) Electrolytic conversion of BeO to Be metal in a fused-salt electrolyte at a temperature below the melting point of Be metal.

2) Separation of the Be metal deposit from the adhering fused electrolyte. This is accomplished by coalescing the Be metal at temperatures above its melting point [3, 4].

For satisfactory electrodeposition and optimal coalescence of Be metal, the electrolyte must possess a reasonable electrical conductivity, dissolve enough BeO to meet the electrolysis demand, and be composed of salts not appreciably volatile at the process temperature. It should not react rapidly with Be metal at the coalescence temperature (1300 °C). From numerous tested fluoride mixtures, $MF+MF_2$, selected from MF (M=Li, Na, K) and MF_2 (M=Be, Mg, Ca, Sr, Ba), only BeF_2 containing electrolytes proved applicable. The largest quantity and best quality of Be metal was attained with mixtures of LiF-BeF_2 and LiF-BaF_2-BeF_2 [3 to 5].

In a method developed by [3], Be metal of ~99% purity was obtained by semicontinuous processing on a technical scale.

Electrolysis

BeO is fed into a fused salt electrolyte composed of (in mol %) 56 LiF, 37 BeF_2, and 7 BeO, and electrolyzed at 780 °C in an externally heated graphite crucible with an Mo cathode and a special carbon anode. The electrolysis is carried out at 3.3 V cell voltage with an anode current density ~16 A/dm^2 and an optimal cathode current density of 100 to 125 A/dm^2. The electrode current efficiency (in %) is about 50 for the cathode and 60 to 70 for the anode. The developing cell off-gases (CO, CO_2) serve as a protective atmosphere and are then scrubbed in an Na_2CO_3 solution. The loaded cathode is exchanged at predetermined intervals for a new unloaded one. The particle size (mainly <1 mm, maximum 3 mm) of the deposited Be powder can be controlled via cell current. Increasing cell current decreases the particle size.

Satisfactory Be powder deposits are produced at initial cathode current densities of ~75 A/dm^2. The electrolysis cell is described in [3].

Isolation of Beryllium Metal

Separation of the Be metal deposit from the adhering fused electrolyte is accomplished as follows: After completion of an electrolysis run, the loaded cathode is stripped off. The deposited Be metal is recovered as small beads or in powder form. To get metal beads, the cathode load is briefly heated to 1300 °C in order to coalesce the Be metal into beads, then cooled to 900 °C, and poured through a steel screen to recover the beads. The beads are then leached in 5% aqueous HNO_3. To produce powdered metal, the cathode load is mixed with twice its weight of an NaCl-KCl eutectic, heated to 700 °C, stirred, cooled, crushed, and leached, first with water, then with dilute nitric acid [3].

Other Processes

Previous attempts to develop an electrolytic process by which Be metal could be attained from BeO were performed on a small scale by the same group [4, 5].

In a process described by [4], Be beads of ~98.9% purity were obtained by semi-continuous or by batch operations.

The electrolysis process was carried out under an He atmosphere by feeding commercial BeO into an electrolyte[1)] composed of (in mol %) 60 LiF (reagent-grade) and 40 BeF_2 (techni-

---

[1)] The solubility of BeO in this electrolyte has been reported to be 0.09 mol BeO/kg melt at 700 °C [6], cited in [4].

cal grade). The process temperature was 700 °C, and the cell voltage was 2.6 V. For semicontinuous processing, the average current was 3.8 A, and the cathode current density was 18 A/dm^2. The anode current density was 4.5 A/dm^2. Cathode current efficiency was 74% [4]. This efficiency, however, couldn't be obtained when the cell was in daily operation [3].

For batch processing, the average current was higher (4.9 A). The cathode current density was 26 A/dm^2, and the anode current density was 3 A/dm^2. The cathode current efficiency, however, was lower (~51%) [4].

Another batch process utilized an electrolyte composed of (in g) 169 LiF, 196 BeF_2, 135 BaF_2, and 30 BeO.The process temperature was ca. 740 °C; the cell voltage was 4.8 V. The average current was 15 A, and the cathode current density was 84.39 A/dm^2 (784 A/ft^2). The anode current density was 9.042 A/dm^2 (84 A/ft^2). The Be content of the coalesced and leached metal product was 99.2%. The electrolysis cell was equipped with an Ni or Fe cathode and a graphite anode [5].

References:

[1] K. Illig, H. Fischer, W. Birett (in: G. Eger, Die technische Elektrolyse im Schmelzfluß, 2nd Ed., Leipzig 1955, pp. 571/641, 599).
[2] G. Jaeger (Metall **4** [1950] 183/91, 188).
[3] D. A. O'Keefe, D. E. Couch (U.S. Bur. Mines Rept. Invest. No. 7347 [1970] 1/11; C.A. **73** [1970] No. 101193).
[4] M. M. Wong, D. E. Couch, D. A. O'Keefe (J. Metals **21** No. 1 [1969] 43/5).
[5] M. M. Wong, United States Dept. of the Interior (U.S. 3666444 [1968/72]; C.A. **77** [1972] No. 82990).
[6] A. L. Mathews, C. F. Baes Jr., Oak Ridge National Laboratory (ORNL-TM-1129 [1965] 101/3; N.S.A. **19** [1965] No. 30155).

1.4.5.1.2 From Beryllium Fluoride and Oxide Fluoride

The first successful attempt to obtain Be metal from the electrolysis of BeF_2 was achieved by Lebeau [1] in 1898/99. This method was later studied by another group of workers [2 to 4], but has only attained laboratory scale production [5, 6]. A great step forward in the electrowinning of Be metal from BeF_2 was developed by Stock and Goldschmidt [7]. This process, developed for commercial scale production by the firm of Siemens and Halske [8] in Germany, was operated up to the 1950's [6]. Today, these [7, 8] and similar processes given by Vivian [9] and Dickinson [10] are no longer in use [11].

For the production of Be metal by electrolysis of the highly covalent BeF_2, an ionic salt must be added to produce a conducting electrolyte. The fluoride electrolysis processes may be divided into two main classes according to the electrolyte compositions. In the first one, the chief addition is NaF. The process temperature is below the melting point of Be, and solid Be metal (flake form) is deposited. In the second one, BaF_2 is added. The process temperature lies at or just above the melting point of Be, and fused Be metal is deposited [5, 6].

Electrolysis of beryllium fluoride-sodium fluoride mixtures has been developed by Lebeau [1], Fichter, Jablczynski [2], Hopkins and Meyer [4], using electrolyte compositions of 2 NaF · BeF_2 [1], NaF · BeF_2 [1, 4] or NaF · 2 BeF_2 [2] at a process temperature of about 600 °C. A nickel crucible serves as the cathode and a carbon rod as anode [1, 2, 4]. The electrolysis conditions vary from ~100 V at 6 A [4] to 15 V at 10 A [2]. Only low current efficiencies are attained and the electrolysis has to be terminated prematurely because of the anode effect [1, 2, 4]. The low efficiencies are partly the result of the formation of a stable sodium fluoride-beryllium fluoride complex from which only sodium metal is deposited [6, 8]. The cathode deposit is leached with water to free it from the fluoride salt, and then the Be metal is purified by a sink-float process using dense organic liquids [2, 4]. Be metal of 98% purity is claimed by [4]. However, a commercial method based on this electrolysis process failed because of the low yields [6].

Electrolysis of beryllium fluoride (or oxide fluoride)-barium fluoride mixtures was first carried out by Stock and Goldschmidt [7]. Later developments were added by Vivian [9] and by the firm of Siemens and Halske [8].

The Be metal is produced at temperatures from ~1350 to 1400 °C; thus simplifying separation from the electrolyte and avoiding the anode effect. Barium fluoride is the most suitable electrolyte addition [7 to 9]. BeF_2 was used as cell feed [7, 9] as well as beryllium oxide fluoride of the approximate composition $5\,BeF_2 \cdot 2\,BeO$, which forms on dehydration of the usual moisture-containing BeF_2 [8]. The success obtained with the oxide fluoride additions is due to some decomposition of BeO in the bath, but it cannot be obtained by adding BeO [6, 8, 12].

In the commercially operated Siemens and Halske process, the beryllium oxide fluoride-barium fluoride-natrium fluoride mixture is electrolyzed in a graphite crucible serving as the anode. The cathode is a water-cooled iron rod. The system is operated at 1400 °C, and as a rule at ~55 V at 600 A. The fused Be metal is deposited and freezes on the cooled cathode, which is continuously withdrawn carrying a compact rod-shaped Be deposit. The volatilized fluorides are recovered from a condensing system. The current efficiency is 75 to 80%. The purity of the Be obtained is ~99.5% [8], cf. [6].

This process was eventually abandoned because of the high costs and the great technical difficulties connected with the toxic effluent gases and the necessity of condensing and reclaiming volatilized fluorides [12, 13].

References:

[1] P. Lebeau (Ann. Chim. Phys. [7] **16** [1899] 495/503; Compt. Rend. **126** [1898] 744/6).
[2] F. Fichter, K. Jablczynski (Ber. Deut. Chem. Ges. **46** [1913] 1604/11).
[3] F. Fichter, E. Brunner (Z. Anorg. Allgem. Chem. **93** [1915] 84/94).
[4] B. S. Hopkins, A. W .Meyer (Trans. Am. Electrochem. Soc. **45** [1924] 475/80).
[5] G. Jaeger (Metall **4** [1950] 183/91, 188).
[6] K. Illig, H. Fischer, W. Birett (in: G. Eger, Die Technische Elektrolyse im Schmelzfluß, 2nd Ed., Leipzig 1955, pp. 571/641, 605/18, 612, 616).
[7] A. Stock, H. Goldschmidt (Ger. 375824 [1921/24]).
[8] K. Illig, M. Hosenfeld, H. Fischer (Wiss. Veröffentl. Siemens-Konzern **8** No. 1 [1929] 42/58, 44, 47, 53, 55).
[9] A. C. Vivian (Trans. Faraday Soc. **22** [1926] 211/25).
[10] J. Dickinson (U.S. 1511829 [1921]).
[11] G. Petzow, F. Aldinger (Ullmanns Encykl. Tech. Chem. 4th Ed. **8** [1974] 451).
[12] G. E. Darwin, J. H. Buddery (Beryllium, Butterworth, London 1960, pp. 92/3).
[13] C. W. Schwenzfeier (Kirk-Othmer Encycl. Chem. Technol. 2nd Ed. **3** [1964] 459).

1.4.5.1.3 From Beryllium Chloride

The fusion electrolysis of $BeCl_2$ was one of the earliest attempts at producing Be metal and has since developed into a significant process of commercial applicability. Although the Be metal is obtained in high purity, the process is less economical than the thermal reduction of BeF_2 with metallic Mg [1, 2, 38, 39].

1.4.5.1.3.1 Basic Process Design

The $BeCl_2$ electrolysis process is normally carried out in two major steps:

(1) Conversion of BeO to $BeCl_2$ according to the reaction

$$BeO + C + Cl_2 \rightarrow BeCl_2 + CO.$$

(2) Conversion of $BeCl_2$ to Be metal by electrolysis according to the reaction

$$BeCl_2 \rightarrow Be + Cl_2 \quad [4, 12].$$

The raw material for this process is usually BeO and sometimes $Be(OH)_2$, either dry or as a moist cake, or $BeSO_4 \cdot 4\,H_2O$. One can also start directly from the beryl ore [4].

The gaseous chlorine and carbonaceous material in step (1) can be substituted by carbon tetrachloride according to the reaction $2\,BeO + CCl_4 \rightarrow 2\,BeCl_2 + CO_2$ [9].

At the applied electrolysis temperatures (see below), no melting and essentially no oxidation of the Be metal occurs [2]. Since $BeCl_2$ is extremely hygroscopic and hydrolyzable, moisture and air must be carefully excluded in all operations [1].

Alkaline chlorides or alkaline earth chlorides are added to the electrolyte because $BeCl_2$ is a nonconductor of electricity. The electrolysis temperature is dependent on the electrolyte composition. Improvements in the electrolyte compositions and the corresponding lowering of the electrolysis temperature improve the industrial practicability of the process [1, 5, 6].

The process was first developed between 1894 and 1896 by Borchers [7] using an electrolysis temperature just reaching the melting point (~1280 °C) of beryllium (see "Beryllium" 1930, p. 42). Other processes were subsequently developed using temperatures in the range of ~730 to 930 °C. These are collectively known as "high temperature" electrolysis processes, see, e.g. [4, 8]. So-called "low temperature" electrolysis processes have also been developed using temperatures in the range of ~320 to 400 °C [4, 9].

The resulting beryllium metal is cathode deposited as solid flakes, which are recovered from the adhering electrolyte by leaching [4]. Another method separates the electrolyte from the metal by vacuum distillation [9].

The "High Temperature" Electrolysis Process (730 to 930 °C)

Early attempts to produce Be metal at "high temperatures" were carried out between 1924 and 1930 by Cooper [10]. This process has been applied on a large scale (batch process) by the Beryllium Corporation, New York. The $BeCl_2$ starting material is produced by passing chlorine over a mixture of BeO and carbon at about 1000 °C. The $BeCl_2$ sublimate is collected in a condenser maintained below 400 °C [8, 10]. The electrolysis process is carried out under inert gas in a tightly covered, corrosion resistant pot serving as the cathode and equipped with a graphite anode. The electrolyte consists of (in parts by weight) 1 $BeCl_2$ and 5 NaCl. It is electrolyzed at a temperature of ~730 °C, which rises to ~820 °C at the end of the Be metal deposition. The initial current is 600 A, and the cell voltage is 5 to 8 V. The deposited Be flakes are leached with ice water until free from soluble material, then washed with anhydrous alcohol, and air-dried at low temperatures. The Be metal obtained is said to be free from oxides and other impurities [5, 10, 11]. The chlorine formed is reused for $BeCl_2$ production [10, 12].

In the 1950's another "high temperature" electrolysis process was carried out on an industrial scale (batch process) at Clifton Products, Inc., USA. For the cross section of the electrolysis building and the assembly of the electrolysis cell with its specific ventilation and hooding see in the paper. The electrolysis cell for this process consists of a tightly covered heavy alloy steel crucible serving as cathode and a graphite anode. The electrolyte composition (in %) is 14 $BeCl_2$ ($BeCl_2$ production not described), 43 NaCl, and 43 KCl. The fusion temperature of the electrolyte, 760 to 790 °C, increases during electrolysis to a final temperature of 900 to 925 °C at completion of the Be deposition. Electrolysis is carried out to a point just short of completion at ~6 V, with an anode current density not exceeding 1.008 A/cm^2 (6.5 A/in^2). Upon completion, the cell content is poured and scraped into another container where it is permitted to settle and solidify. The Be flake produced separates to the lower third of the solidified mass and is freed by leaching in cold water. The upper two-thirds of the solidified mass is crushed and used as cell feed. The leached flake is divided into two size grades by screening; the first screen is 60 mesh, and the second is 150 mesh. Only the coarse fraction is further refined to pure beryllium metal. This fraction is first washed in dilute nitric acid ($\leqq$8%) at ~70 °C for 3 h to remove surface oxides, floated in a bromoform-carbon tetrachloride mixture having a specific gravity of 1.94 to 1.96, washed in methanol, dried, and packaged [4].

No details concerning the purity of the thus obtained Be metal are available, but in view of the density of the medium used for the sink-float purification, it would seem that the Be metal is comparatively impure. This Be production process is no longer used. Its greatest deficiencies are cell corrosion, the high vapor pressure of $BeCl_2$ and consequent toxicity of $BeCl_2$ at these temperatures, and difficulties in separating the Be metal [13].

The "Low Temperature" Electrolysis Process (320 to 400 °C)

The principal advantage of this process is that at the applied low temperatures rapid surface oxidation of the Be metal is avoided [4]. Furthermore, a minimum amount of $BeCl_2$ is lost by vaporization, and no cell corrosion occurs [15].

The low temperature electrolysis process has been used for commercial Be production in Germany, the USA, France, Great Britain, Japan, and the USSR [2 to 4].

One of the first low temperature electrolysis processes was developed by the German firm Degussa and was operated on a commercial scale from 1938 to 1944 at Frankfurt/Main and later, in part, at Rheinfelden [3, 5, 12, 14, 16, 17]. After 1945, the Degussa plant was removed by the French, reassembled at Praz near Paris, and operated commercially for a number of years by the French firm Pechiney Company [13, 17, 25]. After 1958, the Pechiney production was transferred to the Calypso Plant in the French Alps [24]. Slightly differing modifications of the same process have been commercially operated in Great Britain since ~1957 by Murex Ltd. at Milford Haven [26] and even earlier in the USSR [9, 27 to 29]. In the USA, a process almost identical with the Degussa process has been commercially used since 1953 [4]. From the 1950's until well into the 1970's the low temperature process has been carried out in France by Trefimetaux G.P., a joint subsidiary company of Pechiney Company, France, and Kawecki Berylco Ind. Inc., USA [2], in Japan by NGK Insulators [2, 30], and in the USSR [2, 9, 27]. At present, however, none of these low temperature electrolysis processes is employed commercially [40, 41].

1.4.5.1.3.2 The "Standard Process" and Modifications

The first commercial low temperature electrolysis process to be described in detail was that of Degussa. Because of the many subsequent modifications by others it may be called the "Standard Process" for this technology.

The process of Degussa in Germany consists of five stages. Three are concerned with the preparation of the starting material $BeCl_2$ for electrolysis: 1) Production of Be oxide-carbon briquettes. — 2) Chlorination of the briquettes. — 3) Purification of the crude $BeCl_2$. — 4) Electrolysis. — 5) Separation of Be metal flakes from the melt [16, 17].

A flow sheet for the process as operated by Degussa is given in **Fig. 1-9**.

Production of Be Oxide-Carbon Briquettes. Technical Be oxide (obtained from a dry product composed of $Be(OH)_2$ containing 50% BeO by calcination at 1000 °C) is mixed with charcoal, wood tar, and water in the weight ratio (in kg) of 100[1]:60:5:51, respectively. The mixture is extruded into cylindrical briquettes, 1.75 cm in diameter and 15 cm long. These are packed in crucibles with charcoal and fired at 1000 °C whereby the wood-tar partly distills and cokes. A strong, porous, gray briquette is formed [12, 16 to 18].

Fig. 1-9

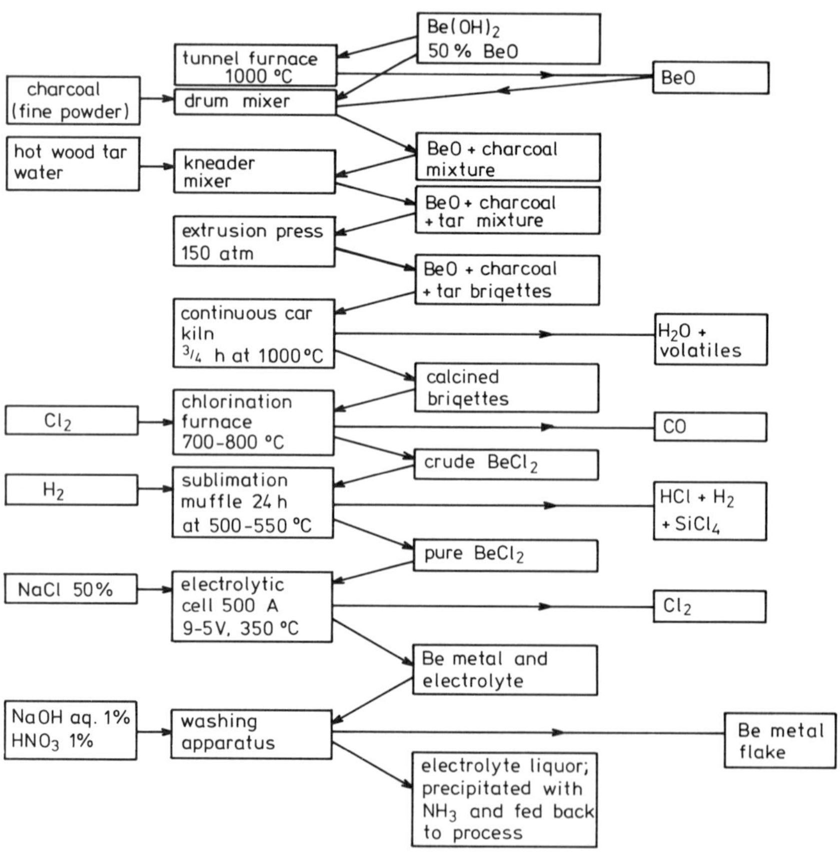

Flow sheet for the production of Be metal from BeO at Degussa [17], cf. [16].

[1] This weight refers to $Be(OH)_2$ with 50% BeO.

Chlorination of the Briquettes. Chlorination is carried out in a heavily insulated furnace consisting of a steel casing lined with several layers of acid-proof bricks. The porous, calcined briquettes are placed between the electrodes and are heated by passing a current through them. Chlorine gas introduction (at the bottom of the furnace) begins at 700 to 800 °C. The temperature is raised to 1000 °C. The volatilized $BeCl_2$ exits through an opening in the top and is collected in an Ni condenser box outside the furnace [12, 16 to 18]. The overall yield of crude $BeCl_2$ from $Be(OH)_2$ is 75 to 78% [13].

Purification of the Crude $BeCl_2$. The crude $BeCl_2$ obtained from the chlorinating furnace is a fluffy material of low density and contains numerous impurities, mainly $SiCl_4$, $AlCl_3$, $FeCl_3$, plus BeO and C, carried over from the chlorinator. It is purified by distillation in a steady H_2 stream in a sublimation muffle. The first condenser is held at 300 to 350 °C and the second at 150 °C. The crude $BeCl_2$ is heated in the hydrogen stream to 500 to 550 °C. Pure $BeCl_2$ condenses mainly in the first condenser as very fine crystals. $FeCl_2$ from the reduction of $FeCl_3$ by hydrogen remains as a nonvolatile substance in the muffle. $AlCl_3$ condenses in the second condenser; $SiCl_4$ and HCl pass out from the condenser system [12, 16, 17].

Electrolysis. The pure $BeCl_2$ is best mixed with an equal weight of NaCl (melting point of the mixture 224 °C [19]) and electrolyzed in an externally heated Ni crucible at 350 °C. The crucible serves as the cathode; a graphite rod centrally located forms the anode (see **Fig. 1-10**, cell A). The electrolysis is carried out at 500 A. The cell voltage drops from 9 to 5 V. The anode current density is 0.775 A/cm² (5 A/in²); the cathode current density is about 0.062 A/cm² (0.4 A/in²), and the current efficiency is about 50%. The developing chlo-

Fig. 1-10

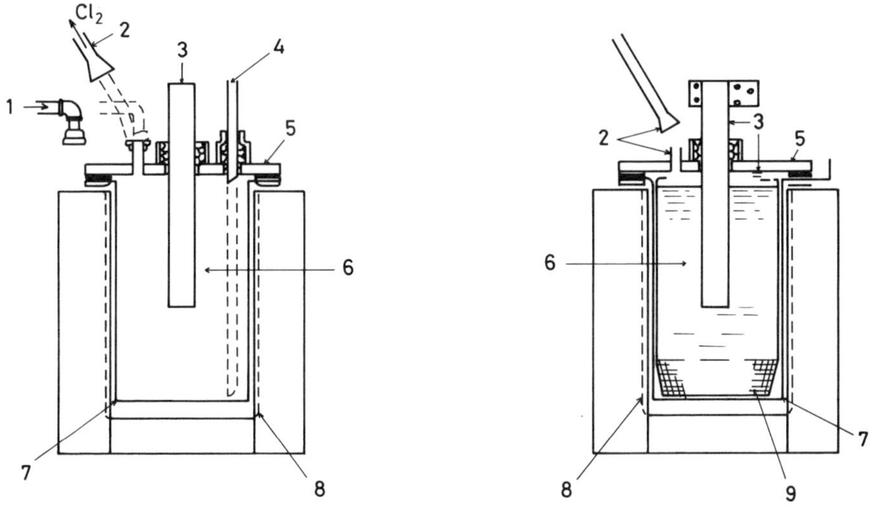

"Low temperature" electrolysis cells A and B for the production of beryllium [4]. 1) Gas inlet; 2) vent; 3) anode; 4) siphon; 5) cover; 6) electrolyte; 7) container cathode; 8) wound resistance heater; 9) perforated insert.

rine serves as a protective atmosphere and is then reused for $BeCl_2$ production (see p. 83). Electrolysis is continued until the $BeCl_2$ is reduced to about 45% of the electrolyte (about 24 h). The electrolyte is then transferred at 500 °C by siphoning to a preheated second cell of the same type. The cell is replenished with fresh $BeCl_2$ up to 50%, and the electrolysis is continued in the new cell while the Be metal flake is collected from the first cell. Interruption periods between two electrolysis runs are very short (~1 min), thus allowing quasi-continuous processing [4, 12, 15 to 17].

The electrolysis cell A is shown in Fig. 1-10, p. 83. The Ni crucible cathode is a flanged cylindrical flat bottom vessel 50 cm in diameter and 80 cm deep. Its cell cover rests on an asbestos insulating gasket. The central anode, insulated from the cover by tightly packed asbestos, is a 10 cm diameter rod made of hard-pressed graphite. Its overall length is 120 cm with the end portion standing about 40 cm from the bottom of the crucible. Ten cells are usually provided in the plant. They are used alternately in groups of five for electrolysis-metal collection [4, 12, 15 to 17].

Separation of Be Flakes from the Fused Electrolyte. After completion of an electrolysis run and removal of the electrolyte by siphoning, the cell is opened, and the Be metal deposit (glittering dendritic flakes [14] along the crucible walls) is ladled out while still hot (300 to 350 °C) with a perforated ladle. It is pressed on the ladle with a small hand-operated, gas-heated press in order to remove the major part of the residual adhering salts. After the metal has reached room temperature, it is plunged into a tank of ice water to absorb the heat of solution of the adhering $BeCl_2$. The resulting $BeCl_2$ containing electrolyte liquor is treated with ammonia and fed back into the process. The last traces of $BeCl_2$ are finally removed by washing with water on screens arranged in a cascade. To remove traces of aluminium oxide and beryllium oxide, the beryllium flakes are washed successively in stoneware vessels with 1% NaOH (solution) and, after a thorough water rinsing, with 1% HNO_3 (solution). After rinsing a second time with water, the flakes are placed in a centrifuge equipped with a bronze screen basket, submitted to a final water rinsing, and then quickly dried. The purity of the flakes at this stage is 99% or more. The major impurities (in %) are: HCl insoluble 0.1 to 0.6, Fe 0.06 to 0.24, Al 0.1 to 0.4 [4, 12, 16, 17].

1.4.5.1.3.3 Various Plant Practices

The low temperature process operated by Pechiney Company in France on a commercial scale is similar to that used earlier by Degussa in Germany. No details of the French technique have been released. The purity of the Be flake is 99.5%, and the main impurities are chlorine and oxygen. Except for the high chlorine content (0.14 wt%), the French Be flake is claimed to be purer than the Be beads made by thermal reduction of BeF_2 with Mg (cf. p. 56) [3, 20 to 24].

To our knowledge, no technical details have been reported on the low temperature process commercially operated by Trefimetaux G.P. in France.

In 1953, an improved commercial scale low temperature process almost identical with the original Degussa process was begun in the USA. The purity of the Be flake produced was 99.48% [4].

A similar earlier process developed by the Brush Beryllium Company in the USA uses a fused salt mixture (in %) of 50 $BeCl_2$, 26.6 KCl, 20.2 LiCl, 3.2 NaCl as the electrolyte and operates at electrolysis temperatures of 260 to 300 °C [6]. The process has not been used in industrial practice [25].

The low temperature process carried out by Murex Ltd. in Great Britain differs only in detail from the original Degussa process. A mixture (in %) of 44.1 BeO, 27.6 carbon black (type P33), and 28.3 coal tar is used for the production of BeO/carbon briquettes. The chlorinator is heated by an external resistance bed instead of using the resistance of the charge. Its temperature is only 850 to 900 °C. In the purification stage of crude $BeCl_2$, particular attention is paid to the production of high density $BeCl_2$. The high density material is easier to handle because it is less hygroscopic than the low density material. Due to its narrow liquid range (405 to 488 °C) and the diluent effects of other gases, most of the crude $BeCl_2$ obtained from the chlorinator condenses directly from the vapor to the solid state producing a very hygroscopic fluffy material. In this purification stage, however, a small quantity of hydrogen is the only diluent and the gas → liquid → solid condensation is more readily achieved resulting in a denser material. A muffle furnace fitted with two condensers, a multiple unit condenser with 13 small cylinders (for pure $BeCl_2$), and a simple secondary condenser (for impurities), is used for the purification operation. After the reduction of $FeCl_3$ with hydrogen at ~275 °C, the temperature is raised to 550 °C. The $BeCl_2$ melts, vaporizes, and passes over to the first condenser, the jacket of which is heated at about 600 °C, while the cylinders are maintained at 300 to 400 °C by blowing air through them. At the end of the operation, the furnace is cooled to 105 °C and the purified $BeCl_2$ adhering to the cylinders in dense layers is removed. A separate cylindrical Ni cathode is used in the electrolysis cell. Instead of the Degussa electrolyte transfer by siphoning, the open-ended cylindrical cathode is withdrawn from the cell, a new one inserted, and the depleted bath replenished [26]. A sketch of an electrolysis cell similar to that used here is shown under "cell B" in Fig. 1-10, p. 83 [4]. The cathode is cooled in a sealed receiving tank and plunged into ice water to recover the beryllium flake [26].

1.4.5.1.3.4 Further Variants of the "Standard Process"

A low temperature process differing in some details from the original Degussa process has been used in the USSR. The chlorination process of the beryllium oxide/carbon briquettes with chlorine gas is carried out at 550 to 650 °C in continuous flow furnaces lined with fused quartz bricks. The crude $BeCl_2$ is distilled and collected at 250 to 280 °C in Ni condensers from which it is intermittently drawn off. The electrolysis is carried out as a two-step, semicontinuous process (very short interruption periods for cathode replacement and electrolyte replenishing) with removable Ni cathodes. In the first step, the $BeCl_2$ is purified by low voltage (~1.5 V) deposition of metals more positive than beryllium (pre-electrolysis step). In the second step the pre-electrolysis cathode is replaced, and high-purity (~99.85%) beryllium is deposited at higher voltages from the purified $BeCl_2$. The electrolysis is performed at 320 to 350 °C. The temperature of the fused electrolyte is raised to 370 to 380 °C before the cathode is replaced. A variation of the Degussa method has been developed to separate the Be flake from the bulk of the adhering fused electrolyte. The adhering electrolyte is directly distilled in a vacuum furnace at 650 to 700 °C. The distilled chlorides and the fused electrolyte are collected in a cooled Ni container from which they are returned to the electrolysis cell. This method simplifies the treatment of wastes and considerably increases the direct yield of beryllium [9, 27 to 29].

This low temperature process is carried out in the USSR on an industrial scale (ton amounts) with a modification to produce highly pure $BeCl_2$. Pure BeO is briquetted with a solution of sizing or dextrin. The briquettes are calcined at 600 °C and then chlorinated with technical carbon tetrachloride at 650 to 700 °C in a quartz-lined furnace. The bulk of the $BeCl_2$ is condensed at 300 to 320 °C [9].

The pre-electrolysis is effected with a cathode current density of 2.4 to 2.8 A/dm^2 (2 h). The electrolysis is carried out at 320 to 340 °C (20 to 22 h) with a cathode current density of 6.5 to 7.5 A/dm^2. The current yield during electrolysis is >80%, and the direct extraction of Be metal within a cycle time of 90 d is 66.7% [9].

The separation of the Be flake from the adhering electrolyte is carried out either by washing off the electrolyte with water (Degussa process) or by removing it by vacuum distillation at 650 to 700 °C [9].

High purity Be flake is continuously produced from crude $BeCl_2$ in a modified low temperature process that is commercially practiced at NGK Insulators in Japan. The process is less expensive than conventional quasi- or semicontinuous processes. The purity of the Be flake equals or is superior to that obtained from distilled and refined $BeCl_2$ [30].

The electrolysis is carried out in several stages with Ni cathodes, each with an axial graphite anode. The resulting Be flake is washed with water, alcohol, or acetone and dried [30], cf. also [44].

Another modification of the standard low temperature process was developed from 1952 to 1958 at Sylvania Electric Products, Inc. in the USA. The continuously operating process involves electrolytic deposition of Be metal from the fused $BeCl_2$/NaCl electrolyte into a mercury cathode. A quasi-amalgam is formed which can be removed without interrupting the electrolysis. The fine particle size of the Be metal in the quasi-amalgam permits direct hot pressing or powder preparation. The process is described for a relatively small scale operation, but is very promising for commercial development [31, 33, 35].

$BeCl_2$ is produced by chlorinating a briquetted mixture of BeO with ~15% excess acetylene black and water with chlorine gas at about 1000 °C. The $BeCl_2$ is condensed at about 300 °C in a dilute chlorine atmosphere, which provides an effective separation from iron impurities [31].

The electrolysis is carried out with a fused $BeCl_2$-NaCl mixture (mole ratio 60:40 or 50:50) at 300 to 350 °C under a purified Ar atmosphere with an Hg pool cathode and a special carbon anode. The resulting Be quasi-amalgam (mushy or pasty semisolid consistency) is continuously circulated to an exchangeable separation unit in the mercury circulation system where the excess Hg is separated from the quasi-amalgam by a screen (200 mesh) held in a metal container. Two disadvantages of the process are that the Be content of even the concentrated amalgam is low (2 wt%), and that both the amalgam and the fine Be powder are very susceptible to oxidation [31 to 34].

This process was later studied by Kawecki Berylco Ind., USA, using an Hg cathode and a Be anode. Using this method, ultrafine Be particles can be obtained [36].

Another modification of the standard low temperature process, developed by Sheer and Korman of Light Metals Refining in the USA, is claimed to produce the starting material, $BeCl_2$, in a low cost process by the direct chlorination of the raw ore in a high intensity arc. In the Sheer-Korman process, the ore (e.g. beryl) is mixed with ~30% soft coal and formed into anodes which are continuously consumed by the arc [37, 42]. The cathode may be of solid carbon or neutral, soft-cored carbon. The temperature in the high intensity arc is estimated to be between 8300 and 11000 °C. Chlorine gas is fed into the furnace to convert the metallic constituents of the ore to chlorides, which are condensed as pure chlorides in a succession of cooling chambers. Electrolysis of the obtained $BeCl_2$ gives Be that is said to be nearly pure and nonbrittle [37, 42], cf. also [43].

References:

[1] W. W. Beaver (Progr. Nucl. Energy V **1** [1956] 277/97, 285; Met. Abstr. **24** [1956/57] 963).

[2] G. Petzow, F. Aldinger (Ullmanns Encykl. Tech. Chem. 4th Ed. **8** [1974] 451).

[3] W. D. Jamrack (Rare Metal Extraction by Chemical Engineering Techniques, Vol. 2, Pergamon, New York 1963, pp. 280/1; N.S.A. **18** [1964] No. 13952).
[4] C. E. Windecker (in: D. W. White Jr., J. E. Burke, The Metal Beryllium, ASM, Cleveland 1955, pp. 102/24, 107/9, 111/5, 118, 119/23).
[5] K. Illig, H. Fischer, W. Birett (in: G. Eger, Die Technische Elektrolyse im Schmelzfluß, 2nd Ed., Leipzig 1955, pp. 600/4).
[6] B. R. F. Kjellgren, C. B. Sawyer, Brush Beryllium Company (U.S. 2188904 [1936/40]; Can. 371194 [1936/38]).
[7] W. Borchers (Z. Elektrochem. **1** [1894/95] 362/420, **2** [1895/96] 39).
[8] W. Hessenbruch (Metall Erz **32** [1935] 234/7).
[9] I. E. Vil'komiskii, G. F. Silina, A. S. Berengard, V. N. Semakin (At. Energiya SSSR **11** [1961] 233/9; Soviet At. Energy **1961** 882/8, 887; C.A. **56** [1962] 5702).
[10] H. S. Cooper, Beryllium Corporation of America (U.S. 1775589 [1924/30]).

[11] H. S. Cooper, Beryllium Development Corp. (U.S. 1805567 [1924/31]).
[12] G. Jaeger (Metall **4** [1950] 183/91, 187/9).
[13] G. E. Darwin, J. H. Buddery (Beryllium, Butterworth, London 1960, pp. 80, 83/4, 86/7).
[14] G. Jaeger (Z. Metallk. **41** [1950] 243/6).
[15] G. Jaeger, Deutsche Gold- und Silber-Scheideanstalt vorm. Roessler, Frankfurt a.M. (Ger. 646088 [1935/37]; U.S. 2041131 [1933/36]; Fr. 785072 [1935]).
[16] R. Potvin, G. S. Farnham (Can. Mining Met. Bull. **39** [1946] 525/38, 529/35).
[17] H. W. West, W. F. Randall, G. L. Miller, R. Turner, E. M. Foster, D. W. Crossley (BIOS-FR-550 [1945] 1/81, 9/10, 12/3, 63, 80).
[18] W. B. C. Perrycoste, M. A. P. (British Intelligence Objectives Subcommittee BIOS Target No. 21/8, cited in [4, p. 107]).
[19] J. M. Schmidt (Bull. Soc. Chim. France **39** [1926] 1686/703, 1691).
[20] R. Paris (Chimia [Zürich] **15** [1961] 443/9).

[21] M. P. Vachet (Chim. Ind. [Paris] **81** [1959] 64/77, 67).
[22] J. Williams (Progr. Nucl. Energy V **1** [1956] 300/4).
[23] L. R. Williams, P. B. Eyre (Nucl. Eng. **3** [1958] 9/18).
[24] F. W. Starratt (J. Metals **13** [1961] 139).
[25] B. R. F. Kjellgren (in: C. Hampel, Rare Metals Handbook, 2nd Ed., New York 1961, pp. 32/57, 38).
[26] P. S. Bryant (Extr. Refining Rarer Metals Proc. Symp., London 1956 [1957], pp. 310/22, 317/21).
[27] G. A. Meyerson (Proc. 1st Intern. Conf. Peaceful Uses At. Energy, Geneva 1955, Vol. 8, pp. 587/9).
[28] G. A. Meerson, D. D. Sokolov, N. F. Mironov, N. M. Bogorad, I. D. Pokhomov, D. S. Lvovsky, E. S. Ivanov, V. M. Shmelev (At. Energiya SSSR **5** [1958] 624/30; Soviet At. Energy **5** [1958] 1555/62, 1556; C.A. **1961** 7218).
[29] C. A. Zapffe (Nucleonics **11** No. 11 [1953] 84/8).
[30] T. Nakamura, NGK Insulators, Ltd. (U.S. 3278402 [1961/66]; C.A. **66** [1967] No. 7911).

[31] M. C. Kells, R. B. Holden, C. Whitman (SEP-207 [1956] 1/43, 6/25, 28/34; C.A. **1957** 4174).
[32] M. C. Kells, R. B. Holden, C. I. Whitman (J. Am. Chem. Soc. **79** [1957] 3925).
[33] R. B. Holden, M. C. Kells, C. I. Whitman (Proc. 2nd Intern. Conf. Peaceful Uses At. Energy, Geneva 1958, Vol. 4, pp. 306/8; C.A. **1960** 7375).
[34] Sylvania Electric Products, Inc. (SEP-107 [1952/57] 1/15, 14; N. S. A. **11** [1957] No. 12499).
[35] T. T. Magel (in: D. W. White Jr., J. E. Burke, The Metal Beryllium, ASM, Cleveland 1955, pp. 124/35, 132/3).

[36] W. G. Lidman, V. Griffiths, Kawecki Berylco Ind., Inc., Reading, Pa. (N-72-27577 [1972] 1/14; C.A. **80** [1974] No. 77459).
[37] Anonymous (Chem. Eng. News **32** No. 24 [1954] 2382).
[38] J. Ballance, A. J. Stonehouse, R. Sweeney, K. Walsh (Kirk-Othmer Encycl. Chem. Technol. 3rd Ed. **3** [1978] 803/23, 809).
[39] D. A. Everest (Top. Inorg. Gen. Chem. **1** [1964] 1/158, 115).
[40] K. A. Walsh (in: D. R. Floyd, J. N. Lowe, Beryllium Science and Technology, Vol. 2, Plenum, New York-London 1979, pp. 1/10, 9).

[41] F. Aldinger, W. C. Heraeus GmbH, Hanau (private information 1984).
[42] G. H. Cleaver (Eng. Mining J. **155** No. 7 [1954] 98/9).
[43] G. F. Silina, J. I. Sarembo, G. J. Kaplan (Tsvetn. Metal. **30** No. 1 [1957] 66/71; C.A. **1958** 5501).
[44] T. Nakamura, N. G. K. Insulators, Ltd. (U.S. 3296107 [1962/67]; C.A. **66** [1967] No. 61309).

1.4.5.1.4 From Beryllium Sulfide

According to [1], Be metal can be obtained by electrolyzing anhydrous beryllium sulfide (for its preparation, cf. p. 74) dissolved in a fused normal anhydrous fluoride or a mixture of anhydrous fluorides whose melting point is substantially below 1900 °C (decomposition temperature of beryllium sulfide), e.g., in fused BeF_2, cryolite, BaF_2, or a mixture of BeF_2 and BaF_2. The electrolysis should be carried out in an atmosphere of a reducing and (or) inert gas at ordinary or reduced pressure and under exclusion of moisture, nitrogen, and oxygen or oxygen-containing gases, e.g., in hydrogen or argon. The beryllium sulfide is wholly decomposed by the electrolysis process and pure Be metal is obtained. The electrode material may be carbon. The reaction vessel is preferably lined with carborundum or a stable carbide such as zirconium or tungsten carbide. It must not be lined with any substance liable to evolve oxygen or oxygen-containing gases at the electrolysis temperature [1], cf. [2].

References:

[1] D. Gardner (Brit. 482468 [1936/38]; C.A. **1938** 7009).
[2] G. Jaeger (Metall **4** [1950] 183/91, 188).

1.4.5.2 Electrolysis in Solutions with Various Solvents

In the electrolysis of Be compounds in aqueous solutions no deposition of Be metal occurs because $Be(OH)_2$ forms at the cathodic discharge of the hydrated Be ion. No Be amalgam, but only $Be(OH)_2$ is obtained when Hg is used as the cathode [1, 2].

The electrodeposition of Be metal from solutions of its compounds in organic solvents was studied by [3 to 7, 16]. Several workers have claimed successful Be deposition from solutions of Be compounds; for example the sulfate in acetamide [4], the chloride, nitrate, sulfate, acetylacetonate, and others in aniline, pyrrole, pyridine, and piperidine [5]. More recent work, however, indicates that many of these early findings cannot be confirmed [6, 7], cf. [14, 15]. The deposition of Be from $Be(ClO_4)_2$ solutions in acetonitrile with tetraethylammonium perchlorate as the supporting electrolyte was reported by [16]; some hydrogen evolution also occurred. Further, metallic beryllium is deposited from solutions of beryllium halides or beryllium dialkyl compounds in ethers. The best results are obtained by electrolyzing mixtures of beryllium chloride and dimethyl beryllium in diethyl ether at 20 °C and by electrolyzing beryllium chloride etherate in diethyl ether at 150 °C [6, 7], cf. [15].

The electrodeposition of Be metal from solutions of its compounds in liquid ammonia has been claimed by several authors [8 to 11]. A more recent study, however, did not confirm these claims [7, 12], cf. [1, 6, 14, 15].

The electrodeposition of traces of Be metal from $BeSO_4$ dissolved in phosphorus oxychloride was reported by [4].

No Be metal deposition occurs on electrolyzing Be compounds (nitrate, halides, sulfate) in fused ethyl pyridinium bromide at ~135 °C [6, 13], cf. [14].

All studies in nonaqueous solvents necessitate rigorously anhydrous and oxygen-free conditions, see e.g. [5, 6, 12, 14].

References:

[1] G. Jaeger (Metall **4** [1950] 183/91, 188).
[2] K. Illig, H. Fischer, W. Birett (in: G. Eger, Die Technische Elektrolyse im Schmelzfluß, 2nd Ed., Leipzig 1955, pp. 571/641, 596/618).
[3] J. M. Tien (Trans. Electrochem. Soc. **89** [1946] 237/45, 238/9).
[4] T. P. Dirkse, H. T. Briscoe (Metal Ind. **36** [1938] 284/5).
[5] H. S. Booth, G. G. Torrey (J. Phys. Chem. **35** [1931] 2465/77, 2492/7).
[6] G. B. Wood, A .Brenner (J. Electrochem. Soc. **104** [1957] 29/37, 30/1).
[7] I. A. Menzies, D. L. Hill, L. W. Owen (Nature **183** [1959] 816/7).
[8] H. S. Booth, G. G. Torrey (J. Phys. Chem. **35** [1931] 3111/20, 3115/9).
[9] H. S. Booth, G. G. Torrey (J. Am. Chem. Soc. **52** [1930] 2581/2).
[10] H. S. Booth, M. Merlub-Sobel (J. Phys. Chem. **35** [1931] 3303/21, 3314).
[11] H. S. Booth, G. G. Torrey, M. Merlub-Sobel (U.S. 1893221 [1933]).
[12] I. A. Menzies, D. L. Hill, L. W. Owen (J. Less-Common Metals **1** [1959] 321/30, 325, 328).
[13] F. H. Hurley, T. P. Wier Jr. (J. Electrochem. Soc. **98** [1951] 203/6).
[14] A. Brenner (J. Electrochem. Soc. **103** [1956] 652/6).
[15] A. Brenner (Advan. Electrochem. Electrochem. Eng. **5** [1967] 205/48, 214/5).
[16] J. F. Coetzee (Diss. Univ. Minnesota 1956, pp. 1/141, 95/6, 107/9; Diss. Abstr. **16** [1956] 1071).

1.4.6 Thermal Decomposition of Beryllium Compounds

All thermal decomposition methods, referred to in the following paragraphs have been studied only in the laboratory and have not attained technical importance.

1.4.6.1 Beryllium Iodide

A method, first described in 1925/26 by van Arkel and his group [1, 2], for the preparation of certain metals (e.g., Ti, Zr, Hf, Th [3]) directly, in a pure form, by thermal decomposition of their iodides was tried for Be metal preparation [3, 4], cf., e.g. [5, 6, 9, 14, 15].

In this method, the thermal decomposition is usually carried out at the surface of a hot wire in order to obtain compact metal without the risk of contamination during melting. The method is based on the fact that the equilibrium in the reaction $M + nI_2 \rightleftharpoons MI_{2n}$ lies more to the left at high temperatures than at low temperatures. A necessary condition for success is that the metal forms a relatively volatile iodide with sufficient thermal instability to decompose at low pressures and at temperatures below the melting point of the metal [1 to 5, 10, 15].

The process is carried out in a sealed, evacuated vessel equipped with a heatable metal wire. The metal iodide is volatilized at the appropriate temperature and pressure to deposit the pure metal on the hot wire. The deposited metal is free from nonvolatile impurities and impurities which do not form volatile iodides under the operating conditions, e.g. carbon, oxides, nitrides, etc. To attain satisfactory deposition rates, the vapor pressure of the volatile metal iodide at the process temperature should be at least 10^{-2} Torr, preferably 10^{-1} or even 1 Torr [1 to 5, 15].

The preparation of Be metal directly in a pure state by thermal decomposition of the iodide has been studied by several workers [5 to 7, 9, 12 to 14].

The Sloman process [5] uses an evacuated Pyrex vessel in which a thin (~0.1 mm) tungsten wire is maintained at about 700 to 900 °C. The apparatus is a copy of that used by van Arkel (see [2]). The beryllium iodide is prepared in situ by reacting iodine with powdered or lumps of crude Be metal at about 350 to 450 °C. The BeI_2 readily sublimes at these temperatures (melting point of BeI_2 480 °C [11]). The Pyrex vessel becomes opaque within a short time, and a coarsely crystalline, brittle deposit forms. In addition to beryllium, the deposit contains appreciable quantities of silicon and boron, arising from beryllium iodide attack on the glass. Attempts to use other glasses with different compositions for the vessel or a tungsten lining were unsuccessful [5], cf. also [10, 15].

Studies of the thermal decomposition of beryllium iodide between 576 and 1200 °C in evacuated Pt containers carried out by Kopelman, Bender [6] indicate that beryllium iodide also reacts with Pt to form a Be-Pt alloy. No practical solution of this corrosion problem was proposed. Thermodynamic calculations based on equilibrium measurements of [6] show that the equilibrium vapor pressure of iodine in the reaction $BeI_2 \rightleftharpoons Be + I_2$ is very low at temperatures up to 1200 °C; this means that the fraction of BeI_2 decomposed is very small. Thus, kinetic rather than thermodynamic factors control the decomposition [6], cf. [7].

Kellog [9] proposed that decomposition of the beryllium iodide vapor in an arc at temperatures above the melting point of Be metal would establish a more favorable equilibrium [4, 9].

Further attempts by Johnson [12] and Magel [13] to prepare Be metal using the "hot-wire" method are thought to have failed due to reactions with the vessel material [8].

The thermal decomposition of beryllium iodide on a tungsten wire in an evacuated glass ampule deposits a silicon precipitate on the wire containing 2 to 5% Be metal, Izhvanov [14]. The temperature of the wire varies from 650 to 1300 °C [14]. The same authors unsuccessfully attempted to obtain Be metal in a metallic apparatus by passing beryllium iodide vapor over a heated metal strip [14].

In experiments, Izhvanov [14] studied the possibility of obtaining Be metal by the iodide decomposition process using liquid beryllium iodide in a Be-lined pressurizable[1)] stainless steel apparatus. The Be was contaminated with iron and traces of nickel and chromium, probably stemming from vessel material components [14].

References:

[1] A. E. van Arkel, J. H. de Boer (Z. Anorg. Allgem. Chem. **148** [1925] 345/50).
[2] J. H. de Boer, J. D. Fast (Z. Anorg. Allgem. Chem. **153** [1926] 1/8).

1) Pressurization of the system is required because of the very narrow temperature range over which beryllium iodide is liquid.

[3] A. E. van Arkel (Metallwirtsch. Metallwiss. Metalltech. **13** [1934] 405/8).
[4] G. E. Darwin, J. H. Buddery (Beryllium, Butterworth, London 1960, pp. 79/80, 97/8).
[5] H. A. Sloman (J. Inst. Metals **49** No. 2 [1932] 365/90, 377/8).
[6] B. Kopelman, H. Bender (J. Electrochem. Soc. **98** [1951] 89/93).
[7] B. Kopelman, Sylvania Electric Products (unpublished information from [8]).
[8] T. T. Magel (in: D. W. White Jr., J. E. Burke, The Metal Beryllium, ASM, Cleveland 1955, pp. 124/35, 133).
[9] H. H. Kellogg (in: D. W. White Jr., J. E. Burke, The Metal Beryllium, ASM, Cleveland 1955, pp. 49/62, 58/9).
[10] R. A. J. Shelton (Metallurgia **56** [1957] 283/9; C.A. **1958** 2711).
[11] D. R. Stull (JANAF Thermochemical Tables: NSRDS-NBS-37 [1971]).
[12] O. Johnson, J. Powell, A. S. Newton, Iowa State College (unpublished information from [8]).
[13] T. T. Magel, Massachusetts Institute of Technology (unpublished information from [8]).
[14] L. A. Izhvanov, V. I. Artem'yev, V. I. Pankratov (Izv. Akad. Nauk SSSR Metally **1971** No. 5, pp. 109/10; Russ. Met. **1971** 82/4; C.A. **76** [1972] No. 16876).
[15] Anonymous (Metallbörse **23** [1933] 18/9, 50/1; C. **1933** I 1344).

1.4.6.2 Beryllium Nitride, Beryllium Carbide, Diethylberyllium

The preparation of Be metal by thermal decomposition of beryllium nitride and condensation of the Be vapor on a cooled surface is described by [1]. The decomposition is performed at temperatures above ~2000 °C, under vacuum [1], cf. [2].

Be metal can also be obtained by thermal decomposition of beryllium carbide at >2100 °C, under vacuum, and condensing the Be vapor on a cooled surface, held slightly above the melting point of the Be metal. The rate of decomposition increases rapidly with increasing temperature. Within a heating period of 30 min, 75% of the Be_2C is decomposed at 2150 °C, 89% at 2200 °C, and 97% at 2250 °C. The Be metal obtained is free from oxidic impurities from the carbide because of the nonvolatility of the Be oxide at the decomposition temperatures [1].

In a more recent study, the thermal decomposition of technical beryllium carbide was carried out at 1500 °C and 10^{-5} Torr. The substantially pure Be metal vapor was condensed on a surface (preferably tantalum) held at 1080 to 1280 °C [3].

Thermal decomposition of diethylberyllium deposits metallic beryllium mirrors [4, 5].

References:

[1] J. Kielland, Heraeus-Vacuumschmelze A.G. (Ger. 635701 [1936]; C.A. **1937** 646).
[2] W. J. Kroll (U.S. Bur. Mines Inform. Circ. No. 7326 [1945] 1/15, 2).
[3] H. Kahan, A. Pattoret, R. Platzer (Brit. 958820 [1961/64]; N.S.A. **18** [1964] No. 25912).
[4] F. A. Paneth, H. Loleit (J. Chem. Soc. **1935** 366/71, 369).
[5] T. T. Magel (in: D. W. White Jr., J. E. Burke, The Metal Beryllium, ASM, Cleveland 1955, pp. 124/35, 133).

1.5 Separation of Metal from Slag and Refining of Metal

Review

Crude Be metal commercially produced either in pebble form by the magnesium reduction of Be fluoride or formerly also in flake form by Be chloride electrolysis is comparatively impure and requires further purification before it can be used in fabrication operations [1 to 3, 6, 9].

The literature before 1931 is reviewed in "Beryllium" 1930, pp. 45/6.

Generally, the flakes from electrolysis are of higher purity than the pebbles from metal reduction. The impurities in the crude Be flakes consist mainly of sodium and chlorine; those of the crude Be pebbles are mainly magnesium and Mg fluoride. In addition, both products contain beryllium oxide, carbon, and many metallic impurities, the most important of which being aluminium, iron, and silicon [1 to 3, 10].

The impurities have a detrimental influence on the mechanical properties and corrosion behavior of beryllium or act as poisons in nuclear reactor applications [2 to 4], cf. also [5].

The brittle intermetallic compound $MgBe_{13}$ may be formed [2, 6] from the magnesium content of the pebbles (up to 1.5 or 2.0% [4, 6]). The chlorine content of the electrolytic flakes (up to 0.15% [2, 7]) is particularly undesirable in nuclear technology as about 0.1 wt% doubles the cross section of beryllium. In addition, it impairs the mechanical properties and corrosion resistance of fabricated beryllium [2, 10]. A series of impurity imparting elements which are highly neutron absorbing and/or generate long half-life, highly radioactive isotopes (e.g., cobalt) on irradiation must be avoided for any nuclear application. Representative chemical purity specification data, setting the corresponding requirements, are given in [4].

Several processes for producing different grades of Be metal of various purities are available.

Large quantities of impurities can be removed from the crude metal by vacuum melting and casting, by melting under a flux, or by oxalic acid leaching. Today, only the vacuum melting and casting method is used for the production of Be metal of technical purity on a commercial scale [1 to 3, 8, 11, 12].

For producing Be metal of much higher purity grades, the following techniques have been tried: re-electrolysis (=soluble anode electrolysis), vacuum distillation or sublimation, transport reaction, zone melting, and liquid metal extraction [2, 3].

Currently only re-electrolysis is used commercially [1, 3]. The other purification techniques are employed only in laboratories for specific purposes. They are generally too expensive or too difficult to operate on a commercial scale [1, 3], cf. [11].

Numerous attempts have been undertaken to attain very high purity Be metal, mainly by re-electrolysis, vacuum distillation/sublimation, and zone melting. The highly purified metal has been sought because it is thought that below a critical impurity concentration, additional planes of slip would become available in the lattice of the Be metal, thereby increasing its ductility. None of these processes, however, has succeeded in producing beryllium with the predicted ductility [2, 3, 8, 9, 13], cf. [5].

All purification techniques must be carried out under rigorous safety precautions, see, e.g. [14, 15].

References:

[1] S. B. Roboff (in: H. H. Hausner, Beryllium, Its Metallurgy and Properties, Univ. California Press, Berkeley 1965, pp. 17/30, 28/29; C.A. **64** [1966] 10879).
[2] G. E. Darwin, J. H. Buddery (Beryllium, Butterworth, London 1960, p. 99).
[3] G. Petzow, F. Aldinger (Ullmanns Encycl. Tech. Chem. 4th Ed. **8** [1974] 444, 452).
[4] C. W. Schwenzfeier Jr. (J. Metals **12** [1960] 793/7, 794).

[5] R. F. Bunshah, California Univ., Livermore, Lawrence Radiation Lab. (UCRL-7289 [1963] 1/24, 2, 4, 7, 15/20; N.S.A. **17** [1963] No. 25784).
[6] C. W. Schwenzfeier Jr. (in: D. W. White Jr., J. E. Burke, The Metal Beryllium, ASM, Cleveland 1955, pp. 71/101, 97/101; N.S.A. **11** [1957] No. 4473).
[7] J. P. Denny (in: H. H. Hausner, Beryllium, Its Metallurgy and Properties, Univ. California Press, Berkeley 1965, pp. 55/67, 55; C.A. **64** [1966] 10879).
[8] J. Ballance, A. J. Stonehouse, R. Sweeney, K. Walsh (in: Kirk-Othmer Encycl. Chem. Technol. 3rd Ed. **3** [1978] 803/23, 810, 811).
[9] G. Jaeger (Metall **4** [1950] 183/91, 190).
[10] J. Williams (Progr. Nucl. Energy V **1** [1956] 300/4, 301; Met. Abstr. **24** [1956/57] 964/5).
[11] A. E. Röllig (in: H. Spindler, H. E. Röllig, H. Steinkopff et al., Werkstoffe der Kerntechnik, Pt. II, VEB Deutscher Verlag der Wissenschaften, Berlin 1964, pp. 104/29, 106/7).
[12] L. R. Williams, P. B. Eyre (in: A. B. McIntosh, T. J. Heal, Materials for Nuclear Engineers, Interscience, New York 1960, pp. 269/318, 278/9; N.S.A. **15** [1961] No. 11485).
[13] R. G. Bellamy, N. A. Hill (Extraction and Metallurgy of Uranium, Thorium and Beryllium, Macmillan, New York 1963, pp. 72/84, 79).
[14] H. T. Sumsion, C. O. Matthews (Trans. 3rd Vacuum Met. Conf., New York 1959 [1960], pp. 114/29, 115).
[15] A. J. Breslin, W. B. Harris (Air Eng. **2** No. 7 [1960] 34/6; 53; N.S.A. **15** [1961] No. 1626).

1.5.1 Milling and Leaching

Crude beryllium material can be purified by milling and leaching the powder in an ammonium carbonate solution [1] or various dilute acids [1, 7, 8]. The method is currently not used in commercial production [1].

Leaching of comminuted, thermally reduced beryllium pebble material is unsuccessful because of the inert nature of the MgF_2 impurity [2]. The method is, however, suitable for purifying electrolytically produced beryllium flake [3, 4]. In addition, it is applicable to recycled beryllium scrap (saw sludges), mill chips, skinning products, machine turnings, etc. [7, 8].

Although vacuum melting (see p. 95) is a successful purification method for electrolytic flake, the milling and leaching method, processing directly from flake to powder, seems economically more attractive. This is because the preferred beryllium fabrication route is powder metallurgy, and because vacuum cast ingots require machining and comminution [2 to 4].

Purification of electrolytic flake is chiefly aimed at lowering the undesirably (see p. 92) high chlorine content (400 to 1000 ppm [3]) [3, 4]. Vacuum heating to just below the melting point of beryllium does not effectively reduce the chlorine content [2]. According to a microscopic study [3], flake beryllium consists of single crystal aggregates with their basal planes all in the plane of the flake. The impurities are concentrated in layers approximately 10 μm apart. The flakes are therefore ball-milled to a powder in order to expose these impurities to attack. The chlorine is, however, not present as a simple chloride because heating the milled powder in vacuum to a temperature near the melting point of beryllium does not affect the chlorine content. Leaching the powder in water or in reagents like methyl and ethyl alcohol, nitric, hydrochloric, or citric acid also has little effect [3, 4]. The chlorine impurity content is reduced to about 200 ppm when the flake is ball-milled in 10% aqueous ammonium carbonate [5]. This technique has been adopted for some time, but difficulties are encountered with corrosion of the mill and balls by the leaching liquid. The corrosion contaminates the beryllium powder surfaces with basic iron compounds [3, 4]. These compounds can be removed by leaching with nitric acid [2, 4] or 10% citric acid [4]. The latter yields beryllium with 200 ppm Cl and 400 ppm Fe [3]. A disadvantage of the process is that the thus obtained beryllium powder has poor sintering properties compared to those of beryllium powder from thermal reduction [1, 2, 4].

Sintering is a process that is generally sensitive to surface conditions [2, 4]. Poor sintering properties are related to a surface film formed by anion adsorption during leaching [6]. Therefore a leaching reagent was sought whose beryllium salt was not hydrolyzed in aqueous solution and which attacked the impurities in the metal. Oxalic, succinic, malonic, and acetic acids and pyrocatechol are suitable reagents. A comparison of the sintering properties of beryllium powders treated with these reagents showed oxalic acid to be the most efficient [4, 6].

For oxalic acid leaching, electrolytic flake beryllium is dry-milled to −200 mesh in a porcelain mill with steel balls; the powder is leached in a tumbling barrel type tank for 1 h with 10% oxalic acid (4 to 6 L/kg Be powder). The temperature is held at 14 to 16 °C with a cooling coil. Leached powder and liquid are poured onto a filter equipped with an agitator to prevent the fine powder from settling and blocking the filter pores. The powder is washed with cold, distilled water and acetone and is then vacuum dried at 10^{-4} Torr, usually at 500 °C. This serves predominantly to anneal and degas the powder [2, 3]. Some beryllium oxide surface contamination occurs when exposing the mill to air, and some iron contamination occurs during milling. Chlorine, oxygen, and iron contents are reduced by oxalic acid leaching and subsequent vacuum drying [2]. A residual chlorine content (200 to 300 ppm), however, remains unchanged by repeated strenuous leaching treatments and does not vary with beryllium particle sizes of <5 to 75 μm [3], cf. [1].

Beryllium powder produced in this way and high quality (Q.M.V.), thermally reduced beryllium powder (Brush Beryllium Company product) possess similar properties. Its slightly higher (~10%) weight loss on sintering and slightly greater grain growth indicate that the electrolytic powder has a cleaner surface [4].

In a further investigation, milling and leaching was used to purify particulate beryllium such as recycled beryllium scrap (saw sludges), mill chips, skinning products, machine turnings, and others. Processing a beryllium powder of generally finer than 100 mesh with an aqueous solution of sulfamic, oxalic, sulfuric, phosphoric, or preferably nitric acid at pH 0 to 1.5 under agitation for about 4 h at up to 50 °C partially removes BeO, Fe, Mn, Cr, Si, Mg, Al, Ti, and C impurities. The concentration of the slurry of beryllium powder in the aqueous acid is generally 200 to 450 g Be/L, and in the case of nitric acid (1 N) 350 to 450 g Be/L. The preferred grain size is from −15 μm to −200 mesh, but some purification is also accomplished with particles as coarse as 1/8 in. After completion of the purification procedure, the beryllium is separated from the dissolved impurities by filtration. It is then washed and dried by conventional procedures. The loss of beryllium metal in the acid is generally ≦2% [7].

Instead of acids, solutions of salts such as beryllium or aluminium nitrate that hydrolyze to provide high acidity, can be used effectively [7].

Results from leachings of beryllium powders with different particle sizes under various treatment conditions with the five previously mentioned acids and two salts are given in [7]. The weight loss of beryllium in the acid is 2.17% [7].

An improved method for the removal of the iron content from beryllium powder (particle size −15 μm to −200 mesh) involves preheating the powder, preferably for 16 to 120 h at 537.8 to 1037.8 °C before acid leaching. Significant amounts of the iron content are thus converted to the soluble compounds $AlFeBe_4$ and $FeBe_{11}$. The process is especially effective for purifying virgin beryllium derived from ore, but may have little effect on powdered scrap that has previously been heated [8].

The preferred leaching agents are aqueous solutions of mixed acids or hydrolyzable salts of sulfamic or oxalic acid, H_3PO_4, HNO_3, or H_2SO_4 at pH 0 to 3.5; the slurries contain 200 to 650 g solids/L. Heating the slurries causes beryllium losses. Analysis data for, e.g., virgin beryllium containing 1028 ppm Fe and 430 ppm Al, preheated 65 h at 854.1 °C, show Fe and Al contents of 757 and 90 ppm, respectively, after leaching with sulfamic acid. Without preheating, 1074 and 76 ppm Fe and Al, respectively, remain [8].

References:

[1] S. B. Roboff (in: H. H. Hausner, Beryllium, Its Metallurgy and Properties, Univ. California Press, Berkeley 1965, pp. 17/30, 28; C.A. **64** [1966] 10879).
[2] G. E. Darwin, J. H. Buddery (Beryllium, Butterworth, London 1960, pp. 101/2).
[3] R. G. Bellamy (AERE-M-619 [1960] 1/17, 2/4, 6/9; N.S.A. **14** [1960] No. 12883).
[4] J. Williams (Progr. Nucl. Energy V **1** [1956] 300/4; Met. Abstr. **24** [1956/57] 964/5).
[5] T. L. Johnston, J. W. S. Jones, A. Blainey (AERE Rept. No. M/R 1442 [1954] from [2]).
[6] J. Williams, W. Munro, J. W. S. Jones (AERE Rept. No. M/R 1679 [1955] from [2]).
[7] K. A. Walsh, A. J. Sandor, The Brush Beryllium Company (U.S. 3642463 [1966/72]; C.A. **76** [1972] No. 102456).
[8] K. A. Walsh, A. J. Sandor, The Brush Beryllium Company (U.S. 3713810 [1970/73] from C.A. **79** [1973] No. 44697).

1.5.2 Melting and Casting

Vacuum melting and subsequent casting is currently the standard purification method for the large-scale isolation of technically Be metal [1 to 4].

This method is successful with crude thermally reduced pebble as well as with crude electrolytic flake [1 to 3, 9], beryllium scrap or recycle metal [5, 6]. In the processing of thermally reduced Be pebbles a purity corresponding to that of electrolytic flake is attained, in that of electrolytic flake a further lowering of halogen and low boiling metal content is achieved [3].

Basic Process Design

The vacuum melting of crude beryllium under oxygen-free conditions effects the vaporization of volatile impurities, mainly of free magnesium and of entrapped excess Be fluoride from the pebble, or of entrapped electrolyte from the flake. Nonvolatiles, such as beryllium oxide, beryllium carbide, and magnesium fluoride, separate from the molten metal as a dross which sinks and adheres to the bottom of the crucible [6 to 9, 11], cf. [5]. The dross is removed after the casting operation (see below) by racking [7].

The vacuum melted and (to a certain extent) purified metal is cast by pouring the liquid metal into an ingot mold with a capacity of between 75 and 200 kg Be metal [6 to 8]. Most of the remaining metallic impurities tend to segregate during solidification and remain within the ingot [10].

The melting step is usually performed in a vacuum induction furnace [4, 6, 8] by heating the crude Be metal under vacuum (0.1 to 0.5 Torr [19, 21]) at or above (superheating) the melting point. Superheating may also be done while back-filling the furnace with argon to a lower vacuum level in order to minimize evaporation losses of the beryllium [4, 5, 9, 21, 27].

Both bottom-pour and lip-pour furnaces have been used [4, 5]. Bottom pouring permits closer control of the melt composition, since contamination by condensed volatiles is avoided during tapping. Lip pouring is simpler from a mechanical standpoint, and is economically the most attractive method for the processing of large quantities of material [2, 4, 5].

The most suitable crucible material for the melting step is Be oxide, because its solubility in molten beryllium is quite low. Other oxide refractories are slowly reduced by molten beryllium. Graphite is unsuitable because beryllium carbide readily forms at temperatures above the melting point of Be metal [2, 4, 5, 27]. For a type of Be oxide crucible, developed at Massachusetts Institute of Technology, see [27, p. 138].

The casting step is usually performed in a vacuum, generally in a low pressure argon atmosphere [2, 4]. However, casting quality is said to be improved by pouring under argon at atmospheric pressure [2, 4, 12, 27].

The most commonly used mold material is graphite which requires a fairly massive mold construction imparting a high heat capacity to prevent excessive Be carbide formation during casting and solidification [2, 5]. A refractory wash-lining in the mold, e.g., from Be oxide, should serve the same purpose, but is of doubtful value [2, 4], cf. [27]. Cast iron, in conjunction with BeO and Al_2O_3 washes, is also widely used as a mold material [4], cf. [21].

Whereas most experience has been accumulated with respect to the vacuum induction melting technique, more recently, vacuum arc or vacuum electron-beam melting techniques have also been reported [1, 5].

Arc melting is performed under vacuum (10^{-2} to 10^{-3} Torr), thereby partial pressures of helium and argon are applied successfully. A consumable-electrode from, e.g., electrolytic flake (Pechiney), or from already vacuum induction-melted and extruded Be metal is preferred. Nonconsumable electrodes (e.g., from tungsten) prove unsatisfactory because of melt contamination. In general, vacuum arc melting is preferred over vacuum induction melting only for reducing the Be oxide level. However, controlled compositions are here more difficult to obtain [1, 5], cf. [13].

Electron-beam melting is carried out in a high vacuum (10^{-4} to 10^{-5} Torr [14]) on pressed beryllium flake, compacted chip and extruded vacuum-cast ingots [5, 16]. Thermally reduced crude pebble cannot be satisfactorily melted because of outgassing of magnesium and excessive splattering [14 to 17]. Two to five melting runs are required for purification. The Be oxide level is thereby reduced by segregation near the ingot surface. No significant decrease in metallic impurity content is observed [5, 15, 16]. Necessary multiple melting, slow melting rates, and relatively high beryllium evaporation losses render electron-beam melting less attractive than the other melting methods [5], cf. [15].

A schematic drawing of the electron-beam melting apparatus is shown in [14], cf. [18]. Heat for melting is provided by an electron gun which serves as the cathode and bombards the melt stock (the anode) with electrons until the melting temperature is reached. The melt stock passes the electron-beam and drops into a specifically designed water-cooled copper crucible [14], cf. [16, 18].

Beryllium ingots are classified as prime virgin beryllium, if the castings are made exclusively from beryllium pebble or flake products. When the casting is made from scrap or recycled metal, or from a mixture of scrap with pebbles or flakes, it is virgin beryllium [6].

A modified vacuum induction melting/casting technique has been developed, which avoids use of a crucible. This technique is applied for purification of crude beryllium of reasonably high purity, such as electrolytic flake. Any contamination of beryllium metal with impurities stemming from the crucible material, from BeO itself or impurities contained therein, or from reactions between beryllium melt and crucible material is avoided [28, 29]. For results see p. 98, for description of an apparatus, see Fig. 1-11, p. 98.

Industrial and Pilot Plant Practices

In the commercial process at Brush Beryllium Company the magnesium content of the beryllium pebble is reduced to 300 to 500 ppm. Magnesium levels of approximately 100 ppm are obtainable in the laboratory. The fluorine content is reduced to 50 ppm [2, 4]. The chlorine content of electrolytically produced flake is usually lowered by vacuum melting and casting to 50 ppm [2] or 20 ppm [4, 31], respectively.

At Brush Beryllium Company the vacuum furnace used for purification of Be pebble is an induction-heated Ajax tilting type furnace, containing both the melting crucible and the casting mold within the evacuated space, see photographs in [19], cf. also [22, 27]. It is rated at 3000 cycles, 100 kW, and 400 V. The operating pressure is approximately 0.5 Torr, since at lower pressure (0.05 to 0.03 Torr) the magnesium content is not substantially reduced and danger of undesirable silent discharge increases [2, 19, 27].

The Be oxide melting crucible is packed in Be oxide powder as insulation between crucible and high frequency coils. This type of construction allows patching of cracked and eroded crucibles and normal crucible life therefore ranges from 100 to 140 melts. No graphite heating sleeve (susceptor) is employed, and therefore heating is by direct induction into the charge. The graphite casting mold is set at right angles to the crucible near its lip, and is covered during melting to prevent contamination by vaporized impurities [2, 19].

A high-frequency, induction-heated vacuum furnace with a beryllium capacity of 4.53 kg (10 lb) rated at 50 kVA, 3200 cycles, and a coil voltage of 200 to 250 V was used in a pre-commercial pilot plant run at Imperial Smelting Corporation. A movable water-cooled dome mounted above the crucible reduces the amount of condensate on the rest of the furnace, and a tipping device enables the addition of beryllium without breaking the vacuum seal. Operation is very similar to the Brush Process. No susceptor is necessary for pebble melting, but must be used for swarf melting [2, 11]. The analytical data show a noticeable increase in some impurities. Boron impurities result from contamination in the reduction stage, the silicon and aluminium impurities are due to contamination from crucible sealing cements [11].

For a vacuum melting process used in India, see [23].

Industrial purification of Be pebble or flake in the USSR is performed under an argon atmosphere of 20 Torr and at a temperature of 1500 to 1550 °C [24]. More recently, vacuum melting has been combined with centrifugal casting [25, 26]. The centrifugal technique is said to almost completely eliminate most gases and slags [26]. Centrifugal casting effects directional crystallization. Hot metal is fed into the center of a rotating mold and centrifugal forces during solidification expel the gases, oxides, and other impurities to the surface of the casting [25, 26].

Beryllium melting is begun at a vacuum of 2×10^{-3} Torr and continued under vacuum up to 1×10^{-4} Torr at a temperature of 800 to 900 °C. Adsorbed humidity, gases, and volatile fluorides are eliminated. The furnace is then filled with chemically pure, dry argon up to a residual pressure of 30 to 50 Torr, and the metal melted and heated to 1450 to 1470 °C for 5 min. The cover is then lifted and the metal poured into the revolving graphite mold [25, 26]. Rapid crystallization of the surface layer of the casting prevents the metal from becoming carbonated [26]. Be metal yield per casting is 80 to 90% [25].

Crude Be metal may also be melted in a high-vacuum arc furnace with a consumable electrode. The melting is carried out at 1×10^{-4} Torr. The ingots thereby produced have a finer crystal structure than those produced by vacuum induction heating [25].

A crucibleless vacuum induction melting and casting technique using electrolytic flake or other high purity crude Be metal, is at present practiced only on a laboratory scale [28, 29]. The technique consists of placing a cylindrical compact of beryllium flake on a water-cooled copper pedestal and surrounding it with an induction coil. Contact between the compact and pedestal is limited to a few points. During heating prior to melting the beryllium distills off the compact and condenses on the pedestal. The actual melting occurs in a skin of distilled beryllium. The eddy currents induced in the compact by the coil heat the metal and melt it. The molten metal is held on the pedestal in a conical shape by a balance of body forces, surface tension forces, and levitation forces due to opposing currents in the induction coil and melt. The conical shaped melt is frozen in place by slowly decreasing the power. The impurities on the surface are removed mechanically or chemically. The amount of impurities separating to the surface of the melt decreases markedly after several melt cycles [28, 29], cf. [30].

A schematic representation of the crucible-free vacuum-induction melting and casting process is shown in **Fig. 1-11** [28].

In practice, a compact of ~2.54 cm diameter by 10.16 cm in height, weighing about 100 to 130 g is placed on the pedestal. After evacuation of the system to its base pressure of 1×10^{-6} Torr, the induction coil is connected to a 10 kc, 50 kW motor-generator-type power supply [28].

Analysis of the data shows that on vacuum-induction melting the high vapor pressure impurities, e.g., Cl, Na, Zn, and Mg are almost completely removed by vaporization. The concentration of the low vapor pressure, soluble impurities, Fe, Ni, Cu, Cr, Ti, and Si increases after melting because some beryllium (about 25% in 6 melt cycles) is lost from

Fig. 1-11

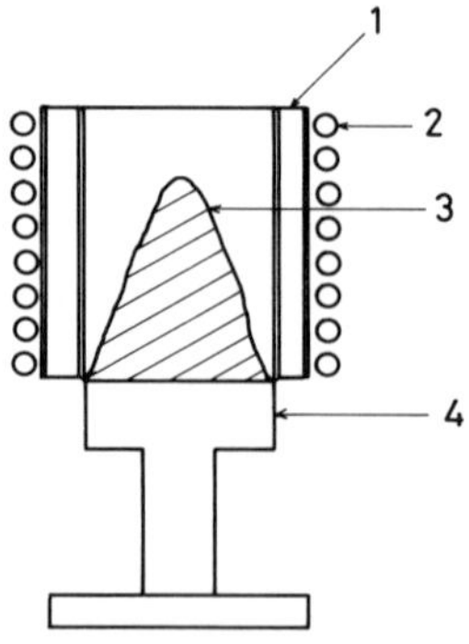

Schematic representation of the crucible-free vacuum-induction melting and casting process. 1) Cu shield, water-cooled with vertical slit; 2) induction coil; 3) molten Be; 4) stool, water-cooled.

the molten pool by vaporization, thus decreasing the total amount of beryllium in the melt. The insoluble impurities, BeO, BeC_2, and other compounds separate to the surface of the melt where they are mechanically or chemically removed [28].

References:

[1] S. B. Roboff (in: H.H. Hausner, Beryllium, Its Metallurgy and Properties, Univ. California Press, Berkeley 1965, pp. 17/30, 28/9; C.A. **64** [1966] 10879).
[2] G. E. Darwin, J. H. Buddery (Beryllium, Butterworth, London 1960, pp. 100/1).
[3] G. Petzow, F. Aldinger (Ullmanns Encycl. Tech. Chem. 4th Ed. **8** [1974] 452).
[4] L. R. Williams, P. B. Eyre (in: A.B. McIntosh, T.J. Heal, Materials for Nuclear Engineers, Interscience, New York 1960, pp. 269/318, 278/80; N.S.A. **15** [1961] No. 11485).
[5] J. P. Denny (in: H.H. Hausner, Beryllium, Its Metallurgy and Properties, Univ. California Press, Berkeley 1965, pp. 55/67, 56/7, 59/61).
[6] K. A. Walsh (in: D.R. Floyd, J.N. Lowe, Beryllium Science and Technology, Vol. 2, Plenum, New York-London 1979, pp. 1/11, 9/10).
[7] C. W. Schwenzfeier Jr. (in: Kirk-Othmer Encycl. Chem. Technol. 2nd Ed. **3** [1964] 450/74, 462).
[8] J. Ballance, A. J. Stonehouse, R. Sweeney, K. Walsh (in: Kirk-Othmer Encycl. Chem. Technol. 3rd Ed. **3** [1978] 803/23, 810).
[9] W. W. Beaver (Progr. Nucl. Energy V **1** [1956] 277/99, 284, 286; Met. Abstr. **24** [1956/57] 963).
[10] R. W. Krenzer (in: D.R. Floyd, J.N. Lowe, Beryllium Science and Technology, Vol. 2, Plenum, New York-London 1979, pp. 31/56, 32).

[11] L. J. Derham, D. A. Temple (Extr. Refining Rarer Metals Proc. Symp., London 1956 [1957], pp. 323/36, 324, 331, 333; C.A. **1958** 19783).
[12] J. G. Kura, J. H. Jackson, M. C. Udy, L. W. Eastwood (Trans. Am. Inst. Mining Met. Eng. **185** [1949] 769/78, 770, 773).
[13] P. D. Kyffin, R. L. Craik (Inst. Metals Monograph Rept. Ser. No. 28 [1963] 677/86, 681/2, 685).
[14] D. Peckner (Mater. Design Eng. **53** No. 3 [1961] 101/6; C.A. **1961** 10244).
[15] S. R. Seagle, R. L. Martin, O. Bertea (J. Metals **14** [1962] 812/20, 815).
[16] H. T. Sumsion, C. O. Matthews (Trans. 3rd Vacuum Met. Conf., New York 1959 [1960], pp. 114/29, 115, 117/8, 121, 125/8; C.A. **1961** 23248).
[17] C. B. Dittmar, S. Abkowitz (NP-7927 [1958] 1/8, 4; N.S.A. **14** [1960] No. 592).
[18] S. H. Dayton (Mining World [San Francisco] **20** No. 8 [1958] 40/3; C.A. **1958** 13578).
[19] C. W. Schwenzfeier Jr. (in: D.W. White Jr., J.E. Burke, The Metal Beryllium, ASM, Cleveland 1955, pp. 71/101, 97/101).
[20] A. R. Kaufmann, B. R. F. Kjellgren (Proc. 1st Intern. Conf. Peaceful Uses At. Energy, Geneva 1955, Vol. 8, pp. 590/99, 593).

[21] J. T. Stacy (AECD-3647 [1955] 55/94, 78/80).
[22] C. W. Schwenzfeier Jr. (J. Metals **12** [1960] 793/7).
[23] K. S. Subbarao, B. P. Sharma, C. M. Paul, C. V. Sundaram (Beryllium 1977 Conf. Prepr. 4th Intern. Conf. Beryllium, London 1977, Paper No. 43, pp. 1/10; C.A. **92** [1980] No. 151299).
[24] G. A. Meyerson (Proc. 1st Intern. Conf. Peaceful Uses At. Energy, Geneva 1955, Vol. 8, pp. 587/9, 588).
[25] E. S. Ivanov, V. M. Shmelev (Proc. 2nd Intern. Conf. Peaceful Uses At. Energy, Geneva 1958, Paper No. 2048, pp. 302/5).

[26] G. A. Meerson, D. D. Sokolov, N. F. Mironov, N. M. Bogorad, Ya. D. Pokhomov, D. S. Lvovskii, E. S. Ivanov, V. M. Shmelev (At. Energiya SSSR **5** [1958] 624/30; Soviet At. Energy **5** [1958] 1555/62, 1560/1; C.A. **1961** 7218).
[27] P. Corzine, A. R. Kaufmann (in: D. W. White Jr., J. E. Burke, The Metal Beryllium, ASM, Cleveland 1955, pp. 136/51, 137/8, 140, 144, 150).
[28] R. F. Bunshah, R. S. Juntz (Met. Soc. Conf. **33** [1966] 1/28, 2/7, 9/11, 13/6).
[29] R. F. Bunshah, R. S. Juntz (J. Metals **16** [1964] 693/9).
[30] R. F. Bunshah, R. S. Juntz (Trans. 5th Vacuum Met. Conf. Am. Vacuum Soc., Boston, Mass., 1962 [1963], p. 110).
[31] R. G. Bellamy (AERE-M-619 [1960] 1/17, 3; N.S.A. **14** [1960] No. 12883).

1.5.3 Melting under a Flux

The method of purifying crude beryllium pebble by melting and casting in open pots under a flux was used before the vacuum melting and casting process came into use [1, 2]. Today this method has been completely replaced by vacuum melting and casting [1 to 3].

Remelting under a flux has been applied to reduce the magnesium fluoride slag content of beryllium pebble material. Fluorides and fluoride-chloride mixtures are recommended for the flux [4, 5], cf. [7]. If fluxes with a high beryllium fluoride content are used, the residual magnesium content can also be reduced [2, 3, 5, 6]. For the disadvantages of this method, see [3, 5].

References:

[1] L. R. Williams, P. B. Eyre (in: A.B. McIntosh, T.J. Heal, Materials for Nuclear Engineers, Interscience, New York 1960, pp. 269/318, 278; N.S.A. **15** [1961] No. 11485).
[2] C. W. Schwenzfeier Jr. (in: D.W. White Jr., J.E. Burke, The Metal Beryllium, ASM, Cleveland 1955, pp. 71/101, 97).
[3] G. E. Darwin, J. H. Buddery (Beryllium, Butterworth, London 1960, p. 99).
[4] G. Jaeger (Metall **4** [1950] 183/91, 190).
[5] J. G. Kura, J. H. Jackson, M. C. Udy, L. W. Eastwood (J. Metals **1** [1949] 769/78, 772/3, 776/7).
[6] B. R. F. Kjellgren (Trans. Electrochem. Soc. **93** [1948] 122/8).
[7] C. H. Monroe, H. C. Claflin, The Beryllium Corporation, New York (U.S. 2051963 [1932/36]).

1.5.4 Vacuum Distillation

Vacuum distillation can be used to obtain highly pure beryllium metal from a prepurified starting material. Because of the very low throughput, the method can only be applied on a small scale. Although pilot plant trials have been reported in Russia, the method is presently normally limited only to research and to special purposes.

General Considerations

According to the equation: $\log(P/atm) = 6.186 + 1.454 \times 10^{-4}\,T - (16734 \pm 80)\,T^{-1}$ (for 1171 to 1552 K) [1], the vapor pressure of solid beryllium is comparatively low. Using this equation, which is confirmed by [2], a vapor pressure of only 4×10^{-2} Torr is extrapolated for beryllium at its melting point (1283 °C). The evaporation rate is therefore expected to be low [3, 4]; a value of $0.6 \times 10^{-6}\,g \cdot cm^{-2} \cdot min^{-1}$ at 850 °C was calculated by [2]. Many of the metallic impurities have vapor pressures on the same order as that of beryllium. Therefore, separation cannot be achieved unless some form of fractional evaporation and condensation is used. However, the vacuum evaporation/condensation method seems attractive for the removal of nonmetals, especially oxygen in the form of beryllium oxide [3, 4]. The vapor pressure [5] of beryllium oxide is extremely low (4×10^{-11} Torr at 1283 °C) and therefore negligible [3, 4]. Beryllium nitride and carbide are less stable than the oxide and sublime under vacuum at 2000 and 2100 °C, respectively. Dissociation occurs at slightly higher temperatures [6].

The surface of the beryllium sample may coat with an oxide and/or nitride film during vacuum evaporation [2, 4]. The investigations of [2] reveal that this film greatly lowers the vapor pressure of beryllium [2], [36, p. 4]. The surface films have to be removed to restore a sufficient evaporation rate [4].

As the evaporation proceeds, the nonvolatile impurities accumulate at the surface of the melt, and a progressive fall in the evaporation rate can be expected [4, 36] .

Experience in Practice

Numerous attempts at vacuum distillation have been made, either to produce pure beryllium foils for X-ray windows [7, 8] or in the search for ductile (see p. 92) beryllium, cf., e.g. [9 to 16].

The evaporation temperature is between 1100 to 1500 °C [34, 35, 42]. The beryllium charge may be heated in a crucible by resistance-, induction-, or electron-beam heating [17, 25, 30, 42], or crucible-free by electron-beam bombardment or induction heating [28, 36]. Fractional condensation can be carried out on a cold surface [14], or more preferably, on a heated surface having a decreasing temperature gradient [17, 21, 25, 42].

Purified beryllium metal in a form suitable for X-ray tube windows is obtained by evaporating thoroughly outgassed lower purity beryllium metal, contained in a heated region of a Pyrex tube under high vacuum ($\sim 5 \times 10^{-4}$ Torr) and condensing it in a cooler region of the tube. Heat is supplied by electron bombardment (voltage 21 kV, overall power 2 kW). Purified foils 30 µm thick and free from pinholes are obtained. By continuing the process for longer periods, homogeneous, malleable plates, 0.16 mm thick for windows of demountable X-ray tubes are obtained [7].

An ordinary low tension furnace under a vacuum of about 5×10^{-5} Torr and a current of 85 A at 6 V is used to evaporate beryllium at $\sim$1400 °C from a beryllium oxide crucible and to deposit it onto a Be-metal surface (25 to 92 °C). Purified foils a few µm thick and up to 1.5 in. in diameter are obtained but are said to contain pinholes [8]. Earlier attempts at producing beryllium X-ray window foils by this method are briefly discussed in [7].

Besides these foils, no significant amount of solidified distillate, applicable to a recognized mechanical test, was obtained [4], e.g. [9, 13, 22, 25].

A few authors [9 to 11], cf. [17, 18] claim to have obtained ductile beryllium foils which were capable of being bent many times before fracture. The others, however, obtained brittle or noncoherent beryllium deposits [4], e.g. [12, 14, 39].

In the search for ductile beryllium, attempts have been made to carefully purify beryllium metal by simple fractional evaporation from a crucible and condensation on a cool surface [14]. Beryllium metal is placed in a tantalum crucible situated in a Pyrex chamber and heated by an electron bombardment furnace. The vacuum distillation/condensation system is outgassed to 2×10^{-9} Torr. The raw material consists of 10 g high purity electrolytic flake (Clifton flake), prepurified by vacuum melting for removal of chlorine and dissolved gases. Several distillation runs were carried out at 5×10^{-8} to 2×10^{-9} Torr at 1400 to

1500 °C. The deposits were 0.010 to 0.040 in thick, and showed well developed hexagonal crystals under low magnification. Residual oxygen within the vacuum distillation device was calculated assuming a total absorption by beryllium of 1×10^{-7} ppm [14]. In a subsequent discussion of this work [14] by [16], however, the calculated oxygen content was quoted as 10 ppm [16].

The distilled beryllium was ~99.92% pure. The impurity separation attained, except for the oxide, was very limited. All deposits were brittle and failed in bending. The bending of individual crystals also showed brittleness under the microscope. The postulated ductility of high purity, oxygen-free beryllium was therefore not made evident [14].

Russian workers [17] claim a greatly improved purification. It is achieved by fractional evaporation of liquid beryllium from less volatile impurities followed by fractional condensation on a heated surface at such a temperature that enables the more volatile impurities to re-evaporate. The method is reported to give beryllium metal of 99.98% purity on a semicommercial scale. The increased purity is chiefly due to the removal of impurities having similar vapor pressures as beryllium [17].

The vacuum unit used has a volume of ~500 L, with a residual pressure of 10^{-5} to 10^{-6} Torr. The BeO crucible is heated by Mo or W coils embedded in lining cylinders of BeO or Al_2O_3. Fractional distillation and condensation (condenser open on top) are achieved by suitably adjusting the temperatures. The metal evaporates at 1300 to 1450 °C, and the metal vapors are condensed at 1200 to 900 °C [17]. The Be condensate, obtained from commercial pure raw material consists of regular hexagonal crystals several millimeters long. The effect of this method on oxygen impurity content is not mentioned. The Be foils obtained are 0.1 to 0.3 mm thick and are not destroyed by repeated bending deformation. Their hardness is also considerably reduced [17].

A modification of the method [17] is reported by [18]. This procedure gives a product with a higher O and C content [18], see also [19]. The distilled beryllium has greater ductility than the raw material [18]. Its hardness, compared with that of purified beryllium from [17], is reduced [18]. Mass spectrometric analysis of the residual gases in the vacuum chamber of a beryllium distillation process indicates that the gas content of the distilled beryllium is lowered. Processing the metal in a vessel, isolated from pumps gives beryllium containing 10 ppm oxygen and nitrogen [20], cf. p. 103. Compare also [42].

The method of [17] was slightly modified by [21]. The apparatus consisted of an induction heated BeO crucible topped by a cylindrical hot surface BeO condenser (closed at the top). The crucible was heated by an induction coil with a frequency of 5 kc/s and a maximum voltage of 200 V. Before using the apparatus, the molybdenum or tantalum liners were pickled in a mixture of nitric and hydrofluoric acid to remove oxide films. The BeO crucible was freed from volatile impurities by preliminary distillation of (25 g) Be metal onto a tantalum lined condenser and replacing the latter by a new one [21].

A beryllium charge (500 to 1000 g) was heated to ~1375 °C and 80% distilled at a pressure of 4×10^{-6} Torr. The evaporation rate at 1375 °C ranged from 0.73 to 1.10 $g \cdot cm^{-2} \cdot h^{-1}$. It approached the value calculated from the Rayleigh equation (1.2 $g \cdot cm^{-2} \cdot h^{-1}$) and was considerably higher than that obtained in similar studies. This is attributed to improved stirring, the low induction frequency, and the use of a thinner susceptor than in comparable work. Condensation took place from ~1200 to ~1020 °C, the purest beryllium depositing in the 1100 to 1050 °C zone [21].

In the fractionation process within the condenser, many impurity elements behave in accordance with their relative vapor pressures, but some apparently, are distilled as compounds more volatile than the elements themselves. Metallic impurities are divided into 4 groups according to their distribution behavior in the condenser: 1) Fe, Si, Al; 2) Mg, Ca, Ba, Sr, Mn, Ga; 3) Na and K; 4) Ni, Cu, Cr. Group 1) behaves as if certain compounds, more or less volatile than beryllium itself, distill. The elements of group 3), in spite of their high vapor pressures, behave anomalously in not migrating to the cooler condenser end, but distribute themselves fairly uniformly between all condenser zones. Groups 2) and 4) behave consistently with their vapor pressures, i.e., the group 2) elements migrate more to the cooler end of the condenser, and the group 4) elements remain in the melt. Most nonmetallic impurities are reduced to a small fraction of their original concentration [21].

The beryllium condensate (coarse and plate-like crystals) is removed from the condenser by cracking the BeO and peeling off the liner [21].

For earlier smaller scale investigations of the same group, see [22 to 24].

A similar purification operation was reported by [25] using a bell jar vacuum distillation apparatus. A BeO crucible containing the charge ($\sim$350 g), placed in a tantalum wire wound furnace, is heated to about 1375 °C for 6 to 9 h. The distillate is collected on a tantalum cone collector (hot surface), open on top and seated on the crucible. The collector is heated by radiation from the crucible to 1090 °C at the top and $\sim$1160 °C at the bottom. During distillation, a vacuum of 2 to 5×10^{-6} Torr is maintained. The shiny and dense beryllium distillate, obtained from vacuum-melted Pechiney flake starting material, is of high purity. The Fe, Ni, Cr, and Mn contents are in the range of 1 to 5 ppm. The analytical data from twice distilled beryllium given in [25] are equally good as those obtained by [21]. The higher silicon content found by [25] is probably due to the use of a glass vacuum vessel, whereas [21] used a steel vessel [21]. The evaluation of the mechanical properties of the distillate [25] is limited by the thinness of the deposits [25].

In another distillation study [26] based on the method of [17], the advantage of purification by condensation on a hot surface is confirmed. In this process the starting material ($\sim$28 g) is contained in a BeO crucible and evaporated at 1500 °C under an ultra-high vacuum (1 to 3×10^{-7} Torr) and deposited on a pyramidal condenser at 1000 °C [26].

Several authors [17, 25, 27 to 29] have investigated the effectiveness of a double vacuum distillation [30], and recently published results report on three to five distillation runs [30].

The vacuum unit consists of a beryllium oxide crucible heated in a resistance furnace, with a conical alundum hot surface column lined with molybdenum, as condenser. The distillation chamber is evacuated with oil-diffusion pumps in the presence of sorption type getters to a residual pressure of 10^{-6} Torr. After removal of adsorbed gases, the metal is melted and the evacuation rate is reduced by means of a diaphragm to establish a residual pressure of 10^{-5} Torr. This method prevents backstreaming of active gases through the pumps into the distillation zone [30, cf. 20].

The evaporation temperature is 1350 °C for a rate of $0.32\ g \cdot cm^{-2} \cdot h^{-1}$. In each run, the impurities are gradually deposited along the condenser at different ratios; 15% of the charge remains in the crucible. Beryllium condenses in the central part of the condenser. Analysis indicates that the oxygen, carbon, and nitrogen contents are unchanged by repeated distillation runs. The purity after five runs is >99.99%. The distribution of Mg, Cu, and Fe contents over the temperature zones of the condenser from the first three runs is plotted in the paper. As was also noted by other authors, cf. [31, 32], no definite relationship can be established for silicon distribution [30].

Crucible-Free Distillation

Substantial reductions in the impurity content are attained by crucible-free vacuum distillation employing a hot condenser technique with electron-beam heating [28].

This technique allows rapid, localized melting and evaporation and minimizes exposure of the melt to residual oxygen in the vacuum system. Focussing of the beam impingement area permits core melting of the beryllium source material. The elimination of the BeO crucible removes a recognized source of oxygen contamination [28], cf. also p. 97.

The starting material is beryllium SR (super refined) grade Pechiney flake, cold pressed to 1 to 2 in. diameter compacts, and single-pass vacuum distilled beryllium compacts (Nuclear Metals Corp.) [28].

The distillation is carried out under a vacuum of 2 to 5×10^{-7} Torr. The beryllium compact is placed on a water-cooled copper base plate. A condenser assembly consisting of a series of thin tantalum strips (total surface area 270 cm^2) is positioned a small lateral distance away to allow passage of the electron beam directed from above down to the center of the beryllium compact. The beam gun is rated at 20000 V, 300 mA, with a minimum beam diameter of 0.010 in. and a focal length of 4 to 12 in. [28].

The beryllium is melted at the center of the beryllium compact, where the beam is focussed. The outer edges of the disk act as the crucible [28]. The condenser temperature is kept at 850 to 900 °C. The evaporation from the molten pool in the core of the beryllium source at 1350 to 1700 °C is continued until 50 to 80% of the beryllium source is consumed. The refined beryllium is collected in the form of 0.140 in. thick sheets [28, 33].

Another attempt at producing more ductile beryllium using crucible-free vacuum distillation employing the hot condenser technique and operating with induction heating was reported by [36, 37].

The beryllium source is a 1 in. diameter by 3 in. long cast rod placed on a water-cooled copper pedestal in the center of a tantalum condenser. The condenser is situated between the beryllium and a water-cooled cylindrical copper shield. The shield is surrounded by an induction coil. The condenser is heated by radiation from the hot beryllium. The base pressure of distillation is 6×10^{-6} Torr [36, 38]. The distilled beryllium is collected mostly on the tantalum condenser. Two different distillation methods are carried out [36]: 1) The beryllium rod is heated to >1200 °C, but below the melting point of the metal[1)] [36, 38]. 2) The beryllium rod is heated above the melting point, collapses into a cone, and evaporates from the molten metal. After every run, the apparatus is cooled under vacuum to room temperature. The beryllium is easily separated from the tantalum collector because a very thin Ta-Be intermetallic compound forms at the Ta-Be interface which serves as an efficient parting compound. The sheet from the solid-state distillation is thicker at the middle than at the top or bottom. The sheet from the liquid-state distillation is thickest at the bottom and tapers down to a ragged edge at the top [36].

Experimental parameters for a typical distillation of a solid beryllium rod ($\sim$50 g; 1 in.; diameter by 2.25 in. high) at $\sim$1175 °C and 6×10^{-6} Torr during 3 h, are: measured condenser temperature, 1010 °C; weight loss in distillation, 22.4 g; beryllium recovery, 86% [36].

The impurity content of metallic elements in the distillate (obtained by method [36]) of four solid beryllium samples of varying compositions is comparable to calculated values. Good agreement results for Fe, Ni, Cr, and Cu, fair agreement for Al, and poor agreement

[1)] Because the melting point of the charge is not exceeded, sublimation rather than distillation occurs.

for Si which apparently exhibits a positive deviation from Raoult's law behavior. No comparisons can be made for Ca, Mg, Na, and Mn because they re-evaporate from the distillate, which is collected at very high temperatures (~1050 °C) [38].

The results of a limited number of bend tests on the beryllium sheets obtained according to method [36] show improved bend ductility over commercial material. It cannot be stated if this improvement results from higher purity and/or less pronounced basal texture [39].

A further effort to produce high purity beryllium metal was carried out by [40, 41]. The crucible-free distillation technique operates with a device similar to that described by [36]. In contrast, an ultra-high vacuum, developed by continuous ion pumping, is applied. Distillation pressures in the range from 5×10^{-9} to 2×10^{-8} Torr are obtained without the back streaming and trapping problems associated with oil type pumping systems [40, 41].

References:

[1] R. B. Holden, R. Speiser, H. L. Johnston (J. Am. Chem. Soc. **70** [1948] 3897/9).
[2] E. A. Gulbransen, K. F. Andrew (J. Electrochem. Soc. **97** [1950] 383/95, 386, 393/4).
[3] R. G. Bellamy, N. A. Hill (Extraction and Metallurgy of Uranium, Thorium and Beryllium, Macmillan, New York 1963, pp. 72/84, 80).
[4] G. E. Darwin, J. H. Buddery (Beryllium, Butterworth, London 1960, pp. 102/4).
[5] N. D. Erway, R. L. Seifert (J. Electrochem. Soc. **98** [1951] 83/8, 86).
[6] W. W. Beaver (in: D. W. White Jr., J. E. Burke, The Metal Beryllium, ASM, Cleveland 1955, pp. 570/98, 572).
[7] H. Smith (J. Sci. Instr. **26** [1949] 378/9).
[8] H. Bradner (Rev. Sci. Instr. **19** [1948] 662/4).
[9] H. A. Sloman (J. Inst. Metals **49** [1932] 365/90, 381/3).
[10] W. Kroll (Metal Ind. [London] **47** [1935] 29/31).

[11] W. Kroll (Metals Alloys **8** [1937] 349/53).
[12] G. E. Claussen, J. W. Skehan (Metals Alloys **15** [1942] 599/603).
[13] A. R. Kaufmann, P. Gordon, D. W. Lillie (Trans. Am. Soc. Metals **42** [1950] 785/844, 789/90).
[14] C. S. Pearsall (MIT-1104 [1952] 1/14; N.S.A. **7** [1953] No. 1435).
[15] G. A. Meyerson (Proc. 1st Intern. Conf. Peaceful Uses At. Energy, Geneva 1955, Vol. 8, pp. 587/9).
[16] A. R. Kaufmann (in: D. W. White Jr., J. E. Burke, The Metal Beryllium, ASM, Cleveland 1955, pp. 367/71).
[17] K. D. Sinelnikov, V. E. Ivanov, V. M. Amonenko, V. D. Burlakov (Proc. 2nd Intern. Conf. Peaceful Uses At. Energy, Geneva 1958, Vol. 4, pp. 295/301, 296/7, 300).
[18] V. E. Ivanov, V. M. Amonenko, G. F. Tikhinskii, A. A. Kruglykh (Fiz. Metal. Metalloved. **10** [1960] 581/5; Phys. Metals Metallog. [USSR] **10** [1960] 88/92; C. A. **1961** 9197).
[19] V. M. Amonenko, L. N. Ryabchikov, G. F. Tikhinskii, V. A. Finkel (Dokl. Akad. Nauk SSSR **125** [1959] 977/8; Proc. Acad. Sci. USSR Phys. Chem. Sect. **124/129** [1959] 825/6).
[20] V. E. Ivanov, B. M. Amonenko, G. F. Tikhinskii, I. I. Papirov, L. N. Ryabchikov, V. N. Grinyuk (3rd Conf. Intern. Met. Beryllium, Grenoble 1965 [1966], pp. 33/7 from C.A. **67** [1967] No. 84259).

[21] N. J. Keen, E. W. Hooper (Inst. Mining Met. Trans. C **75** [1966] 262/4; C.A. **66** [1967] No. 13409).
[22] E. W. Hooper, N. J. Keen (Inst. Metals Monograph Rept. Ser. No. 28 [1963] 579/87, 581, 585; C.A. **60** [1964] 11687).
[23] E. W. Hooper, N. J. Keen (CONF-170 [1965] 1/5; Met. Abstr. [3] **1** [1966] 1774).
[24] E. W. Hooper, N. J. Keen (AERE-R-3321 [1960] from [20]; N.S.A. **15** [1961] No. 11567).

[25] J. P. Pemsler, S. H. Gelles, E. D. Levine, A. R. Kaufmann (Inst. Metals Monograph Rept. Ser. No. 28 [1963] 570/8; C.A. **60** [1964] 10307).
[26] B. L. Blanc (Bull. Inform. Sci. Tech. [Paris] No. 62 [1962] 16/9).
[27] E. D. Levine, J. P. Pemsler, S. H. Gelles (J. Nucl. Mater. **12** [1964] 40/9; C.A. **61** [1964] 5318).
[28] M. J. Hordon, W. T. Hess (Met. Soc. Conf. **33** [1966] 67/86).
[29] V. E. Ivanov, I. I. Papirov, G. F. Tikhinskii, V. M. Amonenko (Chistye Sverkhchistye Metally, Metallurgi, Moscow 1965, pp. 1/263; C.A. **64** [1966] 13879 from [29]).
[30] G. F. Tikhinskii, I. N. Khristenko (Izv. Akad. Nauk SSSR Metally **1969** No. 5, pp. 90/4; Russ. Met. **1969** No. 5, pp. 58/61; C.A. **72** [1970] No. 5341).

[31] K. L. Edwards, A. J. Martin (Inst. Metals Monograph Rept. Ser. No. 28 [1963] 557/69, 561).
[32] B. L. Blanc (3rd Conf. Intern. Met. Beryllium, Grenoble 1965 [1966], pp. 39/62; C.A. **67** [1967] No. 102331).
[33] M. J. Hordon, W. T. Hess (J. Metals **16** [1964] 693/9, 694).
[34] E. H. Hohmann (Chem. Age India **20** [1969] 169/74).
[35] G. F. Silina, Yu. I. Sarembo, G. J. Kaplan (Tsvetn. Metal. **30** No. 1 [1957] 66/71).
[36] R. F. Bunshah, R. S. Juntz (Met. Soc. Conf. **33** [1966] 1/28; C.A. **66** [1967] No. 87975).
[37] R. F. Bunshah (UCRL-7289 [1963] 1/24 from N.S.A. **17** [1963] No. 25784).
[38] R. F. Bunshah (UCRL-12253 [1965] 1/16; C.A. **63** [1965] 10950).
[39] R. F. Bunshah (CONF-170 [1965] 214/20; Met. Abstr. [3] **1** [1966] 1774).
[40] G. J. London, M. Herman (3rd Conf. Intern. Met. Beryllium, Grenoble 1965 [1966], pp. 21/32; C.A. **67** [1967] No. 84263).

[41] G. London, M. Herman (AD-426539 [1963] 1/13; N.S.A. **19** [1965] No. 7872).
[42] R. F. Bunshah, R. S. Juntz (Trans. 8th Vacuum Met. Conf., New York 1965 [1966], p. 200 from [43]).
[43] F. Sperner (Vakuum-Tech. **19** [1970] 42/8, 47; C.A. **73** [1970] No. 28114).

1.5.5 Refining by Transport Reaction

The vapor pressure of beryllium metal is rather low at technically practicable temperatures for a commercial scale use (cf. p. 100) of direct vacuum distillation. For a further purification of prepurified beryllium metal, direct vacuum distillation is therefore only possible with low production rates [1].

In order to provide a distillation process at an increased production rate, methods of indirect vacuum distillation by transport reactions have been proposed in which a more volatile halide is substituted for the beryllium in the vapor state [1], cf. [2, 3].

A reaction between prepurified beryllium and beryllium dihalide, according to $Be(c, l) + BeX_2(g) \rightleftharpoons 2\,BeX(g)$, has been suggested for use in beryllium refining [1, 3]. This reaction is analogous to the reaction $2\,Al(c) + AlCl_3(g) \rightleftharpoons 3\,AlCl(g)$, proposed by [4] for aluminium refinement. The reaction is based on the thermal instability of gaseous beryllium monohalide, BeX (X = halide). Crude beryllium must be contacted with vaporized BeX_2 at elevated temperatures and/or reduced pressure to be partially volatilized as gaseous BeX. On cooling or increasing the pressure, the reaction is believed to reverse, redepositing the beryllium in purified form and regenerating BeX_2 [1 to 3]. No experimental work on this method has been reported [1].

A similar process for indirect distillation of crude beryllium by means of sodium chloride according to: $Be(c) + 2\,NaCl(g) \rightleftharpoons BeCl_2(g) + 2\,Na(g)$ has been patented by [2]. The equilibrium constant K_p (in atm) for this reaction applicable for a reaction temperature of 1150 °C is $\log K_p = -11740/T + 4.59$ (T in K) [2].

This reaction is also strongly endothermic. When crude beryllium is contacted with the vapor of NaCl at high temperatures and reduced pressure, the reaction proceeds to a limited extent to the right of the equation. If the gas mixture is cooled or the pressure increased, the reaction reverses and pure beryllium deposits [2, 3].

The reaction has been investigated by [5]. The $BeCl_2$ partial pressure is estimated to be 3×10^{-1} Torr at the melting point of beryllium and at a total salt vapor pressure of 1 Torr. This is more favorable than that calculated for beryllium monochloride in the previous reaction (see p. 106) [5].

The apparatus consists of an evacuated (10^{-4} Torr) horizontal quartz tube containing NaCl in a Nimonic boat at the closed end and crude beryllium (94% Be) in a molybdenum tube at the center with an extension, acting as the beryllium condenser [5].

About 3 g of beryllium are distilled in 5 h by heating the NaCl to 800 °C and the crude beryllium to 1000 °C. Analysis of the reaction products shows that the distillation rate is tenfold higher than that obtainable by distilling beryllium alone [5]. The beryllium deposits are thin foils, bright, pliable but coarsely crystalline. The tensile test of a 0.005 in. thick piece shows a maximum stress of 13.5 tons/in.^{2} and an elongation of 3.2% [6], cf. [2].

Studies of the thermal vapor phase dissociation of Be iodide (van Arkel technique) for obtaining purified Be metal are described under 1.4.6.1, p. 89.

References:

[1] G. E. Darwin, J. H. Buddery (Beryllium, Butterworth, London 1960, p. 99).
[2] P. Gross, International Alloys Ltd. (U.S. 2607675 [1949/52]).
[3] H. H. Kellogg (in: D. W. White Jr., J. E. Burke, The Metal Beryllium, ASM, Cleveland 1955, pp. 49/62, 61).
[4] P. Gross, C. S. Campbell, P. J. C. Kent, D. L. Levi (Discussions Faraday Soc. **4** [1948] 206/15).
[5] D. R. Levi, D. H. Munro (Fulmer Res. Inst. Unpubl. Rept. No. R 16 [1952] from [1]).
[6] W. Munro (AERE Harwell, private communication from [1]).

1.5.6 Refining by Zone Melting

Floating zone-melting has proved to be an effective means of purification. After ten passes, the oxygen content in the middle of a single crystal rod (150 mm length) was reduced from 200 to 90 ppm, that of Fe from 90 to 36 ppm, Al and Mg from 20 to 0.2 ppm, Cr from 5 to 0.6 ppm, and Co from 5 to 0.2 ppm. The smallest effect was noted for Cu which decreased from 20 to 16 ppm. The degree of purity can also be characterized by the ratio of the residual resistance measured at 298 and 4.2 K ($R_{298}/R_{4.2}$) which increases with the purity of the metal, Jönsson et al. [1]. An ingot with 99.98% Be was purified in twelve zone refining passes to a residual content (atomic fraction) of 5 to 10 ppm each of C and O, 1 to 5 ppm of each Ni and Fe, about 1 to 2 ppm of Cu and <1 ppm of other residual impurities, as shown by mass spectrographic, emission spectrographic, and γ activation analyses, London et al. [2]. If zone melting of a 150 mm rod is done under an alternating current of 4 A/mm^2, the Fe content is decreased after three passes from 100 to 40 ppm, that of Mn from 20 to <2 ppm, Al from 60 to 30 ppm, Mg from 40 to 20 ppm, and Cu from 20 to 15 ppm, according to spectrographic analysis in the middle of the rod. An increase to six zone melting passes did not lead to an appreciably purer crystal, Grinyuk et al. [3], Ivanov et al. [4]. A considerable reduction of the O_2, Al, Mn, Cr, Co, Ta and C content which after six zone melting passes at a speed of 40 mm/h were concentrated in the "tail" of the bar, a less distinct reduction of Fe, no effect for Cu and an inverse effect for Ni were found by Schaub, Cabane [5], Schaub [6]. The purification effect (ratio: initial/final head content of impurity) in vacuum cast bars 15 to 25 cm long and 4.8 to 19 mm in diameter after six to eight refining passes under argon was high for Al(4.6) and Si(4.8). It was, however, less for Fe(1.8) when the melting zone traveled upward and only about half as high when it moved downward in the bars at a speed of 46 mm/h. Reducing the zone traveling speed to 4.6 mm/h essentially did not change the degree of purification. In horizontal zone melting (4.8 mm/h), this effect was much lower for Al(2.1) and Si(1.3), but somewhat higher for Fe(2.3) than in vertical zone melting with upward motion, Edwards, Martin [7]. Horizontal zone melting (up to 12 passes) of a hot-pressed powder ingot (97.84% Be) in BeO boats under Ar ($^1/_6$ atm) with traveling speeds of 38 and 76 mm/h reduced the Al content to only 10 ppm and removed other impurities such as Si, Fe, and BeO, Mitchell et al. [8]. The removal of Fe from beryllium is rather difficult at a transport speed of 40 mm/h because the distribution coefficient of Fe is only $K_{Fe}=0.9\pm0.05$. However, the presence of Al (3 wt%) in the molten zone shifts the effective value to $K_{Fe}=0.75\pm0.05$ thus facilitating the separation of Fe, Schaub, Desre [9], also see Desre et al. [10]. A similar reduction of K_{Fe} to 0.7 ± 0.05 can be achieved (in six passes) by applying a direct electrical current of 200 A/cm² with the positive pole at the head of the bar, Schaub, Potard [11]. Since at the zone traveling speed of 40 mm/h K_{Al} in Be is very low ($K_{Al}=0.01$), the aluminium (3%) added to promote the separation of Fe may be rapidly eliminated (<0.1 ppm) by subsequent zone melting passes, Schaub, Desre [9].

Purification of beryllium is also done by zone melting the bar-shaped metal under electron bombardment (6 kVA) in a high vacuum (10^{-5} to 4×10^{-7} Torr). After degassing at 700 °C, the bar is placed on a water-cooled Cr plated hard copper sole through the melting zone of a focussed electron beam, twice at 200 mm/h and twice at 130 mm/h. Almost all electrolyte ($NaCl+BeCl_2$) and Mg are eliminated by the process. A considerable reduction of the oxygen (from 600 to 25 at-ppm) and carbon (from 75 to 30 at-ppm) content was also reported. The other impurities (Si, Al, Fe, Mn, Ni, Cu, Cr, 6 to 0.6 at-ppm) were left almost unchanged, Schaub, Gallet [12].

References:

[1] S. Jönsson, A. Freund, F. Aldinger (Metall **33** [1979] 1257/61, 1260).
[2] G. London, V. V. Damiano, H. Conrad (Trans. AIME **242** [1968] 979/86, 979).
[3] V. N. Grinyuk, I. I. Papirov, G. F. Tikhinskii, I. G. D'yakov (Izv. Akad. Nauk SSSR Metally **1967** No. 4, pp. 77/84; Russ. Met. **1967** No. 4, pp. 33/6).
[4] V. E. Ivanov, B. M. Amonenko, G. F. Tikhinskii, I. I. Papirov, L. N. Ryabchikov, V. N. Grinyuk (3rd Conf. Intern. Met. Beryllium Commun., Grenoble 1965 [1966], pp. 32/7, 35).
[5] B. Schaub, G. Cabane (Compt. Rend. **257** [1963] 444/7, 446).
[6] B. Schaub (Symp. Zonenschmelzen Kolonnenkristallisieren, Karlsruhe 1963, pp. 227/46, 229, 234; C.A. **63** [1965] 5317).
[7] K. L. Edwards, A. J. Martin (Inst. Metals Monograph Rept. Ser. No. 28 [1963] 557/69, 566/8).
[8] W. R. Mitchell, J. A. Mullendore, S. R. Maloof (Trans. AIME **221** [1961] 824/6).
[9] B. Schaub, P. Desre (3rd Conf. Intern. Met. Beryllium Commun., Grenoble 1965 [1966], pp. 93/4; C.A. **67** [1967] No. 102236).
[10] P. Desre, B. Schaub, E. Bonnier (Thermodyn. Proc. Symp., Vienna 1965 [1966], Vol. 1, pp. 493/500, 499, discussion p. 520; C.A. **65** [1966] 8032).

[11] B. Schaub, C. Potard (3rd Conf. Intern. Met. Beryllium Commun., Grenoble 1965 [1966], pp. 87/91, 89; C.A. **67** [1967] No. 102333).
[12] B. Schaub, J. Gallet (3rd Conf. Intern. Met. Beryllium Commun., Grenoble 1965 [1966], pp. 511/5, 511, discussion p. 516; C.A. **68** [1968] No. 15325).

1.5.7 Refining by Recrystallization from Aluminium Melts

A mixture of Al grains and Be flakes (up to 20 at%) was heated under Ar at 1200 °C until molten. The melt was then cooled to 750 °C at a rate of 50 °C/h and held at that temperature for 10 h. The first crystals appeared at 930 °C and the last ones at 645 °C, a temperature slightly above the eutectic point where the melt contains 2.5 at% Be. The Be crystals appeared as needles several mm long. To avoid contamination, the crystals were decanted and vacuum filtered off at 700 °C through a filter crucible of recrystallized Al_2O_3. Al was removed from the Be crystals by leaching with aqueous 20% NaOH, which at 20 °C dissolves Al readily but not Be. The average yield of recrystallized Be was 70%, but yields up to 90% are possible. After two passes, the foreign metal content shifted from 0.81 to 0.04 at-ppm for Fe, from 4 to 0.45 at-ppm for Mn, from 4.5 to 0.4 at-ppm for Ni (decrease), but from 0.7 to 17 at-ppm for Cu and from 3 to 400 at-ppm for Al. The Al can be removed from Be by Mg which forms a low melting eutectic (melting point 449 °C) containing 52.7 at% Mg and 0.165 at% Be. For this purpose Mg was added to the recrystallized Be up to a 1:1 Al:Mg atomic ratio. The mixture was heated to 700 °C under Ar and kept at that temperature for 10 h after which it was filtered. Two extractions were sufficient to eliminate the Al. The Mg that adhered to the Be crystals was removed by evaporation in vacuum at 700 °C for 3 h. The last traces of Al were removed with dilute aqueous KOH, Potard [1], also see Potard, Schaub [2]. Pure crystals of Be were also obtained by cooling a molten Be alloy containing Si, Al, or both from 1200 to 800 °C for Al (43%) or Al+Si (36+3%) or to about 500 °C for Si. At said temperatures the Be crystals were removed by skimming, centrifuging or filtering in an inert atmosphere in order to avoid coprecipitation of impurities. The melt may also be slowly cooled to allow the Be to crystallize, then quenched, and the solid Be particles were separated from the crushed slag. The recovery of pure Be crystals was 98% with Al and 98.4% with Al+Si. The yield for a Be alloy with 25% Si and no Al at 1090 °C was 78.7%. The higher the Be content and the lower the Si:Al weight ratio in the melt (preferably 1:12 or lower), the greater the crystallization yield. A Be:Si weight ratio of >2:3 was recommended, and the impurities in the melt should be <5%, preferably about 1%, Lerman [3]. Thermal treatment of commercial Be in the presence of Al at 650 to 800 °C shows that impurities like Fe are recovered in the liquid Al phase where they are more soluble than in solid Be. The separation of Fe from commercial Be is controlled by the thermodynamic activity coefficients of Fe in liquid Al and solid Be and by the diffusion of Fe in Be rather than by the precipitation of Fe from a supersaturated solid solution in Be. Thus the content of Fe in the matrix was reduced far below the limit of solubility, from 400 ppm to almost zero, Adda et al. [4]. The relation between the distribution coefficient and the maximum solubility of impurities such as Fe, Co, Zr, Cu, Ag, Al, and N in solid Be was studied and calculated by Vakhobov et al. [5].

References:

[1] C. Potard (3rd Conf. Intern. Met. Beryllium Commun., Grenoble 1965 [1966], pp. 95/9; C.A. **67** [1967] No. 102337).
[2] C. Potard, B. Schaub (Fr. 1435134 [1965/66] 1/6, 4/5; C.A. **65** [1966] 18280).
[3] F. Lerman (U.S. 3719472 [1970/73] 1/12, 3, 7; C.A. **78** [1973] No. 150379).
[4] Y. Adda, N. Azam, G. Donze, J. Mallen, F. Maurice, M. Weisz (Compt. Rend. **254** [1962] 1052/4).

[5] A. V. Vakhobov, V. G. Khudaiberdiev, V. N. Vigdorovich (Dokl. Akad. Nauk Tadzh. SSR **11** No. 8 [1968] 19/22; C.A. **70** [1969] No. 61675).

1.5.8 Refining of Be by Spinel Forming Oxides

Beryllium can be purified from BeO and other spinel forming metals by sintering 1 to 2 vol% of finely powdered (<44 μm) Al_2O_3 at 1149 °C or Cr_2O_3 at 1095 °C for 3 h under argon. To remove the spinels, the metal is melted in a controlled (inert) atmosphere where the spinels separate [1].

Reference:

[1] D. M. Scruggs, Bendix Corp. (Ger. 1188816 [1964/65] 1/3; C.A. **62** [1965] 12858).

1.5.9 Electrolytic Refining with $BeCl_2$ in Eutectic KCl-LiCl Melts

The LiCl-KCl eutectic was chosen because it allows suitable electrolytic refining of Be at a far lower concentration of $BeCl_2$ in the electrolyte than the NaCl-$BeCl_2$ eutectic (10 to 20% instead of 40 to 50%), although it is operated at higher temperatures (500 ± 50 °C instead of 300 to 350 °C). Such preference is due to the fact that this solvent is not depleted of any of its constituents, especially of the expensive $BeCl_2$, by the removal of the metal. This causes serious losses of $BeCl_2$ in the NaCl-$BeCl_2$ electrolyte due to the occlusion of $BeCl_2$ in the cathodic deposit. The melting point of the KCl-LiCl eutectic is unaffected by $BeCl_2$ contents up to 20%. The electrolysis can be carried out at higher operating currents and $BeCl_2$ concentrations down to 3%. However, 10 ± 3% $BeCl_2$ is recommended for Be dendrites and 20% $BeCl_2$ for platelets. On the other hand the higher partial pressure of $BeCl_2$ at 500 °C and the higher cost of electrolytic grade LiCl (as compared with NaCl in the $BeCl_2$-NaCl melts) can not be neglected, Schimmel [1]. Generally, the electrolyte was prepared by melting a eutectic mixture of oven-dried (300 °C) alkali chlorides (60 mol% LiCl) at 400 °C overnight under a vacuum of 10^{-3} to 10^{-4} Torr for complete dehydration. The melt then was treated with gaseous HCl for 1 h to convert any LiOH into LiCl and heated again to 450 °C for 8 h under vacuum to drive off the last traces of water which would cause excessive foaming upon addition of $BeCl_2$, forming a basic chloride, Schimmel [1], Menzies et al. [2]. The $BeCl_2$ solution (10 to 20 wt%) in the KCl-LiCl eutectic (59 mol% LiCl) was also prepared "in situ" either by reacting Be with $SnCl_2$ dissolved in the salt melt, Bilard et al. [3], Chauvin et al. [4], Boisdé et al. [5], with gaseous HCl in a perforated graphite tube which was partially immersed into the salt melt, Wong et al. [6, 7], or with Cl_2, Wong et al. [8, 9]. The electrolyte should contain 10 to 20 wt% $BeCl_2$ (purified by vacuum distillation) since at appreciably lower $BeCl_2$ levels the deposited Be may contain Li. The cell body should be constructed of graphite because it is completely inert under the conditions of electrolysis, and to moist HCl. Stainless and mild steel and Pyrex glass are more or less seriously corroded under these conditions. The only drawback of graphite cells is that they crack sometimes on remelting the solidified electrolyte. Because of the hygroscopicity of the electrolyte it is advisable to work in an inert atmosphere. The most convenient anode is a wire gauze basket (Ni) loaded with chips or pieces of hot-pressed beryllium. The cathode consists of a sheet of nickel, stainless steel, Inconel, or a graphite rod, Schimmel [1]. The electrolysis cell used by Bilard et al. [3] consisted of a cylindrical silica crucible placed in a waterproof container of stainless steel which was provided with a sidetap for the admission of argon and a lid bearing the electrodes. The anode is a rod of commercial Be to be purified and the cathode is a nickel pot with a perforated bottom [3]. A nickel cell fitted with an annular nickel anode basket containing beryllium scraps or pebbles and a cylindrical nickel cathode was used by Ballance et al. [10]. A mild steel cell with a similar nickel anode and an internally water-cooled nickel tube cathode was used by Wong, O'Keefe [8]. For the use of oblong shaped electrodes, see Wong, Klosterman [9]; for iron electrodes, see Wong et al. [6, 7].

Prior to refining, pre-electrolysis was carried out under flowing Ar using a graphite anode and a nickel cathode. The voltage was kept just below the decomposition voltage of the eutectic to remove final traces of moisture and metallic impurities such as iron, Menzies et al. [2], and tin which had been introduced into the melt by the chlorination of Be with $SnCl_2$, see above, Bilard et al. [3]. Pre-electrolysis was continued overnight after which the cell was evacuated and purged with Ar. The refining electrolysis electrodes were then slid down and electrolysis was started under flowing (300 mL/min) argon, Menzies et al. [2]. The operation conditions for electrolysis may vary over a wide range. The purest products were obtained with a current strength of 8 to 20 A (graphite: 110 A), a voltage of 0.44 to 1.25 V, an anodic current density of 15.0 to 38.0 A/dm^2, and electrolyte temperatures between 500 and 550 °C (in graphite 460 °C) under purified argon, Schimmel [1]. Very pure Be was also obtained by electrolysis at 450 to 500 °C and an anodic current density of $\leqq$15 A/dm^2, Bilard et al. [3], 2 to 10 A/dm^2, Chauvin et al. [4], Boisdé et al. [5, 11], Coriou [12], at the cathode $\leqq$30 A/dm^2 [3], 10 to 30 A [4, 5, 11, 12], up to 93 A/m^2 at 550 °C see Ballance [10]. For the refining of Be in electrolytes containing 3.5 to 16.5 mol% $BeCl_2$ at 450 to 550 °C and cell currents of 3 to 53 A, see Wong et al. [6 to 9].

The cathodic deposit (Be) is first quenched with ice to avoid losses of Be due to hydrolysis, then filtered off, washed, and digested with cold 5% HNO_3 for about 8 h to remove the electrolyte salt, then washed with acetone or ethanol and dried, Schimmel [1]. Repeated immersion in aqueous 4 N HNO_3 at 2 to 4 °C and final immersion in 4 to 6 N HNO_3 for several hours alternating with short periodic immersions into 0.5 N HF removed the last traces of salt trapped in the cavities of the metal. It was then rinsed with demineralized water and pure methanol, dried under vacuum, and stored under argon, Boisdé et al. [5]. Leaching with dilute aqueous NaOH to remove Al and washing with dilute HNO_3, water, and acetone was recommended by Menzies et al. [2]. The dried metal was treated by a sink-float method with a mixture of ethylene bromide and acetone at a density of 1.88 g/cm^3 to remove BeO. The floating Be was collected and washed with acetone [1]. The oxide content of the metal was minimized by bubbling dry gaseous HCl through the electrolyte salt melt and subsequent preliminary electrolysis. The amount of surface oxides in the refined metal was reduced when the deposit was separated from the adherent metal chlorides by ultrasonic treatment with ethanol, Schulze [13]. For earlier papers dealing with the electrorefining of Be under varying conditions, see Wong et al. [14], Gurklis et al. [15].

Depending on the $BeCl_2$ content in the electrolyte, the steel gray beryllium consisted of large lamellas (at 16.5 mol% $BeCl_2$) or plate-like crystals (at 10.3 mol% $BeCl_2$ and low current densities). High current densities produced fiber-like crystals, Wong et al. [6, 7], also see [8, 9]. Large, well-crystallized flakes were obtained by Boisdé et al. [5]. Lamellas or dendrites consisting of hexagonal scales were found by Bilard et al. [3], also see Ballance et al. [10]. The cathodic current efficiency ranged from 70 to 80%, Bilard et al. [3], and 70 to 90% of the recovered beryllium, Wong et al. [6, 7]. The procedure adopted by Bilard et al. [3] reduced the main metallic impurities (Al, Cr, Cu, Fe, Mg, Mn, Ni) from 1000 to 4000 ppm to $\leqq$35 ppm and the oxygen content from 1000 to 5000 ppm down to 100 to 300 ppm [3]. For values for the individual metals (<15 ppm except for Al: <25 ppm) and salt ions (K^+, Li^+, Cl^-), see Chauvin et al. [4], Boisdé et al. [5]. The purest product isolated by Schimmel [1] contained <100 ppm of metallic impurities such as Al, B, Cd, Co, Cr, Cu, Fe, Mg, Mn, Ni, Pb, Si or Ti and attained the spectrographic purity of commercial triple-refined super-fine grade beryllium [1]. The refined metal produced by Ballance et al. [10]

assayed the following maximum impurity levels (in ppm): 300 Fe, 300 C, 200 Ni, 100 Al, 100 Si, 60 Mg and 50 Cu [10]. Wong et al. obtained somewhat poorer results using electrolytes with 10.3 and 16.5 mol% Be [6, 7]. The oxygen content was lowered from 3.27% (in the anode feed) to about 0.004% and the magnesium content from 2.66 to 0.0003%. The Al, Ni, Co, and Cr content was reduced to $<0.01\%$, that of Fe and Si became substantially lower, but the purification from Mn, Cu, and Ca was much less effective [6, 7].

References:

[1] F. A. Schimmel (Y-1380 [1961/62] 1/31, 11/3, 21/8; N.S.A. **16** [1962] No. 15127).
[2] A. Menzies, D. L. Hill, L. W. Owen (J. Less-Common Metals **1** [1959] 321/30, 324/6; Nature **183** [1959] 816/7).
[3] J. Bilard, G. Boisdé, M. Broc, G. Chauvin, H. Coriou, J. Hardy (Metaux **42** [1967] 259/69, 261/3).
[4] G. Chauvin, H. Coriou, J. Hure (Metaux **37** [1962] 112/26, 119/21).
[5] G. Boisdé, M. Broc, G. Chauvin, H. Coriou, J. Hure, P. Jarny (J. Nucl. Mater. **6** [1962] 256/64, 257/60).
[6] M. M. Wong, R. E. Campbell, D. H. Baker (U.S. Bur. Mines Rept. Invest. No. 5959 [1962] 1/14, 3/5, 8, 11/2; C.A. **56** [1962] 15222; N.S.A. **16** [1962] No. 15117).
[7] M. M. Wong, R. E. Campbell, D. H. Baker (J. Metals **12** [1960] 786/8).
[8] M. M. Wong, D. A. O'Keefe (U.S. Bur. Mines Rept. Invest. No. 6570 [1964] 1/8, 3/6; C.A. **62** [1965] 8671; N.S.A. **19** [1965] No. 13870).
[9] M. M. Wong, J. E. Klosterman (BM-RI-6489 [1963] 1/19, 2/4, 8, 12, 15; C.A. **62** [1965] 14229).
[10] J. Ballance, A. J. Stonehouse, R. Sweeney, K. Welsh (Kirk-Othmer Encycl. Chem. Technol. 3rd Ed. **3** [1978] 803/23, 810).
[11] G. Boisdé, M. Broc, G. Chauvin, H. Coriou, L. Hardy, P. Jarny (Bull. Inform. Sci. Tech. [Paris] No. 62 [1962] 29/38, 29/32; C.A. **58** [1963] 12172).
[12] H. Coriou (Bull. Inform. Sci. Tech. [Paris] No. 84 [1964] 17/35, 28/31; C.A. **65** [1966] 16397).
[13] K. Schulze (AED-CONF-78-078-002 [1978] 1/16, 7/9; C.A. **91** [1979] No. 42520).
[14] M. M. Wong, F. R. Cattoir, D. H. Baker (U.S. Bur. Mines Rept. Invest. No. 5581 [1960] 1/9, 4/7; C.A. **1960** 10591).
[15] J. A. Gurklis, J. G. Beach, C. L. Faust (BMI-781 [1952/55] 1/26, 4/8; N.S.A. **10** [1956] No. 1367).

1.5.10 Electrolytic Refining in Eutectic $BeCl_2$-NaCl Melts

The anodic dissolution of Be in a eutectic $BeCl_2$-NaCl melt containing 50 mol% $BeCl_2$ and its migration to the cathode were studied at 400 to 500 °C in an inert (Ar) atmosphere. The electrolyte was prepared by melting a mixture of the oven-dried (300 °C) components at about 400 °C under Ar. Treatment with gaseous HCl and electrolysis (anode: Be, cathode: Ni) was carried out as described for the electrolytic refining of Be in eutectic KCl-LiCl melts on p. 110, Menzies et al. [1]. The best results were obtained with an electrolyte composed of a fused mixture of $BeCl_2$ and NaCl or the KCl-NaCl eutectic with a $BeCl_2$ content of 40 to 50 wt% at temperatures between 280 and 340 °C and cathodic current densities of 5 A/dm^2, Silina, Grinberg [2]. The deposited metal was leached with dilute aqueous NaOH to remove Al, washed with dilute HNO_3, water and acetone, and air dried. It consisted of large adherent shiny flakes, Menzies et al. [1], of large needles at 50 wt% $BeCl_2$, of fine needles and scales at 40 wt% $BeCl_2$, Silina, Grinberg [2] which attained a purity of 99.3 to 99.4% Be [1]. The highest current yield (93%) and refining effect was observed at

50 wt% $BeCl_2$ which produced the following impurity reductions: Fe and Al each from 0.05 to 0.003%, Ni from 0.05 to 0.005%, whereas Cu and Mn remained almost constant at 0.02 and 0.001%, respectively. The increase in the anodic current density from 6 to 50 A/dm^2 gained by submerging the anode to various depths afforded essentially no changes of the impurity levels in the cathodic deposit. A reduction of the $BeCl_2$ content to 40% or less, however, strongly increased the amount of impurities in the cathodic deposit. Enlarged pilot plant tests (20 runs) conducted at 350 °C in a 9 L quartz crucible at an average current strength of 25 A and a voltage of 2.5 to 3.5 V yielded a refined metal containing (in %, technical feed in parentheses): Fe 0.004 (0.12), Ni 0.007 (0.14), Mn 0.002 (0.008), Cu 0.0028 (0.02). The current yield fluctuated between 75 and 80%. It became almost constant and reached 85 to 90% after a dense layer of Be had formed at the cathodes. Under these conditions, 93% of the anodic Be feed could be recovered at the cathodes. Similar results were also obtained in factory tests, Silina, Grinberg [2]. According to Logerot [3], electrolytic refining of Be in eutectic $BeCl_2$+NaCl (or KCl+LiCl) melts yielded a metal containing 99.8% Be in the first and 99.9% Be in the second refining pass [3].

References:

[1] A. Menzies, D. L. Hill, L. W. Owen (J. Less-Common Metals **1** [1959] 321/30, 324/6; Nature **183** [1959] 816/7; N.S.A. **10** [1956] No. 1367).

[2] G. F. Silina, L. L. Grinberg (Tsvetn. Metal. **33** No. 12 [1960] 47/53; Soviet J. Non-Ferrous Metals **33** No. 12 [1960] 46/52, 47/9).

[3] J. M. Logerot (3rd Conf. Intern. Met. Beryllium Commun., Grenoble 1965 [1966], pp. 471/87, 471/3, discussion p. 488; C.A. **68** [1968] No. 15444).

1.6 Preparation of Special Forms

1.6.1 Powders

Beryllium powders can be obtained by multiple turning of cast ingots and grinding the swarf in an attrition mill under dry N_2 to a <0.074 mm (−200 mesh) size. The attrition device consists of two counter rotating (800 rpm) water cooled disks with beryllium faces [1, 2]. Prior to grinding the swarf is transferred into a magnetic separator to remove any ferrous impurity [2]. A finer powder (about 50% <0.044 mm) can be obtained by repeated grinding [1]. Fine powders of a 6 to 13.5 µm particle size were obtained by ball-milling of Pechiney or Brush (coarser) beryllium powders for 12 h under Ar with methanol as the grinding fluid, Bufferd et al. [3]. Finely grained flake-shaped particles (splats) are produced by quenching molten droplets of Be against a cooled rotating copper disk or a heavy-walled copper tube. The splats are obtained by arc melting the tip of a Be electrode rotating at a high speed above the face of the copper disk in vacuum or under a static or dynamic inert gas (He) atmosphere. The solidified splats are irregularly shaped platelets about 5 to 10 µm thick and 1000 to 2000 µm long with grain diameters of 1 to 10 µm [4, 5].

Spherical particles can be prepared by induction melting of Be in a vibrating crucible provided with a nozzle at its bottom. The droplets leaving the nozzle fall through a long (5.3 m) tube where they become solid. The grain size can be modified by varying the jet rate with the aid of H_2 or an inert gas but it is rather difficult to obtain particles of uniform size since liquid Be has a low density (1.42 g/cm^3) and a high surface tension (1950 dyn/cm) [6].

Powder particles of 1 to 2 µm size can be produced by evaporating Be from heated tungsten baskets after previous evacuation of the Pyrex vessel to 10^{-5} Torr and re-admission of Xe to a pressure of 2 to 15 Torr. The basket is heated with an energy of 100 W for a few seconds to produce Be vapor which deposits on the walls of the vessel. Finer particles ($^1/_4$ to $^1/_3$ µm) can be obtained using Ar instead of Xe. The particles appear as rhombic dodecahedra (mainly at $p_{Xe}=4$ Torr) or tetragonal plates (prevailing at $p_{Xe}=14$ Torr) [7], also see [8 to 10]. Ultrafine beryllium powder of 0.03 to 0.3 nm particle size was produced by rapid evaporation of beryllium (melt) at 1450 °C and 0.1 to 0.3 Torr (Ar). The Be vapors were condensed in the space adjacent to the rotating drum to yield the desired powder size in a recovery of about 18% of the initial charge or 27 to 61% of the evaporated metal, Raymond [11]. For preparation of Be powder see also Section 1.4.4, pp. 74/5.

References:

[1] W. W. Beavor (Metal Progr. **65** No. 4 [1954] 92/7, 93, 170/3; C.A. **1954** 5762).

[2] R. E. Green (Mach. Prod. Eng. **110** [1967] 1072/9, 1077; C.A. **69** [1968] No. 69257).

[3] A. S. Bufferd, R. Widmer, N. J. Grant (ASD-TDR-62-509-Vol. III [1964] 44/65, 45, 49; N.S.A. **19** [1965] No. 4644).

[4] A. D. Kaufman, W. C. Muller (NMI-6021-13 [1963] 1/52, 3/4; C.A. **63** [1965] 3969; N.S.A. **19** [1965] No. 13995).

[5] L. R. Aronin (3rd Conf. Intern. Met. Beryllium Commun., Grenoble, Fr., 1965 [1966], pp. 547/51, 547; C.A. **67** [1967] No. 84331).

[6] F. Aldinger, F. Linck, N. Claussen (Mod. Dev. Powder Metall. **9** [1976/77] 141/51, 145; C.A. **87** [1977] No. 139607).

[7] N. Wada (Japan. J. Appl. Phys. **7** [1968] 1287/93, 1287/8; C.A. **69** [1968] No. 98864).

[8] R. Uyeda (J. Cryst. Growth **24** [1974] 69/75, 71).

[9] K. Kimoto (Thin Solid Films **32** [1976] 363/5; C.A. **84** [1976] No. 158243).

[10] Y. Fukano, K. Nakao (Japan. J. Appl. Phys. **20** [1981] 477/87, 477/8).

[11] P. L. Raymond (ASD-TDR-62-509-Vol. III [1964] 53/65, 60; N.S.A. **19** [1965] No. 4644).

1.6.2 Beryllium with Low Density

Ultra-light beryllium was produced by milling near spherical particles of NH_4Cl or naphthalene with beryllium powder in a pelletizer. The addition of a paraffin wax binder in a solvent carrier provided a sufficiently adherent and coherent coating of the particles which was consolidated by applying isostatic pressures of 6.4 and 9.6 kg/cm² at low temperature. The solid carrier was then removed by heating the particles overnight in a vacuum at 88 °C in the case of naphthalene or at 143 °C in the case of NH_4Cl. This treatment did not affect the finely porous network structure of the Be coating which was consolidated into a self-supporting body by sintering in an evacuated closed container at 1177 and 1260 °C for 0.5 h. The density of the sintered product varied between 10.3 and 24.7% of the theoretical value (1.848 g/cm³). The density could be lowered somewhat by rounding the corners and dimpling the surface of the pellets, Mueller [1].

Reference:

[1] J. J. Mueller (UCRL-1372 [1976]; SANL-456019 [1976] 1/41, 13/28, 32; C.A. **90** [1979] No. 65818).

1.6.3 Filamentary Crystals

Filamentary crystals on a beryllium rod were observed when the rod was heated in vacuum at 1200 °C. The crystals appeared on the surface of the rod and were orthogonally orientated toward the axis of the rod. The Be vapors were collected in a tantalum collector heated to 1100 °C which surrounded the rod. Comparison with the starting material and the distillate revealed that the filamentary crystals were richer in low-vapor-pressure impurities such as Fe, Ni, and Cu, Bunshah [1].

Reference:

[1] R. F. Bunshah (UCRL-12252 [1955] 1/19, 7; C.A. **63** [1965] 11104).

1.6.4 Films and Foils

Thin self-supporting Be foils 5 to 10 μg/cm² thick and 0.63 cm in diameter were prepared by outgassing Be pellets for 15 min in vacuum (10^{-5} Torr) followed by rapid evaporation of the metal onto microscope slides covered with $Na_5P_3O_{10}$ in a vacuum of $<10^{-5}$ Torr. After reaching the desired thickness the slides were removed from the vacuum system, and the coating was loosened and floated off by slow immersion into distilled water. The foils serve as targets for beams of atomic and molecular ions with energies up to 2 MeV, Oona, Rickel [1]. Evaporation onto glass plates covered with a potassium soap film and recovering the deposited metal film by floating off with distilled water was proposed by Malakhov et al. [2]. Evaporation of Be onto a polished slab of rocksalt covered with a celluloid film and recovering the metal film by first removing the rocksalt with water and then the celluloid with amylacetate was suggested by Lukirskii [3]. Be films up to 0.025 mm thick were obtained by evaporation of Be in the focus of an electron beam under a vacuum of $\leqq 10^{-6}$ Torr. Glass proved to be the best substrate since it allowed the film to be stripped off with some success, Basche, Schetky [4]. The evaporation of Be in a vacuum of $<10^{-10}$ Torr and its deposition and growth on a tungsten {110} surface was studied by Schlenk, Bauer [5]. For deposition onto Ta backing by vacuum evaporation with direct heating see Bondar et al. [6]. Be films on Mo were produced by evaporation of Be at 1300 to 1500 °C in a vacuum of 10^{-6} Torr and precipitation of the vapors on suitably heated and degassed polished Mo plates. The dependence of the physical properties (plasticity) of the films on the thickness and temperature of deposition was studied by Amonenko et al. [7]. The Be films deposited on Fe could not be isolated, Basche, Schetky [4]. Slow evaporation of Be at 1000 °C and precipitation on a helium cooled (4.2 K) substrate (collodion?) yielded a film apparently consisting of a low temperature modification assumed to be stable up to at least 130 K, Fujime [8].

Beryllium coated Formvar plastic films for X-ray windows were produced by slow evaporation of Be in a vacuum onto Formvar films held by frames in a rigid stand. Eight to fourteen runs, each of 7 min followed by cooling for $^1/_2$ h between the runs were required to produce a smooth, reflective Be film coating of up to 60 μg/cm² total weight on both sides of the plastic film. The process was complete within about 8 h, Voigt [9]. Evaporation of Be at 1300 to 1400 °C in a vacuum of about 10^{-4} Torr produced a smooth coating of Be on a Zapon plastic film fastened to a substrate or backed by glass plates covered with a mold release. Be coated Formvar films gave similar results. Higher evaporation temperatures or poor thermal contact with the glass substrate produced a rough sandpaper finish of the Be coating on Formvar and tended to destroy the Zapon films to a larger degree than the Formvar films, Eichelberger [10]. For earlier papers concerning the production of flexible Be foils by vacuum distillation see Sloman [11], Smith [12]; for X-ray windows, see Smith [12], Brandner [13]. Beryllium coatings (>97% Be) up to 25 μm thick on various substrates such as copper or mild steel were produced as smooth gray deposits by thermal decomposition of $Be(t\text{-}C_4H_9)_2$ in a closed vessel at 280 to 305 °C and pressures of 0.7 to 0.8 Torr. The best smooth coatings were obtained at low plating rates of <25 nm/h. Higher plating rates, higher pressure and higher or lower temperature always lead to admixtures of metallic bubbles, Wood et al. [14, 15]. Beryllium coatings on tungsten filaments were obtained by reducing $BeCl_2$ with Zn or a mixture of Zn and H_2 at 732 °C (boiling point of $ZnCl_2$), optimum conditions see Davies et al. [16].

Beryllium foils up to 35 μm thick were obtained by passing Cu strips through a molten mixture of $BeCl_2$ and $Ba(BeF_4)_2$ (weight ratio 6:1), followed by electrolysis in a bath containing (in g/L) 120 $CuSO_4$, 45 H_2SO_4 and 30 mannitol with 40 A/dm², Inoue [17]. Electrodeposition of a compact adherent layer of Be on Cu was achieved by electrolyzing a saturated solution of BeF_2 (35 wt%) in a ternary eutectic mixture of LiF, NaF, and KF. The optimum conditions were around 600 °C with a cathodic current of 2 to 10 A/dm² and an anodic current of 1 to 4 A/dm². A current density of 9 A/dm² and stirring produced a uniform Be layer of 30 to 60 μm on the Cu cathode within 5 h, Binard et al. [18]. A strongly adherent film consisting of Be (80%), and Be_2C (20%) was obtained by electrolyzing a solution of KF · $2Be(C_2H_5)_2$ in a threefold excess of $Be(C_2H_5)_2$ at 75 °C on Cu cathodes, Strohmeier, Gernert [19], also see Strohmeier, Popp [20].

References:

[1] H. Oona, D. G. Rickel (Rev. Sci. Instr. **38** [1967] 980).
[2] I. Ya. Malakhov, A. S. Deineko, G. B. Andreev (Prib. Tekhn. Eksperim. **11** No. 3 [1966] 218/9; Instr. Exptl. Tech. **11** [1966] 741/2).
[3] A. P. Lukirskii (Izv. Akad. Nauk SSSR Ser. Fiz. **25** [1961] 910/2; Bull. Acad. Sci. USSR Phys. Ser. **25** [1961] 923/5).
[4] M. Basche, L. M. Schetky (WADC-TR-Pt. 2-58-457 [1960] 1/48, 8/10, 22/5; N.S.A. **14** [1960] No. 19333).
[5] W. Schlenk, E. Bauer (Surface Sci. **94** [1980] 528/46, 529).
[6] A. D. Bondar, A. S. Emlyaninov, A. P. Klyuchaev, L. G. Lishenko, V. N. Medyanik, A. D. Nikolaichuk, O. E. Shalaeva (Izv. Akad. Nauk SSSR Ser. Fiz. **24** [1960] 929/33; Bull. Acad. Sci. USSR Phys. Ser. **24** [1960] 926/30, 926).
[7] V. M. Amonenko, A. A. Kruglykh, V. S. Pavlov, G. F. Tikhinskii (Zavodsk. Lab. **26** [1960] 625/6; Ind. Lab. [USSR] **26** [1960] 668/9).
[8] S. Fujime (Japan. J. Appl. Phys. **5** [1966] 59/67, 66, 778/87, 778/80).
[9] J. W. Voigt (MLM-1194 [1964] 1/15, 6, 10; N.S.A. **19** [1965] No. 938).
[10] J. F. Eichelberger (MLM-1126 [1961] 1/27, 12/3; N.S.A. **16** [1962] No. 32881).

[11] H. A. Sloman (J. Inst. Metals **49** [1932] 365/88, 381).
[12] H. Smith (J. Sci. Instr. **26** [1949] 378/9).
[13] H. Brandner (Rev. Sci. Instr. **19** [1948] 662/4).
[14] J. M. Wood, F. W. Frey (Proc. Conf. Chem. Vapor Deposition Refract. Metals Alloys Compounds, Gatlinburg, Tenn., 1967, pp. 205/16, 208/10; C.A. **68** [1968] No. 32522).
[15] J. M. Wood, J. B. Chidester, E. Dyble (UCRL-13100 [1963/64] 1/21, 2/4, 7, 10; N.S.A. **19** [1965] No. 39102).
[16] L. G. Davies, J. C. Withers, D. F. Bazzarre (AD-611757 [1965] 1/39, 21/4; C.A. **63** [1965] 9638).
[17] K. Inoue (Japan. 71-16802 [1968/71] 1/3 from C.A. **76** [1972] No. 102457).
[18] M. Binard, G. Boisdé, M. Broc, G. Chauvin, H. Coriou (Fr. 1521522 [1967/68] 1/4; C.A. **71** [1969] No. 18301).
[19] W. Strohmeier, F. Gernert (Z. Naturforsch. **20b** [1965] 829/31).
[20] W. Strohmeier, G. Popp (Z. Naturforsch. **23b** [1968] 38/41).

1.6.5 Single Crystals

1.6.5.1 Growth from Melts and by Evaporation

Large monocrystals of beryllium were grown from a Be melt in a BeO crucible at 10^{-6} Torr. The crystals grew by slow-directional solidification from the bottom of the crucible over a period of $12^1/_2$ to $16^1/_4$ h. The crystals resulting from Pechiney SR grade Be ($12^1/_2$ h) were larger than those obtained from vacuum distilled Be, Gelles [1]. The metal was bottom cast under vacuum at 1325 to 1350 °C into a mold with smooth interior walls. A suitable cooling rate stimulated the growth of a single crystal in each corner of the mold thus producing four large crystals in a cross section. The size of the crystals was limited only by the size of the mold and the melt, Tuer, Kaufmann [2]. Single cylindrical crystals were also grown in vacuum and under inert gas at p=200 Torr using monocrystals 10 mm in diameter as the seed crystals. The seed crystal was fastened to a rod which was drawn from a BeO crucible containing the melt at a rate of 10 to 60 mm/h. A specimen consisting of a few large crystals elongated along the rod axis was obtained in this way, Ivanov et al. [3], Grinyuk et al. [4]. For the growth of single crystals in a vacuum of 10^{-6} Torr, see Gelles [5]. Attempts to grow large monocrystals from melts in BeO crucibles by the Czochralski or Bridgman technique were unsuccessful. Thus, only small crystals were obtained by the Czochralski method because of spurios growth and strong nucleation. The Bridgman method encountered problems in procedure, Kirn [6]. In an earlier attempt single crystals up to 60 mm in diameter and 100 mm long were grown from melts in alundum crucibles lined with BeO using seed crystals. These were drawn out of the melt under Ar gas by slow removal (12 to 100 mm/h) of the induction heating source, Kaufmann, Nowak [7]. Thread shaped single crystals of Be were produced by evaporation of the metal in a vacuum and condensing the vapors on a cylindrical sheet of molybdenum. Growth of Be whiskers was observed at evaporation rates of 0.4 to 0.9 g/h in some ring-like areas. Filar crystals were attained that were around 10 μm in diameter and several mm long, Amonenko et al. [8]. They were single crystals with the c-axis inclined at a small angle to the whisker axis, Bunshah [9]. The filar crystals did not grow in the direction of maximum packing density, but mainly in the [221], [331], and [111] crystal directions [8]. Defect-free single crystals of Be of the cubic modification were obtained by heating the monocrystals at 1260 to 1280 °C in a vacuum of 10^{-6} Torr for 17 h. This phase was cooled to 1250 °C at a rate of 120 °C/h and held at that temperature (transition $\beta \rightleftharpoons \alpha$-Be) for 120 h. The crystals thereafter were cooled to 1000 °C at a rate of 10 °C/h and then to room temperature at a rate of 100 °C/h after which they showed an internal desorientation $<5°$ as required for neutron monochromators, Faure et al. [10].

References:

[1] S. H. Gelles (ASD-TDR-62-509-Vol. III [1964] 34/44, 35/7; N.S.A. **19** [1965] No. 4644).

[2] G. L. Tuer, A. R. Kaufmann (in: D. W. White Jr., J. E. Burke, The Metal Beryllium, ASM, Cleveland 1955, pp. 372/424, 374/6); also see A. R. Kaufmann, P. Corzine (MJT-1086 [1952] 11/2; N.S.A. **11** [1957] No. 8517).

[3] V. E. Ivanov, B. M. Amonenko, G. F. Tikhinskii, I. I. Papirov, L. N. Ryabchikov, V. V. Grinyuk (3rd Conf. Intern. Met. Beryllium Commun., Grenoble, Fr., 1965 [1966], pp. 33/7, 36; C.A. **67** [1967] No. 84259).

[4] V. N. Grinyuk, I. I. Papirov, G. F. Tikhinskii, I. G. D'yakov (Izv. Akad. Nauk SSSR Metally **1967** No. 4, pp. 77/84; Russ. Met. **1967** No. 4, pp. 33/6, 35).

[5] S. H. Gelles (CONF-170 [1965] 184/213, 187; N.S.A. **20** [1966] No. 13174).

[6] J. F. Kirn (TID-21852 [1965] 1/19, 9, 14; C.A. **63** [1965] 15631; N.S.A. **19** [1965] No. 26758).

[7] A. R. Kaufmann, W. B. Nowak (NMI-11399 [1955] 6/13, 8; N.S.A. **11** [1957] No. 12490).

[8] V. M. Amonenko, I. I. Papirov, G. F. Tikhinskii, V. A. Finkel (Fiz. Metal. Metalloved. **13** [1962] 928/30; Phys. Metals Metallog. [USSR] **13** No. 6 [1962] 117/9).
[9] R. F. Bunshah (UCRL-12252 [1965] 1/19, 7; C.A. **63** [1965] 11104; N.S.A. **19** [1965] No. 20654).
[10] J. Faure, Y. Malmejac, B. Schaub (Fr. Demande 2231426 [1973/74] 1/7, 3/5; C.A. **83** [1975] No. 64623).

1.6.5.2 Single Crystals by Vertical Floating Zone Melting

Single crystals were grown by passing a beryllium rod slowly through a fixed induction coil heated by a 20 kW high frequency generator at a frequency of 360 kHz. Because of the high vapor pressure of Be, the growth was carried out in a vacuum chamber filled with high purity argon at a pressure of 50000 Pa. About 12 kW was required to establish a molten zone 20 mm high. This allowed a crystal 32 mm in diameter and 100 mm long to be grown, Freund et al. [1], Jönsson [2]. A moving rate of 40 mm/h allowed to grow crystals with a purity of >99.999% Be and a maximum length of 150 mm, Jönsson et al. [3], see also Aldinger, Freund [4], and Wilhelm [5]. Single Be crystals were also grown in a double ellipsoid mirror furnace in which melting occurs in the common focus of the two ellipsoides, Stiltz, Jönsson [6]. This permits the application of the necking technique and enables crystal growth under vibration free conditions yielding sub-grain free single crystals [1]. The rods were sealed in quartz containers to prevent contamination of the mirror furnace by Be vapors. The diameter of the rods was restricted to about 10 mm and the length to 200 mm because the maximum energy available in the furnace was 0.9 kW [2], also see [3]. The rods used in the mirror furnace technique consisted of a seed crystal joined through a neck with the single crystal and its feed material. If not sufficiently pure (at least 99.8 mass%) the feed material had to be refined by zone melting (up to ten times) prior to crystal growth. The crystals were grown at a zone speed of 1 mm/s and their orientation was determined by that of the seed crystals, Freund et al. [1]. Thus, after one zoning pass the seeded rods yielded single crystals with a 90° angle between the direction of growth and the [0001] crystal axis and a 0° angle relative to the [10$\bar{1}$0] axis. The orientation of crystals grown from unseeded rods differed little for the [0001] axis (82° to 88°) and considerably for the [10$\bar{1}$0] axis (up to 29°). It seemed to depend on the zone migration speed, Jönsson [2]. The slow movement of the melting zone across the rod permits the growth of the crystallites adhering to the liquid-solid interface not only along the migrating crystallization front, but also parallel to the interface. In this manner part of the crystallites are separated from the melt and further growth is stopped. Thus, each melting pass reduces the number of crystallites until the texture of the rod becomes that of a single crystal. The orientation of the single crystals obtained by zone melting depends on one of the orientations of the premelted crystallites in the parent material. This is the source of the divergent and contradictory observations regarding the crystallization behavior of beryllium. In almost all cases, extruded rods were used which had a pronounced deformation texture in which the basal planes of the grains were aligned parallel to the direction of extrusion. The crystallites thus preoriented in the polycrystallinic parent material grew during zone melting and gradually formed crystals with the basal plane parallel to the rod axis, Jönssen [7]. The vertical floating zone melting technique allows zone melting of beryllium rods (bars) up to 32 mm in diameter. The light weight and high surface tension of molten Be holds the melt in place, London, Meakin [8].

In order to produce highly pure single crystals, beryllium rods containing seed crystals of the desired orientation at the top were suspended near the bottom of a Vycor tube together with smaller rods of Ti and Be placed above the specimens. The system was evacuated and heated to red heat for preliminary outgassing. The Ar gas was admitted up to a pressure slightly below atmospheric and circulated while the Ti rod was heated up to 1100 °C until it showed no further change of color. After a second outgassing of the specimen at red heat a melting zone was passed through the smaller Be getter rod for ulterior purification. Then zone melting of the specimen at the desired zone migrating speed was started. After each run the rods were etched with an aqueous solution of HNO_3 (40%) and HF (2%) to remove any scales formed during the run, Carrabine et al. [9]. Purification of the Ar atmosphere by Ti and Be getters see also Kaufman et al. [10], Spangler et al. [11], by liquid Li, heated Ti turnings and Mg vapor see Ivanov et al. [12], Grinyuk et al. [13]. The formation of pure single crystals can be promoted by applying direct or alternate current to the beryllium specimen during zone melting. The current contributes to the mixing of the metal in the molten zone of the cylindrical specimen (diameter 10 to 14 mm) thus increasing the refinement to 99.96% Be and accelerating the growth of single crystals, Ivanov et al. [12], Grinyuk et al. [13].

Single crystals of any desired orientation could be grown by vertical zone melting of bars carrying seed crystals of the desired orientation at their heads. The crystals were purified by 8 to 12 zone melting passes, Spangler et al. [14]. Thus a seed crystal was fused to a bar of 6.4 mm in diameter at the beginning of zone melting (four passes at 12 mm/h) and in most cases the angle between the basal plane of the seed crystal and the bar axis was 45°, Spangler et al. [15], Jenkins et al. [16]. For a similarly seeded bar 32 mm in diameter the desired purity was attained after five zone refining passes, Jenkins et al. [17]. In other cases the angle between the basal plane of the seed crystal and the bar axis was initially 90° and was reduced to 88° after three zone refining passes, Jenkins et al. [18]. In almost all earlier publications the growth of single crystals was achieved by zone melting of extruded rods or bars in an inert atmosphere (He, Ar; $p < 760$ Torr) or under vacuum. The rods were slowly conducted upward through the center of an induction coil of 2, 3 or 4 turns. Melting was induced by a high frequency alternating current of 360 to 450 kHz and 7 to 25 kW. The enclosure vessel was made of glass (Vycor, Pyrex), quartz or silica. The following table gives the origin of the feed material and its diameter in mm, the melting zone length in mm and its migrating speed in mm/h, the number of passes and the angle between the basal plane of the single crystals and the rod axis in degrees [2, 4, 5, 9 to 13, 19 to 29]:

feed metal	rod diameter in mm	zone length in mm	speed in mm/h	passes	angle basis: rod axis in degrees	Ref.
Kawecki Berylco XT-1	25	20	0.5 to 40	up to 10	56	[2, 4]
Pechiney SR	10	15	40	10	<6	[5]
Brush, seeded rods	9.5	—	25 to 100	n	—	[9]
Pechiney SR, vacuum cast	31.8	16 to 19	varying	6	<6	[10, 22]
Pechiney SR, vacuum distilled	6.4	6.4	varying	>4	<1	[10, 22]
Pechiney SR/Brush vacuum cast	25.4	16	1 × 38, then 12.7	5	45	[11]
Pechiney SR flake sinter	6.4	6.4	12.7	5	19	[11, 19]

feed metal	rod diameter in mm	zone length in mm	speed in mm/h	passes	angle basis: rod axis in degrees	Ref.
vacuum distilled Be	10 to 14	~15	20 or 40	2 to 6	—	[12, 13]
Brush vacuum cast	6.4	—	12.7	5	—	[20]
vacuum distilled Be	25.4	—	1 × 38, then 12.7	7	50 to 56	[21]
Pechiney SR grade	25.4	—	2 × 51, then 12.7	3	48 to 52	[21]
Brush vacuum cast	25.4	—	2 × 38, then 12.7	10	74	[21]
vacuum distilled, bundled	31.8	—	12.7	10 to 12	—	[23]
Pechiney SR, vacuum cast	31.8	—	—	6 to 12	—	[24]
vacuum bidistilled Be	12.7	—	12.7 to 38.1	n	—	[25]
Be reduced, rods	12.7	—	—	—	—	[26]
Be electrolytic SSR	25	—	40 to 80	6	75	[27]
Pechiney SR vacuum cast	6.4	—	—	6	—	[28]
vacuum cast, electrolytic	4.8 to 19.1	13 to 20	4.5 to 45	6 to 8	—	[29]

References:

[1] A. K. Freund, S. Jönsson, S. Stiltz, G. Petzow (J. Nucl. Mater. **124** [1984] 215/21, 216).
[2] S. Jönsson (J. Cryst. Growth **63** [1983] 116/24, 118).
[3] S. Jönsson, A. Freund, F. Aldinger (Metall **33** [1979] 1257/61, 1258).
[4] F. Aldinger, A. Freund (Beryllium 1977, Conf. Preprint 4th Intern. Conf., London 1977, Paper No. 55, pp. 1/12, 5; C.A. **92** [1980] No. 66536).
[5] M. Wilhelm (Diss. Univ. Stuttgart 1974, pp. 1/87, 18/9), M. Wilhelm, F. Aldinger (Z. Metallk. **66** [1975] 323/8, 324).
[6] S. Stiltz, S. Jönsson (Metall **38** [1984] 748/53, 751).
[7] S. Jönsson (Naturwissenschaften **69** [1982] 483/90, 488).
[8] G. J. London, J. D. Meakin (NYO-3716-4 [1970] 1/26, 2/4; C.A. **74** [1971] No. 131354).
[9] A. Carrabine, A. J. Stonehouse, D. G. Fetsko, W. W. Beaver (AD-612696 [1965] 1/70, 3; Met. Abstr. [2] **32** [1964/65] 1287; AFML-TR-64-388 [1964] 1/67, 3/4; N.S.A. **19** [1965] No. 20544).
[10] D. F. Kaufman, J. J. Pickett, L. R. Aronin (NMI-1266 [1961/65] 1/111, 7/11; N.S.A. **19** [1965] No. 34686).
[11] G. F. Spangler, M. W. Herman, E. J. Arndt (NP-11591 [1961] 1/66, 3/4, 39; N.S.A. **16** [1962] No. 13515; NASA Doc. N 62-12044 [1961] 1/66; C.A. **60** [1964] 6275).
[12] V. E. Ivanov, B. M. Amonenko, G. F. Tikhinskii, I. I. Papirov, L. N. Ryabchikov, V. N. Grinyuk (3rd Conf. Intern. Met. Beryllium Commun., Grenoble, Fr., 1965 [1966], pp. 33/7, 35).

[13] V. N. Grinyuk, I. I. Papirov, G. F. Tikhinskii, I. G. D'yakov (Izv. Akad. Nauk SSSR Metally **1967** No. 4, pp. 77/84; Russ. Met. **1967** No. 4, pp. 33/6, 33).
[14] G. E. Spangler, M. Herman, D. B. Hoover, V. V. Damiano (AD-620264 [1963/64] 1/87, 34; N.S.A. **19** [1965] No. 46909).
[15] G. E. Spangler, M. Herman, E. J. Arndt, D. B. Hoover, V. V. Damiano, C. H. Lee (AD-413120 [1963] 1/39, 1; N.S.A. **17** [1963] No. 34318).
[16] R. G. Jenkins, D. F. Kaufman, E. D. Levine, A. R. Lumbert, J. J. Pickett, L. R. Aronin (NMI-2098 [1961] 13/4; N.S.A. **16** [1962] No. 5715).
[17] R. G. Jenkins, D. F. Kaufman et al. (NMI-2105 [1962] 11/3; N.S.A. **16** [1962] No. 20955).
[18] R. G. Jenkins, D. F. Kaufman et al. (NMI-2102 [1962] 17/20; N.S.A. **16** [1962] No. 10534).
[19] M. Herman, G. E. Spangler, E. Hein (NP-9871 [1960] 1/26; N.S.A. **15** [1961] No. 13247; PB-171951 [1960] 1/24; C.A. **58** [1963] 1206).
[20] M. Herman, G. E. Spangler (Inst. Metals Monograph Rept. Ser. No. 28 [1963] 75/83, 76; Met. Abstr. [2] **29** [1961/62] 419).
[21] G. E. Spangler, M. Herman, V. V. Damiano, E. J. Arndt, D. B. Hoover (NP-11159 [1961] 1/14, 2/3; N.S.A. **16** [1962] No. 4541).
[22] D. F. Kaufman, E. D. Levine, J. J. Pickett, L. R. Aronin (NMI-1256 [1963] 1/67, 5/9; N.S.A. **17** [1963] No. 30976).
[23] D. F. Kaufman, E. D. Levine, J. J. Pickett, L. R. Aronin (NMI-2112 [1962/63] 1/12, 2/3; N.S.A. **17** [1963] No. 11093).
[24] D. F. Kaufman, L. R. Aronin (NMI-1265 [1965] 1/29, 4/5; N.S.A. **19** [1965] No. 18460; NMI-1257 [1964] 1/28, 3/4; N.S.A. **18** [1964] No. 24134).
[25] J. H. Keith (Y-DA-196 [1962] 1/43, 15/8, 30; N.S.A. **18** [1964] No. 32062).
[26] J. L. Lukesh, L. McD. Schetky, H. S. Spacil, M. Basche (WADC-TR-58-457-Pt. 1 [1959] 1/38, 8, 19/20; NP-7708 [1958] 1/16; N.S.A. **13** [1959] No. 16989).
[27] B. Schaub, G. Cabane (Compt. Rend. **257** [1963] 444/7, 446), B. Schaub (Symp. Zonenschmelzen Kolonnenkristallisieren, Karlsruhe 1963, pp. 227/46, 229, 233/4; C.A. **63** [1965] 5317).
[28] C. E. R. Tristam, A. Moore (CONF-729-7 [1964] 29/41, 31; N.S.A. **19** [1965] No. 16193).
[29] K. L. Edwards, A. J. Martin (Inst. Metals Monograph Rept. Ser. No. 28 [1963] 557/69, 566).

1.6.5.3 Single Crystals by Horizontal Zone Melting

In earlier work single crystals were grown by horizontal zone melting of beryllium bars (19.1 mm diameter) in BeO boats. A 51 mm long melting zone, produced by a suitable inductor like tantalum, was drawn through the bar by moving the boat at a speed of 0.3 mm/min. The bar then consisted of long crystals parallel to the bar axis which were identical to those grown from the melt at the same speed and the impurities were concentrated in the tail of the bar, Pearsall [1]. Enclosure of the boat with the Be ingot in a Pyrex or quartz tube and use of several high-frequency induction heating coils arranged in series was recommended by Lange [2].

References:

[1] C. S. Pearsall (MIT-1103 [1952] 3/12, 7; C.A. **1954** 93).
[2] A. Lange (Neue Hütte **2** [1957] 56/9, 57).

1.6.5.4 Bicrystal Grown by Zone Melting

So-called bicrystals were grown by parallel seeding and subsequent zone melting of bars resulting from several zone melting passes of SR grade flake ingots. Such seeding from the floating zone minimizes contamination and produces clear boundaries in the bicrystal. This allows the growth of bicrystals with the desired orientation, Aronin [1].

Reference:

[1] L. R. Aronin (3rd Conf. Intern. Met. Beryllium Commun., Grenoble, Fr., 1965 [1966], pp. 309/18, 311; C.A. **68** [1968] No. 15425).

2 Uses

This chapter summarizes the actual and the proposed uses of beryllium. Developmental work on the use of Be as a reflector and moderator in nuclear technology has slowed because the limits of its applicability (damage by irradiation) have apparently been reached. On the other hand, its low density, high thermal conductivity, and heat permeability make it an ideal material for the construction of the heat shields and heat collectors of space vehicles. It is also used in aeronautical applications such as aircraft brakes, antennas, and rocket heads, in inertial guidance components which require light-weight, stiffness, and dimensional stability, in scanning and large mirror components of satellite optical systems, and for window materials in X-ray tubes (high transmission of low energy X-rays). To the extent possible only reviews are cited, completeness is not intended.

A synopsis of the technically important properties of beryllium and their application is given in the tabular survey:

density strength melting point resistance to oxidation	Young's modulus of elasticity micro strain limit ($\sigma \times 10^{-6}$)	rigid low-weight constructions	aircrafts, missiles, satellites, orbiters, orbital stations, space shuttles, submarines, automotives, rotor blades for jet engines and gas centrifuges, computers, sound recording systems, high speed film transporting systems
	reflection power	nondeforming parts	optical devices (mirrors, casings for lenses and prisms of telescopes, spectroscopes, scanning devices), control systems based on inertia, seismographs, antennas
	heat conductivity, specific heat	heat sinks	brakes, heat collectors, heat shields, cooling systems
	nuclear properties	nuclear reactor construction parts	reflectors, moderators
		weapons	
		instruments	X- and γ-ray technique systems, scintillators, neutron monochromators
	others	carriers for radioactive isotopes	radiation sources for X- and γ-rays energy sources

Survey taken from F. Aldinger, G. Petzow (Radex Rundschau **1972** 280), also see F. Aldinger (Haus Tech. Essen Vortragsveröff. No. 376 [1977] 70), D. Lupton, F. Aldinger (Radex Rundschau **1983** 43/51, 49).

General References:

D. Lupton, F. Aldinger, Radex Rundschau **1983** 43/51; Kirk-Othmer Encycl. Chem. Technol. 3rd Ed. **3** [1978] 803/23, 815.

F. Aldinger, Haus Tech. Essen Vortragsveröff. No. 376 [1977] 64/71, 67.

G. Petzow, F. Aldinger, Ullmanns Encykl. Tech. Chem. 4th Ed. **8** [1974] 442/61, 453; F. Aldinger, G. Petzow, Radex Rundschau **1972** 275/83, 279/82.

G. Dressler, M. Rühle, in: K. Winnacker, L. Küchler, Chemische Technologie, Vol. 6, Seltene Metalle, München 1973, pp. 549/601, 590/4.

H. H. Hausner, Beryllium, Its Metallurgy and Properties, Berkeley-Los Angeles 1965, pp. 1/6.
W. Schreiter, Seltene Metalle, 2nd Ed., Vol. 1: Beryllium, Bor, Cäsium, Gallium, Germanium, Hafnium, Leipzig 1963, pp. 91/162, 143/54.
J. P. Pemsler, in: A. R. Kaufmann, Nuclear Reactor Fuel Elements Metallurgy and Fabrication, Chapter 7: Cladding Materials, New York-London 1962, pp. 231/56, 248/50.
G. E. Darwin, J. H. Buddery, Metallurgy of the Rarer Elements, Vol. 7: Beryllium, London 1960, pp. 1/392, 8/10, 353/65.
S. J. Polkin, Flotatsiya Rud Redkikh Metallov i Olova, Chapter IV: Flotatsia Beryllievykh Rud, Moscow 1960, pp. 387/407, 387/8.

2.1 Beryllium Metal

2.1.1 Nuclear Technique

Beryllium is a highly efficient generator of neutrons when bombarded with α particles. It may serve as a source of neutrons to initiate the nuclear fission within the fuel nucleus of a reactor (according to: ${}^{9}_{4}Be + {}^{4}_{2}He \rightarrow {}^{12}_{6}C + {}^{1}_{0}n$) [1]. Beryllium discs serve as the artificial neutron sources of neutrons in cyclotrons. A mixture of powdered Be and salts of Ra, Ac, or Pu can be used as a simple and comparatively cheap source of neutrons, Schreiter [1], also see Darwin, Buddery [3], Lübke [19].

Mixtures of Be and Po are strong emitters of α-radiation and are almost free of γ-radiation. A mixture of Be and Sb yields low energy neutrons after being irradiated in a reactor; the isotope ^{124}Sb, a γ-emitter (171 MeV), is formed. High energy neutrons (7.8 MeV), are obtained from a Be powder exposed to the α-radiation of Rn [1], also see [3, 4]. For the use of Be as a target for neutron activation by deuterons irradiation (conversion into ^{10}Be), see [5]. Beryllium foils used as targets for neutron bombardments were obtained by vacuum evaporation of Be onto a Ta foil [6]. For the use of Be foils as proton stoppers (more efficient than Al), see [7]. Beryllium is an especially good moderator, that is, it is able to retard the speed of nuclear fission neutrons to that of thermal neutrons, because the amount of kinetic energy released increases as the mass of the material involved becomes less and approximates that of the neutron. The same moderator effect is also obtained with BeO. In some nuclear reactions the moderator envelopes the fissionable fuel whereas in others, moderator and fuel may be used as a mixture or an alloy such as Be and U [1]. However, the high cost and large amount of nuclearly pure Be required for the moderator equipment of a nuclear reactor [1], as well as the damage of the equipment upon irradiation, Petzow, Aldinger [8] has prevented its use in commercial reactors. A comparison of the moderator properties of Be (energy decrement (ξ), slowing down and diffusion areas (L_s^2, L^2), moderating ratios ($\xi \Sigma_s/\Sigma_a$), slowing down power ($\xi \Sigma_s$), and scattering and thermal absorption cross sections (σ_s, σ_a)), with those of BeO, Be_2C, graphite, D_2O, and H_2O is given by Darwin, Buddery [3], also see Aldinger, Petzow [4]. The use of Be as a moderator in nuclear reactors because of its low capture cross section for thermal neutrons has been discussed by Meyer [9], Barrett [10], Boland [11], Onitsch-Modl [12].

Beryllium can be used as a reflector which envelopes the fissionable fuel material. Such envelopment reduces the evasion of neutrons by reflecting them into the core. In the case of a thermal reactor, the reflector box consists of the same material as the moderator, i.e., Be. The reflected neutrons are slowed by collisions with the Be atoms of the moderator. These lower energy neutrons are energetically more favorable for nuclear chemical reactions than the initial high energy neutrons [1]. An efficient reflector such as Be has a low $\Sigma_a L$ value and a high $\xi \Sigma_s$ value. Thus, a large proportion of the neutrons are reflected into the core rather than absorbed in the reflector and the leakage of fast neutrons is minimized. For mobile reactors which require a minimum size it becomes particularly important to balance the cost of a beryllium reflector against the increase in efficiency and decrease in size [3]. The use of a combined beryllium and graphite reflector for a materials testing reactor in the USA was reported by Huffmann [13] and Schreiter [1]. For the use of Be as a neutron reflector in high-flux test reactors and nuclear weapons see Kirk-Othmer [14].

Beryllium can be used as a canning or cladding material in gas-cooled nuclear reactors at elevated (200 to 600 °C) temperatures because of its good compatibility with gases and fuels. Its low neutron absorption cross section allows its application up to a considerable wall thickness, Schreiter [1], also see [15, 16, 20]. However, at operational temperatures below 500 °C, embrittlement occurs due to the presence of He atoms in solution which precipitate to form bubbles at temperatures above 600 °C. Although the swelling is unlikely to exceed 30%, the occurrence of large bubbles on the grain boundaries may cause permeability to gases [3].

The use of finned Be tubes as cladding devices for reactor fuel elements was mentioned by Blainey [17]. Since beryllium is almost unaffected by neutrons and shows only little absorption of thermal neutrons it may be used as a solid metallic carrier (matrix) of the uranium fuel in a fine and homogeneous dispersion, like zirconium [1]. Swelling of beryllium, however, is expected in the region of maximum fast particle flux irradiation at sufficiently high temperatures. Swelling of Be and ThO_2 mixtures starts at 700 °C and attains 160% at 1100 °C within 2 h whereas of Be and UO_2 mixtures start to swell at 600 °C, but attain only 120% after 2 h of irradiation at 1100 °C. Therefore from considerations of swelling due to diffusional porosity, chemical reaction (reduction), and irradiation, the fuel temperatures in systems involving beryllium metal and U, Th, or Pu and their compounds should be kept below 600 °C. A higher thermal stability is expected with BeO as the fuel matrix since UO_2 and BeO are mutually insoluble and BeO is more resistant to irradiation than the metal [3].

Beryllium can be used as a sensitive threshold indicator of γ-radiation which causes emission of neutrons. Beryllium monocrystals may be applied as effective neutron monochromators [1, 4], also see Jönsson [18]. Beryllium is a convenient photoneutron source because of its low photoneutron threshold [3, p. 363].

References:

[1] W. Schreiter (Seltene Metalle, 2nd Ed., Vol. 1: Beryllium, Bor, Cäsium, Gallium, Germanium, Hafnium, Leipzig 1963, pp. 91/162, 148/51).
[2] G. Dressler, M. Rühle (in: K. Winnacker, L. Küchler, Chemische Technologie, Vol. 6, Seltene Metalle, München 1973, pp. 591/4, 593).
[3] G. E. Darwin, J. H. Buddery (Metallurgy of the Rarer Elements, Vol. 7: Beryllium, London 1960, pp. 1/392, 353/64).
[4] F. Aldinger, G. Petzow (Radex Rundschau **1972** 275/82, 280), F. Aldinger (Haus Tech. Essen Vortragsveröff. No. 376 [1977] 64/71, 68, 70).
[5] G. J. Atchison, W. H. Beamer (Anal. Chem. **28** [1956] 237/43).
[6] A. D. Bondar, A. S. Emlyaninov, et al. (Izv. Akad. Nauk SSSR Ser. Fiz. **24** [1960] 929/33; Bull. Acad. Sci. USSR Phys. Ser. **24** [1960] 926/30).
[7] V. C. Burkig, K. R. McKenzie (Phys. Rev. [2] **106** [1957] 848/51).
[8] G. Petzow, F. Aldinger (Ullmanns Encykl. Tech. Chem. 4th Ed. **8** [1974] 442/61, 453).
[9] R. Meyer (Rev. Met. [Paris] **58** [1961] 871/85, 879/80).
[10] T. R. Barrett (Engineering [London] **187** [1959] 412/4).

[11] L. F. Boland (J. Metals **10** [1958] 401/3).
[12] E. M. Onitsch-Modl (Berg-Hüttenmänn. Monatsh. Leoben **101** [1956] 363/9, 366).
[13] J. R. Huffmann (Nucleonics **12** No. 4 [1954] 20/6, 21).
[14] Kirk-Othmer Encycl. Chem. Technol. 3rd Ed. **3** [1978] 803/23, 815.
[15] J. P. Pemsler (in: A. R. Kaufmann, Nuclear Reactor Fuel Elements Metallurgy and Fabrication, Chapter 7: Cladding Materials, New York-London 1962, pp. 231/56, 248/50).
[16] Anonymous (Ind. Chemist **36** [1960] 10/2).
[17] A. Blainey (Metal Progr. **77** No. 1 [1960] 104/8).
[18] S. Jönsson (Naturwissenschaften **69** [1983] 483/90, 483).
[19] A. Lübke (Metall **10** [1956] 152/4).
[20] D. T. Livey, J. Williams (Proc. 2nd Intern. Conf. Peaceful Uses At. Energy, Geneva 1958, Vol. 5, pp. 311/8, 317; C.A. **1960** 22249).

2.1.2 Aircrafts and Space Vehicles

Since beryllium is the only light metal with a high melting point (1285 °C) and an extremely high elasticity modulus (28000 ± 3000 kp/mm^2) it is specially suitable as a structural material for jet propelled long range aircrafts and guided missiles (rockets), Schreiter [1]. It is used in the structural components of leading edges, secondary wing skins, fuselage plating, and control surfaces of aircraft. Because of the excellent heat capacity and conductivity it is used in heat sink devices in fuselage noses, inlet ducts, brakes and multiwall enclosures for the crew. Missile and space craft applications include the interconnecting structures of upper stages, the vehicle nose shell, the internal structure in the final stage, heat sink devices for multiwall enclosures of payloads and heat shields for various parachute-type recoveries, Hausner [2], also see [3]. Application in equipment systems such as gears, gyros, insulation, brackets, linkage and bellcranks of spacecrafts (missiles) is covered in [2]. The heat absorbing properties of beryllium enable it to withstand the temperatures produced by the friction of high speed reentry of spacecrafts. The heat absorption ability of beryllium is higher than that of aluminium (twice), steel (3.8 times), and copper (5 times). The far lower density of Be (1.85 g/cm^3) as compared with Al (2.7 g/cm^3) allows for a weight reduction in carrier planes of 40 to 60% and consequently a considerably extended flying range, Schreiter [1]. The use of Be in the construction of long range aircraft was also suggested by Darwin, Buddery [4]. For information concerning Be as a heat sink material in light weight high performance aircraft brakes, see Kirk-Othmer [5], Petzow, Aldinger [3], Boland [6]. For its use in space stations and space shuttles, see Aldinger [7], Barrett [8], Mathews [10], Gassner [9], Logerot, Syre [11], Dunmur [12]. Ultrafine Be powder can be used as a solid fuel in missiles because its heat of combustion (6000 kcal/kg) is rather high [3, p. 281].

References:

[1] W. Schreiter (Seltene Metalle, 2nd Ed., Vol. 1: Beryllium, Bor, Cäsium, Gallium, Germanium, Hafnium, Leipzig 1963, pp. 91/162, 151).
[2] H. Hausner (Beryllium, Its Metallurgy and Properties, Berkeley-Los Angeles 1965, pp. 1/6, 3/4).
[3] F. Aldinger, G. Petzow (Radex Rundschau **1972** 275/83, 280).
[4] G. E. Darwin, J. H. Buddery (Metallurgy of the Rarer Elements, Vol. 7: Beryllium, London 1960, pp. 1/392, 9).
[5] Kirk-Othmer Encycl. Chem. Technol. 3rd Ed. **3** [1978] 813/23, 815.
[6] L. F. Boland (J. Metals **10** [1958] 401/3).
[7] F. Aldinger (Haus Tech. Essen Vortragsveröff. No. 376 [1977] 64/71, 70).
[8] J. C. Barrett (J. Metals **15** [1963] 349/51).

[9] R. H. Gassner (Metal Progr. **78** No. 3 [1960] 88/92).
[10] C. O. Matthews (J. Metals **12** [1960] 780/5).
[11] J. M. Logerot, R. Syre (Metaux **44** [1969] 169/86, 171, 185/6).
[12] I. W. Dunmur (in: D. R. Floyd, J. N. Lowe, Beryllium Science and Technology, Vol. 2, New York-London 1979, pp. 135/75,170).

2.1.3 Other Uses

Beryllium is used extensively as a window material for X-ray tubes. It has a permeability to low-energy X-rays [1, 2], which is 17 times that of aluminium [2, 3, 11] and six to ten times that of Lindeman glass, a lithium beryllium borate glass, Schreiter [2]. The difference in mass absorption of Be and Al increases with increasing wavelength λ of the X-rays from the metal source (MoKα with $\lambda = 0.71$ Å $<$ CuKα with $\lambda = 1.54$ Å $<$ FeKα with $\lambda = 1.93$ Å). The mass absorption coefficients for these wavelengths are for Be 0.30, 1.35, and 3.24 respectively, however for Al 5.30, 48.7, and 92.8 respectively, Darwin, Buddery [4]. At the same total absorption, Be windows are much less sensitive to mechanical damage than Al windows but vacuum tight brazing joints of Be are difficult to achieve without cracking the window [4, 5]. The application of Be for X-ray tube windows was also suggested by Griffith [6], Taylor, Parish [7], Pol'kin [8], Aldinger, Petzow [9]. For its use in X-ray and high speed cameras, see Boland [10]; in Geiger proportional and scintillation counters see Darwin, Buddery [4, p. 365]. Beryllium can be applied as an object carrier in electronic microscopy, Schreiter [2], Dressler, Rühle [3], as a scavenger in heated liquid metals, as a deoxidation additive for other metals and alloys, Boland [10], Schreiter [2], and in high frequency electronics, Boland [10]. It can also be used in anion safety grids of mercury discharging tubes and for cold cathodes in gas discharging tubes [2]. An important application of Be is in inertial guidance components where its high elasticity modulus coupled with its excellent dimensional stability and machining characteristics are utilized. Its light weight, stiffness, dimensional stability and ability to accept an optical polish make it very suitable for scanning mirrors and large mirror components of optical systems in satellites, Kirk-Othmer [1]; also see the survey of Aldinger, Petzow [9] on p. 124 which gives further uses.

References:

[1] Kirk-Othmer Encycl. Chem. Technol. 3rd Ed. **3** [1978] 813/23, 815.
[2] W. Schreiter (Seltene Metalle, 2nd Ed., Vol. 1: Beryllium, Bor, Cäsium, Gallium, Germanium, Hafnium, Leipzig 1963, pp. 91/162, 151/2).
[3] G. Dressler, M. Rühle (in: K. Winnacker, L. Küchler, Chemische Technologie, Vol. 6: Seltene Metalle, München 1973, pp. 590/4, 593).
[4] G. E. Darwin, J. H. Buddery (Metallurgy of the Rarer Elements, Vol. 7: Beryllium, London 1960, pp. 1/392, 364/5).
[5] J. T. Perry (Rev. Sci. Instr. **27** [1956] 759/62, 760).
[6] R. F. Griffith (Metal Progr. **65** No. 4 [1954] 81/5, 84).
[7] J. Taylor, W. Parish (Rev. Sci. Instr. **26** [1955] 367/73, 370).
[8] S. I. Pol'kin (Flotatsiya Rud. Redkikh Metallov i Olova, Chapter IV: Flotatsia Beryllevykh Rud, Moscow 1960, pp. 387/407, 388).
[9] F. Aldinger, G. Petzow (Radex Rundschau **1972** 275/82, 280).
[10] L. F. Boland (J. Metals **10** [1958] 401/3).
[11] A. Lübke (Metall **10** [1956] 152/4).

2.2 Alloys and Compounds of Beryllium

A tabular survey of the most important beryllium-containing alloys and compounds and their use is given by Lupton, Aldinger [1, p. 46]:

materials	examples	composition	admixtures	use
beryllium basis materials	technical beryllium	Be + impurities	0.8 to 8% BeO	nuclear technology
	Lockalloy	BeAl	38% Al	constructions (air and
	Matex	BeAl	24.8% Al	space aeronautics)
	composite materials	Be/Al, Be/Ti	—	compressor blades
beryllium compounds	beryllium oxide	BeO	—	electrotechnology, heat resistent materials for high temperatures, crucibles, nuclear technology
	beryllides (X = Nb, Ta, Zr)	$Be_{12}X$, $Be_{17}X_2$	—	highly heat resistent materials
hardening materials with beryllium additive	conductivity bronze	CuCoBe	0.5% Be	engine constructions
	alloys of high solidity	CuBe	1.7 to 2% Be	electrotechnology
	(strength, resistency)	NiBe	2% Be	control devices
	casting materials	CuCoBe	1.6 to 2.8% Be	clock manufacture
		NiBe	2 to 2.8% Be	die (chill) materials
	Nivarox	FeNiCrFeBe	0.8% Be	tools
	Duratherm	CoNiCrFeBe	0.5% Be	medical equipments
	Thermelast	FeNiMoBe	0.5% Be	
	Nivaflex	CoNiCrFeWMoBe	0.3% Be	
others	Al-casting alloy		0.1 to 0.5% Be	improved casting, deoxidant
	Mg-casting alloy		0.005% Be	improved resistance
	steels		1% Be	to corrosion

Alloys

There are only two beryllium based alloys with a heterogeneous texture, Lockalloy (38% Al) and Matex (24.8% Al), the latter containing some Si and Mg to strengthen the Al phase. In both alloys the plasticity of the beryllium is enhanced by the comparatively ductile inclusions of Al without losing the favorable properties of beryllium (low density, high strength, and excellent elasticity) which are desired in light-weight constructions of high strength such as in aircrafts, missiles, and satellites, Lupton, Aldinger [1], also see Petzow, Aldinger [2], Denny [3]. An alloy based on Be with 40% Al serves as a film former for copy (reprint) films in electron microscopy because of its low density, high electrical conductivity and insolubility in almost all usual solvents, Schreiter [4]. Several composites of beryllium and Al, Ag, Mg, or Ti were also studied, but only those with Ti have been found suitable for technical use such as in the compressor blades of gas centrifuges, Lupton, Aldinger [1], also see Petzow, Aldinger [2].

Far more important are alloys where beryllium is the minor component. A copper alloy which contains 2% Be has great economic and strategical significance. These beryllium bronzes can be hardened while warm because copper dissolves up to 2.1% Be at 964 °C but almost no Be at 20 °C. High strength Be bronzes contain 2% Be and 0.5% Co; those with high conductivity contain less Be (0.5%) and more Co (2.5%). Beryllium bronzes are nonmagnetic, resistant to corrosion and fatigue, and easily malleable when heated. A bronze with 2% Be and 0.25% Co can be used in flat and spiral springs found in relays, switching and controlling devices, vibrators, membrane (diaphragm) equipped instruments, and snap switches. Sparkless tools of beryllium bronze are used in dangerous gas atmospheres in industrial settings. Since such tools are resistent to salt (sea) water, they are also used in navigation. Beryllium bronze tools are preferred also if great hardness and a cutting edge are required, Schreiter [4], also see Denny [3], Darwin, Buddery [5], Kirk-Othmer [6], Dressler, Rühle [7], Griffith [11], Lübke [15], Clemmons, Browning [16].

Beryllium bronze is used in dies for drawing metals and pressing plastics because of its considerable hardness and high thermal conductivity. These properties and its high electrical conductivity also make it suitable for electrodes and other components of resistance welding machines [5 to 7]. Beryllium nickel alloys containing 1.7 to 2.4% Be have a resistance to corrosion similar to that of stainless steel and are therefore used in drill chucks (bits), matrices of diamond boring crowns, [4, 16], as stainless ball bearings, heat-resistant springs, stirrer paddles, valves of lye pumps [4], injection needles and other surgical instruments [4, 7]. Their hardness is conserved up to 400 to 500 °C and they are impermeable to hydrogen even under pressure and at red heat [4]. Nonsparking chisels of a nickel alloy containing 2% Be preserve their cutting power even better than those of the Cu-Be alloy [4, 15]. A hardenable, completely stainless alloy with 2% Be and 1% Ti attains a maximum Brinell hardness of 480 and is heat-resistant up to 400 °C; its suitability for shafts of vibration meters has been approved, Schreiter [4], Lübke [15]. For the use of Be-Ni alloys as castings for drill bits and pump impellers, see Denny [3], Darwin, Buddery [5]; as strips or wires for springs such as watch main springs [5, 7] and balance wheels [16], heat-resistant springs, diaphragms [3] and electrical connectors [6] where hardness, high tensile strength, good resistance to corrosion, heat and wear are required, see Denny [3], Darwin, Buddery [5], Lübke [15], Clemmons, Browning [16]. Alloys of Ni-Be are used as dyes for glass molding because of their resistance to heat shocks, crazing, and corrosion, and the antigalling characteristics of the alloy, by Kirk-Othmer [6].

Iron-beryllium alloys are mainly used in the manufacture of springs, valves, controlling devices, nozzles (jets), and welding electrodes for the construction of engines and motor vehicles. Steels containing 1% Be, 11% Ni, and 12% Cr preserve their high strength even at elevated temperatures. Invar, Elinvar (Nivarox), Contracid, and Triconium are beryllium steel alloys used in the manufacture of hair springs in clockworks and watches, for surgical and dental purposes, and for heat-resistant membranes and valve springs, Schreiter [4], also see Darwin, Buddery [5].

Beryllium forms alloys with aluminium in any ratio, but the reciprocal solid solubility is very low. However, even small admixtures of Be greatly increase the tensile strength and resistance to heat and oxidation. The castability and heat resistance of Al is notably increased even by 0.01 to 0.5% Be, which yields a finer grain, Schreiter [4], also see Kirk-Othmer [6]. The addition of Be (0.001 to 0.005%) to magnesium-based aluminium alloys prevents the segregation of Mg, the formation of hard magnesium nitride which hampers rolling, the deflagration of Mg from such alloys above the melting point (~800 °C) and enhances the resistance to corrosion in water. A 0.004% content of Be prevents the absorption of H_2 by the alloy and reduces its porosity. The oxide film formed on Be-containing light metal alloys does not fade. The casting of magnesium-rich alloys in green sand molds without relevant detrimental reactions becomes possible upon addition of 0.001% Be. The casting fluidity (mold filling ability) of aluminium-based alloys is improved by adding 0.01 to 0.05% Be for difficult castings, Schreiter [4], also see Denny [3], Darwin, Buddery [5], Kirk-Othmer [6]. The casting of Mg-based alloys at 760 °C without protection in the presence of at least 0.042% Be was reported by Burns [8], the grain growth during casting is reduced by Ti or Zr [3, 5]. Magnox alloys of Mg containing Be (0.05%), Ca (0.1%) and Al (1%) are much more resistant to corrosion by wet CO_2 than pure Mg. They are used for cans in the Calder Hall reactors and in gas-cooled nuclear reactors, Darwin, Buddery [5]. Standard silver (92.5% Ag, 7.5% Cu) alloyed with 0.4% Be is much more resistant to wet H_2S and cooling liquors (vinegar, fruit juices) than standard silver without Be [5]. A zinc-based alloy containing 0.1% Be and 2 to 2.5% Cu has about the same properties as a brass containing 70% Cu and 30% Zn [4, 5]. A zirconium based alloy containing 0.5% Be or 0.1% Be + 0.5% Cu shows less CO_2 corrosion at 450 °C than unalloyed Zr. The ternary alloy shows no signs of brake-away after 6000 h and its corrosion rate of only 1.25×10^{-4} mg · cm^{-2} · h^{-1} is rather low. Beryllium is the most efficient hardening and grain refining alloy component for molybdenum [5].

A copper base alloy containing 30% Ni and 30% Mn is further hardened upon addition of 0.4% Be. Beryllium can replace iridium in the hardening of platinum. The hardness and tensile strength of platinum with 0.3% Be is increased by 5 to 20% [4]. The hardening effect of Be is 125 times that of Ir and the alloys are malleable up to 0.3% Be. For most applications a Be content of 0.06 to 0.1% is sufficient and does not affect the chemical or catalytic properties of platinum, Darwin, Buddery [5]. For the use of beryllium (in small amounts) as a deoxidizing agent (oxidation inhibitor) in metals such as Mg, Al, and alloys see Boland [9], Schreiter [4, p. 152], Darwin, Buddery [5, p. 9], Lupton, Aldinger [1, p. 51], Petzow, Aldinger [2, p. 71].

Beryllia

Beryllia (BeO) is an excellent refractory of high melting point (2570 °C), extreme chemical stability, high thermal conductivity (better than some metals), good thermal shock resistance, and high electrical resistance. It is a useful component of ceramic systems such as crucibles, rocket nozzle liners, electrical porcelaines and of ultraviolet permeable glasses. It is occasionally used as a component of phosphors (luminophores) in fluorescent lighting, gas mantles, welding rod coatings and porcelain enamel fluxes, Darwin, Buddery [5, pp. 9, 37], also see Dressler, Rühle [7], Ryshkevitch [10, p. 277], Lupton, Aldinger [1, p. 51], Griffith [11], Pol'kin [12]. Because of their moderate and steady thermal expansion, high heat conductance, thermal shock resistance, and high electrical resistance at high temperatures, beryllia crucibles and tubes can be used in an inert atmosphere or high vacuum at temperatures up to 2000 °C. Sintered beryllia vessels are used for distillation of metals in a high vacuum because they do not contaminate melts and are gas tight. Beryllia also serves as a fire-proof (refractory) material for spark plugs in aircrafts, ultrahigh frequency radar insulators, protective tubes in transistors and thermocouples, lining of electrical high temperature furnaces, embedding of heating filaments, and as a coating for incandescent filaments. Its resistance to strong heat and flame shocks allows its application in combustion chambers of missiles and turbines of jet motors, Schreiter [4], also see Clemmons, Browning [16]. The mechanical strength, light weight, and high transparency of sintered beryllia to microwaves makes it an ideal material in high-power radio-tube envelopes, thermoionic units, and magnetohydrodynamic power generators, Ryshkevitch [10, p. 277]. Beryllia can be used as a neutron reflector and moderator in nuclear reactors because of its low neutron capture and high neutron scattering cross section, its high melting point and stability to heat (super refractory class) and radiation, Darwin, Buddery [5], also see Schreiter [4, p. 153], Dressler, Rühle [7], Boland [9], Finniston [13, p. 459], Buresch [14, p. 139]. It is stable in most gaseous coolants even at high temperatures if grain growth inducing radiation is absent [4]. Radiation at an intensity of 10^{20} n/cm^2 is believed to reduce the compressive strength and heat conductance by up to 60% [5, p. 350]. Mixtures of BeO and UO_2 are stable to oxidation by the coolant gas and, thus, suitable as fuels in gas-cooled high temperature reactors where highly refractory materials are required, Schreiter [4], also see Ryshkevitch [10]. For use of BeO in nuclear technology, see also "Uran" Erg.-Bd. A 3, 1981, pp. 186, 233/5.

Other Compounds

The very light (D = 0.58 g/cm^3) beryllium hydride (BeH_2) may be used as a solid propellant for missiles, Petzow, Aldinger [2]. The nitrate ($Be(NO_3)_2 \cdot 4H_2O$) serves to harden the incandescent mantles of gas incandescent lamps [2], also see Griffith [11]. Beryllium carbide (Be_2C) is used in crucibles. It is applied in nuclear reactors because of its excellent heat resistance and nuclear properties. The glassy metaphosphate ($Be(PO_3)_2$) has a high permeability to UV radiation. Beryllium boratsilicate glasses containing BaO, CaO, and ZnO are used as high melting coatings of heat-resistant alloys, Schreiter [4]. Sodium fluoroberyllates (Na_2BeF_4, $NaBeF_3$) are used as fluxes for the soldering and welding of light metals, Dressler, Rühle [7]. Distillable organic beryllium compounds, such as the basic acetate ($Be_4O(CH_3COO)_6$) and acetylacetonate, are used in the production of high purity beryllium, Petzow, Aldinger [2].

References:

[1] D. Lupton, F. Aldinger (Radex Rundschau **1983** 43/51, 46, 51).
[2] G. Petzow, F. Aldinger (Ullmanns Encykl. Tech. Chem. 4th Ed. **8** [1974] 442/61, 453/5).
[3] J. P. Denny (in: H. H. Hausner, Beryllium, Its Metallurgy and Properties, Berkeley - Los Angeles 1965, pp. 179/90, 181/2, 186/9).
[4] W. Schreiter (Seltene Metalle, 2nd Ed., Vol. 1: Beryllium, Bor, Cäsium, Gallium, Germanium, Hafnium, Leipzig 1963, pp. 143/8, 152/4).
[5] G. E. Darwin, J. H. Buddery (Metallurgy of the Rarer Elements, Vol. 7, Beryllium, London 1960, pp. 8/10, 37, 321/39, 322, 328, 331/6).
[6] Kirk-Othmer Encycl. Chem. Technol. 3rd Ed. **3** [1978] 803/23, 816/22.
[7] G. Dressler, M. Rühle (in: K. Winnacker, L. Küchler, Chemische Technologie, Vol. 6: Seltene Metalle, München 1973, pp. 590/4, 593).
[8] J. R. Burns (Trans. Am. Soc. Metals **40** [1948] 143/60).
[9] L. F. Boland (J. Metals **10** [1958] 401/3).
[10] E. Ryshkevitch (in: H. H. Hausner [3, pp. 267/78, 277]).
[11] R. E. Griffith (Metal Progr. **6** No. 4 [1954] 81/5, 84).
[12] S. I. Pol'kin (Flotatsia Rud Redkikh Metallov i Olova, Chapter IV: Flotatsia, Beryllevykh Rud, Moscow 1960, pp. 387/407, 388).
[13] H. M. Finniston (Research [London] **5** [1952] 456/63, 459).
[14] F. E. Buresch (Radex Rundschau **1983** 133/45, 139/40).
[15] A. Lübke (Metall **10** [1956] 152/4).
[16] B. H. Clemmons, J. S. Browning (J. Metals **5** [1953] 1433/4).

3 Nuclides

3.1 Atomic Weight. Isotopic Abundance

The atomic weight of Be on the unified mass scale ($A_r(^{12}C)=12$) as proposed by the IUPAC Commission on Atomic Weights and Isotopic Abundances is $A_r(Be)=9.01218(1)$ [1, 2]. This is consistent with the nuclidic mass $A_r=9.0121822(4)$ [4] of ^9Be, the only stable isotope of beryllium (isotopic abundance 100%) [1, 3]. Peiser et al. [1] have recently reviewed how the atomic weight of Be has developed since 1961.

The A=10 isotope, the longest-lived unstable Be nuclide ($T_{1/2}=1.6\times10^6$ a), is continuously produced in the earth's atmosphere by cosmic ray-induced spallation from N_2, O_2, Ne, and Ar, and is thus distributed over the earth's surface in equilibrium concentrations of $<10^{-9}$ at%. This, however, is too low to be of significance for the atomic weight of the element [1].

References:

[1] H. S. Peiser, N. E. Holden, P. De Bièvre, I. L. Barnes, R. Hagemann, J. R. De Laeter, T. J. Murphy, E. Roth, M. Shima, H. G. Thode (Pure Appl. Chem. **56** [1984] 695/768, 720).
[2] N. E. Holden, R. L. Martin (Pure Appl. Chem. **56** [1984] 653/74, 663, 665).
[3] N. E. Holden, R. L. Martin, I. L. Barnes (Pure Appl. Chem. **56** [1984] 675/94, 680).
[4] A. H. Wapstra, G. Audi (Nucl. Phys. A **432** [1985] 1/54, 14).

3.2 Properties of Nuclides

This section presents the ground state properties of beryllium isotopes in order of increasing mass number A. Only the A=9 isotope is stable, whereas the A=7, 10 to 12, and 14 isotopes are β unstable. Among these, the A=10 isotope is the longest-lived. The A=5, 6, 8, and 13 isotopes are particle unstable.

Evaluated nuclear data rather than data from original papers are presented wherever possible in the following. In particular, extensive use has been made of the most recent compilations of energy levels of light nuclei published by Ajzenberg-Selove in three parts: A=5 to 10 [1] (literature closing date mid 1983), A=11 and 12 [2] (literature closing date mid 1984), and A=13 to 15 [3] (literature closing date mid 1980). Generally, only recent papers not registered in any of these compilations are cited in the following text. Except for ^5Be, atomic masses and decay energies are taken from the first part of: "The 1983 Atomic Mass Evaluation" of Wapstra and Audi [4]; binding energies are taken from the second part of this evaluation [5].

All data are presented in a short-hand notation which is essentially that used in [1 to 4] and reads as follows:

M; ΔM	atomic mass; mass excess; $\Delta M=M-A$
(syst)	value obtained by interpolating or extrapolating systematic trends
β^+; β^-	positron emission; negatron emission
EC	electron capture
γ	γ-ray emission
$I(\beta^{\pm})$, $I(\gamma)$	intensity of β or γ radiation
α	α particle
$T_{1/2}$	half-life
Q(EC)	total β^+ or EC decay energy of nuclide (A, Z); $Q(EC)=\Delta M(A,Z)-\Delta M(A,Z-1)$
$Q(\beta^-)$	total β^- decay energy of nuclide (A, Z); $Q(\beta^-)=\Delta M(A,Z)-\Delta M(A,Z+1)$
E(γ)	energy of a γ transition

J^{π}	nuclear spin of ground state, in units of $\hbar$, with superscript " + " or " − " sign for parity π
μ	nuclear magnetic dipole moment in nuclear magnetons (nm)
Q	nuclear electric quadrupole moment in barns (b)

Beryllium-5

Mass. $\Delta M > 33700$ keV [1].

Decay Properties. ^{5}Be is particle unstable with respect to ^{3}He$+2\,^{1}$H by >4.2 MeV [1].

Spin, Parity. $J^{\pi} = 1/2^{+}$ (syst) [1].

Beryllium-6

Mass. $M = 6.019725 \pm 0.000006$ u, $\Delta M = 18374 \pm 5$ keV [4].

Decay Properties. ^{6}Be is particle unstable with respect to ^{4}He$+2\,^{1}$H by 1371 ± 5 keV [1, 5]. The width of the ground state for this decay is 92 ± 6 keV in the center-of-mass system [1]. Analysis of the energy spectrum of α-particles following the reaction ^{6}Li(^{3}He, ^{3}H)^{6}Be showed that about 50% of the decays of the ^{6}Be ground state can be described by a sequential proton decay through the ^{5}Li system [6].

Spin, Parity. $J^{\pi} = 0^{+}$ [1].

Beryllium-7

Mass. $M = 7.0169283 \pm 0.0000008$ u, $\Delta M = 15768.7 \pm 0.8$ keV [4].

Decay Properties. ^{7}Be → ^{7}Li EC 100%, $T_{1/2} = 53.29 \pm 0.07$ d [1].

Q(EC) = 861.90 ± 0.04 keV [4]. The decay proceeds to the first excited state of ^{7}Li at 477.61 keV and directly to the ground state with a branching ratio ^{7}Li(477)/^{7}Li(ground state) of (10.39 ± 0.06)% [1]. A much larger value of $(15.4 \pm 0.8\%)$ was reported by one group of workers [7, 8] which stimulated a series of redeterminations of this branching ratio during the period 1982 to 1984. The values obtained in the course of this redetermination are between 9.8 and 11.4%, most of them between 10.3 and 10.7%. The smaller value has thus been confirmed [9 to 19]. The results from these papers (except [19]) are summarized in [18].

The branching ratio in the EC decay of ^{7}Be is of certain astrophysical relevance, as outlined, e.g., in [10]. It is often used for calibrating cross section measurements of the reactions ^{3}He(α,γ)^{7}Be and ^{7}Be(p,γ)^{8}B, which are members of the so-called proton-proton chain producing solar neutrinos. Their cross sections are important input parameters for a calculation of the flux of solar neutrinos with energies above 810 keV, to which a ^{37}Cl detector is sensitive. There exists a striking discrepancy between the solar neutrino flux measured with such a detector and the calculated one, see the review of Bahcall et al. [20], and references therein.

The energy of the γ-ray following the EC decay of ^{7}Be is 477.605 ± 0.003 keV and may serve as a calibration standard [1, 21].

Spin, Parity. $J^{\pi} = 3/2^{-}$ [1].

Beryllium-8

Mass. M = 8.00530512 ± 0.00000012 u, ΔM = 4941.73 ± 0.11 keV [4].

Decay Properties. ^{8}Be is particle unstable with respect to $^{4}He + ^{4}He$ by 91.88 ± 0.05 keV [5]. The width of the ground state for this decay is 6.8 ± 1.7 eV in the center-of-mass system [1].

Spin, Parity. $J^{\pi} = 0^{+}$ [1].

Beryllium-9

Mass. M = 9.0121822 ± 0.0000004 u, ΔM = 11347.7 ± 0.4 keV [4].

Stable Isotope. Natural isotopic abundance is 100% [22, 23].

Spin, Parity, Nuclear Moments. $J^{\pi} = 3/2^{-}$, $\mu = -1.1778 \pm 0.0009$ nm [1] (μ is from NMR measurements on free Be atoms [24]). $\mu = -1.17745$ nm has been selected from spectroscopic papers published up to 1974 [25]. Q = +0.053 ± 0.003 b is from atomic beam-magnetic resonance spectra [1, 25]. The magnetic moments are corrected for diamagnetic shielding [24, 25], and the quadrupole moment is corrected for Sternheimer polarization [25].

Beryllium-10

Mass. M = 10.0135341 ± 0.0000004 u, ΔM = 12607.0 ± 0.4 keV [4].

Decay Properties. $^{10}Be \rightarrow ^{10}B$ β^{-}, $T_{1/2} = (1.6 \pm 0.2) \times 10^{6}$ a [1].

$Q(\beta^{-}) = 556.2 \pm 0.5$ keV [4]. The β^{-} spectrum consists of a single branch to the ground state of ^{10}B, i.e., decay of ^{10}Be is not accompanied by γ radiation [1].

Spin, Parity. $J^{\pi} = 0^{+}$ [1].

Beryllium-11

Mass. M = 11.021658 ± 0.000007 u, ΔM = 20174 ± 6 keV [4].

E(γ) in keV	E(^{11}B level) in keV	I(γ)
692.31 ± 0.10	7977.84 ± 0.42	0.85 ± 0.04
1771.31 ± 0.30 [a)]	6791.80 ± 0.30	4.0 ± 0.3
2124.473 ± 0.027	2124.693 ± 0.027	100
2895.30 ± 0.40	5020.31 ± 0.30	14.4 ± 0.6
4443.90 ± 0.50	4444.89 ± 0.50	100
4665.90 ± 0.40	6791.80 ± 0.30	28.5 ± 1.1
5018.98 ± 0.40	5020.31 ± 0.30	85.6 ± 0.6
5851.47 ± 0.42	7977.84 ± 0.42	53.2 ± 1.2
6789.81 ± 0.50	6791.80 ± 0.30	67.5 ± 1.1
7282.92	7285.51 ± 0.43	87.0 ± 2.0
7974.73	7977.84 ± 0.42	46.2 ± 1.1

[a)] E(γ) = 1171.31 keV for this transition appears to be a misprint in [2], and also in the original paper [26], on which data evaluation in [2] is based. The transition is between the 6792- and 5020-keV levels of ^{11}B, cf. [2], and also Fig. 2 and Table 1 in [26].

Decay Properties. $^{11}Be \rightarrow {}^{11}B$ β^-, $T_{1/2} = 13.81 \pm 0.08$ s [2].

$Q(\beta^-) = 11506 \pm 6$ keV [4]. The β^- decay feeds several levels of ^{11}B up to the 9.88-MeV level which decays via α emission to ^{7}Li. The main β^- branch with $I(\beta^-) = 54.7\%$ is to the ^{11}B ground state [2].

$E(\gamma)$ from ^{11}Be β^- decay. All lines are listed on p. 134. $I(\gamma)$ is absolute intensity per 100 ^{11}Be decays [2].

Spin, Parity. $J^\pi = 1/2^+$ [2].

Beryllium-12

Mass. $M = 12.026921 \pm 0.000016$ u, $\Delta M = 25077 \pm 15$ keV [4].

Decay Properties. $^{12}Be \rightarrow {}^{12}B$ β^-. The upper limit of a branch involving delayed neutrons is 1% [2, 27]. $T_{1/2} = 24.4 \pm 3.0$ ms [2], $T_{1/2} = 24 \pm 1$ ms [27]. $Q(\beta^-) = 11707 \pm 15$ keV [4].

Spin, Parity. $J = 0^+$ [2].

Beryllium-13

Mass. $M = 13.037520 \pm 0.000540$ u (syst), $\Delta M = 34950 \pm 500$ keV (syst) [4].

Decay Properties. ^{13}Be is particle unstable with respect to $^{12}Be + n$ by 1800 ± 500 keV as obtained from systematic trends [5] and from $^{14}C(^{7}Li, ^{8}B)$ reaction data [28].

Spin, Parity. $J^\pi = 1/2^+$ or $5/2^+$ (syst) [3].

Beryllium-14

Mass. $M = 14.044040 \pm 0.000320$ u, $\Delta M = 41020 \pm 300$ keV [4]. $\Delta M = 40100 \pm 160$ keV was obtained from $^{14}C(\pi^-, \pi^+)$ reaction data [29].

Decay Properties. ^{14}Be is particle stable with respect to $^{13}Be + n$ [3] by 2000 ± 580 keV (syst) [5] and to $^{12}Be + 2n$ [3] by 200 ± 300 keV (syst) [5]. Since earlier theoretical mass predictions, on which the evaluation of ^{14}Be decay data in [3] is based, predict even higher one- and two-neutron separation energies, a β^- decay $^{14}Be \rightarrow {}^{14}B$ was adopted [3]. $Q(\beta^-) = 17360 \pm 300$ keV (syst) [4].

Spin, Parity. $J^\pi = 0^+$ [3].

References:

[1] F. Ajzenberg-Selove (Nucl. Phys. A **413** [1984] 1/214, 3/4, 22, 45/7, 62, 65/7, 80/1, 109, 132/4).

[2] F. Ajzenberg-Selove (Nucl. Phys. A **433** [1985] 1/158, 2, 5/6, 8, 21/2, 43/5).

[3] F. Ajzenberg-Selove (Nucl. Phys. A **360** [1981] 1/186, 2, 9, 56).

[4] A. H. Wapstra, G. Audi (Nucl. Phys. A **432** [1985] 1/54, 14/5).

[5] A. H. Wapstra, G. Audi (Nucl. Phys. A **432** [1985] 55/139, 58/61).

[6] O. V. Bochkarev, A. A. Korsheninnikov, E. A. Kuz'min, I. G. Mukha, A. A. Ogloblin, L. V. Chulkov, G. B. Yan'kov (Pis'ma Zh. Eksperim. Teor. Fiz. **40** [1984] 204/6; JETP Letters **40** [1984] 969/72).

[7] C. Rolfs et al. (personal communication [1982] to F. Ajzenberg-Selove [1]).

[8] H. P. Trautvetter, H. W. Becker, L. Buchmann, J. Görres, K. U. Kettner, C. Rolfs, P. Schmalbrock, A. E. Vlieks (Verhandl. Deut. Physik. Ges. [6] **18** [1983] 1141/2).

[9] D. P. Balamuth, L. Brown, T. E. Chapuran, J. Klein, R. Middleton, R. W. Zurmühle (Phys. Rev. [3] C **27** [1983] 1724/7).

[10] E. B. Norman, T. E. Chupp, K. T. Lesko, J. L. Osborne, P. J. Grant, G. L. Woodruff (Phys. Rev. [3] C **27** [1983] 1728/31, Erratum: C **28** [1983] 1409; Bull. Am. Phys. Soc. [2] **28** [1983] 713).

[11] C. N. Davids, A. J. Elwyn, B. W. Filippone, S. B. Kaufman, K. E. Rehm, J. P. Schiffer (Phys. Rev. [3] C **28** [1983] 885/7; Bull. Am. Phys. Soc. [2] **28** [1983] 714).

[12] T. R. Donoghue, E. Sugarbaker, M. Wiescher, T. C. Rinckel, K. E. Sale, C. P. Browne, E. D. Berners, R. W. Tarara, R. W. Warner (Phys. Rev. [3] C **28** [1983] 875/8; Bull. Am. Phys. Soc. [2] **28** [1983] 713/4).

[13] G. J. Mathews, R. C. Haight, R. G. Lanier, R. M. White (Phys. Rev. [3] C **28** [1983] 879/84; Bull. Am. Phys. Soc. [2] **28** [1983] 660/1).

[14] S. A. Fisher, R. L. Hershberger (Nucl. Phys. A **423** [1984] 121/9; Bull. Am. Phys. Soc. [2] **28** [1983] 713).

[15] D. A. Knapp, A. B. McDonald, C. L. Bennett (Nucl. Phys. A **411** [1984] 195/8; Bull. Am. Phys. Soc. [2] **28** [1983] 713).

[16] T. N. Taddeucci, J. Rapaport, C. D. Goodman, C. C. Foster, C. A. Goulding, C. Gaarde, J. Larsen, D. J. Horen, T. Masterson, E. Sugarbaker, P. Koncz (Bull. Am. Phys. Soc. [2] **28** [1983] 714).

[17] R. T. Skelton, R. W. Kavanagh (Nucl. Phys. A **414** [1984] 141/50).

[18] H. C. Evans, I. P. Johnstone, J. R. Leslie, W. McLatchie, H. B. Mak, T. Alexander (Can. J. Phys. **62** [1984] 1139/44).

[19] R. D. von Dincklage, C. Garcia-Recio, P. Graller, P. Hoff, B. Jonson, J. J. Simpson (Verhandl. Deut. Physik. Ges. [6] **19** [1984] 985/6).

[20] J. N. Bahcall, W. F. Huebner, S. H. Lubow, P. D. Parker, R. K. Ulrich (Rev. Mod. Phys. **54** [1982] 767/99).

[21] W. Seelmann-Eggebert, G. Pfennig, H. Münzel, H. Klewe-Nebenius (Karlsruhe Chart of Nuclides, 5th Ed., Kernforschungszentrum Karlsruhe, 1981).

[22] N. E. Holden, R. L. Martin, I. L. Barnes (Pure Appl. Chem. **56** [1984] 675/94, 680).

[23] H. S. Peiser, N. E. Holden, P. De Bièvre, I. L. Barnes, R. Hagemann, J. R. De Laeter, T. J. Murphy, E. Roth, M. Shima, H. G. Thode (Pure Appl. Chem. **56** [1984] 695/768, 720).

[24] E. W. Weber, J. Vetter (Phys. Letters A **56** [1976] 446/7).

[25] G. H. Fuller (J. Phys. Chem. Ref. Data **5** [1976] 835/1092, 845, 907).

[26] D. J. Millener, D. E. Alburger, E. K. Warburton, D. H. Wilkinson (Phys. Rev. [3] C **26** [1982] 1167/85, 1170/1).

[27] J. P. Dufour, S. Beraud-Sudreau, R. Del Moral, et al. (Z. Physik A **319** [1984] 237/8).

[28] D. V. Aleksandrov, E. A. Ganza, Yu. A. Glukhov, V. I. Dukhanov, I. B. Mazurov, B. G. Novatskii, A. A. Ogloblin, D. N. Stepanov, V. V. Paramonov, A. G. Trunov (Yadern. Fiz. **37** [1983] 797/9; Soviet J. Nucl. Phys. **37** [1983] 474/5).

[29] R. Gilman, H. T. Fortune, L. C. Bland, R. R. Kiziah, C. F. Moore, P. A. Seidl, C. L. Morris, W. B. Cottingame (Phys. Rev. [3] C **30** [1984] 958/61).

4 Atom and Atomic Ions

4.1 The Be^- Ion

Although Be^- ions have been produced by several means (described below), it is generally agreed that a stable Be^- ion, i.e., an anionic state below the ground state of neutral Be, does not exist. Rather, Be^- is believed to be metastable.

The lowest orbital outside the $1s^2\,2s^2\,{}^1S$ ground state of neutral Be is the 2p orbital. Addition of an electron to it gives an anionic $1s^2\,2s^2\,2p\,{}^2P^o$ state, which from ab initio calculations is known to be unstable. It should be observable in electron-atom collisions as a shape resonance (unbound state in the Coulomb potential of the atom, but separated from the continuum by a centrifugal barrier, see, e.g., [1]). The position and widths of this resonance have been calculated in several theoretical papers, among which only a few of the more recent ones [2 to 5] are cited here. Earlier results are discussed in references presented in [2]. It does not appear that this resonance has been experimentally observed.

These findings are in accord with semiempirical estimates of a negative electron affinity for Be, based on extrapolations from a series of atoms and ions of the first period. The estimates involve either a comparison of isoelectronic series ("isoelectronic extrapolation") [6 to 11], or of a series of ions with the same degree of ionization ("horizontal extrapolation", see [1]) [12 to 16]. On the basis of isoelectronic [6] and horizontal extrapolations [13, 16] it has been suggested that Be^- $1s^2\,2s^2\,3s\,{}^2S$ might be stable [6, 13, 16], but this was not confirmed by HF calculations including correlation effects [17, 18].

Experimentally, Be^- ions have been produced by several groups of workers, mostly for use in tandem accelerators. They were obtained either by extraction from a Penning ion source [19], or by charge exchange of Be^+ ion beams ($\leq$100 keV) with sodium [20], magnesium [21], or caesium vapors [22], or in a so-called caesium sputter source (sputtering with caesium ions on a Be or Be alloy surface, overlayed with caesium to reduce the work function) [23 to 27], or in a vacuum spark ion source [28]. It appears very likely that Be^- is produced in a metastable state, i.e., one which is lower in energy than a particular excited state of the neutral atom and can autodetach to the ${}^1S_0$ ground state of neutral Be only through magnetic interactions violating the spin selection rule $\Delta S=0$ for the autodetachment process (the same rule as for autoionization) as outlined on p. 166 [18, 29 to 32].

The most probable candidate is Be^- $1s^2\,2s\,2p^2\,{}^4P$ [29] which from ab initio calculations is predicted to be bound by $\leq$0.5 eV with respect to Be $1s^2\,2s\,2p\,{}^3P^o$, the lowest excited state of the neutral atom [18, 30 to 34]. Be^- ions produced by charge exchange of a Be^+ ion beam (3.5 keV) with caesium vapor were indeed metastable, and their autodetachment lifetime was measured by a time-of-flight method. Different components with lifetimes between 10^{-4} and 10^{-5} s were assigned to the autodetachment decay to the $1s^2\,2s^2\,{}^1S_0$ ground state of neutral Be of substates of the metastable ${}^4P$ state with different values of total angular momentum J. The $J={}^5/_2$ state would presumably have the longest lifetime and the $J={}^3/_2$ and ${}^1/_2$ states would have shorter lifetimes because of their spin-orbit coupling to autodetaching Be^- ${}^2P$ states [22]. An energy of the ${}^4P$ state of 2.53 ± 0.09 eV above the ground state of neutral Be was determined by measuring the energy of electrons autodetaching from fast Be^- ions in the center-of-mass system. These ions were produced in sequential charge exchange collisions between lithium vapor and Be^+ ions accelerated to fixed energies between 55 and 60 keV. The ${}^4P$ state is thus bound by 0.195 eV with respect to Be $1s^2\,2s\,2p\,{}^3P^o$ [32].

Another possible metastable state is Be^- $1s^2\,2p^3\,{}^4S^o$, which is theoretically predicted to be bound by ≤ 0.3 eV with respect to Be $1s^2\,2p^2\,{}^3P$ and to decay radiatively to Be^- ^4P. The predicted wavelength of the transition is between 2638 and 2671 Å [30, 31, 33]. Attempts to observe this transition have been unsuccessful [35]. Furthermore, there may be a core excited metastable state of Be^-, $1s\,2s\,2p^3\,{}^6S^o$. This state is theoretically predicted to be bound with respect to the core excited Be $1s\,2s\,2p^2\,{}^5P$ state and to possess two decay channels, relativistic autodetachment (i.e., violating $\Delta S = 0$) to Be $1s^2\,2s\,2p\,{}^3P^o$ [30, 36] and to Be $1s^2\,2p^2\,{}^3P$ [36], and radiative decay to Be^- $1s^2\,2s\,2p^2\,{}^4P$ [30].

References:

[1] H. S. W. Massey (Negative Ions, 3rd Ed., Cambridge University, Cambridge 1976, pp. 48/60, 97/104).
[2] M. Mishra, O. Goscinski, Y. Öhrn (J. Chem. Phys. **79** [1983] 5505/11).
[3] D. T. Chuljian, J. Simons (Intern. J. Quantum Chem. **23** [1983] 1723/38).
[4] B. R. Junker (Phys. Rev. [3] A **27** [1983] 2785/9).
[5] R. A. Donnelly, J. Simons (J. Chem. Phys. **73** [1980] 2858/66).
[6] B. L. Moiseiwitsch (Advan. At. Mol. Phys. **1** [1965] 61/83, 63).
[7] C. W. Scherr, J. N. Silverman, F. A. Matsen (Phys. Rev. [2] **127** [1962] 830/7).
[8] B. Edlén (J. Chem. Phys. **33** [1960] 98/100).
[9] H. O. Pritchard (Chem. Rev. **52** [1953] 529/63, 547/8).
[10] H. A. Skinner, H. O. Pritchard (Trans. Faraday Soc. **49** [1953] 1254/62).

[11] S. Geltman (J. Chem. Phys. **25** [1956] 782/3).
[12] S. Fraga (Can. J. Phys. **58** [1980] 544/5).
[13] R. J. Zollweg (J. Chem. Phys. **50** [1969] 4251/61).
[14] R. J. S. Crossley (Proc. Phys. Soc. [London] A **83** [1964] 375/89).
[15] J. W. Edie, F. Rohrlich (J. Chem. Phys. **36** [1962] 623/7).
[16] A. P. Ginsberg, J. M. Miller (J. Inorg. Nucl. Chem. **7** [1958] 351/67).
[17] K. D. Jordan, J. Simons (J. Chem. Phys. **67** [1977] 4027/37).
[18] A. W. Weiss (Phys. Rev. [2] **166** [1968] 70/4).
[19] H. J. Kaiser, E. Heinicke, H. Baumann, K. Bethge (Z. Physik **243** [1971] 46/59), E. Heinicke, K. Bethge, H. Baumann (Nucl. Instr. Methods **58** [1968] 125/33), K. Bethge, E. Heinicke, H. Baumann (Phys. Letters **23** [1966] 542/3).
[20] J. Heinemeier, P. Hvelplund (Nucl. Instr. Methods **148** [1978] 425/9), J. Heinemeier, P. Tykesson (Nucl. Instr. Methods **141** [1977] 183/4; Rev. Phys. Appl. **12** [1977] 1471/5).

[21] J. Heinemeier, P. Hvelplund (Nucl. Instr. Methods **148** [1978] 65/75).
[22] Y. K. Bae, J. R. Peterson (Phys. Rev. [3] A **30** [1984] 2145/7).
[23] R. Middleton (Nucl. Instr. Methods Phys. Res. **214** [1983] 139/50).
[24] W. Kreisman (private communication to R. Middleton, Nucl. Instr. Methods **122** [1974] 35/43).
[25] G. Braun-Elwert, J. Huber, G. Korschinek, W. Kutschera, W. Goldstein, R.L. Hershberger (Nucl. Instr. Methods **146** [1977] 121/38, 126, 133).
[26] C. Lukner (Nucl. Instr. Methods **167** [1979] 249/54, 250).
[27] A. Kh. Ayukhanov, M. K. Abdullaeva (Izv. Akad. Nauk SSSR Ser. Fiz. **30** [1966] 2000/7; Bull. Acad. Sci. USSR Phys. Ser. **30** [1966] 2083/90, 2088).
[28] T. Makita, H. Kishi, K. Kodera (Shitsuryo Bunseki **21** [1973] 293/301).
[29] H. Hotop, W. C. Lineberger (J. Phys. Chem. Ref. Data **14** [1985] 731/50, 749; **4** [1975] 539/76, 564, 572).
[30] D. R. Beck, C. A. Nicolaides, G. Aspromallis (Phys. Rev. [3] A **24** [1981] 3252/4).

[31] C. F. Bunge, M. Galán, R. Jáuregi, A. V. Bunge (Nucl. Instr. Methods Phys. Res. **202** [1982] 299/305).
[32] T. J. Kvale, G. D. Alton, R. N. Compton, D. J. Pegg, J. S. Thompson (Phys. Rev. Letters **55** [1985] 484/7).
[33] D. R. Beck, C. A. Nicolaides (Intern. J. Quantum Chem. Symp. No. 18 [1984] 467/81).
[34] L. A. Cole, J. P. Perdew (Phys. Rev. [3] A **25** [1982] 1265/71).
[35] T. A. Andersen (private communication to D. R. Beck et al. [33] and to T. J. Kvale et al. [32]).
[36] G. Aspromallis, C. A. Nicolaides, Y. Komninos (J. Phys. B **18** [1985] L545/L549).

4.2 The Be Atom

4.2.1 Ground State. Ionization Energies

Observed values of the valence and core level ionization energies relative to the Be $1s^2\,2s^2\,{}^1S_0$ ground state are compiled in Table 4/1. Most were obtained from series limits, either in optical spectra for valence levels (2s shell) or in soft-X-ray spectra for core levels (1s shell).

Valence level ionization limits given in the Landolt-Börnstein Tables [9] and in the "Tables of Atomic Energy Levels" of Moore [10] are 75194.3 cm^{-1} and 75194.29 cm^{-1}, respectively. References to spectroscopic papers prior to 1950, from which these values have been selected, may be found in these tables.

Table 4/1
Ionization Limits and Ionization Energies of Neutral Be.
Values placed in parentheses have been converted from the unit used in the cited reference by 1 eV $\triangleq$ 8065.479 cm^{-1}.

state of Be^+ ion	ionization limit in cm^{-1}	ionization energy in eV	method	Ref.
$1s^2\,2s\,{}^2S_{1/2}$	75192.64 ± 0.06	(9.323)	series nd $^1D_2$	[1]
(ground state)	75192.5 [a]	(9.323)	series ns $^{1,3}S$, np $^{1,3}P^o$, nd $^{1,3}D$, nf $^{1,3}F^o$	[2]
	75192.07	(9.323) [b]	series nf $^1F^o$, ns ^3S, nd ^3D [3], np $^3P^o$, nf $^3F^o$, np $^1P^o$ [4]	[3, 4]
$1s^2\,2p\,{}^2P^o$	107125.22 [c]	(13.282)	series 2p np ^3P	[3]
$1s^2\,3s\,{}^2S_{1/2}$	163400 [d]	(20.26)	series 3s np $^1P^o_1$	[5]
$1s\,2s^2\,{}^2S_{1/2}$	(996900)	123.6 ± 0.1	123.6 eV $= E_A + E_i(Be^+) + E_i(Be)$ [e]	[6]
	994900	(123.4)	series $1s\,2s^2$ np $^1P^o_1$	[7]

[a] Reevaluation from term energies given in [3, 4], based on semiempirical calculations of quantum defects.
[b] Moore [8] cites the value from [3] as 9.322 eV, based on the conversion factor 1 eV $\triangleq$ 8065.73 cm^{-1}.
[c] Value refers to center of gravity of $Be^+\ {}^2P^o_{1/2,3/2}$ doublet.
[d] Original reference [5] gives 612 Å.
[e] E_A is the energy of the Auger transition $Be^+\ 1s\,2s^2\,{}^2S \rightarrow Be^{2+}\ 1s^2\,{}^1S$ for the free ions, measured by the authors [6], $E_i(Be^+)$ and $E_i(Be)$ are ionization energies taken from the literature.

The core level ionization energy of elemental Be in its standard state has been frequently measured by X-ray and photoelectron spectra. This energy refers to the Fermi level of the solid and is substantially lower than that of the free atom: Reported values are between 111.4 and 112.2 eV from X-ray spectra, see p. 142, and 110.5 and 111.3 eV from photoelectron spectra, see p. 145. Attempts have been made (see, e.g., [11, 12]) to obtain the core level ionization energy of the free atom from that of the element in its standard state by applying semiempirical corrections for environmental effects (work function correction, core level shifts in the metal relative to the free atom). In view of the accurate core level ionization energies for beryllium and boron free atoms, it appears that such procedures underestimate solid state effects [6].

References:

[1] R. Beigang, D. Schmidt, P. J. West (J. Phys. Colloq. [Paris] **44** [1983] C7-229/C7-237).
[2] M. J. Seaton (J. Phys. B **9** [1976] 3001/7).
[3] L. Johansson (Arkiv Fysik **23** [1963] 119/28).
[4] J. E. Holmstroem, L. Johansson (Arkiv Fysik **40** [1969] 133/8).
[5] J. M. Esteva, G. Mehlman-Balloffet, J. Romand (J. Quant. Spectrosc. Radiat. Transfer **12** [1972] 1291/303).
[6] P. Bisgaard, R. Bruch, P. Dahl, B. Fastrup, M. Roedbro (Phys. Scr. **17** [1978] 49/52).
[7] G. Mehlman, J. M. Esteva (Astrophys. J. **188** [1974] 191/5).
[8] C. E. Moore (NSRDS-NBS-34 [1970] 1/8).
[9] Landolt-Börnstein 6th Ed. **1** Pt. 1 [1950] 73.
[10] C. E. Moore (NBS-C-467-Vol. I [1949] 12/3).
[11] D. A. Shirley, R. L. Martin, S. P. Kowalczyk, F. R. McFeely, L. Ley (Phys. Rev. [3] B **15** [1977] 544/52).
[12] J. C. Slater (Phys. Rev. [2] **98** [1955] 1039/45).

4.2.2 Electron Affinity A

Hotop and Lineberger [1] and Bunge et al. [2] have presented strong evidence showing that a stable Be^- ion does not exist ($A<0$), i.e., Be^- has no stable states below the ground state of neutral Be. These authors conclude that Be^- ions which have been observed so far are metastable and are bound with respect to an excited state of the neutral atom. A full discussion of the existence of Be^- with literature references is presented on pp. 137/139.

References:

[1] H. Hotop, W.C. Lineberger (J. Phys. Chem. Ref. Data **14** [1985] 731/50, 742, 749; **4** [1975] 539/76, 564, 572).
[2] C. F. Bunge, M. Galán, R. Jáuregui, A. V. Bunge (Nucl. Instr. Methods Phys. Res. **202** [1982] 299/305, 301/3).

4.2.3 Atomic Structure Calculations

The Be atom and its uni- and dipositive ions are members of the simplest few-electron systems which serve as probes for testing atomic structure theories, see, e.g., [1]. Their ubiquitous occurrence in the literature on atomic structure calculations is well documented in a bibliography on ab initio quantum chemical calculations covering the period 1978 to 1980 [2] and from 1981 covering each year by annual supplements, up to 1984 at present [3]. For the seven-year period now accessible this bibliography cites, e.g., more than 200 papers dealing with neutral Be (compared with a total of 7824 literature references for quantum chemical calculations in that bibliography).

In view of the huge abundance of papers on atomic structure calculations for Be and its positive ions, a complete documentation of all of them would be beyond the scope of the present volume. Rather, for the most recent literature the reader is referred primarily to the "Quantum Chemical Literature Data Base" [2, 3] mentioned above. Access to some earlier references for special topics of atomic structure is provided by bibliographies published by the U.S. National Bureau of Standards: The "Bibliography on Atomic Energy Levels and Spectra", covering 1968 through 1971 [4], 1971 through 1975 [5], and 1975 through 1979 [6], and the "Bibliography on Atomic Transition Probabilities", covering 1914 through 1977 [7] and 1977 through (March) 1980 [8]. Some earlier papers devoted specifically to ab initio calculations for Be are discussed in a textbook by Froese Fischer [9] and in a review article by Hibbert [1].

References:

[1] A. Hibbert (Rept. Progr. Phys. **38** [1975] 1217/338, 1266/8).
[2] K. Ohno, K. Morokuma (Quantum Chemistry Literature Data Base, Elsevier, Amsterdam 1982).
[3] K. Ohno, K. Morokuma (J. Mol. Struct. **91** THEOCHEM 8 [1982] 1/252; **106** THEOCHEM 15 [1983] 1/215; **119** THEOCHEM 20 [1984] 1/229; **134** THEOCHEM 27 [1985] 1/298).
[4] L. Hagan, W. C. Martin (NBS-SP-363 [1972] 1/102, 9).
[5] L. Hagan (NBS-SP-363-Suppl. 1 [1977] 1/182, 13).
[6] R. Zalubas, A. Albright (NBS-SP-363-Suppl. 2 [1980] 1/115, 13).
[7] J. R. Fuhr, B. J. Miller, G. A. Martin (NBS-SP-505 [1978] 1/270, 18).
[8] B. J. Miller, J. R. Fuhr, G. A. Martin (NBS-SP-505-Suppl. 1 [1980] 1/112, 9/10).
[9] C. Froese Fischer (The Hartree-Fock Method for Atoms, Wiley, New York 1977, 1/308, 113, 191/2).

4.2.4 X-Ray Spectra

Several K X-ray series have been observed in absorption spectra for the free atoms in vacuum spark: $1s^2\,2s^2 \rightarrow 1s\,2s^2\,np$, n=2 to 5, 107.38 to 100.9 Å [1, 2]; $1s^2\,2s^2 \rightarrow 1s\,2s\,2p\,nl$, n=3, 4, l=s or d, 99.19 to 89.16 Å; $1s^2\,2s\,2p \rightarrow 1s\,2s\,2p\,np$, n=2, 3, 107.2 to 100.86 Å [1].

The X-ray spectrum of elemental Be in its standard state has been measured repeatedly, both in emission [3 to 20, 35] and in absorption [5, 12, 19, 21 to 26]. References to literature up to 1950 may be found in [27]. The K emission band of Be was among the first to be examined experimentally, though much of the preliminary work appears to have suffered from the presence of target impurities [28]. The K emission and absorption spectra of elemental Be are continuous and bandlike (overall width ~14 eV). This reflects the splitting of atomic valence levels involved in the absorption or emission process into a band of finite width in the solid. The emission and absorption profiles are similar in shape to the density-of-states curve, but are not a direct mapping of it, since they also depend on the transition probabilities between the Bloch states of the valence electrons and the 1s core states, see, e.g., [28, 29].

The sharp rise in the intensity at the high energy (short wavelength) side of the emission spectrum, or at the low energy (high wavelength) side of the absorption spectrum, is referred to as the K edge, i.e., the binding energy of a 1s electron in solid Be with respect to the Fermi level, cf. p. 140. Derived values are compiled in Table 4/2.

Table 4/2
Position of K Edge in Elemental Be.
First and second row: Values in parentheses have been converted from the unit used in the cited reference by E (in eV) · λ (in Å) = 12398. Third row: Em = emission, Ab = absorption.

K edge in eV	112.00	111.80	112.2	(111.63)	111.45	112.1	(112)	(111.8)
K edge in Å	(110.70)	(110.89)	110.85	111.06	(111.24)	(110.6)	111	110.9[a)]
spectrum . .	Em [3, 5], Ab [5]	Em	Em	Em	Em	Ab	Ab	Ab
Ref	[3, 5]	[6]	[17]	[15]	[19]	[22, 23]	[24, 26]	[30]

a) "Best value", selected in [30] from three primary sources prior to 1950.

A high-energy satellite is observed beyond the K edge in the emission spectrum of elemental Be at 143 to 146 eV. It is assigned as a double-ionization satellite, i.e., X-ray transitions involving initial states with a double 1s core hole [8, 9, 15, 31]. This peak was also observed after ion impact on a Be target (ions H^+, He^+, N^+, Ar^+ at energies $\leqq$1.2 MeV [32, 33], ions He^+, Ne^+, Ar^+ at 4.5 MeV, Kr^+ at 3.0 MeV [34]).

References:

[1] G. Mehlman, J. M. Esteva (Astrophys. J. **188** [1974] 191/5).
[2] J. M. Esteva, G. Mehlman-Balloffet, J. Romand (J. Quant. Spectrosc. Radiat. Transfer **12** [1972] 1291/303).
[3] J. A. Tagle, E. T. Arakawa, T. A. Callcott (Phys. Rev. [3] B **21** [1980] 4552/7).
[4] O. Aita, K. Ichikawa, H. Nakamura, Y. Iwasaki, K. Tsutsumi (Japan. J. Appl. Phys. **17** [1978] 595/6).
[5] T. A. Callcott, E. T. Arakawa, D. L. Ederer (Japan. J. Appl. Phys. **17** [1978] Suppl. 2, pp. 149/53).
[6] R. S. Crisp (Phil. Mag. [8] **36** [1977] 609/28).
[7] A. I. Kozlenkov, Yu. I. Belov, V. G. Bogdanov (Pribory Tekhn. Eksperim. **1976** No. 1, pp. 202/5; Instr. Exptl. Tech. [USSR] **19** [1976] 247/50).
[8] E. T. Arakawa, M. W. Williams (Phys.Rev. [3] B **8** [1973] 4075/8).
[9] T. Sagawa (J. Phys. Colloq. [Paris] **32** [1971] C4-186/C4-192).
[10] K. Feser, J. Müller, G. Wiech, A. Faessler (J. Phys. Colloq. [Paris] **32** [1971] C4-331/C4-333; DESY-70-59 [1970] 1/11; C.A. **75** [1971] No. 103332).
[11] O. Aita, T. Sagawa (J. Phys. Soc. Japan **27** [1969] 164/75).
[12] T. Sagawa (Soft X-Ray Band Spectra Electron. Struct. Metals Mater., Strathclyde,Scotl., 1967 [1968], pp. 29/43; C.A. **71** [1969] No. 130289).
[13] L. M. Wason, R. K. Dimond, D. J. Fabian (Soft X-Ray Band Spectra Electron. Struct. Metals Mater., Strathclyde, Scotl., 1967 [1968], pp. 45/58; C.A. **71** [1969] No. 130288).
[14] G. Wiech (Soft X-Ray Band Spectra Electron. Struct. Metals Mater., Strathclyde, Scotl., 1967 [1968], pp. 59/70; C.A. **71** [1969] No. 130303).
[15] Y. Hayasi (Sci. Rept. Tohoku Univ. [1] **51** [1968] 1/8), T. Hayasi, Y. Hayasi (Sci. Rept. Tohoku Univ. [1] **50** [1967] 228/35).
[16] D. W. Fischer, W. L. Baun (Norelco Reporter **14** [1967] 92/8; C.A. **69** [1968] No. 6411).
[17] J. E. Holliday (in: E. F. Kaelble, Ed., Handbook of X-rays, Chapter 38, McGraw Hill, New York 1967, pp. 1/41, 35, 38).
[18] R. C. Ehlert, R. A. Mattson (Advan. X-Ray Anal. **9** [1966] 456/70).
[19] A. P. Lukirskii, I. A. Brytov (Fiz. Tverd. Tela [Leningrad] **6** [1964] 43/53; Soviet Phys.-Solid State **6** [1964] 33/41).

[20] R. S. Crisp, S. E. Williams (Phil. Mag. [8] **6** [1961] 365/9).

[21] C. Kunz, R. Hänsel, G. Keitel, P. Schreiber, B. Sonntag (NBS-SP-323 [1971] 275/7).
[22] R. Hänsel, G. Keitel, B. Sonntag, C. Kunz, P. Schreiber (Phys. Status Solidi A **2** [1970] 85/90).
[23] N. Swanson, K. Codling (J. Opt. Soc. Am. **58** [1968] 1192/4).
[24] T. Sagawa, Y. Iguchi, M. Sasanuma et al. (J. Phys. Soc. Japan **21** [1966] 2602/10).
[25] A. P. Lukirskii (Izv. Akad. Nauk SSSR Ser. Fiz. **25** [1961] 910/2; Bull. Acad. Sci. USSR Phys. Ser. **25** [1961] 923/5).
[26] R. W. Johnston, D. H. Tomboulian (Phys. Rev. [2] **94** [1954] 1585/9).
[27] Landolt-Börnstein 6th Ed. **1** Pt. 4 [1955] 775, 807/8.
[28] D. H. Tomboulian (in: S. Flügge, Encyclopedia of Physics, Vol. 30, Springer, Berlin-Göttingen-Heidelberg 1957, pp. 246/304, 273/4).
[29] G. A. Rooke (Soft X-Ray Band Spectra Electron. Struct. Metals Mater., Strathclyde, Scotl., 1967 [1968], pp. 3/27; C.A. **71** [1969] No. 130290).
[30] Landolt-Börnstein 6th Ed. **1** Pt. 1 [1950] 216/7, 229/32.

[31] Y. Hayasi (Sci. Rept. Tohoku Univ. [1] **52** [1969] 41/4).
[32] K. Kawatsura, K. Ozawa, F. Fujimoto, M. Terasawa (Phys. Letters A **60** [1977] 327/9), K. Kawatsura (Rev. Phys. Chem. Japan **47** [1977] 69/79).
[33] K. Fujimoto, K. Kawatsura, K. Ozawa, M. Terasawa (Phys. Letters A **57** [1976] 263/4), K. Kawatsura (Rev. Phys. Chem. Japan **47** [1977] 53/68).
[34] D. J. Nagel, A. R. Knudson, P. G. Burkhalter (J. Phys. B **8** [1975] 2779/86).
[35] A. I. Kozlenkov, A. I. Shulgin, A. V. Postnikov, E. Z. Kurmaev, A. I. Ivanovskii, V. A. Gubanov (J. Phys. C **18** [1985] 3581/9).

4.2.5 Auger Spectra

The ejection of Auger electrons from neutral free Be atoms has been observed after passage of fast Be^+ ions (500 keV) either through He or CH_4 gas targets under single collision conditions [1 to 3], or through thin carbon foils [4, 5]. The core excited initial states are formed by both core excitation and capture of target electrons [1]. The observed Auger transitions are listed in Table 4/3. A comparison of observed transition energies with theory and references for theoretical work are presented in [1, 2].

Table 4/3
Observed Auger Transitions of Neutral Be.

Auger energy in eV	initial Be state	final Be^+ state	Ref.
101.0 ± 0.1	$1s\,2s^2\,2p\,^3P^o$	$1s^2\,2p\,^2P^o$	[1]
102.1 ± 0.1	$1s\,2s^2\,2p\,^1P^o$	$1s^2\,2p\,^2P^o$	[1]
102.2 ± 0.3	$1s\,2s\,2p^2\,^5P$	$1s^2\,2p\,^2P^o$	[3]
102.7 ± 0.5	$1s\,2s\,2p^2\,^5P$	$1s^2\,2p\,^2P^o$	[5]
104.9 ± 0.1	$1s\,2s^2\,2p\,^3P^o$	$1s^2\,2s\,^2S$	[1, 2]
106.1 ± 0.3	$1s\,2s\,2p^2\,^5P$	$1s^2\,2s\,^2S$	[3]
106.7 ± 0.5	$1s\,2s\,2p^2\,^5P$ [a)]	$1s^2\,2s\,^2S$	[5]

[a)] Lifetime observed by Auger electron detection is 132 ± 50 ns [5].

A peak at 102 eV in the Auger spectrum of solid Be, excited by 4.5 keV Ar^+ ions, was also assigned to Auger transitions from neutral free Be atoms. These were assumed to escape from the surface as sputtered recoils [17].

By the method of ion-atom collisions, Auger electron emission from free Be^{+} and Be^{2+} ions has been observed as well. The spectra are presented under the respective ions on pp. 160 and 174. Nevertheless, the Auger spectrum of elemental Be in its standard state, although arising from Be^{+} as the autoionizing species, is referred to here, because it has a close resemblance to the X-ray emission spectrum. The Auger spectrum of solid Be is excited by electron impact [6 to 16] or ion impact (Ar^{+}, Kr^{+}) [15, 17] on a Be surface. Like the X-ray emission spectrum, the Auger spectrum of solid Be is broad and bandlike. Its profile has a close resemblance to the density-of-states curve, and even to the total density of states in the valence band. This is in contrast to the X-ray spectrum which, due to selection rules, yields information only about p-like valence band levels [14]. However, spectral resolution is usually worse than in the X-ray spectrum. Furthermore, the Auger spectrum appears to strongly depend on uncontrollable surface conditions.

There is agreement about an Auger peak at ~104 eV [6, 8 to 15, 18], which is assigned as an Auger emission from a $1s\,2p^2$ initial state [8, 14]. The energy of ~104 eV corresponds to the adopted peak position in the differential spectrum, dN/dE, where N is the counting rate and E is the energy. The position shifts to 98.5 eV in an N(E) spectrum [16, 17]; it is not clear, however, whether the work function correction is the same in these two papers. A second strong peak near 92 eV was observed by some authors [6, 8, 14], but has also been reported [6, 8 to 13, 18] to be conditionally absent or only weak. It is unclear whether this is an intrinsic property of pure crystalline Be. The peak was assigned by some authors as a transition from a $1s\,2s^2$ initial state of Be^{+} in crystalline Be [6, 8, 14], which should broaden and merge into the line at ~104 eV for a disordered Be phase produced, e.g., by evaporation of Be onto a cold substrate [6, 8]. Other authors, however, relate the occurrence of this peak with the presence of oxygen [9, 13] or silicon [12]. Likewise, a high energy satellite at ~143 eV was assigned to an Auger process initiated from a Be^{+} ion with two core holes [10, 11], but was claimed to be completely absent and ascribed to an impurity by others [6].

References:

[1] M. Roedbro, R. Bruch, P. Bisgaard (J. Phys. B **12** [1979] 2413/47).
[2] M. Roedbro, R. Bruch, P. Bisgaard (J. Phys. B **10** [1977] L275/L279).
[3] R. Bruch, M. Roedbro, P. Bisgaard, P. Dahl (Phys. Rev. Letters **39** [1977] 801/4).
[4] R. Bruch, G. Paul, J. Andrä, B. Fricke (Phys. Letters A **53** [1975] 293/4).
[5] R. Bruch, G. Paul, J. Andrä (J. Phys. B **8** [1975] L253/L258).
[6] H. G. Maguire (Solid State Commun. **45** [1983] 71/3).
[7] D. R. Jennison, H. H. Madden, D. M. Zehner (Phys. Rev. [3] B **21** [1980] 430/5).
[8] H. G. Maguire, P. D. Augustus (Phil. Mag. [8] **30** [1974] 95/103).
[9] S. Thomas (Solid State Commun. **13** [1973] 1593/4).
[10] L. H. Jenkins, D. M. Zehner, M. F. Chung (Surf. Sci. **38** [1973] 327/40).
[11] L. H. Jenkins, D. M. Zehner (Solid State Commun. **12** [1973] 1149/51).
[12] D. M. Zehner, N. Barbulesco, L. H. Jenkins (Surf. Sci. **34** [1973] 385/93).
[13] M. Suleman, E. B. Pattinson (J. Phys. F **3** [1973] 497/504).
[14] R. G. Musket, R. J. Fortner (Phys. Rev. Letters **26** [1971] 80/2).
[15] L. Viel, C. Benazeth, N. Benazeth (Surf. Sci. **54** [1976] 635/46), N. Colombie, C. Benazeth, J. Mischler, L. Viel (Radiat. Eff. **18** [1973] 251/5), F. Louchet, L. Viel, C. Benazeth, B. Fagot, N. Colombie (Radiat. Eff. **14** [1972] 123/30).
[16] T. Sekine, A. Mogami, M. Kudoh, K. Hirata (Vacuum **34** [1984] 631/6).
[17] O. Grizzi, R. A. Baragiola (Phys. Rev. [3] A **30** [1984] 2297/303).
[18] E. Jensen, R. A. Bartynski, T. Gustafsson, E. W. Plummer, M. Y. Chou, M. L. Cohen, G. B. Hoflund (Phys. Rev. [3] B **30** [1984] 5500/7, 5501).

4.2.6 Fluorescence Yield

These quantities (both experimental and theoretical) are presented under the Be^+ ion on p. 161.

4.2.7 Photoelectron Spectra

Photoelectron spectra excited with Al Kα or Mg Kα radiation have been used to obtain a core level binding energy of elemental Be in its standard state of 111.3$\pm$0.3 eV [1] and 110.5$\pm$0.3 eV [2]. These values refer to the Fermi level of the solid, as do the binding energies obtained from X-ray spectra of the solid, see p. 140. The X-ray excited valence band photoelectron spectrum [3] is an intrinsic property of the crystalline state and thus is not outlined here.

References:

[1] P. Steiner, H. Höchst, S. Hüfner (Z. Physik B **30** [1978] 129/43), H. Höchst, P. Steiner, S. Hüfner (Phys. Letters A **60** [1977] 69/71).
[2] K. Hamrin, G. Johansson, A. Fahlman, C. Nordling, K. Siegbahn (UUIP-548 [1967] 1/9; N.S.A. **22** [1968] No. 13211).
[3] H. Höchst, P. Steiner, S. Hüfner (Z. Physik B **30** [1978] 145/54; J. Phys. F **7** [1977] L309/L314).

4.2.8 Atomic Energy Levels. Optical Spectra (Be I)

Levels. A survey of atomic energy levels of neutral Be found by various spectroscopic methods is presented in Table 4/4. This table is arranged by term systems belonging to the same "running" electron. The energy of only the lowest and highest member of each system is given numerically. Papers reporting lifetime measurements are cited in the column next to the references for the term values.

Most of the terms compiled in Table 4/4 were obtained by Johansson et al. [1 to 3] who studied atomic emission spectra from 1400 to 35000 Å using a hollow cathode discharge. Most excited-state lifetimes were measured by Andersen et al. [6, 10], Bergstroem, Bromander et al. [5, 7], and Hontzeas et al. [13, 14]. These authors used a time-of-flight method after beam-foil excitation.

The compilation of atomic energy levels for Be in the well-known tables of Moore [20] is based on several spectroscopic studies up to 1946. The most important changes in the energy levels as given by Moore [20] after the spectrum of neutral Be was reexamined are the following: (1) The quantity "x" in Moore's tables, denoting the then existing uncertainty of $\pm 2\ cm^{-1}$ in the position of the triplet terms relative to the singlets, is $x = -1.18\ cm^{-1}$. This was determined from a precision measurement of the very weak intercombination line $2s^2\ ^1S_0 - 2s\,2p\ ^3P^o_1$ at 4548.538$\pm$0.002 Å [21]. Same wavelengths within the limits of error were obtained later by others [2, 22]. (2) The energy values of $2s\,np\ ^3P^o$, n=3 to 6, $2s\,4p\ ^1P^o$, $2p^2\ ^1S$, $2p^2\ ^1D$, and $2p\,3p\ ^3P$ have been replaced by entirely new values [1 to 3, 12]. (3) The term $2p\,3d\ ^3P^o$ has been removed [1, 3].

Table 4/4
Term Systems of Be I and Source List for Lifetime Measurements.

term system	energy E of lowest member in cm^{-1}	energy E of highest member in cm^{-1}	references for energy	references for lifetime τ	mode of excitation; comment
2s ns $^1S_0$, $2 \le n \le 11$	0	74163.4	[1 to 3]	—	hollow cathode
2s ns $^1S_0$, $10 \le n \le 38$	~73929[a)]	~75114[a)]	[4]	—	two-step laser excitation
2s np $^1P^o_1$, $2 \le n \le 6$	42565.35	71746.09	[1, 2]	n=2 [5 to 7, 13, 52, 53][b)]	E: hollow cathode; τ: beam foil
2s nd $^1D_2$, $3 \le n \le 12$	64428.31	74443.3	[1 to 3]	n=3, 4 [5 to 7, 9, 52] $3 \le n \le 6$ [10]	E: hollow cathode; τ: beam foil [5 to 7, 10, 52], level crossing [9]
2s nd $^1D_2$, $10 \le n \le 38$	~73868[a)]	~75113[a)]	[4]	—	two-step laser excitation
2s nf $^1F^o_3$, $4 \le n \le 7$	68241.17	72931.64	[1]	—	hollow cathode
$2p^2\ ^1S_0$	76189.5±5[c)]	—	[12]	—	see footnote[c)]
2p np $^1P_1$, $3 \le n \le 6$	89121.08	103529.1	[3]	n=3 [5 to 7][d)] n=3, 4, 5 [13, 14]	E: hollow cathode; τ: beam foil
$2p^2\ ^1D_2$	56882.43	—	[1, 2]	[10]	E: hollow cathode; τ: beam foil
2p nd $^1P^o$, $3 \le n \le 16$	95672[e)]	106740[e)]	[12, 28]	—	E: three-photon absorption +resonance ion mass spectrometry [12][e)], vacuum spark [28]
2p nd $^1D^o_2$, n=3, 4, 5	93393.98	102441.9	[1, 3]	n=3 [13, 14]	E: hollow cathode; τ: beam foil
2p nf $^1F_3$, n=4, 5	100092.42	102640.99	[3]	—	hollow cathode
2s ns $^3S_1$, $3 \le n \le 8$	52080.94	73088.5	[1, 2]	n=3, 5 [6] n=3 [5, 7, 17]	E: hollow cathode; τ: beam foil [5 to 7], electron impact+laser excitation [17]

2s np $^3P^o_{0,1,2}$, $2 \leq n \leq 8$	21980.15 [f]	73309.15	[1, 2]	—	hollow cathode
2s nd ^3D, $3 \leq n \leq 12$	62053.72	74415.3	[1, 2]	n = 3, 4 [5, 6, 18] n = 3 [7, 37]	E: hollow cathode; τ: beam foil [5 to 7, 18], electron impact + laser excitation [17]
2s nf $^3F^o$, $4 \leq n \leq 7$	68241.02	72931.60	[2]	—	hollow cathode
2p 3s $^3P^o_{0,1,2}$	85558.11 [f]	—	[1, 2]	—	hollow cathode
2p np $^3P_{0,1,2}$, $2 \leq n \leq 5$	59696.03 [f]	102198.59 [f]	[1, 3]	n = 2 [5 to 7] n = 3, 4 [15, 18]	E: hollow cathode; τ: beam foil
2p nd $^3D^o_{1,2,3}$, n = 3, 4, 5	94189.79 [f]	102587.1	[1, 3]	n = 3 [6, 13, 14]	E: hollow cathode; τ: beam foil
2p 4f $^3F_3$, 2p 4f $^3F_4$	100230.53, 100231.82	— —	[3]	—	E: hollow cathode; τ: beam foil

[a] Calculated from Rydberg-Ritz formula with quantum defect δ given in a diagram in the original reference [4]. The Rydberg constant for ^9Be is 109730.63 cm^{-1}, see, e.g. [11].

[b] Among the measured lifetimes of the 2s 2p $^1P^o_1$ state, $\tau = 1.85 \pm 0.07$ ns [52] and 1.80 ± 0.15 ns [13] are regarded as the most accurate [52]. $\tau = 1.85$ ns implies an oscillator strength f = 1.34 for the resonance transition $2s^2\ ^1S_0 - 2s\,2p\ ^1P^o_1$ [52]; thus, f = 2.5 from flame spectrometry [8] appears to be in error. Theoretical work for f up to 1973 is reviewed in [52].

[c] Energy: 997 ± 5 cm^{-1} above Be^+ 2s ^2S, this taken as 75192.5 cm^{-1}, see p. 139 [12].
Mode of excitation: Three-photon absorption from Be $2s^2\ ^1S_0$ ground state. Light source was frequency-doubled output of dye laser, 2620 to 2630 Å. Detection of Be^+ from autoionization of Be $2p^2\ ^1S_0$ by resonance ion mass spectrometry. Observed level width was 74.9 cm^{-1} [12].
Comment: A previous spectroscopic level energy of 71498.9 cm^{-1} [20] (adopted in [1, 26]), i.e., below Be^+ 2s ^2S, was based on classification of a line at 3455 Å as $2s\,2p\ ^1P^o - 2p^2\ ^1S$, now reclassified, however, as $2s\,3p\ ^1P^o - 2p\,3p\ ^1P$ [14, 15]. Prior to the experimental detection [12], the position and width of the $2p^2\ ^1S_0$ level was predicted by numerous theoretical calculations which are reviewed in [12]. The origin of the value 64460.6 cm^{-1} for the energy [16, 25] is unclear.

[d] Observed line was 3455 Å previously classified as $2s\,2p\ ^1P^o - 2p^2\ ^1S$ [6, 7], but now reclassified as $2s\,3p\ ^1P^o - 2p\,3p\ ^1P$, cf. footnote [c].

[e] Energies: Term value given here for either lowest or highest member is the inverse of the wavelength reported for the respective transition from the Be $2s^2\ ^1S_0$ ground state. Disposable single-photon wavelength for three-photon absorption was 3135.7 Å for n = 3 and 2810.58 Å for n = 16 [12].
Mode of excitation, comment: Excitation and Be^+ detection as in the case of $2p^2\ ^1S_0$ [12], cf. footnote [c], or excitation in a vacuum spark [28]. Wavelength reported in [28] may be deteriorated by plasma effects [12].

[f] Center of gravity of fine structure components.

Fig. 4—1

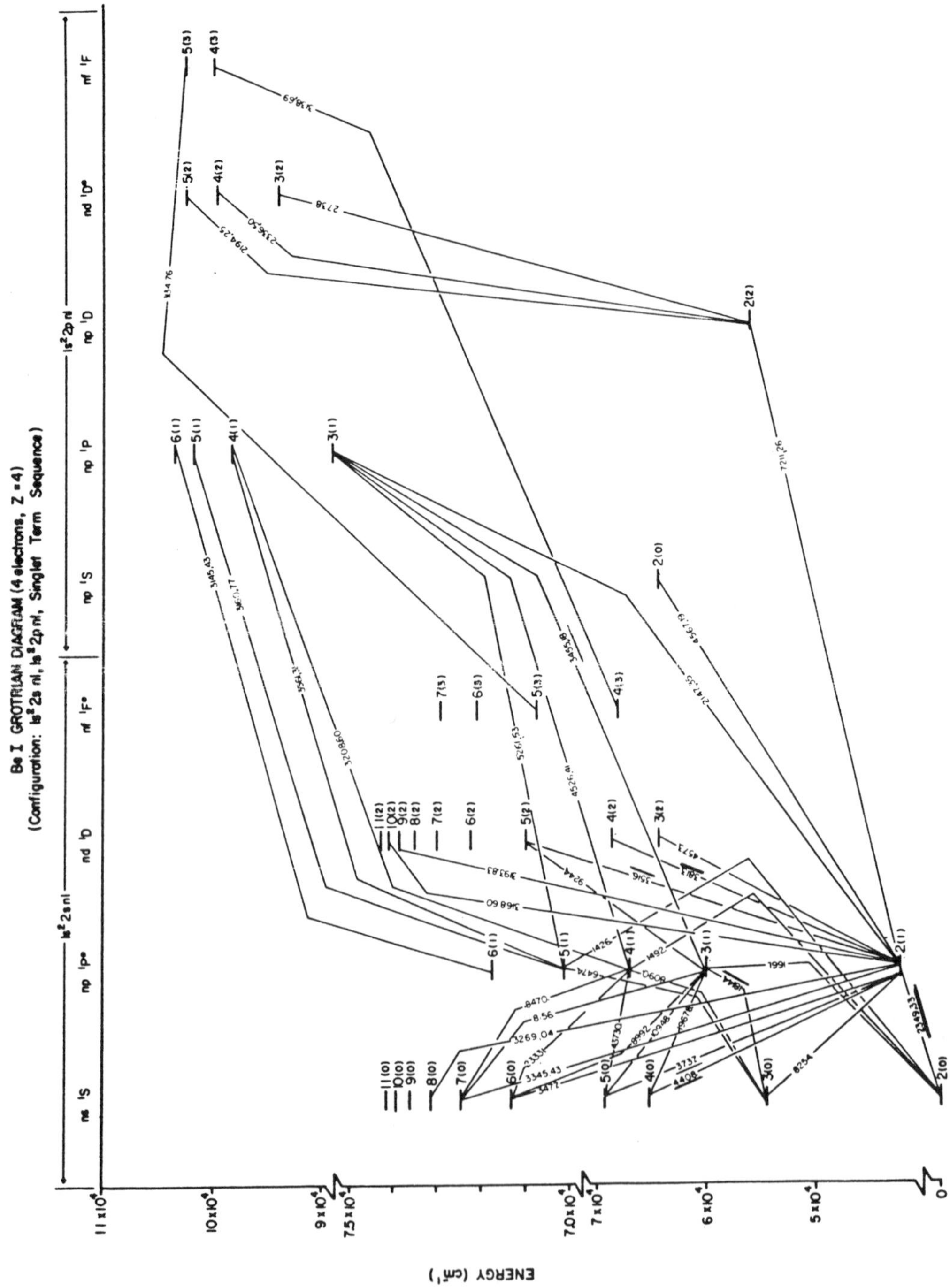
Be I GROTRIAN DIAGRAM (4 electrons, Z = 4)
(Configuration: 1s²2s nl, 1s²2p nl, Singlet Term Sequence)
1s²2s nl
1s²2p nl
ns ¹S
np ¹P°
nd ¹D
nf ¹F°
np ¹S
np ¹P
np ¹D
nd ¹D°
nf ¹F
ENERGY (cm⁻¹)
11×10⁴
10×10⁴
9×10⁴
7.5×10⁴
7.0×10⁴
7×10⁴
6×10⁴
5×10⁴
0

Fig. 4-2

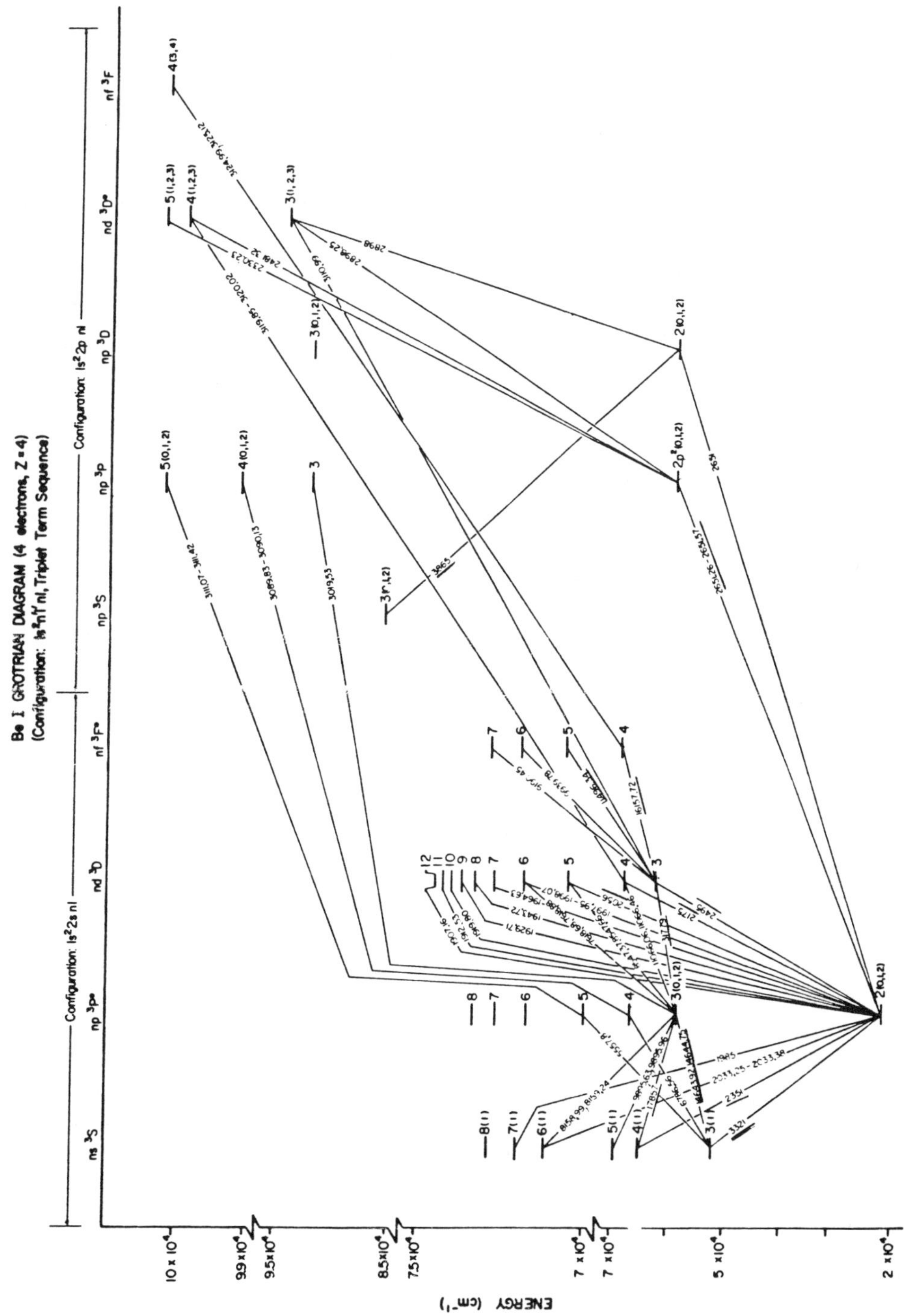
Be I GROTRIAN DIAGRAM (4 electrons, Z = 4)
(Configuration: 1s²nʹlʹ nl, Triplet Term Sequence)
Configuration: 1s²2s nl
Configuration: 1s²2p nl
ENERGY (cm⁻¹)

Fig. 4-1, p. 148

Grotrian diagram of the singlet system of neutral Be (Be I) [25]. Transitions are indicated by wavelengths in Å. The most intense transitions are underlined, cf. text on this page.

Note: The wavelength of the $2s\,2p\ ^1P^o_1-2p^2\ ^1D_2$ transition should be replaced by 6983 Å, cf. [1]. The term $2p^2\ ^1S$ should be placed at 76189.5 cm^{-1} and the transition to $2s\,2p\ ^1P^o$ deleted, cf. footnote c) under Table 4/4, p. 147.

Fig. 4-2, p. 149

Grotrian diagram of the triplet system of neutral Be (Be I) [25]. Transitions are indicated by wavelengths in Å. The most intense transitions are underlined, cf. text on this page.

Note: At the top of the figure, under "configuration $1s^2\,2p\,nl$", the term symbol "$np\ ^3S$" should be replaced by "$ns\ ^3P^o$". The terms $2p^2\ ^3D$ and $2p\,3p\ ^3D$ appear to be redundant ($2p^2\ ^3D$ is physically unreasonable). The 3865 Å-line attached to the term labeled "$2p\,3p\ ^3S(0, 1, 2)$" (to be read as $2p\,3s\ ^3P^o(0, 1, 2)$, see above) should be connected downwards with $2p^2\ ^3P(0, 1, 2)$. The set of quantum numbers for the term $2p\,3p\ ^3P$ should read "3 (0, 1, 2)".

The position of the $1s\,2s\,2p^2\ ^5P$ core excited state has been determined at 115.4 ± 0.3 eV by adding the first ionization energy of Be to the energy of the Auger transition of this state to $Be^+\ 1s^2\,2s\ ^2S_{1/2}$ [23]. Its lifetime has been determined by Auger spectroscopy [24].

Wavelengths. Transitions. Fig. 4-**1** and **Fig.** 4-**2** on pp. 148 and 149 are Grotrian diagrams for the singlet and triplet systems of neutral Be. The diagrams are photographically reproduced from the revised edition [25] of Bashkin and Stoner's book [26]. These diagrams were compiled on the basis of literature published up to 1975. Some obvious errors in the diagrams are indicated in the figure captions on this page. The transitions shown include nearly all of the strong lines from the vacuum UV to IR region which are listed in the compilation of Reader and Corliss [27] (cf. Table 4/5, p. 151) with intensities ≥ 100 on a linear scale. These transitions have been supplementarily underlined in Fig. 4-1 and Fig. 4-2. The strongest line, the $2s^2\ ^1S_0-2s\,2p\ ^1P^o_1$ resonance transition, is given an intensity of 950 in [27]. Further strong transitions (intensity ≥ 100 in [27]) not presented in Fig. 4-1 and Fig. 4-2 include a group at 4253 to 4254 Å, $2s\,3d\ ^3D-2p\,3s\ ^3P^o$, 7209.13 Å, $2p^2\ ^1D_2-2s\,5f\ ^1F^o_3$, and 8801.37 Å, $2p^2\ ^1D_2-2s\,4f\ ^1F^o_3$ (wavelengths from [27], classifications from [1]). The selection of the remaining lines drawn by Bashkin and Stoner [25] in Fig. 4-1 and Fig. 4-2 appears to be arbitrary.

A key list for papers and monographs containing wavelength measurements or literature evaluations of such measurements is presented in Table 4/5, p. 151.

Critically evaluated transition probabilities (mostly theoretical) have been compiled for several transitions of Be by Wiese and Martin [39]. The literature base up to 1980 used for this compilation is collected in two bibliographies published by the U.S. National Bureau of Standards [49, 50].

Some transitions have been observed merely for the purpose of studying their excitation mechanism: beam foil excitation [40]; excitation by collisions with He atoms [41], by ion sputtering from elemental Be [42, 43, 54], or by electron impact on a Be atomic beam [44].

Table 4/5
Key List for Wavelengths of Neutral Be from the Vacuum UV to IR Region (Ab = absorption, Em = emission).

wavelength range in Å	mode of excitation; spectrum	comment	Ref.
618.5 to 2348.6	vacuum spark; Ab	40 classified lines, series $2s^2-2s\,np$, $2p\,ns$, $2p\,nd$, $3s\,np$. For wavelengths of excitation of $2p\,nd\,^1P^o$ series by three-photon absorption, see [12], cf. Table 4/4, p. 146	[28, 29]
1908 to 4525	beam foil; Em	4 classified lines, 1 unclassified; study of $2p\,np\,^1P$ terms	[19]
1426.117 to 18143.54	hollow cathode; Em	120 classified lines	[1]
2033 to 4572	laser-produced plasma; Em	19 classified lines; plasma diagnostics	[51]
2125.57 to 4572.66	inductively coupled plasma; Em	25 unclassified lines	[30], see also [31]
2147.35 to 16956.68	hollow cathode; Em	27 classified lines <5262 Å, 20 unclassified lines >12580 Å	[3]
2337.0 to 5557.8	beam foil; Em	7 classified lines	[13, 14]
2350 to 4647	(Be in Ar or Kr matrices); Em	3 classified lines	[32]
3865.130 to 31778.70	hollow cathode; Em	32 classified lines	[2]
1426.117 to 1998.07	–	14 classified lines; literature evaluation	[33]
1426.12 to 31778.70	–	130 unclassified lines; literature evaluation	[27]
2033.254 to 3477.559	–	76 partly classified lines; literature evaluation	[34]
2174.99 to 8254.10	–	17 unclassified lines; literature evaluation	[35]
2349.331 to 4567.189	–	9 classified lines (transitions $2s^2-2s\,2p$ and $2s\,2p-2p^2$); literature evaluation	[16]
2348.612 to 2986.09	–	9 classified lines; literature evaluation (prior to 1947)	[36]
1426.1167, 1491.7647, 1641.4790	–	3 recommended reference lines for wavelength calibration	[37]
10331.06 to 233444.83	–	~550 classified lines; literature evaluation (unobserved IR lines calculated from term values)	[38]

Some emission lines due to transitions between quintet levels of core excited neutral Be (transitions $1s\,2s\,2p^2\,{}^5P - 1s\,2p^3\,{}^5S^o$, $1s\,2s\,2p\,3p\,{}^5P - 1s\,2p^2\,3p\,{}^5S^o$, and $1s\,2s\,2p\,4p\,{}^5P - 1s\,2p^2\,4p\,{}^5S^o$) have been observed in the region 1900 to 2310 Å [45, 46] and classified by comparison with ab initio calculations [45, 47, 48].

References:

[1] L. Johansson (Arkiv Fysik **23** [1963] 119/28).
[2] J. E. Holmstroem, L. Johansson (Arkiv Fysik **40** [1969] 133/8).
[3] L. Johansson (Phys. Scr. **10** [1974] 236/40).
[4] R. Beigang, D. Schmidt, P. J. West (J. Phys. Colloq. [Paris] **44** [1983] C7-229/C7-237).
[5] J. Bromander (Phys. Scr. **4** [1971] 61/3).
[6] T. Andersen, K. A. Jessen, G. Soerensen (Phys. Rev. [2] **188** [1969] 76/81).
[7] I. Bergstroem, J. Bromander, R. Buchta, L. Lundin, I. Martinson (Phys. Letters A **28** [1969] 721/2).
[8] B. V. L'vov (J. Quant. Spectrosc. Radiat. Transfer **12** [1972] 651/81, 669; Methodes Phys. Anal. **8** No. 1 [1972] 3/25, 15).
[9] O. Poulsen, T. Andersen, N. J. Skouboe (J. Phys. B **8** [1975] 1393/405).
[10] T. Andersen, J. R. Roberts, G. Soerensen (Phys. Scr. **4** [1971] 52/4).

[11] M. J. Seaton (J. Phys. B **9** [1976] 3001/7).
[12] C. W. Clark, J. D. Fassett, T. B. Lucatorto, L. J. Moore, W. W. Smith (J. Opt. Soc. Am. B **2** [1985] 891/6).
[13] S. Hontzeas, I. Martinson, P. Erman, R. Buchta (Phys. Scr. **6** [1972] 55/60).
[14] S. Hontzeas, I. Martinson, P. Erman, R. Buchta (Nucl. Instr. Methods **110** [1973] 51/4).
[15] A. W. Weiss (Phys. Rev. [3] A **6** [1972] 1261/6; Nucl. Instr. Methods **90** [1970] 121/31).
[16] B. C. Fawcett (At. Data Nucl. Data Tables **16** [1975] 135/64, 139, 152).
[17] H. Kerkhoff, M. Schmidt, P. Zimmermann (Phys. Letters A **80** [1980] 11/3).
[18] J. Bromander, R. Buchta, L. Lundin (Phys. Letters A **29** [1969] 523/4).
[19] H. G. Berry, J. Bromander, I. Martinson, R. Buchta (Phys. Scr. **3** [1971] 63/7).
[20] C. E. Moore (NBS-C-467-Vol. I [1949] 12/3).

[21] W. R. Bozman, C. H. Corliss, W. F. Meggers, R. E. Trees (J. Res. Natl. Bur. Std. **50** [1953] 131/2).
[22] K. V. Subbaram, R. Vasudev, W. E. Jones (J. Opt. Soc. Am. **65** [1975] 318/9).
[23] R. Bruch, M. Roedbro, P. Bisgaard, P. Dahl (Phys. Rev. Letters **39** [1977] 801/4).
[24] R. Bruch, G. Paul, J. Andrä (J. Phys. B **8** [1975] L253/L258).
[25] S. Bashkin, J. O. Stoner Jr. (Atomic Energy Levels and Grotrian Diagrams, Vol. I, Addenda, North Holland, Amsterdam 1978, pp. 8, 10).
[26] S. Bashkin, J. O. Stoner Jr. (Atomic Energy Levels and Grotrian Diagrams, Vol. I, North Holland, Amsterdam 1975, pp. 28, 30).
[27] J. Reader, C. H. Corliss (NSRDS-NBS-68 [1980] 1/357, 15/6).
[28] J. M. Esteva, G. Mehlman-Balloffet, J. Romand (J. Quant. Spectrosc. Radiat. Transfer **12** [1972] 1291/303).
[29] G. Mehlman-Balloffet, J. M. Esteva (Astrophys. J. **157** [1969] 945/56).
[30] T. A. Anderson, A. R. Forster, M. L. Parsons (Appl. Spectrosc. **36** [1982] 504/9).

[31] R. K. Winge, V. J. Peterson, V. A. Fassel (Appl. Spectrosc. **33** [1979] 206/19).
[32] J. M. Brom Jr., W. D. Hewett, W. Weltner Jr. (J. Chem. Phys. **62** [1975] 3122/30).
[33] R. L. Kelly (ORNL-5922 [1982] 1/404, 9; C.A. **98** [1983] No. 151855).
[34] R. L. Kelly (NASA-TM-80268 Sect. 1 [1979] 1/400, 27; C.A. **94** [1981] No. 38748).
[35] W. F. Meggers, C. H. Corliss, B. F. Scribner (NBS-MON-145-Pt. 1 [1975] 1/387, 8).

[36] C. E. Moore (NBS-C-488-Sect. 1 [1950] 5).
[37] V. Kaufman, B. Edlén (J. Phys. Chem. Ref. Data **3** [1974] 825/95, 866).
[38] E. Biemont, N. Grevesse (At. Data Nucl. Data Tables **12** [1973] 217/310, 227/30).
[39] W. L. Wiese, G. A. Martin (NSRDS-NBS-68 [1980] 359/406, 368).
[40] B. Dynefors, I. Martinson, E. Veje (Phys. Scr. **12** [1975] 58/62).

[41] N. Andersen, K. Jensen, J. Jepsen, J. Melskens, E. Veje (Z. Physik A **273** [1975] 1/8).
[42] M. Braun, B. Emmoth, I. Martinson (Phys. Scr. **10** [1974] 133/8).
[43] N. Andersen, W. S. Bickel, R. Boleu, K. Jensen, E. Veje (Phys. Scr. **3** [1971] 255/6).
[44] I. S. Aleksakhin, V. A. Zayats (Opt. Spektroskopiya **36** [1974] 1229/30; Opt. Spectrosc. [USSR] **36** [1974] 717).
[45] M. Agentoft, T. Andersen, C. Froese Fischer, L. Smentek-Mielczarek (Phys. Scr. **28** [1983] 45/50).
[46] S. Mannervik, I. Martinson, B. Jelenkovic (J. Phys. B **14** [1981] L275/L278).
[47] R. L. Brooks, J. E. Hardis, H. G. Berry, L. J. Curtis, K. T. Cheng, W. Ray (Phys. Rev. Letters **45** [1980] 1318/22).
[48] A. V. Bunge, M. F. Rubio (Phys. Rev. [3] A **30** [1984] 1747/51).
[49] J. R. Fuhr, B. J. Miller, G. A. Martin (NBS-SP-505 [1978] 1/270, 18).
[50] B. J. Miller, J. R. Fuhr, G. A. Martin (NBS-SP-505-Suppl. 1 [1980] 1/112, 9/10).

[51] A. Zago, G. Tondello (Nuovo Cimento B [11] **85** [1985] 59/78).
[52] I. Martinson, A. Gaupp, L. J. Curtis (J. Phys. B **7** [1974] L463/L465).
[53] N. P. Penkin, L. N. Shabanova (Opt. Spektroskopiya **26** [1969] 346/50; Opt. Spectrosc. [USSR] **26** [1969] 191/3).
[54] G. M. Mladenov, M. Braun (Phys. Status Solidi A **53** [1979] 631/40).

4.2.9 Interaction Constants

Electron-Electron. Slater-Condon parameters F and G for various electron configurations, calculated from spectroscopic term values as presented in the 1949 published Tables of Moore (ref. [20] on p. 152) may be found in [1, 2].

Electron-Nucleus. The following values for hyperfine structure (hfs) constants A (magnetic hfs) and B (electric quadrupole hfs) have been determined (nuclear spin I (^{9}Be) = $^3/_2$):

^{9}Be state	A in MHz	B in MHz	method	Ref.
$1s^2\,2s\,2p\ ^3P^o_1$	-139.373 ± 0.012	-0.753 ± 0.44	atomic beam	[3]*)
$1s^2\,2s\,2p\ ^3P^o_2$	-124.5368 ± 0.0017	1.429 ± 0.008	magnetic resonance	
$1s^2\,2s\,3d\ ^1D_2$	4 ± 1		level crossing	[5]

*) For preliminary results, see [4].

References:

[1] H. Hosoya (J. Chem. Phys. **47** [1967] 4190/8).
[2] J. Hinze, H. H. Jaffe (J. Chem. Phys. **38** [1963] 1834/47, 1838).
[3] A. G. Blachman, A. Lurio (Phys. Rev. [2] **153** [1967] 164/76).
[4] A. Lurio, A. G. Blachman (Bull. Am. Phys. Soc. [2] **5** [1960] 344/5, **6** [1961] 142).
[5] O. Poulsen, T. Andersen, N. J. Skouboe (J. Phys. B **8** [1975] 1393/405).

4.2.10 Electric Dipole Polarizability α

The polarizability does not appear to have been experimentally determined so far. Recent Hartree-Fock CI calculations [1] show that α is probably close to the rigorous lower bound, $\alpha \geq 5.297 \times 10^{-24}$ cm^3, derived from a rigorous upper bound to the second-order energy calculated by a variational procedure [2]. Thus, the value $\alpha = 5.60 \times 10^{-24}$ cm^3, which, as a result of a finite-field perturbation calculation [3], was recommended by Miller and Bederson [4], is likely to be too large [1]. For a comprehensive discussion of other quantum mechanical calculations, see [1, 2].

References:

[1] G. H. F. Diercksen, A. J. Sadlej (Chem. Phys. **65** [1982] 407/15).
[2] J. S. Sims, J. R. Rumble Jr. (Phys. Rev. [3] A **8** [1973] 2231/5).
[3] H. J. Werner, W. Meyer (Phys. Rev. [3] A **13** [1976] 13/6).
[4] T. M. Miller, B. Bederson (Advan. At. Mol. Phys. **13** [1977] 1/55, 32, 49).

4.2.11 Photon Cross Sections. Mass Attenuation Coefficients

The total photon cross section per atom is the sum of the contributions from four interaction processes: coherent (Rayleigh) scattering, incoherent (Compton) scattering, photoelectric absorption, and e^+e^- pair production (threshold 1.022 MeV). The mass attenuation coefficient μ/ϱ is related to the total cross section per atom by $\mu/\varrho = (N_A/A) \cdot \sigma_{TOT}$, where N_A is Avogadro's number and A is the relative atomic weight. For elemental Be, $\mu/\varrho = 0.066821 \cdot \sigma_{TOT}$, for μ/ϱ in cm^2/g and σ_{TOT} in barn, see, e.g., [15].

Several comprehensive tabulations of total photon cross sections and mass attenuation coefficients in extended energy regions are available, and an allocated bibliography of some of them is presented in Table 4/6. The cited tabulations have been selected by their significance as assessed by Henke et al. [2] and in particular by Pratt et al. [17], and by availability. For completeness, some recent tabulations devoted to the energy region above 1 MeV have been included. The data from McMaster et al. [8] in the energy region 1 keV to 1 MeV are presented graphically for demonstration purposes only in **Fig. 4-3**, p. 156.

Table 4/6
Survey of Publications Presenting Photon Cross Sections and Mass Attenuation Coefficients of Be.
"σ" without subscript denotes the total cross section and partial cross sections to be found in the cited reference. "Bibl" denotes a bibliography of experimental measurements, if present in the cited reference.

energy E in eV	number of data points	comment	Ref.
30 to 6000	91	σ_{TOT} and μ/ϱ vs. E from experiment+theory Bibl: 6 refs. 1930 to 1969	[1]
30.5 to 9886.4	50	μ/ϱ for K, L, M lines of selected elements, Be to Ba, from experiment+theory Bibl: 3 refs. 1964 to 1970 [2], 3 refs. 1930 to 1964 [3]	[2], see also [3]

Table 4/6 [continued]

energy E in eV	number of data points	comment	Ref.
10^2 to 10^6	38	σ vs. E from experiment + theory; assessed accuracy 2 to 5% Bibl: 13 refs. 1935 to 1970	[4]
1.82×10^2 to 2.52×10^4	46	μ/ϱ for Kα lines of elements B to Sn	[5], see also [6]
2.29×10^3 to 1.42×10^4	53	μ/ϱ for Lα lines of elements Mo to Pu Both sets interpolated from preliminary version of [4]	
10^3 to 10^6	25	σ and μ/ϱ from experiment + theory [8] Bibl: 11 refs. 1961 to 1969 [7, 9]	[7 to 9]
4.509×10^3 to 24.942×10^3	24	σ_{TOT} and μ/ϱ for Kα and Kβ lines of selected elements, Ti to Ag, interpolated from [8]	[10, 11]
10^3 to 2×10^7	35	μ/ϱ vs. E from theory; assessed accuracy within spread of experimental uncertainties	[12]
10^3 to 10^8	41	σ vs. E from theory; assessed accuracy 10% for $E \leq 10^4$ eV, 3% for 2×10^5 eV $\leq E \leq 5 \times 10^5$ eV, 10% for $E > 10^6$ eV	[13]
10^4 to 10^{11}	57	σ and μ/ϱ vs. E from experiment + theory Bibl: 5 refs. 1951 to 1960 for $E \geq 10^8$ eV	[14]
10^6 to 10^{11}	46	σ and μ/ϱ vs. E from theory; assessed accuracy ≤0.5% by comparison with experiment (1 ref. 1975)	[15], see also [16]

In the tabulations cited in Table 4/6, the cross sections have been either calculated from theory or have been evaluated by a combination of experimental and theoretical methods. In the energy region where the photoelectric effect dominates, "experimental" photoelectric cross sections may be derived from total attenuation coefficients measured at discrete energy points by subtracting calculated coherent and incoherent scattering cross sections. The "experimental" photoelectric cross sections so obtained are then smoothed and interpolated by assuming them to be continuous functions of both energy and atomic number. For Be, this method is well applicable for photon energies between 1 and 10 keV, as can be seen in a diagram presented by Hubbell and Veigele [18].

Papers reporting measured cross sections are not individually cited here. Whenever a tabulation of cross sections or absorption coefficients contains a bibliography of evaluated experimental data for Be, this is indicated in the column headed "comment" in Table 4/6. An allocated bibliography, of experimental data for Be for the energy region 10 to 10^9 eV containing 33 papers between 1935 and 1970, has been compiled by Hubbell [19].

Fig. 4-3

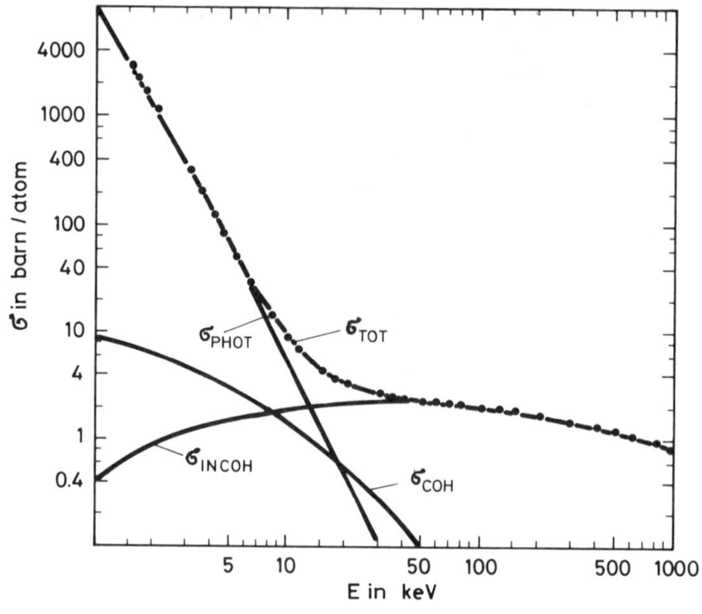

Log-log plot of photon cross sections σ of Be for coherent (COH) and incoherent (INCOH) scattering, photoelectric absorption (PHOT), and total cross section (TOT). Individual points on the σ_{TOT} curve are experimental [8].

References:

[1] B. L. Henke, R. L. Elgin (Advan. X-Ray Anal. **13** [1970] 639/65, 647/8, 656/7, 659).

[2] B. L. Henke, P. Lee, T. J. Tanaka, R. L. Shimabukuro, B. K. Fujikawa (At. Data Nucl. Data Tables **27** [1982] 1/144, 7, 25, 141/4).

[3] B. L. Henke, E. S. Ebisu (Advan. X-Ray Anal. **17** [1974] 150/213, 179, 184/5, 187).

[4] W. J. Veigele (At. Data **5** [1973] 51/111, 66/7, 102/10).

[5] B. L. Bracewell, W. J. Veigele (Develop. Appl. Spectrosc. **9** [1971] 357/400).

[6] B. L. Bracewell, W. J. Veigele (Advan. X-Ray Anal. **15** [1972] 352/64).

[7] W. H. McMaster, N. Kerr Del Grande, J. H. Mallett, J. H. Hubbell (UCRL-50174-Sect. I [1970] 1/16, 12/6).

[8] W. H. McMaster, N. Kerr Del Grande, J. H. Mallett, J. H. Hubbell (UCRL-50174-Sect. II Rev. 1 [1969] 1/350, 15/8).

[9] W. H. McMaster, N. Kerr Del Grande, J. H. Mallett, J. H. Hubbell (UCRL-50174-Sect. III [1969] 1/193, 48/9).

[10] W. H. McMaster, N. Kerr Del Grande, J. H. Mallett, J. H. Hubbell (UCRL-50174-Sect. IV [1969] 1/39, 2, 21).

[11] J. H. Hubbell, W. H. McMaster, N. Kerr Del Grande, J. H. Mallett (in: J. A. Ibers, W. C. Hamilton, International Tables for X-Ray Crystallography, Vol. 4, Birmingham 1974, pp. 47/70, 54/5, 61).

[12] J. H. Hubbell (Intern. J. Appl. Radiat. Isotop. **33** [1982] 1269/90, 1271).

[13] E. Storm, H. I. Israel (Nucl. Data Tables **7** [1970] 565/681, 575).

[14] J. H. Hubbell (NSRDS-NBS-29 [1969] 1/80, 14, 42, 78/80).
[15] J. H. Hubbell, H. A. Gimm, I. Oeverboe (J. Phys. Chem. Ref. Data **9** [1980] 1023/147, 1043, 1047, 1145).
[16] H. A. Gimm, J. H. Hubbell (NBS-TN-968 [1978] 1/72, 16).
[17] R. H. Pratt, A. Ron, H. K. Tseng (Rev. Mod. Phys. **45** [1973] 273/325, 317/22; Erratum: 663/4).
[18] J. H. Hubbell, W. J. Veigele (NBS-TN-901 [1976] 1/43, 18).
[19] J. H. Hubbell (At. Data **3** [1971] 241/97).

4.2.12 Interaction with Electrons

Recommended cross sections for electron impact ionization of Be are empirical estimates due to a lack of reliable experimental and theoretical data. They are based on an analysis of scaled cross sections in the Be isoelectronic series in which experimental values are available for C^{2+}, N^{3+}, and O^{4+}. The scaling law used relates the first ionization energy I and the cross section σ as a function of energy by $\sigma_{sc} = I^2 \cdot \sigma(E)$, where the scaled cross section σ_{sc} is a function of E/I. Cross sections are given for $1.25 \leq E/I \leq 100.00$ [1]. For earlier semiempirical estimates, see [2]; they come reasonably close to the recommended values [1].

References:

[1] K. L. Bell, H. H. Gilbody, J. G. Hughes, E. A. Kingston, F. J. Smith (J. Phys. Chem. Ref. Data **12** [1983] 891/916, 904, 910, 914; CLM-R-216 [1982] 1/120, 31, 93, 95).
[2] W. Lotz (Z. Physik **216** [1968] 241/7).

4.2.13 Thermodynamic Functions

For the ideal monoatomic gas, values for the heat capacity C_p°, entropy S°, Gibbs free energy $(G^\circ - H^\circ_{298})/T$, in $cal \cdot mol^{-1} \cdot K^{-1}$, and for the enthalpy $H^\circ - H^\circ_{298}$ in kcal/mol from the JANAF Tables [1] are given below. The electronic partition function was calculated [1] with energies and statistical weights taken from Moore's tables [2]. Below 2000 K, only translational and no electronic degrees of freedom contribute to the thermodynamic functions, thus, for instance, $C_p^\circ = {}^5/_2\, R = 4.968\ cal \cdot mol^{-1} \cdot K^{-1}$ [1].

T in K	2000	3000	4000	5000	6000
C_p°	4.969	5.021	5.378	6.215	7.340
S°	42.001	44.021	45.503	46.785	48.015
$-(G^\circ - H^\circ_{298})/T$	37.773	39.541	40.583	41.913	42.828
$H^\circ - H^\circ_{298}$	8.455	13.438	18.599	24.359	31.126

Similar values of $(G^\circ - H^\circ_{298})/T$ had been calculated earlier [7] with the same data base for the electronic partition function. Values in another tabulation extending to 10000 K were obtained by summing the partition function over a restricted set of electronic energy levels [3].

The electronic partition function Q has been fitted in the form of polynomial expansions of log Q in powers of log (5040/T) [4], of ln Q in powers of ln T [5], and of Q in powers of T/1000, but taking into account only levels below 30000 cm^{-1} [6].

References:

[1] D. R. Stull, H. Prophet (JANAF Thermochemical Tables, 2nd Ed., NSRDS-NBS-37 [1971]).
[2] C. E. Moore (NBS-C-467-Vol. I [1949]).

[3] J. Hilsenrath, C. G. Messina, W. H. Evans (AD-606163 [1964] 1/434, 45/6; C.A. **62** [1965] 7177).
[4] A. J. Sauval, J. B. Tatum (Astrophys. J. Suppl. Ser. **56** [1984] 193/209, 207/8).
[5] A. W. Irwin (Astrophys. J. Suppl. Ser. **45** [1981] 621/33, 623).
[6] L. De Galan, R. Smith, J. D. Winefordner (Spectrochim. Acta B **23** [1968] 521/5).
[7] H. G. Kolsky, R. M. Gilmer (J. Chem. Phys. **27** [1957] 494/5).

4.3 The Be^+ Ion

4.3.1 Ground State. Ionization Energies

1 eV ≙ 8065.479 cm^{-1}

The g factor of the 2s $^2S_{1/2}$ ground state of $^9Be^+$ is $g_J = 2.00226263(33)$ [1] (preliminary result: 2.00226206(42) [2]), as determined by measurement of ion cyclotron resonance frequencies in a Penning trap by a laser fluorescence technique [1, 2].

A selected transition within the Zeeman-split hyperfine structure manifold of the ground state, namely $(M_I, M_J) = (-3/2, 1/2) \leftrightarrow (-1/2, 1/2)$, was measured with Be^+ ions stored in a Penning ion trap. The ion kinetic energy was reduced with radiation pressure from a laser ("laser cooling") in order to reduce the second-order Doppler shift. The transition frequency at a field of 0.8194 T, where the frequency is field-independent to first order, was determined to be 303016377.265070(57) Hz. The accuracy of a frequency standard referenced to this transition was comparable to the best frequency standards which are based on caesium atomic beams [3].

An ionization limit of 146882.86 cm^{-1} (18.211 eV) was obtained for Be^+ $1s^2 2s$ $^2S_{1/2} \rightarrow$ Be^{2+} $1s^2$ $^1S_0$ (ground states of both ions), by fitting observed term values of nd, nf, and 5g levels to a formula for hydrogenic terms [4]. Another spectroscopic ionization limit given in the Landolt-Börnstein Tables [5] and in Moore's Tables of Atomic Energy Levels [6] is 146881.7 cm^{-1}. References for spectroscopic papers prior to 1950, from which this value has been derived, may be found in these tables.

The K shell ionization energy of Be^+ may be obtained from the above ionization energy (limit) [4] and the term energy of Be^{2+} 1s 2s ^1S, 981187 cm^{-1}, and of Be^{2+} 1s 2s ^3S, 956509 cm^{-1}, both relative to the Be^{2+} $1s^2$ ^1S ground state, as given in [7]. This yields 139.86 and 136.80 eV for the energies to reach the Be^{2+} 1s 2s ^1S and Be^{2+} 1s 2s ^3S ionic states, respectively, also see Fig. 12 in [8].

References:

[1] J. J. Bollinger, D. J. Wineland, W. M. Itano, J. S. Wells (Springer Ser. Opt. Sci. **40** [1983] 168/72).
[2] D. J. Wineland, J. J. Bollinger, W. M. Itano (Phys. Rev. Letters **50** [1983] 628/31; Erratum: 1333).
[3] J. J. Bollinger, J. D. Prestage, W. M. Itano, D. J. Wineland (Phys. Rev. Letters **54** [1985] 1000/3).
[4] L. Johansson (Arkiv Fysik **20** [1962] 489/98).
[5] Landolt-Börnstein 6th Ed. **1** Pt. 1 [1950] 53.
[6] C. E. Moore (NBS-C-467-Vol. I [1949] 14).
[7] S. Bashkin, J. O. Stoner Jr. (Atomic Energy Levels and Grotrian Diagrams, Vol. I, North Holland, Amsterdam 1975, p. 37).
[8] M. Roedbro, R. Bruch, P. Bisgaard (J. Phys. B **12** [1979] 2413/47, 2431).

4.3.2 Atomic Structure Calculations

For Be and its uni- and dipositive ions, atomic structure calculations are not documented in this volume for reasons outlined in Section 4.2.3 on pp. 140/1. Consult that section for bibliographies devoted to the topic.

4.3.3 X-Ray Spectra

Transitions involving a 1s electron have been observed in absorption spectra through laser-produced plasmas [1 to 3, 5] or high current discharges [6]. Emission spectra were taken from laser-produced plasmas [4] and by beam foil spectroscopy [7]. Most transitions may be perceived as resonance transitions $1s^2-1s\,np$ of Be^{2+} perturbed by the screening action of an additional 2s, 2p, or 3p electron, see, e.g., [5]. Some two-electron transitions were observed as well [1, 6]. Classified lines in the region 105 to 95 Å, which has been extensively scanned, are presented in Table 4/7. Wavelengths are from the most recent absorption spectrum [1], and references to other work are also given. The classifications of the lines are the same as in the cited references except where indicated.

Lines in the region below 95 Å were classified as series $1s^2\,2s-1s\,2s\,np$, n=3, 4, 5 [6], $1s^2\,2p-1s\,2p\,np$, n=3, 4, and as series with screening electrons other than 2s, 2p, or 3p [1]. Some lines were classified as two-electron transitions $1s^2\,2s-1s\,2p\,ns$, n=3 to 6, and $1s^2\,2s-1s\,3s\,3p$ [6]. Classifications of lines at 94.79, 93.93, and 93.42 Å differ [1, 6].

The optical spectra of beam foil excited Be^+ (incident energy 100 to 1000 keV) show a line at 80.85±0.03 Å, which is assigned as a $2p^3\,{}^4S^o-1s\,2p^2\,{}^4P$ transition from triply excited Be^+ since a calculated wavelength for this transition coincides with the measured one. Also, other calculations indicate that this line does not originate from the decay of a doubly excited state in Be^{2+}, Be^+, or neutral Be [8].

Table 4/7
Absorption Lines in Soft X-Ray Spectra of Be^+ [1] and References for Other Work.

wavelength in Å [1]	classification [1]	further references
104.68 ±0.02	$1s^2\,2p\,{}^2P^o-1s\,2p^2\,{}^2D$	[2 to 5]
104.42 ±0.03	$1s^2\,2s\,{}^2S-1s(2s\,2p\,{}^3P^o)\,{}^2P^o$	[2, 3, 5, 6]
104.05 ±0.02	$1s^2\,2p\,{}^2P^o-1s\,2p^2\,{}^2P$	[2, 3, 5, 7]
102.51 ±0.01	$1s^2\,2s\,{}^2S-1s(2s\,2p\,{}^1P^o)\,{}^2P^o$	[2, 3, 5, 6]
101.76 ±0.01	$1s^2\,2p\,{}^2P^o-1s\,2p^2\,{}^2S$	[2 to 5]
100.949±0.006	$1s^2\,3p\,{}^2P^o-1s\,2p({}^1P)\,3p\,{}^2D$	
100.778±0.002	$1s^2\,3p\,{}^2P^o-1s\,2p({}^1P)\,3p\,{}^2P$	
100.576±0.008	$1s^2\,3p\,{}^2P^o-1s\,2p({}^1P)\,3p\,{}^2S$	
97.88 ±0.04	$1s^2\,2p\,{}^2P^o-1s\,2s({}^3S)\,3d\,{}^2D$	
96.27 ±0.04	$1s^2\,2p\,{}^2P^o-1s\,2p({}^3P)\,3p\,{}^2P$	[7]
96.02 ±0.07	$1s^2\,2p\,{}^2P^o-1s\,2p({}^3P)\,3p\,{}^2D$	
95.78 ±0.01	$1s^2\,2s\,{}^2S-1s\,2s({}^3S)\,3p\,{}^2P^o$	[6]
95.734±0.005	$1s^2\,2p\,{}^2P^o-1s\,2p({}^3P)\,3p\,{}^2S$ *)	[2, 3, 5]
95.34 ±0.04	$1s^2\,2p\,{}^2P^o-1s\,2s({}^1S)\,3d\,{}^2D$	

*) Other classification: $1s^2\,2p\,{}^2P^o-1s\,2p\,3p\,{}^2D$ [2, 3, 5].

References:

[1] E. Jannitti, M. Mazzoni, P. Nicolosi, G. Tondello, Wang Yongchang (J. Opt. Soc. Am. B **2** [1985] 1078/83).
[2] E. Jannitti, P. Nicolosi, G. Tondello (Physica B+C **124** [1984] 139/47).
[3] E. Jannitti, P. Nicolosi, G. Tondello (AIP [Am. Inst. Phys.] Conf. Proc. No. 90 [1982] 243/53).
[4] P. Nicolosi, G. Tondello (J. Opt. Soc. Am. **67** [1977] 1033/9).
[5] E. T. Kennedy, P. K. Carrol (J. Phys. B **11** [1978] 965/74; Phys. Letters A **64** [1977] 37/8).
[6] G. Mehlman, J. M. Esteva (Astrophys. J. **188** [1974] 191/5).
[7] M. Agentoft, T. Andersen, C. Froese Fischer, L. Smentek-Mielczarek (Phys. Scr. **28** [1983] 45/50).
[8] M. Agentoft, T. Andersen, K. T. Chung (J. Phys. B **17** [1984] L433/L438).

4.3.4 Auger Spectra

Emission of Auger electrons from free Be^+ ions has been observed after passage of fast Be^+ ions ($\leq$500 keV) either through He or CH_4 gas targets under single-collision conditions [1, 2, 4] or through thin carbon foils [3, 5], in the same manner as was observed for neutral Be, see p. 143. The Auger transitions are listed in Table 4/8. A comparison of observed transition energies with theory and references for theoretical work are presented in [1, 2].

Auger widths have been derived for two of the autoionizing Be^+ states listed in Table 4/8 by observing line profiles of optical transitions populating these states. The results were as follows [8]:

Be^+ state	Auger width in meV	transition observed
$1s(2s\,2p\,^3P^o)\,^2P^o$	4.58 ± 0.13	$1s\,2p^2\,^2P-1s(2s\,2p\,^3P^o)\,^2P^o$
$1s\,2p^2\,^2D$	$30.3\ \pm1.1$	$(1s\,2p\,^3P^o)\,3d\,^2D^o-1s\,2p^2\,^2D$

Measured line widths were corrected for upper state lifetime broadening and for the partial level widths for the radiative decays $1s(2s\,2p\,^3P^o)\,^2P^o-1s^2\,2s\,^2S$ and $1s\,2p^2\,^2D-1s^2\,2p\,^2P^o$, respectively. These corrections were found negligible for the width of the $1s\,2p^2\,^2D$ state [8]. A comparison between measured and calculated widths and references for recent theoretical work may be found in a recent review [9].

Auger transitions in the region of electron energies 126 to 144 eV are due to the decay of autoionizing states of triply excited Be^+ ions with two K holes, i.e., with electron configurations $2s^2\,2p$, $2s\,2p^2$, $2p^3$, and $2s^2\,3s$ [1, 6, 7]. Calculations using the "saddle point technique" served to identify nine groups of lines, where each group contains four lines corresponding to the four possible decay channels to the states $1s\,2s\,^{1,3}S$ and $1s\,2p\,^{1,3}P^o$ of Be^{2+} [6, 7].

Table 4/8
Auger Transitions of Core Excited Be^+ to Be^{2+} $1s^2\,^1S_0$.

Auger transition energy in eV[a)]	initial Be^+ state	Ref.
96.1 ± 0.1	$1s\,2s^2\,^2S$	[1 to 3]
97.6 ± 0.1	$1s\,2s\,2p\,^4P^o$	[1, 2, 4]
100.5 ± 0.1	$1s(2s\,2p\,^3P^o)\,^2P^o$	[1, 2]

Table 4/8 [continued]

Auger transition energy in eV[a)]	initial Be^+ state	Ref.
102.5	$1s\,2p^2\,{}^4P$ [b)]	[5]
102.7 ± 0.1	$1s(2s\,2p\,{}^1P^o)\,{}^2P^o$	[1, 2]
104.2 ± 0.1	$1s\,2p^2\,{}^2D$	[1, 2]
107.6 ± 0.2	$1s\,2p^2\,{}^2S$	[1, 2]
110.6 ± 0.2	$1s\,2s({}^3S)\,3s\,{}^2S$ [c)]	[1, 2]
111.2 ± 0.2	$1s\,2s({}^3S)\,3p\,{}^2P^o$	[1, 2]
112.4 ± 0.2	$1s\,2s({}^3S)\,3d\,{}^2D$ [c)]	[1, 2]
113.0 ± 0.2	$1s\,2s({}^1S)\,3s\,{}^2S$	[1]
113.8 ± 0.2	$1s\,2s({}^1S)\,3p\,{}^2P^o$ [c,d)]	[1, 2, 6]
114.6 ± 0.3	$1s\,2p({}^3P^o)\,3s\,{}^2P^o$ [d)]	[1, 6]
114.9	$1s\,2s({}^1S)\,3d\,{}^2D$	[2, 6]

a) Transition energies are from the most recent reference [1] if available. Energies in earlier papers cited together with [1] usually differ by ≤ 0.3 eV.

b) Lifetime observed by Auger electron detection is 3.1 ± 0.4 ns [5].

c) Preliminary assignment [2] was different.

d) Assignment by comparison with calculated transition energy [6].

References:

[1] M. Roedbro, R. Bruch, P. Bisgaard (J. Phys. B **12** [1979] 2413/47).
[2] M. Roedbro, R. Bruch, P. Bisgaard (J. Phys. B **10** [1977] L275/L279).
[3] R. Bruch, G. Paul, J. Andrä, B. Fricke (Phys. Letters A **53** [1975] 293/4).
[4] R. Bruch, M. Roedbro, P. Bisgaard, P. Dahl (Phys. Rev. Letters **39** [1977] 801/4).
[5] R. Bruch, G. Paul, J. Andrä (J. Phys. B **8** [1975] L253/L258).
[6] B. F. Davis, Kwong T. Chung (J. Phys. B **15** [1982] 3113/26).
[7] R. Bruch, Kwong T. Chung (Comments At. Mol. Phys. **14** [1984] 117/26).
[8] H. Cederquist, M. Kisielinski, S. Mannervik, T. Andersen (J. Phys. B **17** [1984] 1969/79).
[9] T. Andersen, S. Mannervik (Comments At. Mol. Phys. **16** [1985] 185/98, 195).

4.3.5 Fluorescence Yield ω_K

Two independent measurements were carried out with elemental Be in its standard state, yielding $\omega_K = (3.6 \pm 1.1) \times 10^{-4}$ [1] and $(3.04 \pm 0.61) \times 10^{-4}$ [2]. The primary 1s vacancy was produced either by synchrotron radiation [1] or by K X-rays of the elements C or Al [2].

Fluorescence yields calculated for a free Be^+ ion [3 to 5] are smaller throughout; the most recent result is 1.2×10^{-4} [3]. However, it does not seem meaningful to compare the experimental results with atomic calculations, since the initial Be^+ states involved in the experiment are in the valence band of solid Be and may include some p character, whereas calculations have been done for the pure $1s\,2s^2\,{}^2S$ initial state of the free ion. A calculation of ω_K for this state is of some interest from a theoretical point of view, since the radiative decay rate of this state would be zero in the absence of electron correlation [5]. For Be^+ $1s\,2s\,2p\,{}^4P^o_{5/2}$, $\omega_K = 2.8 \times 10^{-4}$ has been calculated by a relativistic Hartree-Fock method [6].

References:

[1] K. Feser (Phys. Rev. Letters **28** [1972] 1013/5).
[2] C. E. Dick, A. C. Lucas (Phys. Rev. [3] A **2** [1970] 580/6).
[3] C. A. Nicolaides, Y. Komninos, D. R. Beck (Phys. Rev. [3] A **27** [1983] 3044/52).
[4] F. Bely-Dubau, D. Petrini (J. Phys. B **10** [1977] 1613/24).
[5] H. P. Kelly (Phys. Rev. [3] A **9** [1974] 1582/5).
[6] C. P. Bhalla, T. W. Tunnell (Z. Physik A **303** [1981] 199/201).

4.3.6 Atomic Energy Levels. Optical Spectra (Be II)

4.3.6.1 Be^+ $1s^2$ nl (Singly Excited Be^+)

Levels. A survey of the atomic energy levels of singly excited Be^+ is presented in Table 4/9. This table is arranged by term systems belonging to the same "running" electron, and the energy of only the lowest and highest member of each system is given numerically. Papers reporting lifetime measurements are cited in the column next to the references for the term values.

Most of the terms compiled in Table 4/9 were obtained by Johansson et al. [1, 5], who studied emission spectra from 1500 to 12100 Å using a hollow cathode discharge. Most excited state lifetimes were measured by Hontzeas et al. [6], Bergstroem, Bromander, et al. [7 to 9], and Andersen et al. [10, 11]. These authors used a time-of-flight method after beam foil excitation.

The compilation of atomic energy levels for Be^+ in the well-known tables of Moore [2] is based on spectroscopic work prior to 1949. After reexamination of the Be^+ spectrum, most of the s, p, and d levels have been decreased by about 1 cm^{-1}, and the 3d, 5p, and nf terms have been increased by 1 to 2 cm^{-1} [1, 5] in comparison with the values of Moore [2]. Furthermore, the fine structure splittings of many terms could be resolved [1, 5].

Wavelengths. Transitions. Fig. 4-4, p. 164, is a Grotrian diagram for singly excited Be^+, which is photographically reproduced from the book of Bashkin and Stoner [16]. The diagram is based on literature values up to 1971. Transitions to the lowest term within each system are preferably shown. Wavelengths for transitions between higher members of the term system have been compiled by Johansson et al. [1, 5]. A key list for papers and monographs containing wavelength measurements or literature evaluations of such measurements is presented in Table 4/10, p. 165.

Some transitions have been observed merely for the purpose of studying their excitation mechanism under various conditions: Beam foil excitation [21], excitation by collisions with He atoms [22], by ion sputtering from elemental Be [23 to 25], or by electron impact on atomic or ionic Be beams [26, 27].

Critically evaluated transition probabilities and oscillator strengths (mostly theoretical) have been compiled for several transitions of Be^+ by Wiese and Martin [28, 29]. The literature up to 1980 used in the more recent of these two compilations [28] is collected in two bibliographies published by the U.S. National Bureau of Standards [30, 31].

Stimulated emission of Be^+ on the transitions 4d-3p (4362 Å) [32], 4f-3d (4675 Å), 4s-3p (5272 Å) [32, 33], and 3p-3s (12096 Å) [33] has been observed in Be plasmas produced by a pulsed discharge [33] or by laser heating [32]. Pulsed lasing on the first three transitions was observed in a resonator set up equipped with a special means to prevent plasma expansion [32]. The conditions for population inversion of several transitions of Be^+ have been calculated [33 to 37].

Table 4/9
Terms of Be^+ $1s^2\,nl\,^2L$ and Source List for Lifetime Measurements.

term system	energy E of lowest member in cm^{-1}	energy E of highest member in cm^{-1}	references for energy	references for lifetime τ	mode of excitation
ns ^2S, $2\leq n\leq 7$	0	137226.0	[1, 2]	3s, 4s [6 to 10]	E: hollow cathode [1] τ: beam foil [6 to 10]
np $^2P^o$, $2\leq n\leq 11$	31933.15 a) [1] 31933.1277 a) [15]	143162 b)	[1 to 5, 15, 38]	2p c) [7, 9 to 11] 3p [6] 4p [6, 10]	E: hollow cathode [1, 5]; vacuum spark [3, 4]; tunable laser [15]; laser produced plasma [38] τ: beam foil [6, 7, 9 to 11]
nd ^2D, $3\leq n\leq 8$	98054.90 a)	140020.4	[1, 2]	3d [6 to 8] 4d [6 to 10, 13, 14] 5d, 6d [6, 10, 12, 14]	E: hollow cathode [1] τ: beam foil [6 to 10, 12]; level crossing [13, 14]
nf $^2F^o$, $4\leq n\leq 7$	119446.68 a)	137924.32	[1]	4f [7, 9, 10, 14] 5f [6, 10, 12, 14] 6f, 7f [6, 14]	E: hollow cathode τ: beam foil [6, 7, 9, 10, 12]; level crossing [14]
ng ^2G, $5\leq n\leq 7$	129325.48	137925.13	[1]	—	—

a) Center of gravity of fine structure components.

b) The terms 8p to 11p $^2P^o$ have been observed in absorption spectra of dense plasmas [3, 4, 38] and may be Stark-shifted and -broadened [3, 38]. The term value given here for 11p is the inverse of the reported vacuum wavelength for the 2s–11p transition, 698.5 Å [38].

c) Measured lifetimes of the 2p $^2P^o$ level are 8.1 ns [10, 11] and 9.5 ns [7], implying an oscillator strength $f=0.54$ [10, 11] or 0.47 [7] for the resonance transition 2s ^2S–2p $^2P^o$ (doublet at 3130.42 and 3131.07 Å [1]).

Fig. 4-4

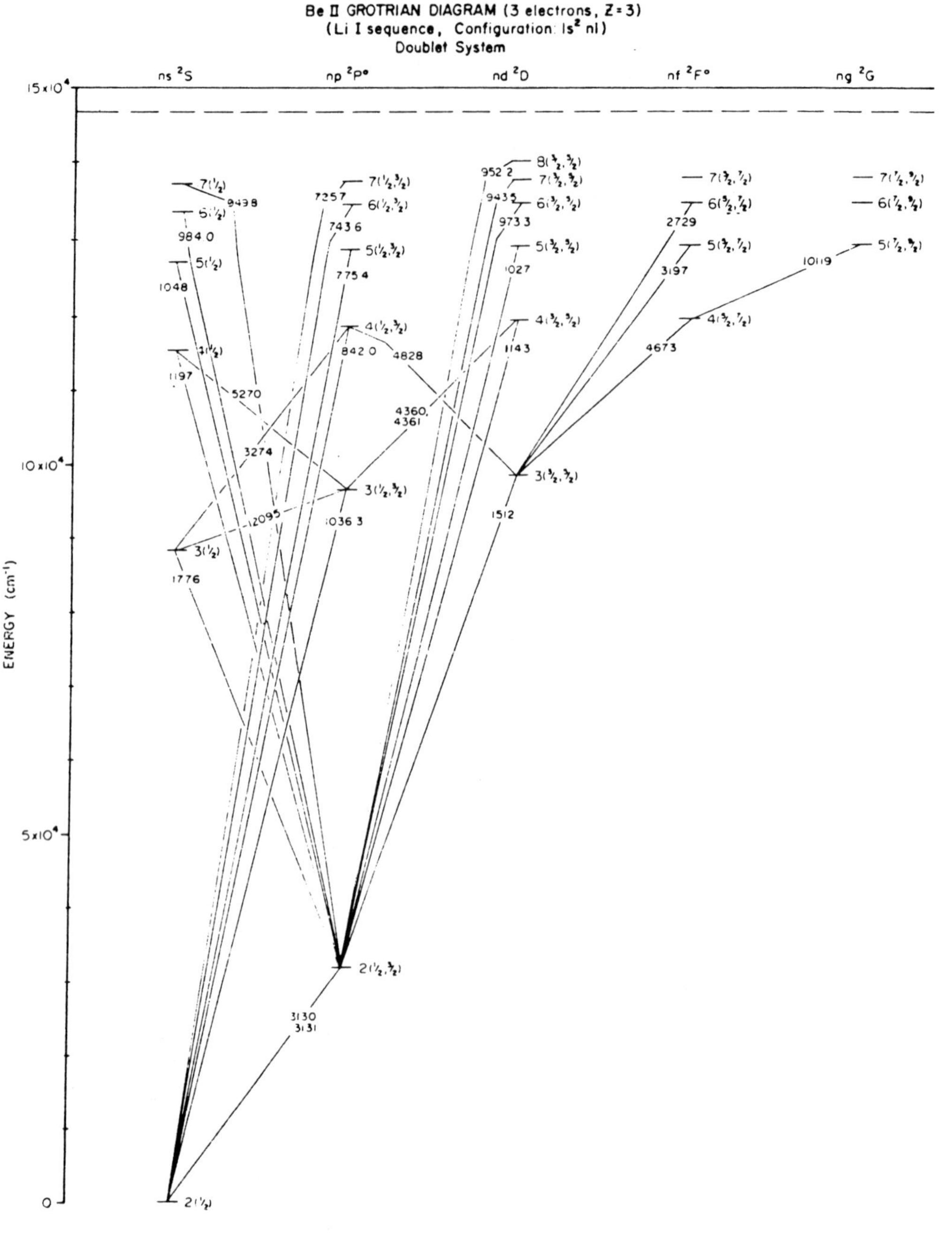

Grotrian diagram of singly excited Be$^+$ [16]. Transitions are indicated by wavelengths in Å.

Table 4/10
Key List for Wavelengths of Singly Excited Be^+ from the Vacuum UV to Near IR Region.

wavelength range in Å	mode of excitation	comment	Ref.
842 to 5271	beam foil	14 classified lines	[6]
1776 to 6279	beam foil	9 classified lines	[10]
1512.269 to 12098.18	hollow cathode	46 classified lines	[1]
4858.22 to 11660.25	hollow cathode	13 classified lines	[5]
3131 to 5270	laser-produced plasma	7 classified lines; plasma diagnostics	[39]
725.71 to 12098.18	—	~55 unclassified lines; literature evaluation a, b)	[17]
725.71 to 1776.307	—	31 classified lines; literature evaluation b)	[18]
2274.2 to 3405.6	—	20 partly classified lines; literature evaluation	[19]
10095.54 to 151501.430	—	94 classified lines; literature evaluation (unobserved IR lines calculated from term differences)	[20]

a) Most of the Be^+ lines listed in [17] without intensity figures appear to belong to doubly excited Be^+.

b) The series $1s^2\,2s\,{}^2S - 1s^2\,np\,{}^2P^o$ and $1s^2\,2p\,{}^2P^o - 1s^2\,nd\,{}^2D$ contained in these compilations have been recently extended to some higher members: $1s^2\,2s\,{}^2S - 1s^2\,np\,{}^2P^o$ to $n=10$ and 11 at 702.2 and 698.5 Å, $1s^2\,2p\,{}^2P^o - 1s^2\,nd\,{}^2D$ to $n=9$ at 912.8 Å [38].

References:

[1] L. Johansson (Arkiv Fysik **20** [1962] 489/98).
[2] C. E. Moore (NBS-C-467-Vol. I [1949] 14).
[3] G. Mehlman-Ballofet, J. M. Esteva (Astrophys. J. **157** [1969] 945/56).
[4] J. M. Esteva, G. Mehlman-Ballofet, J. Romand (J. Quant. Spectrosc. Radiat. Transfer **12** [1972] 1291/303).
[5] J. E. Holmstroem, L. Johansson (Arkiv Fysik **40** [1969] 133/8).
[6] S. Hontzeas, I. Martinson, P. Erman, R. Buchta (Phys. Scr. **6** [1972] 55/60).
[7] J. Bromander (Phys. Scr. **4** [1971] 61/3).
[8] J. Bromander, R. Buchta, L. Lundin (Phys. Letters A **29** [1969] 523/4).
[9] I. Bergstroem, J. Bromander, R. Buchta, L. Lundin, I. Martinson (Phys. Letters A **28** [1969] 721/2).
[10] T. Andersen, K. A. Jessen, G. Soerensen (Phys. Rev. [2] **188** [1969] 76/81).
[11] T. Andersen, K. A. Jessen, G. Soerensen (Phys. Letters A **29** [1969] 384/5), T. Andersen, G. Soerensen (Fys. Tidsskr. **70** [1972] 6/26, 18).
[12] H. G. Berry, J. Bromander, I. Martinson, R. Buchta (Phys. Scr. **3** [1971] 63/7).
[13] B. Denne, H. Dickow, O. Poulsen (Phys. Rev. [3] A **23** [1981] 214/7).

[14] O. Poulsen, T. Andersen, N. J. Skouboe (J. Phys. B **8** [1975] 1393/405).
[15] J. J. Bollinger, J. S. Wells, D. J. Wineland, W. M. Itano (Phys. Rev. [3] A **31** [1985] 2711/4).
[16] S. Bashkin, J. O. Stoner Jr. (Atomic Energy Levels and Grotrian Diagrams, Vol. I, North Holland, Amsterdam 1975, p. 32).
[17] J. Reader, C. H. Corliss (NSRDS-NBS-68 [1980] 1/357, 15).
[18] R. L. Kelly (ORNL-5922 [1982] 1/404, 9/10; C.A. **98** [1983] No. 151855).
[19] R. L. Kelly (NASA-TM-80268-Sect. 1 [1979] 1/400, 28; C.A. **94** [1981] No. 38748).
[20] E. Biemont, N. Grevesse (At. Data Nucl. Data Tables **12** [1973] 217/310, 230).

[21] B. Dynefors, I. Martinson, E. Veje (Phys. Scr. **12** [1975] 58/62).
[22] N. Andersen, K. Jensen, J. Jepsen, J. Melskens, E. Veje (Z. Physik A **273** [1975] 1/8).
[23] M. Braun, B. Emmoth, I. Martinson (Phys. Scr. **10** [1974] 133/8).
[24] N. Andersen, W. S. Bickel, R. Bolen, K. Jensen, E. Veje (Phys. Scr. **3** [1971] 255/60).
[25] G. M. Mladenov, M. Braun (Phys. Status Solidi A **53** [1979] 613/40).
[26] P. O. Taylor, R. A. Phaneuf, G. H. Dunn (Phys. Rev. [3] A **22** [1980] 435/44).
[27] I. S. Aleksakhin, V. A. Zayats (Opt. Spektroskopiya **36** [1974] 1229/30; Opt. Spectrosc. [USSR] **36** [1974] 717).
[28] W. L. Wiese, G. A. Martin (NSRDS-NBS-68 [1980] 359/406, 368).
[29] G. A. Martin, W. L. Wiese (J. Phys. Chem. Ref. Data **5** [1976] 537/70, 540/1, 547).
[30] J. R. Fuhr, B. J. Miller, G. A. Martin (NBS-SP-505 [1978] 1/270, 18).

[31] B. J. Miller, J. R. Fuhr, G. A. Martin (NBS-SP-505-Suppl. 1 [1980] 1/112, 9/10).
[32] V. A. Boiko, F. V. Bunkin, V. I. Derzhiev, G. A. Koldashov, A. Ya. Faenov, A. I. Fedosimov, S. I. Yakovlenko (Pis'ma Zh. Tekhn. Fiz. **9** [1983] 1067/71; Soviet Tech. Phys. Letters **9** [1983] 459/60).
[33] V. V. Zhukov, V. G. Il'yushko, E. L. Latush, M. F. Sem (Kvantovaya Elektron. [Moscow] **2** [1975] 1409/14; Soviet J. Quantum Electron. **5** [1975] 757/60).
[34] L. I. Gudzenko, V. V. Evstigneev, S. I. Yakovlenko (Tr. Fiz. Inst. Akad. Nauk SSSR **90** [1976] 17/38, 27/8; C.A. **87** [1977] No. 125107).
[35] L. I. Gudzenko, V. V. Evstigneev, S. I. Yakovlenko (Kratk. Soobshch. Fiz. **1973** No. 9, pp. 23/7; C.A. **81** [1974] No. 19087).
[36] L. I. Gudzenko, S. I. Yakovlenko (Kratk. Soobshch. Fiz. **1970** No. 7, pp. 3/7; C.A. **74** [1971] No. 70119).
[37] V. I. Kislov (Fiz. Mekhan. **5** [1984] 134/44; C.A. **102** [1985] No. 212281).
[38] E. Jannitti, M. Mazzoni, P. Nicolosi, G. Tondello, Wang Yongchang (J. Opt. Soc. Am. B **2** [1985] 1078/83).
[39] A. Zago, G. Tondello (Nuovo Cimento B [11] **85** [1985] 59/78).

4.3.6.2 Be^+ 1snln′l′ (Doubly Excited Be^+)

Doubly excited (or core excited) Be^+ with two electrons promoted from their ground state orbitals is abundantly produced by beam foil excitation. The quartet states, i.e., those where the three electrons have parallel spin, have attracted the most interest. Most of these states are metastable against autoionization, because selection rules prohibit their autoionization to the $1s^2\,{}^1S_0$ ground state of Be^{2+}, and because the lowest triplet state of Be^{2+}, $1s\,2s\,{}^3S$, is higher in energy than most of the quartet states of Be^+, see Fig. 4-5, p. 169. The selection rules for autoionization, i.e., a decay of a discrete atomic or ionic state to a continuum state under the action of Coulomb forces, are $\Delta J = \Delta\pi = 0$ (no change of total angular momentum and parity) which holds rigorously, and also $\Delta L = \Delta S = 0$ (no change of orbital angular momentum and spin) which is obeyed for pure LS coupling and is violated only by magnetic interactions such as spin orbit coupling [1, 2, 25]. Thus, radiative transitions between core excited quartet levels of Be^+ can be observed which give rise to a great number of lines in the 600 to 5000 Å region.

A bibliography of papers concerning these quartet levels (including their lifetimes) and transitions between them, is presented in Table 4/11. Several controversies about classifications of lines still appear to be unsettled. The status of research up to mid 1985, including the extensive theoretical work devoted to doubly excited states of Be^+, has been reviewed by Andersen and Mannervik [25]. A Grotrian diagram for the quartet levels of Be^+ from [25] is presented in **Fig. 4-5** on p. 169. Other diagrams [2, 4, 5] which were constructed earlier appear to be incomplete or even out of date.

Table 4/11
Reference List for Terms and Optical Spectra of Doubly Excited Be^+ 1snln′l′ ^4L.

observed term	references for term	references for lifetime	comment
2sns ^4S	3s [6 to 8] 4s [8 to 12] 5s [8]	3s [13] 4s [12]	2s 4s ^4S: in [10, 11] through classification of a single line at 3435 Å from [9]
2s np $^4P^o$	2p [6 to 9, 12, 14 to 16] 3p [6 to 9, 12, 17] 4p [8, 12, 14, 18]	4p [12]	2s 3p $^4P^o$: for lines at 3261 and 3530 Å [12], different classifications have been proposed [10, 14, 18, 19]; lines at 1909 and 2273 Å [6, 7] belong to core excited neutral Be, not Be^+ [19]
2s nd ^4D	3d [6 to 9, 12, 17] 4d [6, 7, 9, 12, 17] 5d [12]	3d [6, 7, 13] 4d [6, 7, 12, 13] 5d [12]	2s 4d ^4D: for the line at 3995 Å [9], different classifications have been proposed [10, 14, 19 to 21]
2s nf $^4F^o$	4f [8, 11, 14] 5f [9]	4f [7, 8]	2s 4f $^4F^o$: in [8] through the classification of the line at 4330 Å from [7]; the classification of the line at 3285 Å [14] is questioned [10, 11] Hartree-Fock calculations reveal a strong mutual perturbance of 2s 4f $^4F^o$ and 2p 3d $^4F^o$ terms, making a single-configuration designation less meaningful [20, 22]
2p ns $^4P^o$	2s, see 2s 2p $^4P^o$ 3s [6 to 8, 14] 4s [10, 11, 14]	3s [6 to 8, 13] 4s [9, 14]	2p 4s $^4P^o$: in [10, 11] through classification of a single line at 3435 Å from [9]
2p np ^4S	3p [12, 17] 4p(?) [14, 20]	3p [12] 4p(?) [9]	2p 3p ^4S: for the line at 3261 Å [12] different classification has been proposed [10, 19] 2p 4p ^4S: in [14, 20] based on the classification of the line at 3435 Å [9], but different classification was proposed in [10, 11]

Table 4/11 [continued]

observed term	references for term	references for lifetime	comment
2p np ^4P	2p [6 to 8, 12, 15 to 17] 3p [6, 7, 12, 14, 17] 4p [10, 14]	2p [6, 7, 15, 16, 23] 3p [12] 4p [14]	$2p^2\ ^4P$: in [23] lifetime of Auger decay to Be^{2+} $1s^2$, cf. p. 161; a line at 1909 Å [6, 7] belongs to core excited neutral Be, not Be^+ [19] 2p 3p ^4P: line at 2273 Å [6, 7] belongs to core excited neutral Be, not Be^+ [19]
2p np ^4D	3p [9, 10, 12, 14, 17] 4p [14]	3p [9, 12] 4p [14]	2p 3p ^4D: for lines at 3435 Å [9] and 3530 Å [12], different classifications have been proposed [10, 11, 14, 20]
2p nd $^4P^o$	3d [10, 17, 19] 4d(?) [20]	4d(?) [9]	2p 4d $^4P^o$: in [20] based on reclassification of the line at 3435 Å [9], but different classifications have been proposed [9 to 11, 14]
2p nd $^4D^o$	3d [6 to 9, 12] 4d [12, 17, 19, 20] 5d [12]	3d [6, 7, 12, 13] 4d [12]	—
2p nd $^4F^o$	3d [8, 14, 17] 4d [9, 17]	3d [7, 8] 4d [9]	2p 3d $^4F^o$: in [8, 17] through classification of line at 3511 Å [7]; for perturbation cf. comment to 2s 4f $^4F^o$ 2p 4d $^4F^o$: for lines at 3435 and 3995 Å [9] different classifications have been proposed [10, 11, 14, 19 to 21]
2p 4f ^4D	[10, 19, 20]	[19]	—
2p nf ^4F	4f [8] 5f [9]	4f [8]	2p 4f ^4F: in [8] based on reclassification of the line at 3405 Å [7, 8]
2p 4f ^4G(?)	[14, 20]	[14]	based on classification of lines at 3285 and 3995 Å; different classifications have been proposed in [10, 11, 14, 19 to 21]

Optical transitions between doublet levels of core excited Be^+ can be expected if the upper levels are ^2P, $^2D^o$, or ^2F. For these, autoionization is prohibited by a combination of selection rules for the total angular momentum and parity [19, 25]. Autoionizing doublet levels have been observed by Auger spectra, see p. 160, and by soft X-ray spectra, see p. 159. Eight transitions between doublet levels of core excited Be^+ have been observed in the 800 to 4000 Å region by beam foil spectroscopy and classified by means of Hartree-Fock calculations, including a line at 3995 Å, formerly attributed to the quartet spectrum of Be^+ (cf. Table 4/11, under, e.g., "2p 4f ^4G") [19].

Fig. 4-5

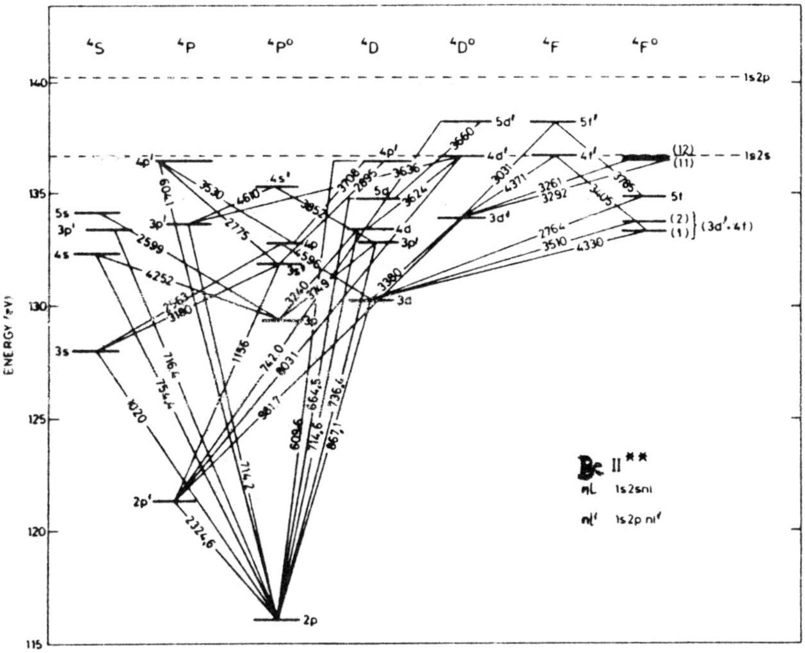

Grotrian diagram of doubly excited Be^+ [25]. The vertical scale is the energy, in eV, above the Be^+ $1s^2\,2s\,^2S_{1/2}$ ground state. Transitions are indicated by wavelengths in Å.

References:

[1] H. G. Berry (Rept. Progr. Phys. **40** [1977] 155/217).
[2] H. G. Berry (Phys. Scr. **12** [1975] 5/20).
[3] M. Roedbro, R. Bruch, P. Bisgaard (J. Phys. B **12** [1979] 2413/47).
[4] K. X. To, E. Knystautas, R. Drouin, H. G. Berry (J. Phys. Colloq. [Paris] **40** [1979] C1-3/C1-5).
[5] S. Bashkin, J. O. Stoner Jr. (Atomic Energy Levels and Grotrian Diagrams, Vol. I, North Holland, Amsterdam 1975, p. 34).
[6] S. Hontzeas, I. Martinson, P. Erman, R. Buchta (Nucl. Instr. Methods **110** [1973] 51/4).
[7] S. Hontzeas, I. Martinson, P. Erman, R. Buchta (Phys. Scr. **6** [1972] 55/60).
[8] S. Mannervik, I. Martinson, B. Jelenkovic (J. Phys. B **14** [1981] L275/L278).
[9] S. M. Bentzen, T. Andersen, O. Poulsen (J. Phys. B **15** [1982] L71/L74).
[10] C. Laughlin (J. Phys. B **16** [1983] 3329/38).
[11] C. Laughlin (Phys. Letters A **91** [1982] 405/6).
[12] S. M. Bentzen, T. Andersen, O. Poulsen (J. Phys. B **14** [1981] 3435/43).
[13] J. Bromander, O. Poulsen, J. L. Subtil (Phys. Scr. **7** [1973] 283/4).
[14] S. M. Bentzen, T. Andersen, O. Poulsen (Phys. Rev. [3] A **26** [1982] 2639/42).
[15] H. G. Berry, J. Bromander, I. Martinson, R. Buchta (Phys. Scr. **3** [1971] 63/7).
[16] I. Martinson (Nucl. Instr. Methods **90** [1970] 81/4).
[17] H. G. Berry, J. L. Subtil (Phys. Scr. **9** [1974] 217/20).
[18] C. Laughlin (J. Phys. B **15** [1982] L67/L70).

[19] M. Agentoft, T. Andersen, C. Froese Fischer, L. Smentek-Mielczarek (Phys. Scr. **28** [1983] 45/50).
[20] C. Froese Fischer (Phys. Rev. [3] A **26** [1982] 2627/38).
[21] C. Froese Fischer (personal communication 1982 to C. Laughlin [10]).
[22] M. Galan, C. F. Bunge (Phys. Rev. [3] A **23** [1981] 1624/31).
[23] R. Bruch, G. Paul, J. Andrä (J. Phys. B **8** [1975] L253/L258).
[24] S. Larsson, R. Crossley, T. Ahlenius (J. Phys. Colloq. [Paris] **40** [1979] C1-6/C1-9).
[25] T. Andersen, S. Mannervik (Comments At. Mol. Phys. **16** [1985] 185/98).

4.3.7 Interaction Constants

Electron-Electron

The fine structure (fs) separation of the $1s^2\,2p\,^2P^o$ doublet has been obtained from a precision measurement of the resonance transitions to $^2P^o_{1/2}$ and $^2P^o_{3/2}$ with tunable lasers [8]. The fs intervals of the doublets $1s^2\,nd\,^2D$, n=4, 5, and $1s^2\,nf\,^2F^o$, n=4, 5, 6, have been determined by level crossing spectroscopy [1, 2]. Similar values (within the limits of error) for $nd\,^2D$ and $nf\,^2F^o$, n=4, 5, were obtained by the zero-field quantum beat method [2, 3]. Less accurate values were obtained from conventional optical spectra [4, 5]. Results from the precision measurements are compiled below.

term	fs splitting in MHz	Ref.	term	fs splitting in MHz	Ref.
$2p\,^2P^o$	197150 ± 64	[8]	$4f\,^2F^o$	3668 ± 5	[1]
				3630 ± 25	[3]
$4d\,^2D$	7293 ± 4	[1]			
	7290 ± 50	[3]	$5f\,^2F^o$	1870 ± 5	[1]
$5d\,^2D$	3730 ± 10	[1]		1810 ± 10	[2]
				1820 ± 50	[3]
			$6f\,^2F^o$	1077 ± 5	[1]

Electron-Nucleus

Magnetic hyperfine structure constants A have been determined for the ground state $1s^2\,2s\,^2S_{1/2}$ and the excited states $1s\,2p\,^2P^o_{1/2,3/2}$ of ^{9}Be as follows:

state	A in MHz	method	Ref.
$2s\,^2S_{1/2}$	-625.00883748 (10)	ion cyclotron resonance detection by laser fluorescence	[6]
$2s\,^2S_{1/2}$	$(\pm)625.009$ (3) *)	radiofrequency-induced transitions detected by optical pumping of $2p\,^2P^o$	[7]
$2p\,^2P^o_{1/2}$	-118.6 ± 3.6	optical double resonance in $2s\,^2S_{1/2}-2p\,^2P^o_{1/2,3/2}$ transitions	[8]
$2p\,^2P^o_{3/2}$	<5	zero-field quantum beats	[2]

*) Hyperfine splitting frequency $\nu_{hfs}=1250.018$ (5) MHz; $A=\nu_{hfs}/J\cdot(2I+1)={}^1/_2\,\nu_{hfs}$ with $I(^9Be)={}^3/_2$ [7].

References:

[1] B. Denne, H. Dickow, O. Poulsen (Phys. Rev. [3] A **23** [1981] 214/7).
[2] O. Poulsen, T. Andersen, N. J. Skouboe (J. Phys. B **8** [1975] 1393/405).
[3] O. Poulsen, J. L. Subtil (Phys. Rev. [3] A **8** [1973] 1181/5).
[4] L. Johansson (Arkiv Fysik **20** [1962] 489/98).
[5] J. E. Holmstroem, L. Johansson (Arkiv Fysik **40** [1969] 133/8).
[6] D. J. Wineland, J. J. Bollinger, W. M. Itano (Phys. Rev. Letters **50** [1983] 628/31; Erratum: 1333), J. J. Bollinger, D. J. Wineland, W. M. Itano, J. S. Wells (Springer Ser. Opt. Sci. **40** [1983] 168/72).
[7] J. Vetter, H. Ackermann, G. zu Putlitz, E. W. Weber (Z. Physik A **276** [1976] 161/5).
[8] J. J. Bollinger, J. S. Wells, D. J. Wineland, W. M. Itano (Phys. Rev. [3] A **31** [1985] 2711/4).

4.3.8 Electric Dipole Polarizability α

The polarizability does not appear to have been experimentally determined so far. The numerous values obtained from theory are not individually cited here for reasons outlined in Section 4.2.3 on pp. 140/1. The most recent from an SCF perturbation calculation is $\alpha = 23.33$ a.u. (3.457×10^{-24} cm^3), G. S. Solov'eva, P. F. Gruzdev, A. I. Sherstyuk (Opt. Spektroskopiya **57** [1984] 776/9; Opt. Spectrosc. [USSR] **57** [1984] 473/4).

4.3.9 Interactions with Photons and Electrons

Values for the photoionization cross sections at the threshold (18.2 eV, see p. 158) of $(1.5 \pm 0.1) \times 10^{-18}$ cm^2 (preliminary result: 1.4×10^{-18} cm^2 [2]) and $(2.5 \pm 0.5) \times 10^{-18}$ cm^2 were derived from intensity distributions in the series $1s^2\,2s\,^2S - 1s^2\,np\,^2P^o$ and $1s^2\,2p\,^2P^o - 1s^2\,nd\,^2D$. Thus, the smaller of the two cross sections corresponds to an outgoing p electron, the larger to an outgoing d or s electron [1].

The cross section for electron impact ionization of Be^+ has been measured by the crossed charged-beam method from threshold to 1600 eV with an absolute uncertainty of 8%. The cross section has a maximum value of 46.5×10^{-18} cm^2 at an impact energy of ~50 eV. A slight discontinuity at ~118 eV is clearly discernible in the high-energy tail of the cross section curve, which is due to resonance excitation and subsequent autoionization of Be^+ $1s\,2s\,2p\,^2P^o$ [3]. These cross sections are recommended in a critical evaluation and assessment of data on electron impact ionization of light atoms and ions [4]. In a preliminary version of that data collection [5] published prior to the above experimental work [3], recent theoretical cross sections [6] had been adopted. These proved to be up to 60% lower than the experimental cross sections [4]. Good agreement was observed, however, with cross sections [3, 4] calculated with the widely used empirical formula of Lotz [7]. The agreement was particularly good when the empirical values were multiplied by a scaling factor of 0.95 and autoionization was taken into account by an additional term [3].

The cross section for electron impact excitation of the $1s^2\,2p\,^2P^o$ term of Be^+ and the polarization of the emitted resonance radiation at 3131 Å have been measured by the crossed charged-beam method from threshold (~4 eV) to 800 eV [8].

References:

[1] E. Jannitti, M. Mazzoni, P. Nicolosi, G. Tondello, Wang Yongchang (J. Opt. Soc. Am. B **2** [1985] 1078/83).
[2] E. Jannitti, P. Nicolosi, G. Tondello, Wang Yongchang (AIP [Am. Inst. Phys.] Conf. Proc. No. 119 [1984] 461/7).

[3] R. A. Falk, G. H. Dunn (Phys. Rev. [3] A **27** [1983] 754/61).
[4] K. L. Bell, H. B. Gilbody, J. G. Hughes, A. E. Kingston, F. J. Smith (J. Phys. Chem. Ref. Data **12** [1983] 891/916, 903, 913).
[5] K. L. Bell, H. B. Gilbody, J. G. Hughes, A. E. Kingston, F. J. Smith (CLM-R-216 [1982] 1/120, 32, 85, 87; C.A. **97** [1982] No. 135245).
[6] S. M. Younger (Phys. Rev. [3] A **22** [1980] 111/7).
[7] W. Lotz (Z. Physik **216** [1968] 241/7).
[8] P. O. Taylor, R. A. Phaneuf, G. H. Dunn (Phys. Rev. [3] A **22** [1980] 435/44).

4.3.10 Thermodynamic Functions

For the ideal monoatomic gas, two groups of workers [1, 2] have calculated thermodynamic functions, namely the heat capacity C_p^o, entropy S^o, Gibbs free energy $(G^o-H_{298}^o)/T$ [1], $(G^o-H_0^o)/T$ [2], all in $cal \cdot mol^{-1} \cdot K^{-1}$, and the enthalpy $H^o-H_{298}^o$ in kcal/mol [1], $(H^o-H_0^o)/T$ in $cal \cdot mol^{-1} \cdot K^{-1}$ [2]. Values from the two sources [1, 2] are in good agreement. In both, the electronic partition function was calculated with energies and statistical weights taken from Moore's Tables [3]. Values from the JANAF Tables [1] are given below for $T \geq 3000$ K. Below 3000 K, only translational and no electronic degrees of freedom contribute to the thermodynamic functions, thus, for instance, $C_p^o = {}^5/_2\,R = 4.968\ cal \cdot mol^{-1} \cdot K^{-1}$ [1, 2].

T in K	3000	4000	5000	6000
C_p^o	4.968	4.976	5.020	5.133
S^o	45.393	46.823	47.937	48.861
$-(G^o-H_{298}^o)/T$	40.918	42.224	43.259	44.118
$H^o-H_{298}^o$	13.423	18.394	23.387	28.457

Values in another tabulation extending to 10000 K were obtained by summing the partition function over a restricted set of electronic energy levels [4].

The electronic partition function Q has been fitted in the form of polynomial expansions of log Q in powers of log (5040/T) [5], of ln Q in powers of ln T [6], and of Q in powers of T/1000, but taking into account only levels below 30000 cm^{-1} [7]. In this latter case the electronic partition function reduces to the constant value 2, the statistical weight of the Be^+ ground state.

References:

[1] D. R. Stull, H. Prophet (JANAF Thermochemical Tables, 2nd. Ed., NSRDS-NBS-37 [1971]).
[2] J. W. Green, D. E. Poland, J. L. Margrave (J. Chem. Phys. **33** [1960] 35/9).
[3] C. E. Moore (NBS-C-467-Vol. I [1949]).
[4] J. Hilsenrath, C. G. Messina, W. H. Evans (AD-606163 [1964] 1/434, 45/6; C.A. **62** [1965] 7177).
[5] A. J. Sauval, J. B. Tatum (Astrophys. J. Suppl. Ser. **56** [1984] 193/209, 207/8).
[6] A. W. Irwin (Astrophys. J. Suppl. Ser. **45** [1981] 621/33, 623).
[7] L. De Galan, R. Smith, J. D. Winefordner (Spectrochim. Acta B **23** [1968] 521/5).

4.4 The Be^{2+} Ion

4.4.1 Ionization Energy E_i

Loefstrand [1] has combined a precision measurement of the wavelength for the Be^{2+} $1s^2\ ^1S_0 - 1s\,2p\ ^1P_1^o$ resonance transition, 100.2552 Å [2], with a calculated value (including relativistic corrections) of the $1s\,2p\ ^1P_1^o$ term value, 243787.7 cm^{-1} [3], to obtain

$E_i = 1241242 \pm 15\ cm^{-1}$. Eidelsberg [4] obtained $E_i = 1241253 \pm 10\ cm^{-1}$ using the same theoretical value, but $\lambda = 100.254$ Å for the resonance transition.

References:

[1] B. Loefstrand (Phys. Scr. **8** [1973] 57/61).
[2] L. A. Svensson (Phys. Scr. **1** [1970] 246).
[3] Y. Accad, C. L. Pekeris, B. Schiff (Phys. Rev. [3] A **4** [1971] 516/36, 524).
[4] M. Eidelsberg (J. Phys. B **5** [1972] 1031/7).

4.4.2 Atomic Structure Calculations

For Be and its uni- and dipositive ions, atomic structure calculations are not documented in this volume for reasons outlined in Section 4.2.3 on pp. 140/1. Consult that section for bibliographies devoted to the topic.

4.4.3 X-Ray Spectra

Preliminary Note. This chapter comprises all of the transitions involving the 1s shell of Be^{2+}. This is the region below 100.2552 Å, which is the wavelength of the Be^{2+} $1s^2\ ^1S_0$–$1s\,2p\ ^1P_1^o$ resonance transition.

X-ray spectra of Be^{2+} were obtained by vacuum sparks [1 to 3], laser-produced plasmas [4, 5], and beam foil excitation [6]. Observed lines with classifications are presented in Table 4/12.

References to spectroscopic work prior to 1950 may be found in [7, 8]. For four levels of doubly excited Be^{2+}, Roedbro et al. [9] and Bruch [10] have used the notation introduced by Lipsky et al. [11] for doubly excited states of He and He-like ions, for the purpose of comparing level energies from X-ray spectra [3, 4], Auger spectra [9], and from calculations [11]. The notation is (N, nα) $^{2S+1}L^{e,o}$, the superscript "e" for "even", besides "o" for "odd", to the right of the quantum number L for the total orbital angular momentum is explicitly added in this notation for clarity. For the states in question, this notation is also given in Table 4/12. The $2s\,2p\ ^1P^o$ [4, 13] state has been identified with (2, 2a) $^1P^o$ [9] by Jannitti et al. [13]. The need for an alternative to a single-configuration classification of the levels of doubly excited He and He-like ions and the meaning of the classification scheme of Lipsky et al. [11] are outlined on p. 177.

Table 4/12
X-Ray Lines of Be^{2+}.

wavelength in Å	spectrum	classification	Ref.
100.2552	emission	$1s^2\ ^1S$ $-1s\,2p\ ^1P^o$	[2]
88.314 to 81.35	absorption	$1s^2\ ^1S$ $-1s\,np\ ^1P^o$, $3 \leq n \leq 9$	[1, 4]
78.94 a)	emission	$1s\,2p\ ^1P^o - 2p^2\ ^1D$ $-(2, 2a)\ ^1D^e$	[3, 5]
78.67 a)	emission	$1s\,2p\ ^3P^o - 2p^2\ ^3P$ $-(2, 2a)\ ^3P^e$	[3, 5, 6]

Table 4/12 [continued]

wavelength in Å	spectrum	classification	Ref.
78.512 a)	emission	1s 2s ^{1}S −2s 2p ^{3}P° −(2, 2a) ^{3}P°	[3, 5]
76.901	emission	1s 3s ^{1}S −2p 3s ^{1}P°	[5]
76.477 a)	emission	{ 1s 3s ^{3}S −2s 3p, 2p 3s ^{3}P° −(2, 3b) ^{3}P° 1s 3d ^{1}D −2p 3d ^{1}D° }	[3, 5]
76.096	emission	1s 3s ^{1}S −2s 3p, 2p 3s ^{1}P° −(2, 3b) ^{1}P°	[3]
44.08 b)	absorption	1s^2 ^{1}S −2s 2p ^{1}P° −(2, 2a) ^{1}P°	[4, 13]
40.93	absorption	1s^2 ^{1}S −2s 3p ^{1}P°	[13]

a) Wavelengths from [5]; those from [3] differ by less than 0.02 Å. In beam foil spectra [6], the transition 1s 2p ^{3}P°−2p^2 (2, 2a) ^{3}P^e was a weak feature located at 78.60 Å.

b) The profile of this line has been fit to a Fano resonance. The wavelength given for this transition is the position of the resonance derived from the fit, 281.25 eV [13], converted to Å.

It has been proposed that lasing of the 1s 2p $^1P^o_1$−1s^2 $^1S_0$ resonance transition of Be^{2+} at 100.25 Å might be achieved if a sufficient number of Be^{2+} ions in the metastable 1s 2s $^1S_0$ state could be accumulated in a cavity and could then be optically pumped with a wavelength of 6141.2 Å from the metastable state to the upper level of the lasing transition [12].

References:

[1] G. Mehlman-Ballofet, J. M. Esteva (Compt. Rend. B **276** [1973] 173/6).
[2] L. A. Svensson (Phys. Scr. **1** [1970] 246).
[3] S. Goldsmith (J. Phys. B **2** [1969] 1075/9).
[4] E. Jannitti, P. Nicolosi, G. Tondello (Physica B+C **124** [1984] 139/47).
[5] P. Nicolosi, G. Tondello (J. Opt. Soc. Am. **67** [1977] 1033/9).
[6] T. Andersen, S. M. Bentzen, O. Poulson (Phys. Scr. **22** [1980] 119/22).
[7] B. Loefstrand (Phys. Scr. **8** [1973] 57/61).
[8] C. E. Moore (NBS-C-467-Vol. I [1949] 14).
[9] M. Roedbro, R. Bruch, P. Bisgaard (J. Phys. B **12** [1979] 2413/47, 2433).
[10] R. Bruch (Phys. Scr. **26** [1982] 381/2).
[11] L. Lipsky, R. Anania, M. J. Conneely (At. Data Nucl. Data Tables **20** [1977] 124/41), L. Lipsky, M. J. Conneely (Phys. Rev. [3] A **14** [1976] 2193/205).
[12] H. Mahr, M. Roeder (Opt. Commun. **10** [1974] 227/8).
[13] E. Jannitti, P. Nicolosi, G. Tondello (Opt. Commun. **50** [1984] 225/30, 229).

4.4.4 Auger Spectra

Emission of Auger electrons from levels of doubly excited Be^{2+} ions has been observed after passage of fast Be^{2+} ions (≤500 keV) through He or CH_4 gas under single-collision conditions, in the same manner as was done for Be and Be^+, see pp. 143 and 160. The observed lines in the region 121 to 155 eV were tentatively assigned [1] by comparison with calculated energy levels [2]. The results [1] are presented in Table 4/13. A preliminary account of this work was given in [3].

Table 4/13
Auger Transitions of Be^{2+} [1].

observed peak energy in eV	assignment of Be^{2+} initial state*)	observed peak energy in eV	assignment of Be^{2+} initial state*)
121.4±0.2	(2, 2a) $^1S^e$	147.3	(2, 3a) $^1S^e$, (2, 3a) $^3D^e$, (2, 3a) $^3P^o$, (2, 3b) $^3P^o$
122.6±0.2	(2, 2a) $^3P^o$		
126.9±0.2	(2, 2a) $^1D^e$	148.5	(2, 3a) $^3F^o$
127.4±0.2	(2, 2a) $^1P^o$	148.7	(2, 3a) $^1D^e$
132.8±0.2	(2, 2b) $^1S^e$	148.9	(2, 3b) $^3D^e$, (2, 3a) $^1P^o$
146.1±0.2	(2, 3a) $^3S^e$	149.9	(2, 3b) $^1D^e$, (2, 3a) $^1F^o$
146.4±0.2	(2, 3b) $^1P^o$	150.8	(2, 3b) $^1S^e$

*) Final state is Be^{3+} $1s_{1/2}$. The notation of Be^{2+} doubly excited states is that used in [2] and outlined on p. 177.

References:

[1] M. Roedbro, R. Bruch, P. Bisgaard (J. Phys. B **12** [1979] 2413/47).
[2] L. Lipsky, R. Anania, M. J. Conneely (At. Data Nucl. Data Tables **20** [1977] 124/41).
[3] M. Roedbro, R. Bruch, P. Bisgaard (J. Phys. B **10** [1977] L275/L279).

4.4.5 Atomic Energy Levels. Optical Spectra (Be III)

Preliminary Note. Within the context of this chapter, the term "optical spectra" is intended to cover all transitions of Be^{2+} except those which are dealt with under "X-ray spectra" on p. 173 involving the 1s shell. Thus, "optical spectra" of Be^{2+} extend down to 378 Å, which appears to be the shortest wavelength observed so far for a transition falling into the above category, namely a transition between states of doubly excited Be^{2+}, see Table 4/16 on p. 178.

Be^{2+} 1s nl (Singly Excited Be^{2+})

Energy Levels. A survey of energy levels of singly excited Be^{2+} is presented in Table 4/14, p. 176. This table is arranged by term systems belonging to the same "running" electron, and the energy of only the lowest and highest member of each system is given numerically.

Table 4/14 is based mainly on the work of Loefstrand [1] and Eidelsberg [2], who took spectra emitted from Be spark sources. The compilation of atomic energy levels of Be^{2+} in the well-known tables of Moore [3] is based on spectroscopic work up to 1947.

Wavelengths. Observed wavelengths of Be^{2+} lines, taken from the above mentioned literature sources [1, 2], are compiled in Table 4/15, p. 176.

Compilations of unclassified observed transitions have been published by Reader and Corliss [7] and of partly classified transitions by Kelly [8, 9] ($\lambda < 2000$ Å, classified lines [8], $\lambda > 2000$ Å, no classifications [9]). Some of the lines listed in Table 4/15 were also

Table 4/14
Terms of Be^{2+} 1s nl (Singly Excited Be^{2+}).

term system			energy of lowest member in cm^{-1}	energy of highest member in cm^{-1}	Ref.
1s ns	^1S,	$1 \leq n \leq 4$	0	1177939 [2]	[1, 2]
1s np	$^1P^o$,	$2 \leq n \leq 9$	997454[a)] [1]	1229256[b)] [4]	[1 to 4]
1s nd	^1D,	n=3, 4, 5	1131462 [1]	1201743 [2]	[1, 2]
1s nf	$^1F^o$,	n=4, 5, 6	1179514 [1]	1213820 [2]	[1, 2]
1s 5g	^1G		1201742 [1] 1201749 [2]		[1, 2]
1s ns	^3S,	$2 \leq n \leq 5$	956502 [1]	1199653 [2]	[1, 2]
1s np	$^3P^o$,	$2 \leq n \leq 5$	983365[a,c,d)] [1]	1201066 [2]	[1, 2, 4]
1s nd	^3D,	$3 \leq n \leq 6$	1131383 [1]	1213791 [2]	[1, 2]
1s nf	$^3F^o$,	$4 \leq n \leq 7$	1179515 [1]	1221098 [2]	[1, 2]
1s 5g	^3G		1201742 [1] 1201749 [2]		[1, 2]

a) For 1s 2p $^1P^o$ and 1s 2p $^3P^o$, calculated term values from [5] have been adopted [1, 2].
b) Inverse of the wavelength reported [4] for the $1s^2$ ^1S − 1s 9p $^1P^o$ transition, 81.35 Å.
c) Center of gravity of fine structure components.
d) Lifetime of 1s 2p $^3P^o$, measured by a time-of-flight method after beam foil excitation, is 22.3 ± 1.0 ns [6].

Table 4/15
Observed Lines of Be^{2+} 1s nl (Singly Excited Be^{2+}).
Vacuum wavelengths below 2000 Å, air wavelengths above.

wavelength in Å [1]	wavelength in Å [2]	classification [1, 2]
6142.01 ± 0.05	6141.2	2s ^1S − 2p $^1P^o$
4497.8 ± 0.3	4498.2 ± 0.1	4f $^{1,3}F^o$ − 5g $^{1,3}G$
—	4495.09 ± 0.1	4d ^1D − 5f $^1F^o$
4487.30 ± 0.1*)	4485.52 ± 0.1	4d ^3D − 5f $^3F^o$
4249.14 ± 0.1	—	4p $^3P^o$ − 5d ^3D
3722.91 ± 0.03	3722.47 ± 0.1	2s ^3S − 2p $^3P^o_1$
3721.31 ± 0.03	3720.91 ± 0.1	^3S − $^3P_0$
3720.85 ± 0.03	3720.36 ± 0.1	^3S − $^3P_2$
—	2191.57 ± 0.1	3p $^1P^o$ − 4s ^1S
—	2137.25 ± 0.1	3d ^3D − 4p $^3P^o$
2127.20 ± 0.04	2127.23 ± 0.05	3p $^3P^o$ − 4s ^3S
2122.27 ± 0.04	2118.56 ± 0.05	3p $^1P^o$ − 4d ^1D
2080.38 ± 0.02	2080.18 ± 0.05	3d ^1D − 4f $^1F^o$
2076.94 ± 0.02	2076.86 ± 0.05	3d ^3D − 4f $^3F^o$
1954.97 ± 0.03	1954.90 ± 0.05	3p $^3P^o$ − 4d ^3D
—	1917.03 ± 0.05	3s ^1S − 4p $^1P^o$
—	1754.80 ± 0.05	3s ^3S − 4p $^3P^o$

Table 4/15 [continued]

wavelength in Å [1]	wavelength in Å [2]	classification [1, 2]
—	1440.77 ± 0.05	3p $^1P^o$ – 5d ^1D
—	1435.17 ± 0.05	3d ^3D – 5p $^3P^o$
—	1422.86 ± 0.05	3d ^1D – 5f $^1F^o$
—	1421.26 ± 0.05	3d ^3D – 5f $^3F^o$
—	1401.52 ± 0.05	3p $^3P^o$ – 5s ^3S
—	1362.25 ± 0.05	3p $^3P^o$ – 5d ^3D
—	1214.32 ± 0.05	3d ^1D – 6f $^1F^o$
—	1213.12 ± 0.05	3d ^3D – 6f $^3F^o$
—	1114.69 ± 0.05	3d ^3D – 7f $^3F^o$
767.75 ± 0.03	—	2p $^1P^o$ – 3s ^1S
746.228 ± 0.008	746.28 ± 0.05	2p $^1P^o$ – 3d ^1D
725.586 ± 0.01	725.77 ± 0.05	2p $^3P^o$ – 3s ^3S
675.593 ± 0.008	675.58 ± 0.05	2p $^3P^o$ – 3d ^3D
661.322 ± 0.01	661.59 ± 0.05	2s ^1S – 3p $^1P^o$
582.078 ± 0.01	582.18 ± 0.05	2s ^3S – 3p $^3P^o$
549.31 ± 0.03	—	2p $^1P^o$ – 4d ^1D
509.99 ± 0.02	—	2p $^3P^o$ – 4d ^3D

*) Improved wavelength is 4486.8 ± 0.3 Å [13].

observed by beam foil excitation [10, 11] and in a laser-produced plasma [23]. The line at 6142 Å (1s 2s ^1S – 1s 2p $^1P^o$) was also observed earlier by Edlén [12] in a vacuum spark.

A Grotrian diagram based on the two main spectroscopic sources [1, 2] has been constructed by Bashkin and Stoner [17]. No intercombination lines have been detected so far.

Two bibliographies of work up to 1980 on transition probabilities (mostly theoretical) have been published by the U.S. National Bureau of Standards [21, 22].

Be^{2+} nln′l′, n ≥ 2 (Doubly Excited Be^{2+})

Using beam foil excitation, Andersen et al. [14] have observed nine transitions in the triplet spectrum of doubly excited Be^{2+}. Their wavelengths and classifications are presented in Table 4/16, p. 178. In addition, a classification scheme introduced by Lipsky et al. [15, 16] is given for levels of doubly excited He and ions in the He isoelectronic sequence. In this notation, any nln′l′ configurational description is completely dropped, since configuration mixing is very strong, as shown by calculations, making such a description useless. Instead, a notation (N, nα) is used which reads as follows: N represents the Nl∞l′ threshold below which the state exists (in the case of Be^{2+}, only states with N = 2 have been observed and the threshold in question is Be^{2+} 2p∞l, 317.19 eV above the ground state of Be^{2+}, cf. [17] and Fig. 14 in [18]), n represents the quantum number of the outer electron, and α = a, b, c, ... labels each series for a given set of quantum numbers L, S, and parity π: a is given to the series to which the lowest level of a given set (L, S, π) belongs, b is given to the series to which the lowest member not belonging to series a belongs, and so on. Assignment of states to one of the series a, b, c is made from their calculated quantum defects [15, 16]. Table 4/16 on p. 178 was compiled as follows: Observed states labeled by single configurations in [14] were identified with calculated states [15] classified using the notation of Lipsky et al. [15, 16] by cross-linking table V in [15] with table II in [14]: The former gives the calculated term energies in the notation of Lipsky et al. [15,

Table 4/16
Observed Lines of Be^{2+} $nln'l'$ (Doubly Excited Be^{2+}).

observed wavelength in Å [14]	classification of transition [14]	[15][a)]
378.2±0.1	2s 2p $^3P^o$ − 2p 4p ^3P	(2, 2a) $^3P^o$ − (2, 4a) $^3P^e$
384.7±0.1	− 2p 4p ^3D	− (2, 4a) $^3D^e$
468.9±0.1	− 2p 3p ^3D	− (2, 3b) $^3D^e$
474.8±0.1	− 2p 3p ^3P	− (2, 3a) $^3P^e$
412.2±0.1	$2p^2$ ^3P − 2p 4d $^3P^o$	(2, 2a) $^3P^e$ − (2, 4c) $^3P^o$
416.3±0.1	− 2p 4d $^3D^o$	− (2, 4a) $^3D^o$
510.2±0.1	− 2p 3d $^3P^o$	− (2, 3c) $^3P^o$
525.8±0.1	− 2p 3d $^3D^o$	− (2, 3a) $^3D^o$
3914 ±1[b)]	2s 2p $^3P^o$ − $2p^2$ ^3P	(2, 2a) $^3P^o$ − (2, 2a) $^3P^e$

a) The superscripts "o" and "e" for "odd" and "even", to the right of the term symbol, are added for clarity [15].
b) A more accurate wavelength for this transition is 3913.4±0.3 Å [20].

16], whereas the latter presents some of these energies as citations and classifies them by single configurations. Also, see the comments of Bruch [19] concerning some doubly excited states of Be^{2+}.

References:

[1] B. Loefstrand (Phys. Scr. **8** [1973] 57/61).
[2] M. Eidelsberg (J. Phys. B **5** [1972] 1031/7).
[3] C. B. Moore (NBS-C-467-Vol. I [1949] 14/5).
[4] G. Mehlman-Ballofet, J. M. Esteva (Compt. Rend. B **276** [1973] 173/6).
[5] Y. Accad, C. L. Pekeris, B. Schiff (Phys. Rev. [3] A **4** [1971] 516/36, 524, 534).
[6] T. Andersen, K. A. Jessen, G. Soerensen (Phys. Rev. [2] **188** [1969] 76/81).
[7] J. Reader, C. H. Corliss (NSRDS-NBS-68 [1980] 1/357, 16).
[8] R. L. Kelly (ORNL-5922 [1982] 1/404, 10/1; C.A. **98** [1983] No. 151855).
[9] R. L. Kelly (NASA-TM-80268 [1979] 1/400, 28; C.A. **94** [1981] No. 38748).
[10] B. Dynefors, I. Martinson, E. Veje (Phys. Scr. **12** [1975] 58/62).
[11] S. Hontzeas, I. Martinson, P. Erman, R. Buchta (Phys. Scr. **6** [1972] 55/60).
[12] B. Edlén (Arkiv Fysik **4** [1951] 441/53).
[13] I. Martinson (personal communication to B. Loefstrand [1]).
[14] T. Andersen, S. M. Bentzen, O. Poulsen (Phys. Scr. **22** [1980] 119/22).
[15] L. Lipsky, R. Anania, M. J. Conneely (At. Data Nucl. Data Tables **20** [1977] 127/41).
[16] L. Lipsky, M. J. Conneely (Phys. Rev. [3] A **14** [1976] 2193/205).
[17] S. Bashkin, J. O. Stoner Jr. (Atomic Energy Levels and Grotrian Diagrams, Vol. I, Addenda, North Holland, Amsterdam 1978, pp. 12, 13, 15).
[18] M. Roedbro, R. Bruch, P. Bisgaard (J. Phys. B **12** [1979] 2413/47, 2434).
[19] R. Bruch (Phys. Scr. **26** [1982] 381/2).
[20] S. Mannervik, I. Martinson, B. Jelenkovic (J. Phys. B **14** [1981] L275/L278).

[21] J. R. Fuhr, B. J. Miller, G. A. Martin (NBS-SP-505 [1978] 1/270, 18).
[22] B. J. Miller, J. R. Fuhr, G. A. Martin (NBS-SP-505-Suppl. 1 [1980] 1/112, 9/10).
[23] A. Zago, G. Tondello (Nuovo Cimento B [11] **85** [1985] 59/78).

4.4.6 Interaction Constants

The hyperfine structure of the 1s 2p $^3P^o$ level of Be^{2+} has been studied by the zero-field quantum beat technique. The transition observed was 1s 2s $^3S-1s\,2p\,^3P^o$ at 3720 Å. The observed structure was analyzed in terms of the coupling constant for the Fermi contact interaction, a (1s) $= -(0.413\pm0.004)$ cm^{-1}, and the radial integral, $\langle r^{-3}\rangle_{2p}=1.0\pm0.2$ a.u., O. Poulsen, J. L. Subtil (J. Phys. B **7** [1974] 31/6).

4.4.7 Electric Dipole Polarizability α

The numerous values obtained from theory are not individually cited here for reasons outlined in Section 4.2.3 on pp. 140/1. The most recent, calculated within the relativistic random-phase approximation, a relativistic generalization of the coupled Hartree-Fock theory, is $\alpha=0.5182$ a.u. (0.0768×10^{-24} cm^3) [1]. An early "experimental" value from ion refractivities is 0.04×10^{-24} cm^3 [2].

References:

[1] W. R. Johnson, D. Kolb, K.-N. Huang (At. Data Nucl. Data Tables **28** [1983] 333/40, 336).
[2] Landolt-Börnstein **1** Pt. 1 [1950] 401).

4.4.8 Interactions with Photons and Electrons

The photoionization cross section of Be^{2+} was measured in a laser-produced Be plasma from threshold at 154 eV to 420 eV. The value at threshold was $(1.4\pm0.2)\times10^{-18}$ cm^2 [1].

Recommended cross sections for electron impact ionization of Be^{2+} are empirical estimates, due to a lack of reliable experimental and theoretical data. They were derived by scaling experimental values for B^{3+}. The scaling law used relates the first ionization energy I and the cross section σ as a function of energy by $\sigma_{sc}=I^2\cdot\sigma(E)$, where the scaled cross section σ_{sc} is a function of E/I. Cross sections are given for $1.25\leq E/I\leq 100.00$ [2]. For earlier semiempirical estimates, which come reasonably close to the recommended values [2], see [3].

References:

[1] E. Janitti, P. Nicolosi, G. Tondello (Opt. Commun. **50** [1984] 225/30).
[2] K. L. Bell, H. H. Gilbody, J. G. Hughes, E. A. Kingston, F. J. Smith (J. Phys. Chem. Ref. Data **12** [1983] 891/916, 901, 910, 913; CLM-R-216 [1982] 1/120, 33, 76, 79).
[3] W. Lotz (Z. Physik **216** [1968] 241/7).

4.4.9 Thermodynamic Functions

Calculated thermodynamic functions (heat capacity C_p^o, entropy S°, Gibbs free energy $(G^o-H_0^o)/T$, and enthalpy $H^o-H_0^o$) up to 10000 K are those of an ideal gas with only translational and no electronic degrees of freedom. Thus, for instance, $C_p^o={}^5/_2R=$ 4.9679 cal·mol^{-1}·K^{-1} applies throughout, J. Hilsenrath, C. G. Messina, W. H. Evans (AD-606163 [1964] 1/434, 49/50; C.A. **62** [1965] 7177).

4.5 The Be^{3+} Ion

Erickson [1] has calculated **atomic energy levels** of hydrogen-like $^9Be^{3+}$ to a high accuracy, including the effects of finite nuclear mass and size, relativistic effects, and quantum electrodynamic corrections, but neglecting hyperfine structure. All levels are given for principal quantum numbers $n \leq 11$, and, additionally, the $s_{1/2}$, $p_{1/2}$, and $j=n-1/2$ levels for $12 \leq n \leq 20$. The energy of the $1s_{1/2}$ level so obtained, i.e., the **ionization limit** of Be^{3+}, is 1756018.824 (8) cm^{-1} [1]. The **Lamb shift** so obtained, i.e., the energy difference between the $2s_{1/2}$ and $2p_{1/2}$ levels, is 6.0006 cm^{-1} [1], or 179.893 GHz, taking $2.99792458 \times 10^{10}$ cm/s for the speed of light. An earlier calculation gave 179.791 GHz [23]. Erickson's calculations [1] update and improve those carried out earlier by Garcia and Mack [2] who obtained an ionization limit of 1756018.67 cm^{-1}. Experimental and theoretical work prior to 1950 is compiled in the "Tables of Atomic Energy Levels" of Moore [3].

The electrostatic potential at the nucleus was calculated from a wave function [14].

The calculated **wavelength** of the first member of the Be^{3+} (Be IV) Lyman series (transition 1s−2p) is 75.92774308 Å (weighted mean over fine structure components $2p_{1/2}$ and $2p_{3/2}$) [2]. The first six lines of the Be^{3+} Lyman series are recommended as calibration standards in the wavelength region 58 to 76 Å [4, 5]. The line profiles and Stark as well as Doppler shifts of the Be^{3+} Lyman series [6 to 11] and also of the $n=5$ to 6 transition [21] were studied repeatedly for use as a tool in plasma diagnostics. A Grotrian diagram based on the calculations of Garcia and Mack [2] was constructed by Bashkin and Stoner [22].

For the metastable $2s_{1/2}$ state, **radiative decay rates** were calculated theoretically. The processes considered were magnetic dipole radiation, two-photon electric dipole radiation, and two-photon magnetic dipole radiation, the latter found to be negligible. The calculated total rate is 3.3684×10^4 s^{-1} [15]. The angular distribution of the radiation accompanying the electric field induced decay of the $2s_{1/2}$ state was calculated in the zero-field limit. A measurement of this distribution could provide a means of determining the Lamb shift of Be^{3+} [16].

The recommended cross sections for **electron impact ionization** of hydrogenic ions, including Be^{3+}, are empirical estimates due to the lack of reliable experimental or theoretical data. They are obtained by scaling theoretical cross sections for C^{5+}; the scaling law is the same as used in case of Be (see p. 157) and Be^{2+} (see p. 179). Cross sections are given for $1.25 \leq E/I \leq 100.00$, where E is the electron kinetic energy and I the ionization energy of Be^{3+} [12]. For earlier semiempirical estimates, which are considerably larger than the recommended values discussed above, see [13].

Phase shifts for **elastic scattering of electrons** (also positrons [17]) have been calculated for impact energies ≤ 148 eV [17] and ≤ 38 eV [18]. Cross sections for electronic **excitation by electron impact** were also calculated theoretically [19, 20].

References:

[1] G. W. Erickson (J. Phys. Chem. Ref. Data **6** [1977] 831/69, 855/8).
[2] J. D. Garcia, J. E. Mack (J. Opt. Soc. Am. **55** [1965] 654/85, 659/60, 670/1).
[3] C. E. Moore (NBS-C-467-Vol. I [1949] 15).
[4] V. Kaufman, B. Edlén (J. Phys. Chem. Ref. Data **3** [1974] 825/95, 866).
[5] B. Edlén, L. A. Svensson (Arkiv Fysik **28** [1965] 427/46, 431).
[6] S. Hashimoto, N. Yamaguchi (Phys. Letters A **95** [1983] 299/302).
[7] S. Hashimoto (Phys. Letters A **105** [1984] 54/6).
[8] E. Jannitti, P. Nicolosi, G. Tondello (J. Phys. [Paris] **43** [1982] 1043/7).

[9] P. Nicolosi, L. Garifo, E. Jannitti, A. M. Malvezzi, G. Tondello (Nuovo Cimento B [11] **48** [1978] 133/51).
[10] G. Tondello, E. Jannitti, A. M. Malvezzi (Phys. Rev. [3] A **16** [1977] 1705/14).

[11] A. M. Malvezzi, E. Jannitti, G. Tondello (Opt. Commun. **13** [1975] 307/10).
[12] K. L. Bell, H. H. Gilbody, J. G. Hughes, E. A. Kingston, F. J. Smith (J. Phys. Chem. Ref. Data **12** [1983] 891/916, 899, 910, 913; CLM-R-216 [1982] 1/120, 34, 66, 70).
[13] W. Lotz (Z. Physik **216** [1968] 241/7).
[14] S. H. Hill, P. J. Grout, N. H. March (J. Chem. Phys. **80** [1984] 3714/9).
[15] W. R. Johnson (Phys. Rev. Letters **29** [1972] 1123/6).
[16] G. W. F. Drake, Chien-Ping Lin (Phys. Rev. [3] A **14** [1976] 1296/8).
[17] I. Shimamura (J. Phys. Soc. Japan **31** [1971] 217/29).
[18] P. Khan, M. Dashkan, A. S. Ghosh, C. Falcon (Phys. Rev. [3] A **26** [1982] 1401/5).
[19] B. K. Thomas (Phys. Rev. [3] A **18** [1978] 452/66).
[20] K. L. Baluja, M. R. C. McDowell (J. Phys. B **10** [1977] L673/L676).

[21] A. Zago, G. Tondello (Nuovo Cimento B [11] **85** [1985] 59/78).
[22] S. Bashkin, J. O. Stoner Jr. (Atomic Energy Levels and Grotrian Diagrams, Vol. 1, Addenda, North Holland, Amsterdam 1978, p. 14).
[23] P. J. Mohr (Beam Foil Spectrosc. Proc. 4th Intern. Conf., Gatlinburg, Tenn., 1975 [1976], Vol. 1, pp. 89/96).

5 Molecules

5.1 The Be_2 Molecule

Occurrence. Detection. The Be_2 molecule has been prepared by deposition of Be vapor into noble gas matrices [1] and, more recently, by pulsed laser vaporization of elemental Be in a stream of He gas cooled to 77 K [2, 3]. It was identified by electronic absorption or emission bands in the UV-visible region, see below.

The equilibrium concentration of Be_2 in the vapor above solid or liquid Be is negligibly small, as was shown by direct mass spectrometric analysis of the vapor composition [4] and by calculating equilibrium constants from parameters of a van der Waals interaction [5]. It is thus assumed that the Be dimer observed in rare gas matrices is formed during condensation of the matrix [1]. Dimeric Be was once claimed to be the predominant species in the vapor just below the melting point of Be [6], but it was later shown that a peak in the mass spectrum at m/e = 18 assigned as $^9Be_2^+$ was mistaken for H_2O^+ [4].

Spectra. Molecular Constants. The products of laser vaporization from a Be surface were cooled in a stream of He at 77 K and their excitation spectrum scanned from 16000 to 32000 cm^{-1} (310 to 620 nm). Two band systems with 0-0 transitions at 21678 and 27860 cm^{-1} were observed and assigned as the $^1\Pi_u - X\,^1\Sigma_g^+$ and $^1\Sigma_u^+ - X\,^1\Sigma_g^+$ transitions of Be_2, respectively. Be_2 was identified as the carrier of the bands and their term assignments were inferred from the rotational structures of the bands: Lines of alternating intensity of odd and even J with an approximate 5:3 ratio suggest a homonuclear diatomic molecule with a nuclear spin of $^3/_2$ [2, 3]. The occurrence of a Q-branch in the lower-energy band [2], and its absence in the other band [3], together with the obvious assumption of a $^1\Sigma_g^+$ ground state, identified the upper states involved in the observed transitions [2, 3]. Furthermore, the radiative lifetime of the $^1\Pi_u$ state is one order of magnitude longer than that of the $^1\Sigma_u^+$ state, consistent with the usual observation of lower transition probabilities for $\Delta\Lambda = 1$ transitions [2]. The $^1\Sigma_u^+ \leftarrow X\,^1\Sigma_g^+$ band was also observed in absorption spectra of Be vapor species trapped in noble gas matrices. It was identified with Be_2 as the carrier by a single vibrational progression with spacing of around 580 cm^{-1} [1].

The following molecular constants of the ground state and the two observed excited states were derived from excitation and resolved emission spectra [2] (preliminary results for $X\,^1\Sigma_g^+$ and $^1\Sigma_u^+$ in [3]):

constant	symbol	unit	value for state $X\,^1\Sigma_g^+$	$A\,^1\Pi_u$	$B\,^1\Sigma_u^+$
term energy	T_e	cm^{-1}	0.0	21468	27738
wave number of 0-0 band	ν_{00}	cm^{-1}	—	21678.38	27860.3
vibrational frequency	ω_e	cm^{-1}	275.8	685.70	511.2
anharmonicity constant	$\omega_e x_e$	cm^{-1}	26.0	4.85	4.69
vibrational spacing 0-1	$\Delta G_{1/2}$	cm^{-1}	233.8	676.01	503.2
dissociation energy	D_e	cm^{-1}	790 a)	21890 b)	15620 b)
rotational constant	B_e	cm^{-1}	0.623	0.938	0.774
internuclear distance	r_e	Å	2.45	1.987	2.19
rotation-vibration constant	α_e	cm^{-1}	0.028	0.012	0.014
centrifugal distortion constant	D	10^{-6} cm^{-1}	14.8	11.0	12.5
radiative lifetime	τ	ns	—	190	15

a) Dissociation limit is Be(^{1}S) + Be(^{1}S).

b) Dissociation limit is Be(^{1}S) + Be(^{1}P°), 42565 cm^{-1} above Be(^{1}S) + Be(^{1}S), see p. 146.

Some constants of the $^1\Sigma_u^+$ state were also derived from absorption spectra of Be in Ne or Ar matrices at 4 K [1] and are given below for comparison:

	Be_2 in Ne	Be_2 in Ar
ν_{00} in cm^{-1}	28551	27670
$\Delta G_{1/2}$ in cm^{-1}	489	474
D_0 in cm^{-1}	~14360*)	~14360*)

*) Dissociation energy of free Be^2 $^1\Sigma_u^+$ molecule, estimated assuming $\nu_{00} \approx 28000\ cm^{-1}$. Dissociation limit is $Be(^1S) + Be(^1P^o)$ [1].

Earlier values of the ground state dissociation energy which are appreciably higher than recent values [2], were derived by interpolations between homonuclear diatomic molecules of the first period [7, 8]. For semiempirical estimates of quadratic, cubic, and quartic force constants from model calculations, see [9].

Bonding. Molecular Structure Calculations. The valence electron configuration of the ground state is $(2s\sigma_g)^2(2s\sigma_u)^2$, resulting in a $^1\Sigma_g^+$ ground term. In a qualitative MO picture there is no net bonding, since the $2s\sigma_g$ orbital is bonding, whereas the $2s\sigma_u$ orbital is antibonding, so that the effects of filling both orbitals cancel [1 to 3]. The question as to whether the ground state of Be_2 is bonded or nonbonded has been the object of numerous ab initio calculations, see below.

Excited states arise from the promotion of one valence electron to the bonding $2p\pi_u$ or $2p\sigma_g$ orbitals as follows [1, 2]:

$(2s\sigma_g)^2\,2s\sigma_u\,2p\pi_u$	$^1\Pi_g$, $^3\Pi_g$
$2s\sigma_g(2s\sigma_u)^2\,2p\pi_u$	$^1\Pi_u$, $^3\Pi_u$
$(2s\sigma_g)^2\,2s\sigma_u\,2p\sigma_g$	$^1\Sigma_u^+$, $^3\Sigma_u^+$
$2s\sigma_g(2s\sigma_u)^2\,2p\sigma_g$	$^1\Sigma_g^+$, $^3\Sigma_g^+$

In a simple MO picture, an Σ_u^+ state with three electrons in bonding orbitals is expected to be lower in energy than a Π_u state with only two electrons in bonding orbitals. Actually, however, the energetic ordering of the $^1\Sigma_u^+$ and $^1\Pi_u$ states as inferred from electronic spectra is just the reverse. Thus, the bonding character of the $2p\pi_u$ orbital appears to be much stronger than that of the $2s\sigma_g$ and $2p\sigma_g$ orbitals. By the same token, the as yet unobserved $^1\Pi_g$ state is likely to be the lowest excited singlet state of Be_2 [2].

The electronic structure and bonding in Be_2 have been dealt with by ab initio HF methods in numerous papers. These are not individually cited here in view of the extensive bibliographies for ab initio quantum chemical calculations available. Specifically, these are the hard-copy forms of the "Quantum Chemistry Literature Data Base" built up in Japan. The hard-copy forms have been published as a bibliography covering the period 1978 to 1980 [10], and as annual supplements to that bibliography, at present available up to 1984 [11]. Literature prior to 1980 is accessible through the bibliographies compiled by Richards and coworkers [12 to 15]. Only one quite recent reference [16] for an ab initio calculation on Be_2 is cited here for the purpose of making available to the reader a discussion and assessment of relevant earlier papers.

Heat Capacity C_p^o. For the ideal gas, C_p^o has been calculated, and presented in graphical form in the temperature interval 50 to 6000 K [17], from a preliminary account of observed spectroscopic constants [18] and theoretical estimates.

References:

[1] J. M. Brom Jr., W. D. Hewett Jr., W. Weltner Jr. (J. Chem. Phys. **62** [1975] 3122/30).
[2] V. E. Bondybey (Chem. Phys. Letters **109** [1984] 436/41).
[3] V. E. Bondybey, J. H. English (J. Chem. Phys. **80** [1984] 568/70).
[4] O. T. Nikitin, L. N. Gorokhov (Zh. Neorgan. Khim. **6** [1961] 224/5; Russ. J. Inorg. Chem. **6** [1961] 111).
[5] R. H. Ewing, A. M. Mellor (J. Chem. Phys. **53** [1970] 2983/4).
[6] V. M. Amonenko, L. N. Ryabchikov, G. F. Tikhinskii, V. A. Finkel' (Dokl. Akad. Nauk SSSR **128** [1959] 977/8; Proc. Acad. Sci. USSR Phys. Chem. Sect. **124/129** [1959] 825/6).
[7] R. T. Sanderson (J. Inorg. Nucl. Chem. **28** [1966] 1553/65).
[8] J. Drowart, R. E. Honig (J. Phys. Chem. **61** [1957] 980/5).
[9] P. Empedocles (J. Chem. Phys. **46** [1967] 4474/81).
[10] K. Ohno, K. Morokuma (Quantum Chemistry Literature Data Base. Bibliography of Ab Initio Calculations 1978 to 1980, Elsevier, Amsterdam 1982).
[11] K. Ohno, K. Morokuma (J. Mol. Struct. **91** THEOCHEM 8 [1982] 1/252, **106** THEOCHEM 15 [1983] 1/215, **119** THEOCHEM 20 [1984] 1/229, **134** THEOCHEM 27 [1985] 1/298).
[12] W. G. Richards, T. E. H. Walker, R. K. Hinkley (A Bibliography of Ab Initio Molecular Wave Functions, Clarendon, Oxford 1971).
[13] W. G. Richards, T. E. H. Walker, L. Farnell, P. R. Scott (Bibliography of Ab Initio Molecular Wave Functions: Supplement for 1970 to 1973, Clarendon, Oxford 1974).
[14] W. G. Richards, P. R. Scott, E. A. Colbourn, A. F. Marchington (Bibliography of Ab Initio Molecular Wave Functions: Supplement for 1974 to 1977, Clarendon, Oxford 1978).
[15] W. G. Richards, P. R. Scott, V. Sackwild, S. A. Robins (A Bibliography of Ab Initio Molecular Wave Functions: Supplement for 1978 to 1980, Clarendon, Oxford 1981).
[16] G. H. F. Diercksen, V. Kellö, A. J. Sadlej (Chem. Phys. **96** [1985] 59/79).
[17] D. J. Frurip, A. N. Syverud, M. W. Chase (J. Nucl. Mater. **130** [1985] 189/98).
[18] V. Bondybey (personal communication 1983 to D. J. Frurip et al. [17]).

5.2 Ions Be_2^+ and Be_2^-. Molecules Be_n, n>2, and Ions

These molecules and ions appear to have been dealt with only theoretically. As for Be_2, the reader is referred to bibliographies of quantum chemical literature, references [10, 11] and [14, 15] above. Species listed there, for which references may be found in these bibliographies, are the ions Be_2^+ and Be_2^-, the molecules Be_3 through Be_7, some larger Be_n clusters up to n=36, and the ions Be_3^- and Be_4^-.

6 Chemical Reactions

6.1 Reactions with Nonmetals

6.1.1 With Helium

The implantation of He in sintered electropolished Be (which was first implanted with an Xe backscatter marker at 1.98 MeV) is studied by irradiation with 5 keV He ions at a rate of 1×10^{16} $cm^{-2} \cdot min^{-1}$ in a vacuum of 7×10^{-4} Pa ($=5.25 \times 10^{-6}$ Torr) at 25 °C. The amount of trapped He atoms in relation to the primary ion flux is shown in Fig. 6-1, p. 189. 100% trapping is observed at a dose up to 2×10^{18} He atoms per cm^2. The decrease of trapping above this flux is indicative of blistering, which occurs at about 1.75×10^{17} He atoms per cm^2. The blister diameter increases with increasing implant dose from about 0.8 µm at 10^{18} He/cm^2 to 5.5 µm at 3×10^{18} He/cm^2. During the irradiation Be acts as a getter, being oxidized by the gaseous impurities (H_2O, CO, CO_2, O_2). The BeO layer thickness increases linearly with irradiation time (at an He dose of 3×10^{18} atoms/cm^2 the oxygen content has reached the value of $\sim 2.5 \times 10^{17}$ atoms/cm^2), Langley [1 to 3].

The relationship of the most probable blister diameter d and the blister skin thickness Δd (both in µm) for annealed Be foils which had been irradiated at room temperature with $^4He^+$ ions (5×10^{13} to 10^{14} $cm^{-2} \cdot s^{-1}$; 15 to 300 keV) was found to be $d = 24.6\ \Delta d^{1.25}$, Das et al. [10]. The effect on the surface corrosion of Be of irradiation with 100 keV $^4He^+$ ions at a flux rate of 6×10^{15} ions $\cdot cm^{-2} \cdot min^{-1}$ and a dose of 0.5 and 1 C/cm^2 in vacuum ($\sim 5 \times 10^{-8}$ Torr) has been studied at room temperature and at 600 °C. Vacuum cast hot rolled Be (sample I) has been found to be less resistant than sintered (II) or sintered and hot rolled Be (III) (all samples electropolished). Blisters with diameters ranging from about 5 to 35 µm can be observed on the room temperature irradiated sample I and in few cases the blisters have exfoliated. After irradiation at 600 °C very serious exfoliation occurs. For the samples II and III little exfoliation is observed after irradiation at 600 °C, while at room temperature the average blister size is smaller than for sample I and no exfoliation is observed, Das, Kaminsky [4]. If discs of extruded and cast Be under cooling are injected with He by bombardment with 40 MeV α-particles, the He remained "in solution". Microscopic studies show that heating to about 800 °C, either during or after the bombardment, is necessary to precipitate He as gas bubbles, which are about 4×10^{-6} cm in diameter. Further heating causes the bubbles to enlarge and to reduce in number; this is accelerated at the grain boundaries. The large difference in the grain size between the extruded and cast Be permits a lower temperature of precipitation and more rapid coarsening in the former, Barnes, Redding [5]. For studies on the nucleation and growth of He gas bubbles see also Hickman [6] and Das et al. [7], for example. According to [6] no bubbles could be observed under the optical microscope in samples heated below 700 °C, but the electron microscope showed the presence of many small pits in certain grains, presumably due to He bubbles [6].

The diffusion coefficient of He in Be is assumed to be much smaller than that of H_2. A value of the order of 2×10^{-15} cm^2/s at 700 °C is evolved by Pemsler et al. [8], using the data of [5], who reported that after an 8 h annealing at 700 °C the He must have diffused about 3×10^{-5} cm.

Values for D in cm^2/s (average of two or three values) between 600 and 900 °C:

t in °C	600	700	800	900
$D \cdot 10^{15}$ . . .	4.2	108	1035	8550

The temperature dependence can be expressed by the following equation $D = (3.1^{+2.8}_{-1.5}) \times 10^{-2}$ $\exp(-51300 \pm 1300)/RT$, Bespalov et al. [11].

For the calculation of the energy of He atom interaction (adsorption) with the Be surface by means of the quantum mechanical formalism, see Gerasimenko [9].

References:

[1] R.A. Langley (J. Nucl. Mater. B **85/86** [1979] 1123/6).
[2] R.A. Langley (CONF-790125-59 [1979] 1/4; C.A. **92** [1980] No. 30282).
[3] R.A. Langley (CONF-781053-12 [1978] 1/4; C.A. **91** [1979] No. 147654).
[4] S.K. Das, M. Kaminsky (CONF-760209 [1976] 1/17; C.A. **85** [1976] No. 181165; J. Nucl. Mater. **63** [1976] 292/8; Proc. Symp. Eng. Probl. Fusion Res. **6** [1975] 1151/3).
[5] R.S. Barnes, G.B. Redding (J. Nucl. Energy A **10** [1959] 32/5).
[6] B.S. Hickman (J. Australasian Inst. Metals **5** [1960] 173/81).
[7] S.K. Das, M. Kaminsky, G. Fenske (J. Nucl. Mater. **76/77** [1978] 215/20).
[8] J.P. Pemsler, R.W. Anderson, E.J. Rapperport (ASD-TDR-62-1018 [1963] 1/25, 7; N.S.A. **17** [1963] No. 22157).
[9] V.I. Gerasimenko (Fiz. Tverd. Tela [Leningrad] **19** [1977] 2862/8; Soviet Phys.-Solid State **19** [1977] 1677/81).
[10] S.K. Das, M. Kaminsky, G. Fenske (CONF-780467-4 [1978] 1/18; 3rd Conf. Plasma Surf. Interact. Control. Fusion Devices, Culham, U.K., 1978, Paper No. 4, pp. 1/18).
[11] A.G. Bespalov, V.N. Bykov, L.V. Pavlinov, Yu.V. Shumov (Tr. Fiz. Energ. Inst. **1974** 443/8).

6.1.2 With Hydrogen

6.1.2.1 Adsorption and Chemisorption

Molecular Hydrogen. Measurement of the sticking coefficient shows that H_2 is not adsorbed by a thick Be film (vapor deposited on a glass substrate), Hurd, Adams [1]. The (0001) surface of a Be single crystal does not adsorb H_2, as shown by LEED studies, Adams [2]. The quantities of H_2 taken up by sintered Be with noticeable porosity inside (no data regarding dimensions of specimens!) when subjected to H_2 (300 to 760 Torr) between 500 and 750 °C are measured by desorption isobars. Depending on the sample, the adsorbed quantity is about 0.06 to 0.3 mL H_2/100 g Be (NTP). Samples with the highest BeO content absorb the most H_2 (BeO can take up significant quantities of H_2, as 3.3 mL/100 g, Blanchard, Bochirol [3]. A surface adsorption of about 0.007 mL/cm^2 was found by others (see p. 189). It was concluded from results of ab initio SCF (=self-consistent-field) calculations of the interaction of H_2 with Be_3 and Be_4 clusters for four possible adsorption sites on the (0001) plane that no adsorption occurs; the Be surface strongly repels H_2 in all sites, Garcia-Prieto, Novaro [4]. This result contradicts calculations by Lavery, Hillier [5] (using a semi-empirical tight-binding crystal orbital method) which result in binding energies between 2.3 and 3 eV for chemisorbed H_2 on the Be (100) face; but the results of [4] confirm the assumptions of Bauschlicher et al. [6]; these authors expect that the dissociative adsorption of H_2 on Be (0001) will be exothermic.

Atomic Hydrogen. The chemisorption of atomic hydrogen on the (0001) surface of Be has been studied theoretically by using clusters of up to 22 Be atoms to simulate the substrate. Ab initio molecular orbital Hartree-Fock (SCF) wave functions have been obtained and the interaction energy of H with the Be clusters for four high symmetry adsorption sites is studied as a function of vertical distance from the surface. For the smaller clusters both, minimum ("MB") and extended (double zeta, "DZ") Gaussian basis sets are used for the SCF calculation. For the larger clusters only "MB" sets could be used. Three of the adsorption sites considered are found to have similar binding energies of about

E ≈ 50 kcal/mol (≈ 2.1×10^5 J/mol) and (vertical) equilibrium distances from the surface of r ≈ 0.1 nm. For the fourth site, H directly over a Be atom, the corresponding values are E ≈ 30 kcal/mol (≈ 1.3×10^5 J/mol) and r ≈ 0.14 nm. The bonding of H to Be (0001) is covalent, its nature is analyzed. The vibrational energies for the motion of H atoms normal to the surface are found to be substantially different for sites with different surface coordination, Bauschlicher et al. [6]. For earlier studies see [7 to 11] and for a summarized review [6]. Calculations for Be_nH clusters with n = 36 instead of 22 give changes in the adsorption energies of about 10%, while for clusters with n = 10 or 13 instead of 36 the change in E is about 30%, Cox, Bauschlicher [12]. The calculations are improved using a mixed basis set, consisting of a "DZ" basis set for the H and Be atoms directly involved in the bonding and a "MB" set for all others (Be_nH clusters with n = 6, 7, and 10 are considered). The mixed basis calculations are only slightly more difficult than the all "MB" calculations; however, the results are found to be very similar to the more difficult all "DZ" calculations, Bauschlicher, Bagus [13]. SCF-molecular orbital cluster calculations have also been conducted to study some of the substrate relaxations that would accompany the adsorption of H. It has been shown that H atoms adsorbed on the (0001) surface of Be causes a marked pinching together of the metal atoms nearest to it. Surface bond lengths are effectively shortened by 1 or 2% and large nonuniform tensile stresses have been shown to be induced in the surface layer if the surface fractionally covered by hydrogen, Cox, Bauschlicher [12].

Three semiempirical molecular orbital theories were used by Hoflund, Merrill [14] to study the chemisorption of atomic H on the (0001) surface of Be: the extended Hückel theory [15], Anderson's modification of it [16], and the complete neglect of differential overlap method (CNDO) [17]. The maximum cluster size used with each method is 5, 73, and 10 atoms, respectively. The CNDO method predicts binding energies which are too large. When these energies are scaled by dividing by a constant for all sites, agreement is found with all other molecular orbital methods. General agreement also is found with results of the ab initio calculations (see above [6]) and of CNDO with bicentric energy rescaling, Companion [18]. With a modified CNDO/2 molecular orbital approach [19], potential energy surfaces are computed for the attack of Be clusters simulating "smooth" (0001) and "corrugated" (1010) faces of Be metal. Several stable sites for chemisorption are found with binding energies of 40 to 55 kcal/mol, but penetration of the lattice appears possible at some points [18].

The chemisorption of an H atom on metal surfaces (Be et al.) has also been treated theoretically by using a jellium model for the metal substrate instead of clusters, Wang, Weinberg [20], see [6, p. 227] for a discussion. In a similar manner the chemisorption of a proton on metal (jellium) surfaces is theoretically treated by Wang, Weinberg [21].

References:

[1] J.T. Hurd, R.O. Adams (J. Vac. Sci. Technol. **6** [1969] 229/33).
[2] R.O. Adams (Struct. Chem. Solid Surf. No. 35 [1969] 70.1/70.9; C.A. **75** [1971] No. 134140).
[3] R. Blanchard, L. Bochirol (AEC-TR-6586 [1965] 1/6; C.A. **63** [1965] 12790).
[4] J. Garcia-Prieto, O. Novaro (J. Chem. Phys. **71** [1979] 3137/8).
[5] R. Lavery, I.H. Hillier (Chem. Phys. **16** [1976] 281/6).
[6] C.W. Bauschlicher, P.S. Bagus, H.F. Schaefer (IBM J. Res. Develop. **22** [1978] 213/34; C.A. **89** [1978] No. 49348).
[7] H.F. Schaefer (Accounts Chem. Res. **10** [1977] 287/93).
[8] P.S. Bagus, C.W. Bauschlicher (Proc. 7th Intern. Vacuum Congr., Vienna 1977, Vol. 2, pp. 989/91; C.A. **88** [1978] No. 79608).

[9] C.W. Bauschlicher, C.F. Bender, H.F. Schaefer (Chem. Phys. **15** [1976] 227/35).
[10] C.W. Bauschlicher, D.H. Liskow, C.F. Bender, H.F. Schaefer (J. Chem. Phys. **62** [1975] 4815/25).
[11] C.W. Bauschlicher (Diss. Univ. California 1976, pp. 1/127; C.A. **87** [1977] No. 141715; Diss. Abstr. Intern. B **38** [1979] 705).
[12] B.N. Cox, C.W. Bauschlicher (Surf. Sci. **102** [1981] 295/311).
[13] C.W. Bauschlicher, P.S. Bagus (Chem. Phys. Letters **90** [1982] 355/8).
[14] G.B. Hoflund, R.P. Merrill (J. Phys. Chem. **85** [1981] 2037/41).
[15] R. Hoffman (J. Chem. Phys. **39** [1963] 1397/412).
[16] A.B. Anderson, R. Hoffman (J. Chem. Phys. **60** [1974] 4271/3).
[17] J.A. Pople, D.L. Beveridge (Approximate Molecular Orbital Theory, McGraw Hill, New York 1970).
[18] A.L. Companion (Chem. Phys. **14** [1976] 7/11).
[19] A.L. Companion (J. Phys. Chem. **77** [1973] 3085/8).
[20] S.W. Wang, W.H. Weinberg (Surf. Sci. **77** [1978] 14/28).
[21] S.W. Wang, W.H. Weinberg (Surf. Sci. **77** [1978] 29/39).

6.1.2.2 Permeation, Diffusion, Solubility, Implantation

The permeability P of H_2 (300 Torr pressure) through Be tubes (hot extruded material of 98.4 wt% purity, density 1.86 g/cm^3) with a wall thickness of 0.5 mm, measured between 500 and 650 °C, increases with the temperature T. From the linear plot log P vs. 1/T (see the paper), an activation energy of $E_A = 10.6$ kcal/mol is derived. The log P values (lying between about −2.9 and −2.3 with P in $cm^3 \cdot cm^{-2} \cdot h^{-1}$) are mean values from 3 to 4 measurements with a maximum deviation from the mean value of 20%. The permeability of H_2 (1 atm pressure) between 500 and 700 °C is lower by one to two orders of magnitude than for Fe and stainless steel, still considerably lower than for Mo, at the same order of magnitude as for Al and higher than for W or SiO_2. For hot pressed Be discs (density 1.85 g/cm^3) P and E_A are higher than for the extruded tubes. Measurements at H_2 pressures between 3 and 21 atm at 500 to 600 °C show for the hot pressed discs that P decreases with decreasing grain size (30 to 600 μm) of the starting powder, Al'tovskii et al. [1]. Measurements at 20 °C at $p(H_2) = 4$ bar give $P = 1.3 \times 10^{-5}\ cm^3 \cdot cm^{-2} \cdot h^{-1}$ (from 31.2 $mm^3 \cdot dm^{-2} \cdot d^{-1}$) for a 0.1 mm thick Be foil. For a 4 mm thick sintered Be disc at $p(H_2) = 80$ bar, P increases between 20 and 100 °C from 0.21×10^{-6} to $5.96 \times 10^{-6}\ cm^3 \cdot cm^{-2} \cdot h^{-1}$ (the last value is a mean value from two samples). Between 20 and 130 °C nonlinear log P vs. 1/T curves are obtained. (For He at 20 °C and 80 bar pressure P was found to be about 60% higher than for H_2, Fidelle et al. [2].)

Tritium was used to study the diffusion and solubility of hydrogen in arc-cast Be discs between 200 and 950 °C. The following equations were obtained for the diffusion coefficient D (in cm^2/s) and the solubility S (in $mL \cdot g^{-1} \cdot atm^{-1/2}$, NTP) in dependence of temperature T:

$$D = 3.0 \times 10^{-7} \exp(-4430/RT)$$
$$S = 7.9 \times 10^{-3} \exp[(0.44 \pm 0.97)/RT]$$

The diffusion coefficient is on the order of about 10^{-9} to 5×10^{-8} cm^2/s, and the diffusion activation energy is 4.4 ± 0.5 kcal/mol, Jones, Gibson [3, 4]. Pemsler et al. [5] have attempted to determine the diffusion coefficient of hydrogen in Be using a 7.5 MeV proton beam to load the material which was subsequently sectioned after heat treatment to determine the distribution and size of gas bubbles microscopically. The authors concluded that, to a first approximation, D is 9×10^{-10} cm^2/s at about 800 to 900 °C and assumed a more rapid diffusion of hydrogen along the grain boundaries than through the lattice (see also below and [11]). In a later paper the authors give the approximated value $D = 3 \times 10^{-9}$ cm²/s at 850 to 900 °C, Pemsler, Rapperport [12].

Attempts to determine the solubility of hydrogen in Be are reported by Cotterill et al. [6] and Evans, Herrington [7], and resulted in the conclusion that hydrogen is virtually insoluble in solid Be. The hydrogen liberated from Be during vacuum outgassing measurements was attributed entirely to a surface adsorption effect involving 0.007 mL/cm² (NTP) [6]. This conclusion was supported by [7], who used a (tritium) isotope dilution technique (850 °C), and vacuum extraction (900 °C), and also found a surface adsorption of ≈0.007 mL/cm².

As described for He (see p. 185), deuterium was implanted in Be by irradiation at a rate of 3×10^{16} atoms · cm^{-2} · min^{-1}. The amount of trapped D atoms in dependence of the primary ion flux is shown in **Fig. 6-1**. A trapping coefficient of 100% is observed up to a dose of 2×10^{18} D/cm² and saturation occurs at 5×10^{18} D/cm². In contrast to the implantation with He, no blisters are formed, Langley [8 to 10].

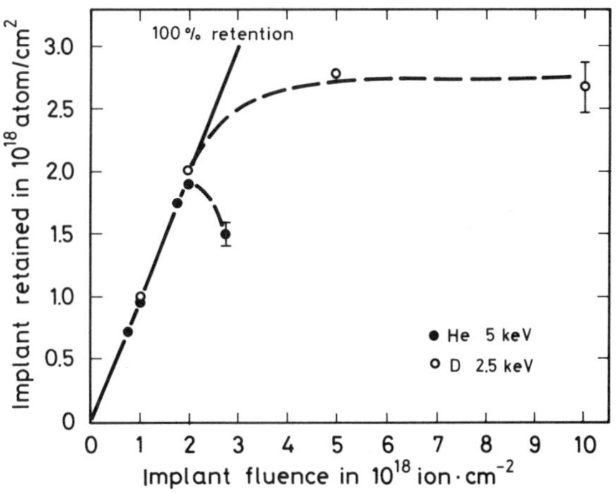

Fig. 6-1

Amount of deuterium and Helium trapped in the surface layer of Be as a function of the implant fluence.

By irradiation of Be (hot pressed or cast) with protons (≈7 MeV) in the cyclotron (at ≦50 °C), the proton beam penetrates the Be to a depth of 0.4 mm and the H-containing layer that is formed is ≈0.04 mm thick [5, 11]. Metallographic studies of samples heated after irradiation show agglomerates of hydrogen after heating for 1 h as low as 320 °C. Grain boundaries are highly preferred points of nucleation for the agglomerates and after a sufficiently severe heat treatment, e.g., 1 h at 1000 °C, extensive grain boundary cracking was observed in most of samples containing 0.5 mL hydrogen per cm³ Be, Ells, Evans [11]. Agglomeration at grain boundaries was stated by [5], who observed at 900 °C evidence of depletion of hydrogen due to dispersal from both the matrix and grain boundaries; after 1 h at 1050 °C no hydrogen agglomerates were visible in the sample.

References:

[1] R.M. Al'tovskii, A.A. Eremin, L.F. Eremina, et al. (Izv. Akad. Nauk SSSR Metally **1981** No. 3, pp. 73/7; Russ. Met. **1981** No. 3, pp. 51/5).

[2] J.P. Fidelle, L.R. Allemand, M. Rapin, B. Hocheid (Hydrogene Metaux Colloq., Valduc, Fr., 1967 [1969], pp. 209/41, 216, 230; C.A. **72** [1970] No. 82202).

[3] P.M.S. Jones, R. Gibson (J. Nucl. Mater. **21** [1967] 353/4).
[4] P.M.S. Jones, R. Gibson (AWRE-0-2-67 [1967] 1/26; C.A. **67** [1967] No. 15122).
[5] J.P. Pemsler, R.W. Anderson, E.J. Rapperport (ASD-TDR-62-1018 [1962] 5/7; N.S.A. **17** [1963] No. 22157).
[6] P. Cotterill, R.E. Goosey, A.J. Martin (Monogr. Rept. Ser. Inst. Metals London No. 28 [1961] 221/36, 232/3).
[7] C. Evans, J. Herrington (Radioisotop. Phys. Sci. Ind. Proc. Conf., Copenhagen 1960 [1962], Vol. 2, pp. 309/16, 315; C.A. **57** [1962] 5295).
[8] R.A. Langley (J. Nucl. Mater. B **85/86** [1979] 1123/6).
[9] R.A. Langley (CONF-790125-59 [1979] 1/4; C.A. **92** [1980] No. 30282).
[10] R.A. Langley (SAND-79 [1978] 1/4; CONF-781053-12 [1978] 1/4; C.A. **91** [1979] No. 147654).

[11] C.E. Ells, W. Evans (AECL-1347 [1961] 1/11; N.S.A. **16** [1962] No. 2171).
[12] J. Pemsler, E.J. Rapperport (Trans. AIME **230** [1964] 90/4).

6.1.2.3 General Reactions

For a nonempirical calculation of the potential surface of the exchange reaction of the Be atom (in the ground state) with hydrogen, $Be(g)+H_2 \rightarrow BeH+H$, see Murav'eva et al. [1]. For the reaction of $Be(g)+H_2 \rightarrow BeH_2$ the perpendicular (C_{2v}) insertion of Be into H_2 was calculated with the multiconfiguration coupled-cluster method within the double-excitation model, Banerjee, Simons [17]. A multireference many-body perturbation theory was discussed and illustrated by the model problem of the vertical insertion pathways of Be in H_2, Lee, Bartlett [18]. Further calculations for this model reaction see Page et al. [19]. A molecular orbital approach has been made and correlation diagrams have been constructed for the reactions $Be(g)+H_2=BeH_2$ and $Be(g)+0.5\ H_2=BeH$. It was concluded that the activation energy for the first reaction is high and LCAO-SCF calculations indicated that BeH_2 is unstable in the triangular form and metastable in the linear form, Griffing et al. [13], see also [14]. For the computer calculation of the activation energy and the rate constant (using a modified Johnston-Parr method) for gas-phase hydrogen-atom transfer reactions as $Be+H_2=BeH+H$ (and others with diatomic hydrogen compounds instead of H_2) see Mayer, Schieler [15]. Pulsed field evaporation of Be in presence of H_2 at 85 to 100 K shows the presence of hydride ions BeH^+ and BeH_3^+ (besides H^+, H_2^+, Be^+, and Be^{2+}), Krishnaswamy, Mueller [2]. By simultaneous quenching of thermally produced Be and H atom beams in argon on a sapphire surface at liquid He temperatures, BeH is formed, which has been studied by ESR investigation, Knight et al. [12]. The charge-transfer process $Be^{2+}+H \rightarrow Be^{+}+H^{+}$ is investigated for the transition probability arising from the pseudo-crossing of the potential-energy curves of the initial and final systems, Bates, Moisewitsch [16].

The equilibrium constant K of the reaction $Be(s)+H_2(g)=BeH_2(g)$ is calculated to be $\log K=-22\pm7$ at 298 K and -3.5 ± 1.3 at 1500 K, Lao, Riter [3].

No appreciable reaction is noted by heating thin plates of Be (from sintered bars) in H_2 at a pressure of 23 Torr over the temperature range 300 to 780 °C (above 780 °C the evaporation of Be becomes appreciable), Gulbransen, Andrew [4].

No reaction between powdered Be and H_2 at pressures of about 100 to 190 atm was observed between 200 and 530 °C; even when catalytic additives (such as Hg, iodine, $BeCl_2$) are used, no BeH_2 was formed, Powers et al. [5]. There was no evidence of the formation of BeH_2 at 600 °C and H_2 pressures up to 1000 atm starting with Be powder (1 to 50 μm particle size), Holley, Lemons [11]. No reaction occurs between Be and H_2 at 1200 or 1500 °C in the pressure range 1 to 500 Torr, Blumenthal, Santy [6, 7], or between Be vapor and H_2 at 1700 °C and 1 to 10 Torr [11]. The pyrolytic decomposition temperature of BeH_2 (prepared by other methods than from the elements) is reported to be as low as 125 °C by Barbaras et al. [8] or at 240 to 300 °C by Coates, Glockling [9].

The white reaction product, which was formed in small amounts by the reaction of powdered Be with atomic hydrogen for 7 h between 170 and 260 °C, was believed to be beryllium hydride, Pietsch [10], but this was not confirmed by Holley, Lemons [11].

References:

[1] N.N. Murav'eva, V.I. Baranovskii, A.I. Panin (Zh. Strukt. Khim. **22** [1981] 13/7; J. Struct. Chem. [USSR] **22** [1981] 8/12).
[2] S.V. Krishnaswamy, E.W. Mueller (Z. Physik. Chem. [N.F.] **104** [1977] 121/30).
[3] R.C.C. Lao, J.R. Riter (J. Phys. Chem. **71** [1967] 2737/9).
[4] E.A. Gulbransen, K.F. Andrew (J. Electrochem. Soc. **97** [1960] 383/95).
[5] J.C. Powers, D.W. Vose, E.A. Sullivan (PB-171489 [1960] 1/34, 16/22; C.A. **58** [1963] 219).
[6] J.L. Blumenthal, M.J. Santy (Symp. Intern. Combust. Proc. **11** [1966/67] 417/25; C.A. **67** [1967] No. 110222).
[7] J.L. Blumenthal, M.J. Santy (WSS-CI-65-5 [1965] 1/56, 6; C.A. **63** [1965] 7684).
[8] G.D. Barbaras, C. Dillard, A.E. Finholt, et al. (J. Am. Chem. Soc. **73** [1951] 4585/90).
[9] G.E. Coates, F. Glockling (J. Chem. Soc. **1954** 2526/9).
[10] E. Pietsch (Z. Elektrochem. **39** [1933] 577/86, 582).
[11] C.E. Holley, J.F. Lemons (LA-1660 [1954] 1/72, 44/56; N.S.A. **11** [1957] No. 9329).
[12] L.B. Knight, J.M. Brom, W. Weltner (J. Chem. Phys. **56** [1972] 1152/5).
[13] V. Griffing, J.P. Hoare, J.T. Vanderslice (J. Chem. Phys. **24** [1956] 71/6).
[14] V. Griffing, J.S. Dooling (J. Phys. Chem. **61** [1957] 11/9).
[15] S.W. Mayer, L. Schieler (J. Phys. Chem. **72** [1968] 236/40).
[16] D.R. Bates, B.L. Moisewitsch (Proc. Phys. Soc. [London] A **67** [1954] 805/12).
[17] A. Banerjee, J. Simons (Chem. Phys. **81** [1983] 297/302).
[18] Y.S. Lee, R.J. Bartlett (Intern. J. Quantum Chem. Quantum Chem. Symp. No. 17 [1983] 347/56).
[19] M. Page, P. Saxe, G.F. Adams, B.H. Lengsfield (J. Chem. Phys. **81** [1984] 434/9).

6.1.3 With Oxygen and Air

Since the corrosion behavior of Be in oxygen and air depends on many factors (purity, fabrication technique, and surface treatment of Be, traces of H_2O vapor in the gases, test time), the results of different authors and the reproducibility of oxidation rate studies show little agreement.

For review see, e.g., W.B. Jepson (in: H.P. Godard, W.B. Jepson, M.R. Bothwell, R.L. Kane, The Corrosion of Light Metals, Wiley, New York 1967, pp. 1/372, 219/52, 226/32) and G.E. Darwin, J.H. Buddery (Beryllium: Metallurgy of Rarer Metals No. 7, Academic, New York 1960, pp. 1/392, 236/8, 251/2).

6.1.3.1 Adsorption and Solubility of Oxygen

LEED (low energy electron diffraction) studies indicate that O_2 is adsorbed on the (0001) face of Be single crystals, but does not change the surface structure of Be, Adams [1]. The total coverage and the sticking coefficient of O_2 on thick Be films (glass substrate) were measured at room temperature; a coverage of 8.4×10^{14} molecules of O_2 per cm^2 and an initial sticking probability of 0.4 were obtained. The chemisorbed oxygen probably forms a monolayer, Hurd, Adams [2]. From secondary ion emission of a polycrystalline Be surface in the presence of O_2 it was stated that for a submonolayer coverage of oxygen, the adsorption follows site-exclusion statistics. The decrease of the sticking coefficient with increasing coverage Θ (from 0 to 1 monolayer) measured by Krauss, Gruen [6] (using the value of 0.4 at $\Theta = 0$ taken from [2]) is in good accordance with [2]. The sticking coefficient for the O_2 adsorption on Be(0001) was found to be 0.010 at $\Theta = 1$ to <0.5 monolayers, <0.005 at Θ about 1 to 3 and <0.0005 at $\Theta > 3$. At 700 K the initial value is 0.025 (see the curves for 300 and 700 K up to 6 monolayers in the paper), Fowler, Blakely [5]. For an estimation of the chemical affinity of the chemisorption of oxygen on Be in comparison with other metals see Lange, Rädlein [3]. The solubility of oxygen in Be is believed to be between 20 and 200 ppm. It was not feasible to obtain data regarding the solubility and diffusion of oxygen in Be, Pemsler [4].

References:

[1] R.O. Adams (Struct. Chem. Solid Surf. No. 35 [1969] 70.1/70.9; C.A. **75** [1971] No. 134140; RFP-1148 [1969] 70.1/70.9; N.S.A. **25** [1971] No. 29894).
[2] J.T. Hurd, R.O. Adams (J. Vac. Sci. Technol. **6** [1969] 229/33).
[3] E. Lange, G. Rädlein (Z. Elektrochem. **61** [1957] 724/6).
[4] J.P. Pemsler (ASD-TDR-62-1018 [1963] 7/8).
[5] D.E. Fowler, J.M. Blakely (Surf. Sci. **148** [1984] 265/82).
[6] A.R. Krauss, D.M. Gruen (Surf. Sci. **90** [1979] 564/78).

6.1.3.2 Reactions with Oxygen

For the ignition and combustion of Be in O_2 and in O_2 containing gases and flames, see p. 204.

With Dry Oxygen below Atmospheric Pressure

The kinetics of the initial stages of the oxidation (up to about 10 monolayers BeO) of Be single crystals at very low O_2 pressure (2×10^{-6} to 3×10^{-8} Torr) between 300 and 770 K was studied by AES (Auger electron spectroscopy), LEED, and ELS (energy loss spectroscopy). The results are shown in several graphs (see the paper). There is evidence that bulk-like BeO nuclei form at an average oxide coverage of <1 monolayer, Fowler, Blakely [23] (for epitaxial growth at increasing oxide thickness see p. 203), see also [24] for earlier AES studies at 300 K.

Electron diffraction studies of Be after exposure to O_2 at 1 Torr pressure for 1 h at 300, 400, 500, 600, or 700 °C show the presence of BeO. The structure of the thin film is identical with that of bulk BeO as determined by X-ray diffraction, Hickman [1]. Weight gain-time curves for the reaction of polished Be plates (prepared from sintered bars containing 1.34% BeO) with dry O_2 at 76 Torr up to 2 h between 350 and 950 °C are shown in **Fig.** 6-**2** and up to 55 h at 500 °C in **Fig.** 6-**3**. Weight gain Δw, thickness d and color of the formed BeO films after a 2 h exposure at the temperature t are shown in the following table:

t in °C	350	600	700	750	800	825	900	950	800
Δw in $\mu g/cm^2$	1.65	4.81	10.78	21.8	28.2	43.7	135	177	77.6
d in Å	85	249	557	1128	1510	2260	6980	9150	4010
color	no	no	straw to blue	blue	straw	straw	violet-gray	gray	straw to violet

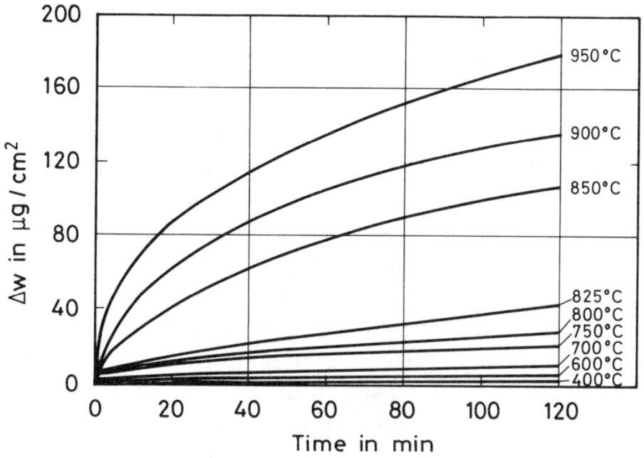

Fig. 6-2

Weight gain Δw of Be in O_2 (76 Torr) vs. time at various temperatures.

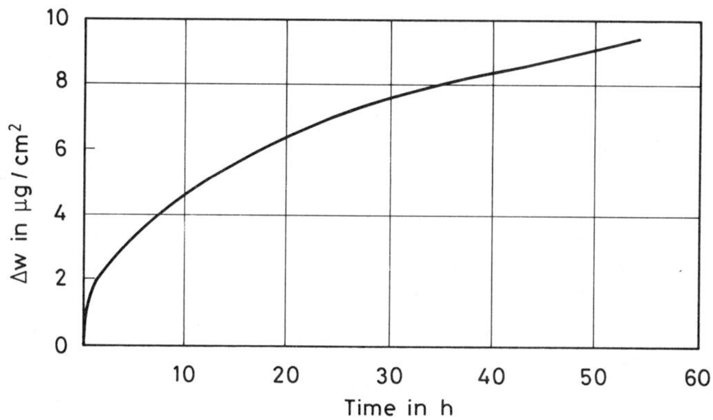

Fig. 6-3

Weight gain Δw of Be in O_2 (76 Torr) vs. time at 500 °C.

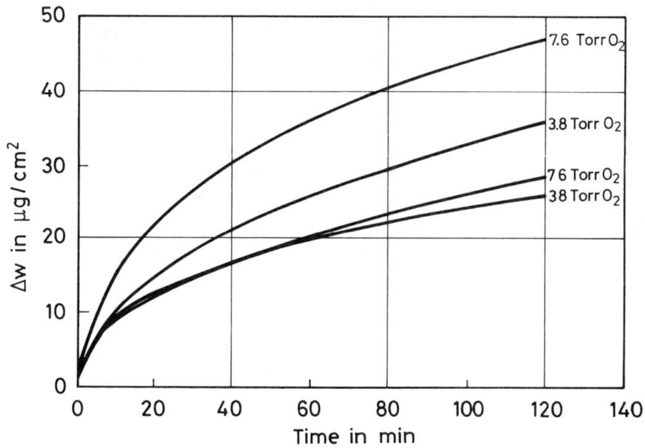

Fig. 6-4

Weight gain Δw of Be vs. time at 800 °C and various O_2 pressures.

Since, after a (very short) initial period, the oxidation rate becomes parabolic with time, the oxidation is protective. The activation energy was found to be 8.5 kcal/mol between 350 and 625 °C, and 50.3 kcal/mol between 625 and 950 °C. No simple relationship for the pressure effect studied at 800 °C at O_2 pressures between 3.8 and 76 Torr was obtained, see **Fig.** 6-**4**, p. 193. The rate of reaction at 750 °C, $p(O_2)=76$ Torr, was lower for specimens which had been pretreated in a high vacuum at 900 °C for several hours before the runs, to remove volatile impurities, Gulbransen, Andrew [2]. For a discussion of the oxidation mechanism with respect to the negative entropies of activation (−51.8 and −6.6 $cal \cdot mol^{-1} \cdot K^{-1}$ for below and above 625 °C, respectively, see above), see Gulbransen [3].

Short time studies of the oxidation of mechanically polished (reactor grade) Be sheet at 840 to 970 °C at $p(O_2)=100$ Torr showed the reaction rate to be parabolic over at least the first 100 min (pressure measurements). The obtained activation energy of 62 kcal/mol was believed to correspond to a diffusion controlled rate. Variation of O_2 pressure between 100 and 720 Torr produced no change in the reaction rate at 885 °C, Cubicciotti [4].

Measurements of the weight gain Δw of various Be samples in O_2 (P_2O_5 dried) at 76 Torr between 500 and 900 °C up to several days, showed initially protective oxidation with a parabolic rate law followed by catastrophic oxidation after a critical weight gain of 0.05 to 0.1 mg/cm^2. The activation energy for the parabolic period was 28.9 kcal/mol for Be powder and 30.7 kcal/mol for mechanically polished Be over the whole temperature range studied. The amount of time required for the accelerated rate to begin depended on the temperature and on the microstructure and purity of the Be. The duration of the initial parabolic period (t_i) decreased with increasing temperature and was shorter for small single crystals and powders than for polished specimens. At 700 to 750 °C, t_i was 7 h for small single crystals (vacuum distilled Be), 30 h for powder (with 3000 to 5000 cm^2/g surface area from hot pressed Be), and 150 h for polished (hot pressed or vacuum cast) Be. For these three cases, t_i was 3, 3, and 70 h, respectively, at 800 to 850 °C and 1, 0, and 30 h at 900 to 950 °C. The pressure dependence of the rate constant, studied at 750 °C and $p=0.06$, 0.20, and 40 Torr O_2 pressure, was found to vary with $p^{1/4.6}$, so that the weight gain Δw (in mg/cm^2) of Be powder and polished Be for a given time t (in min) in dependence of the temperature T (in K) and O_2 pressure p (in atm) for the parabolic period can be expressed by the equations $\log(\Delta w^2/t) = -0.6 - 28900/4.59\,T + 1/4.6 \cdot \log p$ and $\log(\Delta w^2/t) = 0.3 - 30700/4.59\,T + 1/4.6 \cdot \log p$. Microscopic observations of oxidized surfaces showed that pits developed in the metal under the thin adherent oxide layer during the initial protective period, Ervin, Mackay [5, 6].

By observation of polished Be during oxidation under a hot-stage microscope it was found that pitting began within a few minutes at ≧600 °C (even at 500 °C pitting was observed) and that the pits increased in size rapidly at first, but more slowly as time went on. As the pit size increased, the metal surface was disrupted and accelerated oxidation occurred resulting in the formation of a mixture of metal and oxide powder. Pitting and "catastrophic oxidation" are attributed to dislocations in the metal which cause the oxidation of Be to appear as a structure-sensitive property. This provides a qualitative explanation for the observed variations in the time of onset of the accelerating oxidation from sample to sample; that is the dislocation structure depends on sample history. The film thickness, calculated from weight gain, is not substantially thicker over the pits than on the smooth metal. After exposure of polished Be in O_2 at 76 Torr and 750 °C for 0.1, 0.8, 8.5, and 80 h, the thickness of the BeO film was found to be 0.10, 0.21, 0.32, and 0.45 μm, respectively. The values agreed very well with those calculated from interference colors, Ervin, Mackay [7]. The thickness of the BeO layer after exposure of Be to an O_2-He mixture with $p(O_2)=8$ Torr and $p(He)=752$ Torr at 650 °C for 289 h was measured by a radioactive tracer method to be 0.3 μm, Condit, Holt [8].

Contrary to [2, 5], Aylmore et al. [9] did not find a parabolic rate law for the reaction of hot rolled, polished Be sheet (prepared from electrolytic flake Be) with dry O_2. The study was carried out between 500 and 750 °C at $p(O_2)=100$ Torr using a vacuum microbalance. For test times up to 300 h at ≦650 °C, the oxidation was protective; the rate decreased continuously with time, reaching a value of 0.02 to 0.04 $\mu g \cdot cm^{-2} \cdot h^{-1}$ after 300 h (total weight gain 20 to 60 $\mu g/cm^2$ after 300 h). At 750 °C, the rate first decreased, but then increased with time, indicating a "breakaway reaction" and subsequent non-protective oxidation. In one run, a weight gain of 200 $\mu g/cm^2$ was reached after 50 h. In all the tests, the samples were outgassed at the reaction temperatures in vacuum before the O_2 was admitted. The curves of weight gain against time show small discontinuities which were ascribed to cracking and heating of the oxide film. Most of the samples exhibit the following interference colors after 300 h oxidation: At ≦550 °C none, but loss of lustre; at 600 °C blue and green; at 650 °C straw; at 700 °C yellow and purple; at 750 °C dark gray and various colors. The colors varied over the sample surfaces, indicating oxide layers of nonuniform thickness. Samples oxidized at 700 °C and particularly at 750 °C, frequently bent during the course of the run again indicating non-uniform oxidation. Microscopic examination of such samples revealed severe blistering of the oxide surface, Aylmore et al. [9], see also Jepson [10]. The oxidation of pressed Be specimens (from commercial powder of 60 μm particle size and 99.4% purity) studied by gravimetric and volumetric methods at $p(O_2)=10$ to 150 Torr between 900 and 1100 °C was protective within the first 6 to 8 h at 900 °C, 2 to 3 h at 950 °C, 1 h at 1000 °C, and a few minutes at 1100 °C. In this protective range, the value of n in the weight gain vs. time relation $\Delta w^n = k \cdot t$ varied from $n=2$ to $n=1.1$ to 1.3 (at the end of this period), and the O_2 pressure had practically no influence on the reaction rate and duration of the protective period. When the film thickness reached 1 to 2 μm, the reaction rate becames linear ($n=1$) and pressure dependent up to 76 Torr. Values for the dependence of the linear rate constant k at 900 °C on the O_2 pressure p are given in the following table:

$p(O_2)$ in Torr	10	38	76	120	150
k in 10^{-7} $g \cdot cm^{-2} \cdot s^{-1}$	0.37	0.5	1.56	1.56	1.66

Ivanov et al. [11].

High temperature studies of the reaction between polished Be sheet (purity 98.9 and 98.2 wt%) and dry O_2 at $p(O_2)=0.1$ and 0.2 atm (or dry air at 0.5 atm pressure) between 1110 and 1210 °C showed a parabolic rate law for the initial phase of the reaction. The reaction rate became approximately linear after a weight gain of 0.41 mg/cm^2 to 0.56 mg/cm^2 (≈21200 to 28950 Å thickness) which was reached in O_2 after 71.5 min at 1175 °C or 66 min at 1210 °C and in air after 73 min at 1190 °C. The total weight gain, in O_2, after about 4 h at 1210 °C was 1.93 mg/cm^2. Values for the linear and parabolic rate constants k(l) and k(p) at 1175 and 1210 °C for the reaction in O_2 are: $k(l)=6.4$ and 8 $\mu g \cdot cm^{-2} \cdot min^{-1}$, respectively, $k(p)=2.3(\pm 0.7)\times 10^{-9}$ and $4.8(\pm 1.0)\times 10^{-9}$ $g^2 \cdot cm^{-4} \cdot min^{-1}$, respectively. Varying the oxygen pressure between 0.1 and 0.2 atm had no effect on the reaction rate. The resulting BeO layer seemed to adhere so strongly to the metal that, at temperatures near the softening point of Be, it buckled without the formation of blisters, Nakata [12]. Short time oxidation studies (pressure measurements) of reactor grade Be rods (98.3% purity), from 900 °C to temperatures above the melting point of Be at $p(O_2)$ between 0.5 and 700 Torr showed that below 1050 °C, a slow protective oxidation occurs which is relatively independent of O_2 pressure. At 1050 °C, however, at $p(O_2)=500$ to 700 Torr, the oxide coating broke down after exposure to O_2 for only a few minutes. At ≧1050 °C catastrophic oxidation occurred which is characterized by a linear oxidation rate, and a nearly first

Fig. 6-5

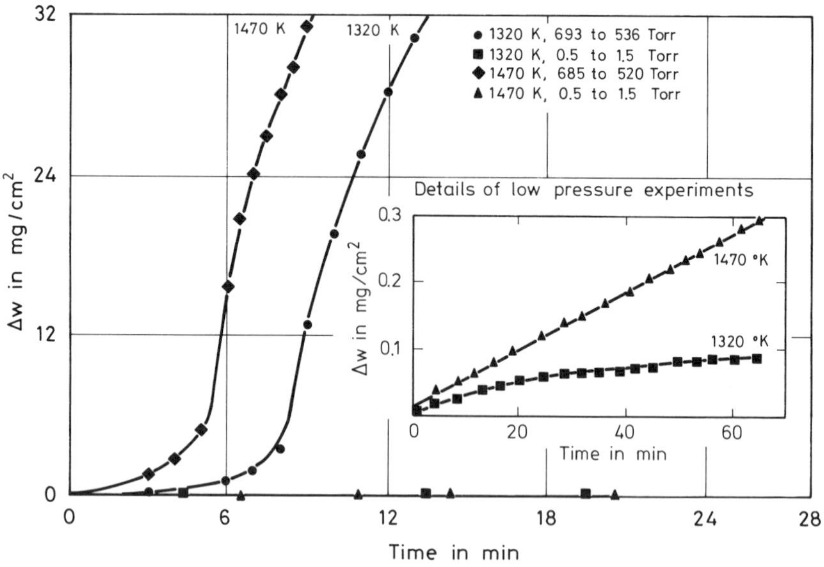

Effect of pressure on the Be–O_2 reaction at 1320 and 1470 K.

Fig. 6-6

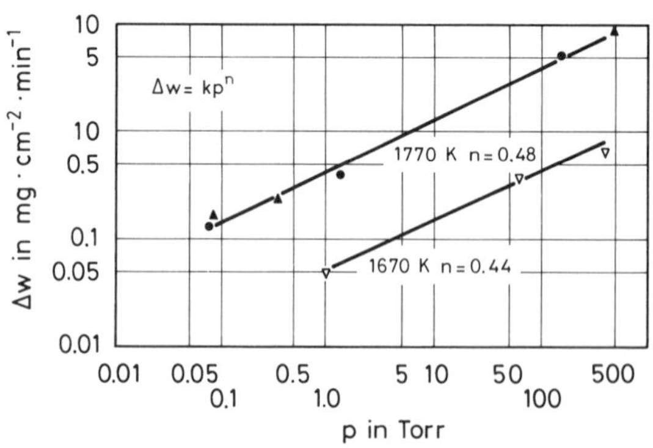

Effect of pressure on the Be–O_2 reaction at 1670 and 1770 K.

order pressure dependence below the Be melting point, and near to half order above the melting point. Weight gain vs. time curves at 1047 and 1197 °C for low and higher O_2 pressures are shown in **Fig.** 6-**5**. The effect of O_2 pressure on the rate constant above the Be melting point is shown in **Fig.** 6-**6**, Blumenthal, Santy [13, 14]. For further Δw values at 1500 °C and 228 to 480 Torr see Fig. 6-24, p. 255.

With Dry Oxygen at Atmospheric Pressure

The behavior of Be sheet in dry O_2 at atmospheric pressure at 600 to 1000 °C varies with the purity of Be and with the method of fabrication of the specimens. As described for the reaction of Be with dry CO_2 (see pp. 264/5) the reaction with flowing O_2 (<20 volppm

H_2O) was studied with the same specimens which had been fabricated in different ways. At 600 °C (studied up to 5000 h), protective films are formed in all cases with values of n between 3.8 and 6, in the weight gain vs. time relation $\Delta w^n = kt$. At 700 and 850 °C, some batches exhibit breakaway and an accelerating rate after 550 or 1500 h (n = 0.3 or 0.4), other batches are resistant up to 5000 or even 8000 h (n values between 3 and 4). At 1000 °C, the reaction rate was found linear in one case with n = 1 and $k = 2.8 \times 10^{-9}$ g · cm^{-2} · s^{-1} in the time range between 250 and 700 h exposure; for another batch, the rate was somewhat lower with n = 1.3 between 250 and 1000 h. The following weight gain data are taken from the weight gain vs. time curves (different batches): <0.05 mg/cm^2 after 5000 h at 600 °C; 0.01 or 0.02 mg/cm^2 after 5000 h at 700 °C, or even 10 mg/cm^2 after 2000 h at 700 °C; ~0.4 mg/cm^2 after 5000 h at 850 °C or 0.5 mg/cm^2 after 1000 h at 850 °C. After 1000 h at 1000 °C, for two batches, 5 or between 15 and 20 mg/cm^2 weight gains were recorded, Higgins, Antill [15], Antill, Higgins [16].

The oxidation of 0.5 mm thick Be sheet (prepared from extruded rods) in dry O_2 (<1 vol ppm H_2O) was studied at 750 and 800 °C by weight gain measurements. At both temperatures, after a small but rapid initial weight gain, the reaction slowed down for some hours, and then accelerated rapidly (after about 40 to 60 h). A number of specimens obtained from different extrusions, behaved in this manner when heated in oxygen. The final weight gain after 100 h exposure at 750 °C was 1.6 mg/cm^2 and after 80 h at 800 °C about 2 mg/cm^2. The sample deformed into a saucer-like shape after 100 h at 750 °C and the surface was extensively cracked, Draycott et al. [17]. Accelerated oxidation of Be was observed at 925 °C after 9 h exposure to flowing O_2 (H_2O content not mentioned) at a pressure slightly above 1 atm. The oxidation was protective at 700 to 800 °C for nearly 100 h. Chromate coatings on Be improved the corrosion resistance, Booker, Stonehouse [18]. Anodic films improved the resistance of Be to dry O_2 at 1 atm pressure considerably. Whilst the weight at 807 °C after 50 h was about 60 mg/cm^2 for uncoated Be, the value is reduced by anodic coating to about 0.33 mg/cm^2 after 50 h at 817 °C (values taken from figures in the paper), Stonehouse, Beaver [19]. Complete oxidation of a 300 to 500 Å thick Be film (prepared by evaporation in a vacuum) in dry O_2 occurred in 10 to 15 s at 800 °C and in 2 to 3 min at 700 °C. Upon further exposure, the size (diameter) of the resulting BeO crystals at 800 °C increased from ≦70 Å to 100 to 150 Å after 30 s and 300 to 350 Å after 1 h, Khorenko et al. [20].

With Moist Oxygen

Studies at a total pressure of 76 Torr (partial pressure of H_2O about 20 Torr) between 600 and 800 °C, Ervin, Mackay [6], at 100 Torr ($p_{H_2O} = 12$ Torr) between 500 and 750 °C, Aylmore et al. [21], and at 1 atm (O_2 with about 2 vol% H_2O) between 650 and 750 °C, Draycott et al. [17], show that the reaction between Be and O_2 is accelerated in the presence of humidity. In dry O_2, the weight gain at the point of transition from a parabolic to an accelerating rate was between 0.05 and 0.1 mg/cm^2. In moist O_2, the critical weight gain was less than 0.01 mg/cm^2 at 600 and 650 °C.

The time required for the weight gain of a polished specimen of commercial Be in moist O_2 at 0.1 atm to reach 0.1 mg/cm^2 at 600, 650, 700, 750, and 800 °C was (in h) 120, 50, 25, 6, and 1, respectively [6]; for weight gain vs. time curves at the 5 temperatures at extended times up to 100 h, see the paper. In tests up to 300 h ($p_{O_2} = 88$ Torr, $p_{H_2O} = 12$ Torr), the oxidation was protective at ≦600 °C, and then at higher temperatures the rate decreased to a constant value of 0.07 µg · cm^{-2} · h^{-1}. The course of the oxidation in moist O_2 was very similar to that in H_2O vapor, see Fig. 6-19, p. 250. At 650 °C and above, breakaway occurred, and both the weight gain and the time at which breakaway occurred diminished with increasing temperature. Weight gain vs. time curves at 600, 650, and 700 °C up to

Fig. 6-7

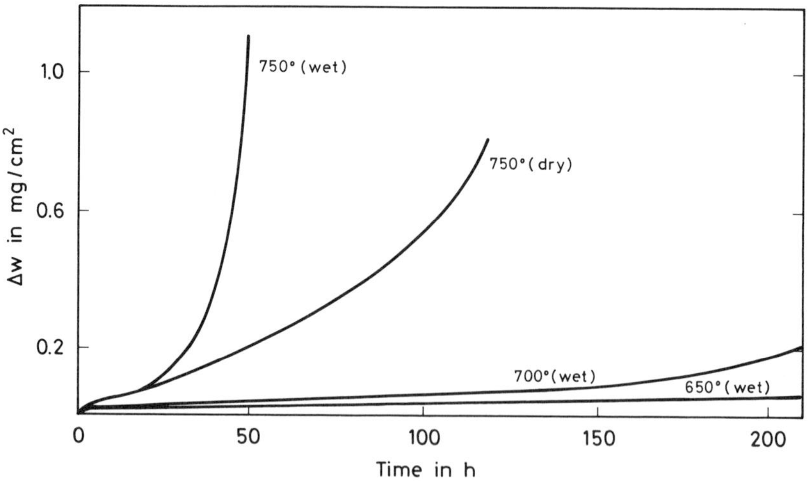

Weight gain Δw vs. time of Be in dry and wet O_2 at atmospheric pressure and various temperatures.

30 h and at 750 °C to 10 h, see in the paper [21]. In **Fig.** 6-**7**, weight gain vs. time curves for the oxidation of extruded Be in moist (2 vol% H_2O) O_2 at 1 atm pressure at 650, 700, and 750 °C are compared with the curve for dry O_2 at 750 °C [17].

With Liquid Ozone

A mixture of powdered Be with 63.9 wt% liquid ozone is highly explosive with a heat of explosion 6.2 times higher than that of TNT (related to 1 g), Brunauer [22].

References:

[1] J.W. Hickman (Am. Inst. Mining Met. Eng. Inst. Metals Div. Metals Technol. **15** No. 8 Tech. Publ. No. 2483 [1948] 1/18, 4, 8; C.A. **1949** 985).
[2] E.A. Gulbransen, K.F. Andrew (J. Electrochem. Soc. **97** [1950] 383/95).
[3] E.A. Gulbransen (Proc. 2nd Intern. Symp. React. Solids, Goeteborg 1952 [1954], pp. 899/908; C.A. **1954** 9175).
[4] D.D. Cubicciotti (J. Am. Chem. Soc. **72** [1950] 2084/6).
[5] G. Ervin, T.L. Mackay (J. Nucl. Mater. **12** [1964] 30/9).
[6] G. Ervin, T.L. Mackay (NAA-SR-9672 [1964] 1/46; N.S.A. **18** [1964] No. 35501).
[7] G. Ervin, T.L. Mackay (React. Solids 5th Intern. Symp., München 1964 [1965], pp. 290/301; C.A. **65** [1966] 14997).
[8] R.H. Condit, J.B. Holt (React. Solids 5th Intern. Symp., München 1964 [1965], pp. 334/41; UCRL-7631-T [1963] 1/11; C.A. **63** [1965] 3967).
[9] D.W. Aylmore, S.J. Gregg, W.B. Jepson (J. Nucl. Mater. **2** [1960] 169/75).
[10] W.B. Jepson (Research [London] **15** [1962] 288/94).
[11] B.E. Ivanov, I.I. Papirov, I.A. Taranenko, et al. (3rd Conf. Intern. Met. Beryllium Commun., Grenoble, Fr., 1965 [1966], pp. 163/9; C.A. **67** [1967] No. 111034).
[12] M.M. Nakata (NAA-SR-4737 [1960] 1/27; N.S.A. **15** [1961] No. 3998).

[13] J.L. Blumenthal, M.J. Santy (Symp. Intern. Combust. Proc. **11** [1966/67] 417/25; C.A. **67** [1967] No. 110222).
[14] J.L. Blumenthal, M.J. Santy (WSS-CI-65-5 [1965] 1/56, 3/5, 25/9; C.A. **63** [1965] 7684).
[15] J.K. Higgins, J.E. Antill (J. Nucl. Mater. **5** [1962] 67/80).
[16] J.E. Antill, J.K. Higgins (TID-7597 [1960] 796/805; N.S.A. **15** [1961] No. 19779).
[17] A. Draycott, F.D. Nicholson, G.H. Price, W.I. Stuart (AAEC-E-83 [1961] 1/34, 6; N.S.A. **16** [1962] No. 12049).
[18] J. Booker, A.J. Stonehouse (Mater. Prot. **8** No. 2 [1969] 43/7; C.A. **70** [1969] No. 80364).
[19] A.J. Stonehouse, W.W. Beaver (Mater. Prot. **4** No. 1 [1965] 24/8 in: H.H. Hausner, Beryllium: Its Metallurgy and Properties, Univ. California Press, Berkeley 1965, pp. 1/322, 155/77; CONF-492-9 [1964] 1/31, 7/9; N.S.A. **19** [1965] No. 20495).
[20] V.K. Khorenko, L.A. Kornienko, B.V. Matvienko (Zh. Fiz. Khim. **40** [1966] 1903/8; Russ. J. Phys. Chem. **40** [1966] 1020/3).

[21] D.W. Aylmore, S.J. Gregg, W.B. Jepson (J. Nucl. Mater. **3** [1961] 190/200).
[22] S. Brunauer (U.S. 3111439 [1949/63]; C.A. **60** [1964] 2720).
[23] D.E. Fowler, J.M. Blakely (Surf. Sci. **148** [1984] 265/82).
[24] D.E. Fowler, J.M. Blakely (J. Vac. Sci. Technol. **20** [1982] 930/3).

6.1.3.3 Reactions with Air

Powder

Ultrafine Be powder (particle size $<0.1\ \mu m$) ignites in air at room temperature, Rhein [1, 2]; see also p. 204 for ignition and combustion of Be. Nuclear grade Be powder of an average particle size of 7 to 14 µm is partially oxidized on heating in air in a closed flask under agitation between 140 and 400 °C. Using an excess of Be, the quantity of oxidized Be is proportional to the increase in temperature and amounts, e.g., at 400 °C the theoretically expected value of about 4.7 wt% BeO (related to the added air) after six air changes in a 550 mL flask containing 30 g Be powder (of 7 µm particle size) with heating times of 20 to 30 min after each change. At 230 °C, after one air change, 0.74 wt% BeO (theoretical 1.04 wt%) and after four air changes 1.02 wt% BeO (theoretical 4.16 wt%) were formed using Be powder with 10 µm particle size, Morana, Koshuba [3]. A somewhat coarser Be powder ($\approx$50 to 60 µm particle size) was oxidized in air at 850 °C to 79% completeness in 6 h and at 1000 °C to 75% completeness in only 15 min, Terem [4]; the reaction started above 500 °C and was slower in dry air, Terem [5].

Thin Layers

At Low Pressures. By reactive evaporation of Be at a partial O_2 pressure of 10^{-2} Pa ($=7.5\times10^{-5}$ Torr) fine crystalline BeO layers are formed on the substrate held at room temperature. If the substrate is kept at liquid air temperature, amorphous BeO layers are obtained which under intensive electron bombardment recrystallizes to thin extended monocrystalline BeO foils, Gruner, Möllenstedt [25]. As shown by electron diffraction, the oxidation of thin films of Be (prepared by evaporation in vacuum from Ta crucibles or from a W spiral) by heating in the diffractometer at an air residue pressure between 10^{-3} and 10^{-5} Torr begins at about 200 °C. The oxide becomes well crystallized at about 550 to 600 °C and only at 850 to 900 °C was the oxidation complete (the thickness of the Be films and the heating times are not mentioned), Trillat et al. [6, 7]. Similar electron diffraction studies of vacuum heated Be films ($\geqq 1\ \mu m$ thick) showed no change of the surface structure of Be at 300 °C. The BeO pattern was observed above 700 °C, it predominated at 800 °C and was the only one at 900 °C. The evaporated Be layers, which exhibit only the Be pattern, showed the same diagram even after having been stored 12 days in open air, Stahl [8].

Be films can be completely oxidized by ionic bombardment (in air or Ar) at 12 keV and a total output of 2×10^{-4} A with a residual vacuum in the diffractograph of 10^{-4} Torr [6], see also [7]. The method of secondary ion-ion emission, with mass spectrometric analysis of the secondary ions, has been used to study the corrosion of 100 μm thick Be films. The films were treated in a residual vacuum of (3 to 5) $\times 10^{-6}$ Torr up to 800 °C and exposed to a primary ion beam of 4 keV He^+ or Ar^+ ions, with a current density of 1×10^{-6} A/cm². The ions BeO^+, Be_2O^+ are observed in the mass spectrum (besides Be^+, Be_2^+, and BeH^+), Kolot et al. [9].

At Atmospheric Pressure. Be films heated under the same conditions as described above (see [6]), except at atmospheric pressure, were totally oxidized at 600 °C [6, 7]. The oxidation of a 200 μm thick, mechanically polished Be plate (99.9% purity, prepared by evaporation in vacuum) in atmospheric air at 500 and 700 °C for 0.5 to 4 h was studied by means of back-back scattered He ions. The calculated weight gain after 4 h was about 5 μg/cm² at 500 °C and 23 μg/cm² at 700 °C. The oxidation rate of the Be layer, which seemed to be initially coated with a 60 Å thick BeO layer, was parabolic for the duration of test time. The rate constants 1.7×10^{-15} and 3.6×10^{-14} $g^2 \cdot cm^{-4} \cdot s^{-1}$ at 500 and 700 °C, respectively, and an apparent activation energy of 22.9 kcal/mol were obtained, Bondarenko et al. [10]. For the formation of oxide layers with preferred orientation, see p. 204.

Compact Beryllium

For the growth of oxide layers with preferred orientation, see p. 203.

In Atmospheric and Moist Air. Compact pure Be specimens are quite resistant to air and can be stored for years at room temperature without changing in appearance. White blisters, sometimes observed on extruded and machined Be at room temperature are caused by the reaction of inclusions, such as Be_2C or chlorine, with humidity in the air, see, e.g., English [11], Darwin, Buddery [12], Beaver [13], Stonehouse, Beaver [14]. Be is protected by forming a very thin uniform nonporous oxide film (~100 Å thick) which reaches a maximum thickness within 2 h at room temperature in air [11]. The oxidation of Be in air at room temperature was not increased by irradiation with UV light, Kerr, Wilman [15].

Bare Be metal coupons ($5.08 \times 2.54 \times 0.635$ cm³) exposed to moist air in a humidity chamber at 38 °C and 95% relative humidity for 30 d showed no evidence of pitting that could be detected microscopically. The surface showed a dull appearance due to the formation of a very thin uniform tarnish film and weight changes were <0.01 mg, McKay et al. [16]. Dense and porous Be samples (80% of theoretical density) were tested in a closed system at 75% relative humidity and 60 °C by pre- and post-test weighing of the samples. Corrosion rates of about 0.004%/year (<0.03 μm/year) and 0.7%/year, respectively, were obtained for the two materials. Protective coatings were tested for the case of salt enhanced corrosion, Wheatherill, Loasby [17]. Cyclic humidity tests at 71 °C and 95% relative humidity gave no clear results. The influence of surface treatment of Be, of C and Fe content as well as vacuum deposits of Ag, Ni, Zr, Al, Ti, and SiO on Be were tested, Steele [18].

The layer thickness Δd of BeO formed on a Be plate ($4 \times 20 \times 12$ mm³) after heating in air up to 500 °C was measured by optical methods. The growth of Δd with time was parabolic and the parabolic rate constant increased exponentially with temperature. The layer thickness after 1 h was about 20 Å at 300 °C and 200 Å at 500 °C (values taken from a figure in the paper), Krylowa [26].

Be (vacuum cast, extruded) was not severely corroded in air at 400 °C after 200 h. Noticeable severe corrosion (grain boundary attack) occurred after 60 h at 700 °C, 12 h at 800 °C, or 1 h at 900 °C, Udy et al. [19], Gordon in [11]. Interference colors observed after heating polished Be samples for 1 h in air are: none at 500 °C, blue at 650 °C, grayish blue at 700 °C, mauve-blue at 750 °C (and a white powder at 800 °C) [15]. A hot pressed, extruded, and 2 h vacuum annealed (at 850 °C) sample of Be was heated in air at 400 °C and after 2 h showed no change in appearance (but BeO was detected on the surface by electron diffraction). At 600 °C a surface interference color was observed (but no change in the luster); at 800 °C, a white-lilac surface coat had formed with loss of luster; and at 1000 °C a white nonadherent layer was present. Weight gains of ~10 mg/cm^2 after 1 h at 950 °C, and of ~70 mg/cm^2 after 1 h at 1000 °C were found (values taken from the curves), Mal'tsev et al. [20]. Commercially pure Be (98.6% purity) and Be containing 1% Al remained stable at 600 °C in air over a period of 4000 h, whereas pure Be or specimens containing 0.5 or 1% Si or Zr showed surface cracks after 1150 or 770 h, respectively, Sinel'nikov et al. [27]. Weight gain-time curves for the oxidation of pressed Be specimens (from commercial powder of 60 μm particle size and 99.4% purity) in atmospheric air (30% relative humidity) between 865 and 1200 °C are shown in **Fig. 6-8**. After an induction period (some hours or minutes) of protective oxidation, the reaction rate becomes linear with time with rate constants k (in $\mu g \cdot cm^{-2} \cdot s^{-1}$) of 0.47 at 865 °C, 0.56 at 915 °C, 2.24 at 965 °C, and 18 at 1200 °C. The rate constants in air are of the same order of magnitude, but somewhat higher than those in dry O_2 (see p. 195). An activation energy of about 46 kcal/mol was found from the Arrhenius plot. Metallographic studies of the surface layer formed after 15 h at 865 °C show appreciable intercrystalline corrosion and X-ray studies show that the oxide films always consist of BeO, Ivanov et al. [21]. Uncoated Be is seriously oxidized in less than 1 d at 843 °C in atmospheric air (dew point 1.67 °C, corresponding to 5.2 Torr H_2O or ~30% relative humidity of air at 20 °C); however, the corrosion resistance can be improved by anodized coatings (in a bath with 40% CrO_3). Depending on the humidity, anodized Be is resistant at 843 °C up to 6 d (p_{H_2O}=4.6 Torr) or up to 50 d in air with p_{H_2O}=1.3 Torr, van Thyne et al. [22].

In Dry Air. The oxidation of reactor grade Be over the range 930 to 1295 °C in dry air (near atmospheric pressure) was studied by weight gain and pressure measurements. After an induction period (~11 min at 1115 °C or 18.5 min at 1055 °C), the reaction rate was nearly linear. Activation energies of 114 kcal/mol below 1050 °C, and of 13 kcal/mol above 1050 °C were found from the Arrhenius plot. The following (high) weight gain was obtained from the weight gain vs. time curves: ~30 mg/cm^2 after 70 min at 1035 °C, after 35 min at 1140 °C, or after 11 min at 1255 °C. Corrosion in air was unaffected by pressure

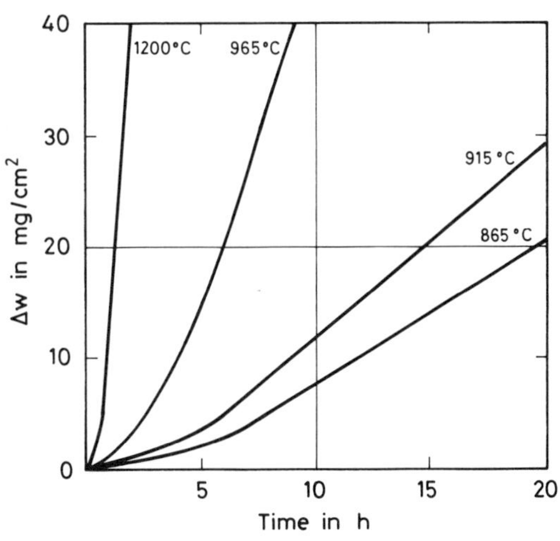

Fig. 6-8

Weight gain Δw of Be in moist air at various temperatures as a function of time.

changes ranging from 760 to 650 Torr. The induction period is a thermally activated process with an activation energy of 30 kcal/mol. The period was increased for Be which had been preoxidized in H_3PO_4-H_2SO_4 mixtures. Metallographic examination of partially oxidized specimens indicated that below 1050 °C, the corrosion was accompanied by intergranular penetration, but that above 1250 °C, the reaction proceeded by total corrosion of the specimen surface. The X-ray diffraction study of the black high temperature corrosion product (containing oxygen and nitrogen) revealed a complex structure, which was different from those of BeO and Be_3N_2, Bradshaw, Wright [23]. An X-ray diffraction pattern of the oxide coating formed in dry air at 1200 °C (at 1 atm pressure) brought out only the Be and BeO patterns; no beryllium nitride was detectable, Nakata [24]. A scale which formed at 1000 °C was analytically shown to contain 0.75% Be_3N_2, however, by electron diffraction, only BeO was detected, Howe in [19].

The oxidation kinetics of Be sheet in dry air at 0.5 atm pressure between 1110 and 1210 °C were similar to those in dry O_2 at 0.1 or 0.2 atm pressure (see p. 195); at 1190 °C a weight gain of 1.08 mg/cm^2 was found after 121 min. After an induction period of 73 min (parabolic rate constant $k=3.6\times10^{-9}\ g^2\cdot cm^{-4}\cdot min^{-1}$) the reaction was accelerated (linear rate constant $k=11.0\ \mu g\cdot cm^{-2}\cdot min^{-1}$) at 1190 °C [24].

References:

[1] R.A. Rhein (Astron. Acta **11** [1965] 322/7).
[2] R.A. Rhein (WSS-CI-64-25 [1964] 1/8; C.A. **62** [1965] 6331).
[3] S.J. Morana, W.J. Koshuba (U.S. 3387966 [1964/68]; C.A. **69** [1968] No. 29793).
[4] H. Terem (Bull. Soc. Chim. France [5] **6** [1939] 664/72).
[5] H. Terem (Istanbul Univ. Fen Fak. Mecm. A **8** No. 1 [1943] 9/22).
[6] J.J. Trillat, L. Tertian, M. Bonnet-Gros (Compt. Rend. **251** [1960] 10/3).
[7] J.J. Trillat, L. Tertian, M. Bonnet-Gros (Mem. Sci. Rev. Met. [Paris] **57** [1960] 845/51).
[8] H.A. Stahl (J. Appl. Phys. **20** [1949] 8/14).
[9] V.Ya. Kolot, V.F. Rybalko, Ya.M. Fogel', G.F. Tikhinskii (Zashch. Metal. **3** [1967] 723/30; Prot. Metals [USSR] **3** [1967] 631/6).
[10] V.N. Bondarenko, L.N. Zvyagintseva, V.Ya. Kolot, et al. (Vopr. At. Nauki Tekhn. Ser. Obshch. Yad. Fiz. No. 2-8 [1979] 42/5; Ref. Zh. Met. **1979** No. 12 I 1030; Ref. Zh. Khim. **1980** No. 1 B 984; C.A. **92** [1980] No. 114983, 136081).

[11] J.L. English (in: D.W. White Jr., J.E. Burke, The Metal Beryllium, ACS, Cleveland 1955, pp. 530/2).
[12] G.E. Darwin, J.H. Buddery (Beryllium: Metallurgy of Rarer Metals No. 7, Academic, New York 1960, pp. 1/392, 251).
[13] W.W. Beaver (Progr. Nucl. Energy V **1** [1956] 277/99, 294).
[14] A.J. Stonehouse, W.W. Beaver (Mater. Prot. **4** No. 1 [1965] 24/8; CONF-492-9 [1964] 1/31, 6/7; N.S.A. **19** [1965] No. 20495).
[15] I.S. Kerr, H. Wilman (J. Inst. Metals **84** [1956] 379/85).
[16] T.L. McKay, C.B. Gilpin, N.A. Tiner (Proc. 1st Jt. Aerosp. Mar. Corros. Technol. Semin., Los Angeles 1968 [1969], pp. 48/53; C.A. **73** [1970] No. 123071).
[17] B.T. Weatherill, R.G. Loasby (Beryllium 1977 Conf. Preprint 4th Intern. Conf. Beryllium, London 1977, pp. 42-1/42-12; C.A. **92** [1980] No. 219538).
[18] J.R. Steele (Mater. Prot. **1** No. 7 [1962] 59/62).
[19] M.C. Udy, H.L. Shaw, F.W. Boulger (Nucleonics **11** No. 5 [1953] 52/9).
[20] M.V. Mal'tsev, L.N. Morozov, K.P. Zverev, Yu.N. Efremov (Izv. Vysshikh Uchebn. Zavedenii Tsvetn. Met. **9** No. 1 [1966] 116/8).

[21] B.E. Ivanov, I.I. Papirov, I.A. Taranenko, et al. (3rd Conf. Intern. Met. Beryllium Commun., Grenoble, Fr., 1965 [1966], pp. 163/9; C.A. **67** [1967] No. 111034).
[22] R.J. van Thyne, J.J. Rausch, J.M. Miller, W.S. Netter (CONF-729-4 [1963] 1/31; N.S.A. **19** [1965] No. 13914).
[23] W.G. Bradshaw, E.S. Wright (Monogr. Rept. Ser. Inst. Metals [London] No. 28 [1963] 314/24; LMSD-288232 [1960]; LMSD-895073 [1961]; N.S.A. **14** [1960] No. 19374, **15** [1961] No. 27994).
[24] M.M. Nakata (NAA-SR-4737 [1960] 1/27, 15; N.S.A. **15** [1961] No. 3998).
[25] H. Gruner, G. Möllenstedt (Thin Solid Films **14** [1972] 43/7).
[26] T.N. Krylowa (Bull. Acad. Sci. URSS Classe Sci. Tech. **1938** No. 10, pp. 89/112, 104; C. **1942** I 2624).
[27] K.D. Sinel'nikov, V.E. Ivanov, B.M. Vasyutinski (Monogr. Rept. Ser. Inst. Metals [London] No. 28 [1963] 355/61; C.A. **60** [1964] 14215).

6.1.3.4 Structural Studies and Oxide Growth in Oxygen and Air

For the growth of BeO whiskers during the combustion of Be droplets, see p. 206.

The oxidation of Be (0001) at very low O_2 pressures (see p. 192) does not show any appreciable epitaxial oxide growth at room temperature. But at $\gtrsim$470 K BeO grows epitaxially on the Be {0001} surface, as evidenced by the appearance of the p (1×1) LEED pattern of BeO {0001}. At a BeO layer thickness of about 6 to 7 monolayers, the LEED pattern changes to p (2×2)-BeO, which implies the occurrence of a surface reconstruction. This reconstruction is believed to be due to the ordering of a surface ionic charge with total charge equal to 1/4 of the net charge per (0001) atomic plane in BeO. The presence of this surface charge may be explained by considering the electrostatic stability of a wurtzite crystal, such as BeO, growing in the ⟨0001⟩ direction, Fowler, Blakely [6].

The structure and growth of BeO on single crystals of Be (zone melted, impurities <0.2%) in air, up to 800 °C, were studied by means of electron diffraction and electron microscopy. The specimens were normally inserted into a hot furnace for 30 min. The diffuse patterns obtained after heating at 300 °C indicated a BeO crystal size on the order of a few tenth Å, whereas the sharp pattern obtained after oxidation at 800 °C corresponded to a crystal diameter of some thousands of Å. Only the normal hexagonal BeO pattern was obtained. An epitaxial growth of BeO on a Be (001) cleavage face between 300 and 400 °C was observed. The epitaxial BeO overgrowth was less well-defined for electropolished (001) surfaces of Be. This disorientation may be due to the formation, during the electropolishing procedure, of surface elements other than (001) upon which the BeO can grow. One-degree (103)-oriented BeO was observed on the (100) and (110) faces of Be. The oxidation rate increased with temperature resulting in an increase of randomly-oriented BeO until, above 600 °C, only random growth was observed in the upper regions of the layer. At about 800 °C, a pronounced (001) habit of randomly oriented BeO crystals developed. The nonprotective nature of the BeO layer formed at $\geqq$600 °C, was evident from electron micrographs which showed the presence of needles of BeO up to 5 μm in length and several hundred Å thickness. Thin distorted BeO flakes up to several μm in diameter were also observed between 600 and 800 °C, Scott [1].

An electron diffraction study of polycrystalline abraded Be which was heated in air between 500 and 800 °C (0.5 to 2 h) showed that the surface film consisted of hexagonal BeO. The crystals were in the form of plates parallel to (001), several hundred Å in diameter, and either randomly disposed or in a one-degree moderate (001) or {100} orientation. On partial removal of the oxide film by abrasion with emery paper, the Be/BeO interface of the heavily oxidized specimens had an etched appearance. The oxide formed on smooth, electropolished, coarsely crystalline Be in air at ≦250 °C (up to 24 h) had a grain size of about 10 Å. That formed at 300 to 600 °C measured 50 to >100 Å with a preferred (001) orientation. After 1 h at 700 °C, the grain size was 500 to 600 Å, Kerr, Wilman [2]. Similar electron diffraction studies on the oxidation of polycrystalline Be (vacuum-melted or hot pressed and abraded) in air between 400 and 800 °C showed the following preferred orientation: ($10\bar{1}3$) at 500 and 600 °C (1 h); ($10\bar{1}0$) at 800 °C (1 h); and (0002) and ($10\bar{1}3$) at 400 °C (0.5 to 20 h).

Textureless, vacuum deposited films (400 to 600 Å) which were heated for 1 h at 400 °C or 0.5 h at 600, 750, or 800 °C, showed oxidation layers with a preferred (0002) and ($10\bar{1}3$) orientation only in patterns taken with an electron beam having an angle of incidence of 45° or 60°. No texture was found from patterns taken perpendicular to the oxide film. The indicated preferred orientation cannot be attributed to epitaxy because the studied compact specimens or films had no texture before oxidation, Kolomiets, Khorenko [3]. Electron diffraction studies of oxides grown at 20 to 400 °C on (gettered) vacuum-deposited Be having {110} orientation on a mica substrate showed a preferred one-degree {103} orientation which was attributed to epitaxy. Preferred (001) orientation of BeO grown at 200 or 400 °C (2 h) on vacuum deposited Be (with (001) planes (preferentially oriented parallel to the glass substrate) was found by electron diffraction studies and also attributed to epitaxy [2]. For qualitative studies of spherical Be particles (105 to 125 μm diameter) during heating and after 35 min exposure to dry and moist air (or O_2) at ~880 °C (near atmospheric pressure) by means of the optical and scanning electron microscope, see Kraeutle [4]. For the change of electronic structure during oxidation of Be, see Madden, Zehner [5].

References:

[1] V.D. Scott (Acta Cryst. **12** [1959] 136/42).
[2] I.S. Kerr, H. Wilman (J. Inst. Metals **84** [1956] 379/85).
[3] L.D. Kolomiets, V.K. Khorenko (Fiz. Metal. Metalloved. **20** [1965] 860/3; Phys. Metals Metallog. [USSR] **20** No. 6 [1965] 59/64).
[4] K.J. Kraeutle (AD-746133 [1972] 46/57; C.A. **78** [1973] No. 113506).
[5] H.H. Madden, D.M. Zehner (SAND-79-0791C [1979]; CONF-7906113-1 [1979], CONF-7906114-1 [1979] 1/7; C.A. **92** [1980] No. 133625).
[6] D.E. Fowler, J.M. Blakely (Surf. Sci. **148** [1984] 283/91).

6.1.3.5 Ignition and Combustion

Despite its high toxicity (see p. 300), the ignition and combustion characteristics of Be have been studied theoretically and experimentally by several authors mainly with respect to the possible use of Be as a propellant in ramjets or rocket motors.

Although the heat of combustion of Be in air is half that of the H_2-O_2 reaction, the adiabatic flame temperature at 1 atm is about 500 °C higher. Computer calculated flame temperatures at various air pressures (298 K) are: 2100 K at 10^{-6} atm, 2670 K at 10^{-3} atm, 3570 K at 1 atm, 3970 K at 10 atm, and 4370 K at 100 atm, Feigenbutz [1].

Ignition of Be foil (~50 μm thick) in a calorimeter bomb under high pressure oxygen produces temperatures so high that the BeO formed in the combustion melts. The found standard heat of formation of BeO is $\Delta H^\circ_{298} = -143.1 \pm 0.1$ kcal/mol, Cosgrove, Snyder [2]. Using C_2H_2 as the ignitor gas, Be powder (~60 μm particle size) burns in O_2 with a white flame, whose temperature has been estimated to be about 4500 K. Snow-white molten BeO was formed when 346 g Be (powder or shaves) was burned in a preheated (1100 °C) pot of BeO in a continuous operation for almost 45 min, Grosse, Stokes [3].

For efficient combustion of Be in propellants, the flame temperature must be kept high enough to insure that any BeO formed remains molten (melting points: Be 1556 K, BeO 2820 K; boiling points: Be 2757 K, BeO 4060 K), Crump et al. [4]. The maximum rocket thrust on combustion of Be in O_2 is obtained by admixture H_2, which takes not place in the initial reaction, Lo [5].

The burning times of spherical Be (and Al) particles (10^{-3} to 1 mm diameter) in rocket motors have been predicted using a computer program based on quasi-steady-state vapor phase combustion. The gas mixtures studied were O_2-Ar, O_2-H_2O, and Ar-H_2O, at a pressure of 20 atm. The minimum conditions for ignition and the ignition delay time have also been calculated as a function of gas temperature and particle size (figures see in the paper), Kuehl [6]. For the theoretical calculation of the ignition and combustion of Be, also see Shevtsov et al. [7] and Gremyachkin et al. [8], respectively.

Theoretical models have been formulated to describe several possible modes of fuel droplet combustion. Surface condensation is found to be an important process that greatly enhances the mass burning rate and increases the flame standoff distance. The burning rate, however, is insensitive to variations in ambient temperatures. The additional gas-phase condensation process has the effect of significantly increasing the outer edge of the extended flame zone. It is also demonstrated that the condensed oxides formed in the gas phase can be effectively treated as gas like in the formulation without losing much accuracy in the quantitative results obtained. Be burns much slower than Al in mixed O_2-inert gas atmosphere and the accumulation of condensed oxides on the droplet surface is also more substantial during Be combustion, Law, Williams [9]. Calculated rate constants for the gas phase reactions of Mg, Ca, Sr, Ba with O_2 are in good accordance with experimental values obtained by a thermocouple variation of the diffusion flame method. Analogously, the rate constant for the reaction $Be(g)+O_2 \rightarrow BeO(g)+O$, was calculated to be $k=1.6\times10^{14}$ $\exp(-21100/RT)$ with k in $cm^3\cdot mol^{-1}\cdot s^{-1}$ (temperature range not mentioned), Kashireninov et al. [10]. A proposed combustion scheme for the gaseous oxidation of Be at 3000 K considers only two-body collisions. The reaction scheme passes through the intermediates BeO, Be_2O, and $(BeO)_2$ and further polymerization reactions to form polymers $(BeO)_n$ with $n=3$ to 6, Henderson [11].

Ultrafine Be powder (<0.1 μm) is said to ignite at room temperature, see p. 199. The ignition and combustion of single phase Be (spherical particles having average diameters of 32 and 25 μm) in dry and moist oxidizing gases at 1 atm pressure have been studied photographically and spectroscopically. Be powder was injected into a hot (2600 to 2960 K) laminar gas stream (~1 cm/ms) generated by combustion in a flat flame burner of CO or propane in O_2. The CO flames consisted of 0.16 to 0.43 atm O_2 and 0.42 to 0.46 atm CO_2 (the remainder being CO and small amounts O atoms) and the propane flames consisted of 0.07 to 0.37 atm O_2 and 0.19 to 0.26 atm H_2O (the remainder being a mixture of N_2, CO_2, CO, O, and OH). For dry gases, temperatures of about 2600 K or higher were needed for ignition, but there was no sharp ignition temperature for any given sample of Be powder. More and more particles ignited with rising temperature. At the higher O_2 partial pressure and the highest temperature (2960 K), nearly all of the particles that passed through the hot gas stream ignited. In moist gases, at applied flame temperatures of 2600 to 2800 K, only 20 to 30% of the particles ignited, i.e., showing a distinct transition from dull glow to bright burning. Ignition efficiences, burning rates, and flame diameters of particles varied both with particle diameters and with ambient gas properties; therefore, Be particles probably burn by several distinct modes. One is rapid vapor-phase diffusion flame, favored by high partial pressure of O_2 and temperatures of a gaseous environment. The burning times corresponding to this mode, ranging from 1.3 to 4.5 ms, are believed to be proportional to d^2 (where d is the diameter of particles) and inversely proportional to a power of O_2 pressure slightly less than unity. As $p(O_2)$ decreased below a certain value, usually lying between 0.1 and 0.2 atm, the metal flame temperature decreased and vapor-flame combustion gradually changed into a slow surface reaction. There is also evidence of a third mode in which combustion is only moderately slower than the vapor phase burning. This mode is favored by low temperatures and by the presence of H_2O vapor in the environment. It may be vapor-phase combustion hindered by a heavy coating of BeO which accumulated on the particles during the pre-ignition surface reaction. The burning rates do not differ much in dry and moist media although H_2O vapor does exert a measurable retarding effect, Maček, Semple [12]; for earlier results see also [13], and for comparison of Al and Be combustion [14]. For a comparison of the combustion of Be with those of B, Mg, and Al, see Barrere [15].

Free falling Be droplets of ~500 μm diameter were prepared from 12.7 μm thick Be foil by ignition with an Xe flash discharge lamp and the combustion of these droplets in dry O_2-Ar mixtures (20 mol% O_2) was studied at 705±5 Torr with scanning electron microscopy. The ignition and burning of Be can occur below its melting point (1556 K) and is accompanied by considerable smoke and a rather massive growth of randomly oriented clusters of BeO needles on the spongy Be substrate. The rising particle temperature produces a molten droplet (after about 30 ms) containing a uniform growth of perfectly hexagonal BeO whiskers. All whiskers disappear when the droplet temperature rises to its maximum and the droplet is encapsulated by molten BeO (melting point 2820 K). Shortly after encapsulation, radiative heat loss from the extensive BeO surface lowers the temperature enough to freeze the oxide (forming platelets) and combustion ends. BeO whiskers can also be grown on burning Be particles in CO_2 and in air; combustion produced in the latter the most perfect crystals, Prentice [16], also see [4] for a comparison with Al combustion. Repetition of such experiments with wet gases (saturated with H_2O vapor at room temperature), results in the prevention of oxide accumulation on the droplet surface and neither whisker formation nor encapsulation occurs (the oxide was removed from the sample surface). Narrower particle tracks (relative to dry gases) and lower color temperatures (and higher burning times) were characteristic of flash-ignited droplets in wet gases. The endothermic reaction $BeO(s)+H_2O(g) \rightleftharpoons Be(OH)_2(g)$ (with $\Delta H^\circ_{298} = +42$ kcal/mol) or a similar reaction with a polymeric species $BeO(s)+H_2O(g) \rightleftharpoons (BeO)_n \cdot H_2O(g)$ should act as a significant energy sink during droplet combustion in the presence of H_2O vapor. The oxide transport process influences flame temperature, flame radius, droplet geometry, burning rate, and burning mechanism, Prentice [17], also see [18 to 20], and for a comparison with the different combustion characteristics of Al, see [21].

In order to understand the reaction mechanism better, the combustion of Be droplets in dry O_2-Ar mixtures near atmospheric pressure (700±7 Torr) using a laser ignition technique was studied photographically and photometrically (scatter typically of ±7%). The droplets burned in free fall and the burning time was proportional to particle diameter (studied between about 200 and 350 μm). It was shown that the vapor phase model is not sufficient to describe the droplet combustion which occurs in 3 distinct stages (following ignition): (1) predominantly vapor phase reaction, but with an attendant solid nonprotective surface oxide (in the form of whiskers), (2) predominantly condensed phase reaction through the molten oxide layer at the droplet surface, and (3) predominantly condensed phase reaction through a solid oxide crust at the particle surface. The experimental results indicate that the dominant mechanism changes from vapor-phase to condensed phase at an O_2 mole fraction near 0.6. When the flame temperature is above the melting point of BeO (2820 K), as is the case for a Be droplet burning in a rocket motor or torch flame, only conditions (1) and (2) apply.

A major reaction product is a large solid BeO sphere. The sphere size increases as the O_2 concentration increases and equals the original droplet size at oxygen mole fractions in the range 0.55 to 0.7. The best theoretical estimates fail to properly account for the back-diffusion of oxide to the droplet surface. Theoretical calculation of droplet burning times are in error by at least a factor of 2, Prentice [18, 19], also see [22]. Analogous systematic quantitative studies of the combustion of single laser-ignited Be droplets (221 to 392 μm diameter) in O_2-N_2 mixtures (instead of O_2-Ar) at 700±7 Torr show that N_2 does not act as an inert gas. The ignition limit is higher in O_2-N_2 than in O_2-Ar, because 272 μm droplets ignite at mole fractions $x(O_2)$ as low as 0.25 in O_2-Ar, whereas they don't ignite below $x(O_2)=0.4$ in O_2-N_2. The presence of N_2 raises the ignition limit and reduces the burning rate (increasing burning time) relative to O_2-Ar. Further, the phenomenon of droplet fragmentation is observed in N_2-O_2 at about $x(O_2)=0.4$ to 0.6; this is possibly due to the formation of unstable BeO_xN_y intermediates which lead to explosions. In explosion debris, $BeO \cdot 3Be_3N_2$ has been found by X-ray diffraction. The time required for a droplet to explode is a linear function of droplet diameter d (≈80 to 180 ms for d≈220 to 320 μm at $x_{O_2}=0.5$). The N_2 induced fragmentation terminates combustion in slightly less than half the "normal" burn time in O_2-Ar. In all cases of droplet combustion in dry O_2-N_2, where fragmentation does not occur ($x_{O_2}<0.4$ and >0.6), a single large BeO sphere remains as the dominant combustion product whose diameter is unaffected by changes in $x(O_2)$. In contrast, in O_2-Ar the diameter increases with $x(O_2)$, Prentice [22, 23].

In the vapor-phase combustion of an electrically heated Be wire in dry or wet O_2-N_2 mixtures (20 to 100 mol% O_2, 1.5 mol% H_2O in the "wet" gases) at 1 to 10 atm pressure, BeO particles are produced in the form of regular hexagonal prisms (length 20 to 600 μm) rather than spheres. O_2 partial pressure, total pressure, or the presence of small amounts of H_2O vapor have no apparent effect on the BeO particle size. Also the particle temperature does not seem to exceed the BeO melting point (2820 K). A set of equations describing the condensation of BeO from an atmosphere containing metal vapor and O_2 has been formulated and programmed on a high-speed digital computer. The time required for BeO condensation and the resulting particle size distribution have been calculated for different

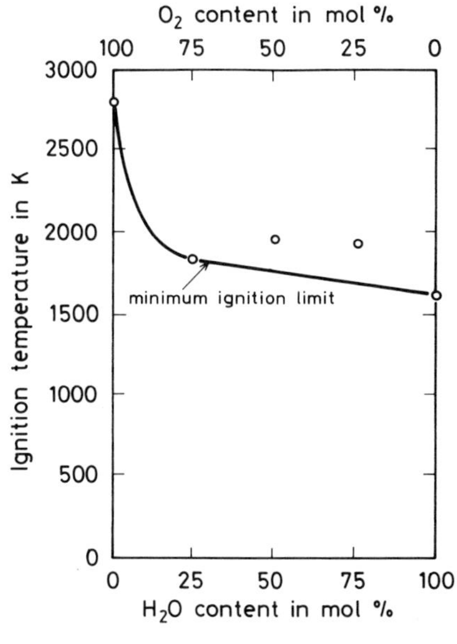

Fig. 6-9

Ignition temperature of a beryllium wire in O_2-H_2O mixtures at 6.8 atm pressure.

initial metal vapor concentrations and temperatures (for comparison with Al see the paper), Hermsen [24], see also [25]. The ignition temperature of Be wire (500 μm diameter) in O_2–H_2O gas mixtures at 6.8 atm (100 psia) pressure, measured under static conditions is shown in **Fig. 6-9**, p. 207. There is no effect of wire size except for wires with much larger diameters. However, the ignition temperatures can vary over a range of about 1000 K due to the fabrication technique of the wire, the pressure, and as shown in Fig. 6-9, p. 207, to the gas composition [6]. The qualitative behavior of bulk Be in H_2–O_2 flames has been studied by photographic and pyrometric techniques. In fuel rich flames in which H_2O vapor was the only reactive species present in significant concentrations, Be samples never reached temperatures of 2000 °C, although rapid reaction of Be occurred. In O_2-rich flames in which both, O_2 and H_2O vapor were present, the reaction was much more violent than in flames with O_2 or H_2O as the only reacting species. When both, O_2 and H_2O were present, the Be ignited and burned with a very hot (>2000 °C) vapor-phase diffusion flame under all but the lowest initial heat flux conditions, Blumenthal, Santy [26, 27]. For the combustion of Be in pyrotechnique mixtures containing $NaNO_3$ or $NaClO_4$, and of 5% of an epoxy binder, see Tanner [28]. For the composition of propellants with high flame temperatures containing an oxygen carrier such as NH_4ClO_4, Be, metal oxides, and inorganic salts in a plastic matrix, see Fortuna [29].

References:

[1] L.V. Feigenbutz (AD-260268 [1961] 1/28; C.A. **58** [1963] 401/2).
[2] L.A. Cosgrove, P.A. Snyder (J. Am. Chem. Soc. **75** [1953] 3102/3).
[3] A.V. Grosse, C.S. Stokes (AFOSR-TR-59-168 [1959] 1/31, 16/8; N.S.A. **14** [1960] No. 18795).
[4] J.E. Crump, J.L. Prentice, K.J. Kraeutle (Combust. Sci. Technol. **1** [1969] 205/23).
[5] R. Lo (Chem. Ing. Tech. **39** [1967] 923/7).
[6] D.K. Kuehl (AIAA [Am. Inst. Aeron. Astron.] J. **3** [1965] 2239/47; C.A. **64** [1966] 11021).
[7] V.I. Shevtsov, V.P. Fursov, E.I. Gusachenko, L.N. Stesik (Gorenie Kondens. Geterog. Sist. Mater. 6th Vses. Simp. Goreniyu Vzryvu, Alma-Ata 1980, pp. 7/10; C.A. **95** [1981] No. 64590).
[8] V.M. Gremyachkin, A.G. Istratov, O.I. Leipunskii (Zh. Prikl. Mekhan. Tekhn. Fiz. **1974** No. 4, pp. 70/8; C.A. **82** [1975] No. 90528).
[9] C.K. Law, F.A. Williams (AIAA [Am. Inst. Aeron. Astron.] Paper **74-147** [1974] 1/18; C.A. **84** [1976] No. 46808).
[10] O.E. Kashireninov, V.A. Kuznetsov, G.B. Manelis (AIAA [Am. Inst. Aeron. Astron.] J. **15** [1977] 1035/7; C.A. **87** [1977] No. 107227).

[11] C.B. Henderson (Combust. Sci. Technol. **1** [1969/70] 275/8).
[12] A. Maček, J. McKenzie Semple (Symp. Intern. Combust. Proc. **12** [1968/69] 71/81; C.A. **74** [1971] No. 130936; AD-669489 [1968] 1/28; C.A. **70** [1969] No. 13116).
[13] A. Maček, R. Friedman, J.M. Semple (CONF-391-2 [1963] 1/6; N.S.A. **18** [1964] No. 10110).
[14] A. Maček (Symp. Intern. Combust. Proc. **11** [1966/67] 203/17; C.A. **68** [1968] No. 52503).
[15] M. Barrere (AGARD Conf. Proc. No. 52-20 [1970] 1/16; C.A. **73** [1970] No. 57636).
[16] J.L. Prentice (J. Am. Ceram. Soc. **52** [1969] 564/5).
[17] J.L. Prentice (J. Electrochem. Soc. **117** [1970] 385/7).
[18] J.L. Prentice (Combust. Sci. Technol. **3** [1971] 287/94).
[19] J.L. Prentice (WSS-CI-71-23 [1971] 1/12; C.A. **75** [1971] No. 89705).
[20] J.L. Prentice (WSS-CI-69-2 [1969] 1/25; C.A. **72** [1970] No. 80993).

[21] J.L. Prentice (Combust. Sci. Technol. **1** [1970] 385/98).
[22] J.L. Prentice (AD-746133 [1972] 1/66; C.A. **78** [1973] No. 113506).

[23] J.L. Prentice (Combust. Sci. Technol. **5** [1972] 273/85).
[24] R.W. Hermsen (AD-670529 [1968] 1/67; C.A. **70** [1969] No. 31265).
[25] R.W. Hermsen, R.W. Woolfolk (AD-617938 [1964] 1/60; C.A. **64** [1966] 400).
[26] J.L. Blumenthal, M.J. Santy (Symp. Intern. Combust. Proc. **11** [1966/67] 417/25; C.A. **67** [1967] No. 110222).
[27] J.L. Blumenthal, M.J. Santy (WSS-CI-65-5 [1965] 1/56; C.A. **63** [1965] 7684).
[28] J.E. Tanner (AD-786216-2GA [1974] 1/24; C.A. **83** [1975] No. 12931).
[29] G. Fortuna (Polim. Vehomarim Plast. [Haifa] **2** No. 3 [1972] 2/4 from C.A. **78** [1973] No. 99933).

6.1.4 With Nitrogen

Measurement of the sticking coefficient shows that N_2 is not adsorbed by a thick Be film (vapor deposited on a glass substrate), Hurd, Adams [1]. The (0001) surface of a Be single crystal does not adsorb N_2, as shown by LEED studies, Adams [2]. As determined by chemical and metallographic analysis, the solubility of nitrogen in beryllium at 1000 °C lies between 48 and 73 ppm, and there is no detectable change in the solubility between 1000 and 500 °C. Using Be-Be_3N_2 diffusion couples, the diffusion coefficient for the diffusion of nitrogen in Be at 1025 °C was found to be approximately $D = 5 \times 10^{-10}$ cm²/s; the result is considered tentative, Pemsler et al. [3]. Rules of the probability of diffusion and dissolution of nitrogen in Be (and other metals) have been derived on the basis of theoretical considerations, Pechenyakov [14].

Theoretical behavior of Be with liquid N_2 is calculated with respect to propulsion in the atmosphere of Mars (95% N_2, 2.5% Ar, 2% CO_2, traces CO, NO_2, O_2, H_2O), Weber, Mueller [4]. Ultrafine Be powder (<0.1 µm particle size) tested as propellant ignites in N_2 at about 515 °C (504 and 527 °C was found in two runs) with an energy gain of about 4.8 kcal/g Be for the reaction $3Be(s) + N_2 \rightarrow Be_3N_2(s)$, Rhein [5, 6]; for a simulated Martian atmosphere ignition temperatures of 553, 762, and 766 °C are found [5].

Due to the exothermic formation of Be_3N_2 the heat of explosion of PbN_6 is higher in mixtures with Be. In this way the calorimetrically determined formation enthalpy of Be_3N_2 is $\Delta H^{\circ}_{298} = -135$ kcal/mol (corresponding to 4.99 kcal/g Be), Apin et al. [7]. In order to determine the heat of formation of Be_3N_2, the degree of short time reaction between Be (roughly powdered in a steel mortar) and N_2 at 25 atm pressure was studied at a heating time up to 2 min at about 1000 °C. The amount of N in the reaction product was low (5 wt% after 1.45 min at 1040 °C), but it can be increased to 30 or 32.6 wt% (theoretical for Be_3N_2: 50.6 wt% N) by addition of 0.25 wt% NaF or 7 wt% BeO, respectively, to the Be, Neumann et al. [8]. The nitriding of roughly powdered Be in a stream of N_2 (atmospheric pressure) was noticeable above 900 °C and more quickly at 1000 to 1100 °C. But a complete reaction was not reached even after heating seven times for 2 h, with intermediate powdering. The nitriding is more complete in a stream of NH_3 at 1000 °C (the reaction products also contained BeO), Fichter, Brunner [13]. The nitriding of Be chips in a stream of N_2 is favored by the presence of 2 to 6 vol% H_2. The reaction began at 870 °C and was continued by heating for about 4 h at 1150 °C and then up to 1250 °C ($\leqq 1$ h). The reaction was completed by refiring the ground and pressed reaction product in the gas mixture at a temperature of about 1200 to 1300 °C, Langsdorf [15].

Nuclear reactor tests (7 d) of the compatibility of compact Be to liquid Na under pressurized N_2 show that at 482 and 593 °C Be is readily nitrided. The Be_3N_2 layers flake off, Brush, Rodd [9].

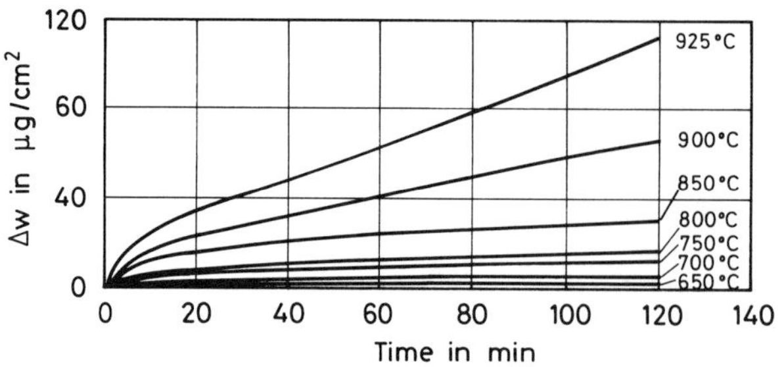

Fig. 6-10

Weight gain Δw of Be in N_2 (76 Torr) in dependence of exposure time at various temperatures.

Weight gain-time curves for the reaction of (abraded and polished) Be plates (prepared from sintered bars containing 1.34% BeO) with purified lamp grade N_2 at 76 Torr pressure between 650 and 925 °C is shown in **Fig. 6-10**. The reaction rate at ≦850 °C, studied for 2 h, follows a parabolic law; at 900 and 925 °C it does so only for the first 60 min. This may indicate that the nitride film formed is protective only below 900 °C. An activation energy of 75 kcal/mol was obtained from the temperature dependence of the parabolic rate constant over the temperature range of 725 to 925 °C. At low N_2 pressures (3.8 and 1.5 Torr) the samples show a weight loss, because Be evaporates faster than nitride was being formed (the rate of Be evaporation at 850 °C was found to be approximately 0.6×10^{-6} g · cm^{-2} · min^{-1}). Colored films are observed after 2 h reactions at ≧700 °C: straw colored at 700, 725, and 850 °C, blue-violet at 750 and 775 °C, blue-gray at 875 °C, and blue-white at 900 and 925 °C. The presence of colored films indicates that the nitride is not soluble in the metal, not even after heating between 800 and 970 °C for several hours, Gulbransen, Andrew [10]. A few corrosion tests in N_2 (600 Torr) at 980 to 1070 °C of induction-melted, chemically etched Be up to t = 360 min show a modified parabolic rate law with n > 0.5 in the time dependence of the weight gain, $\Delta w = k \cdot t^n$. k was found to be 9.43×10^{-3} mg · cm^{-2} · $t^{0.76}$ at 980 ± 15 °C and 3.35×10^{-2} mg · cm^{-2} · $t^{0.625}$ at 1070 ± 25 °C. The activation energy observed was 49.6 kcal/mol, Bradshaw, Wright [11]. The reaction rate of reactor grade Be rods (~5 mm diameter) in N_2 was investigated at 1220 to 1500 °C and N_2 pressures between 0.5 and 500 Torr. The weight gain at 1220 °C and $p(N_2)$ between about 1 and 5 Torr is ~0.2 mg/cm² after 30 min and ~0.35 mg/cm² after 60 min (values taken from the curve). The weight gain with time at 1350, 1430, and 1500 °C for $p(N_2) = 1$, 4, and 10.7 Torr is shown in **Fig. 6-11**. The reaction rate was found to be pressure dependent. For the range 0.5 to 12 Torr the values for n in the equation rate $= k \cdot p^n$ at 1220, 1350,

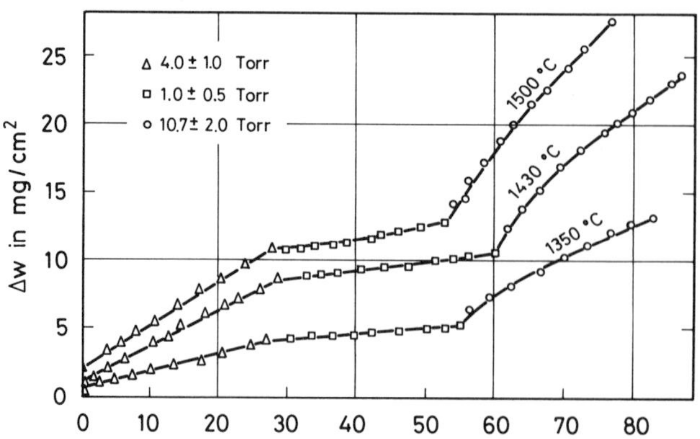

Fig. 6-11

Weight gain Δw of Be in N_2 at various pressures in dependence of time.

1430, and 1500 °C were found to be 0.58, 0,76, 0.83, and 0.88, respectively. Although the Be-N_2 reaction is significant above 1200 °C, there appears to be little possibility that a bulk Be sample would ignite in an N_2 atmosphere below 1500 °C and pressures up to 1 atm, Blumenthal, Santy [12]. Values Δw at 1500 °C and 306 to 387 Torr in comparison with H_2, O_2, NO, CO, and CO_2 are given in Fig. 6-24, p. 255.

References:

[1] J.T. Hurd, R.O. Adams (J. Vac. Sci. Technol. **6** [1969] 229/33).
[2] R.O. Adams (Struct. Chem. Solid Surf. No. 35 [1969] 70.1/70.9; C.A. **75** [1971] No.134140).
[3] J.P. Pemsler, R.W. Anderson, E.J. Rapperport (ASD-TDR-62-1018 [1963] 1/25; N.S.A. **17** [1963] No. 22157).
[4] J.Q. Weber, K.H. Mueller (Chem. Eng. Progr. Symp. Ser. **60** No. 52 [1964] 17/22; C.A. **62** [1965] 14413).
[5] R.A. Rhein (Astronaut. Acta **11** [1965] 322/7).
[6] R.A. Rhein (N65-14807 [1964]; NASA-CR-60125 [1964] 1/29, 7; C.A. **63** [1965] 14628; WSS-CI-64-25 [1964] 1/8; C.A. **62** [1965] 6331).
[7] A.Ya. Apin, Yu.A. Lebedev, O.I. Nefedova (Zh. Fiz. Khim. **32** [1958] 819/23; C.A. **1958** 21108).
[8] B. Neumann, C. Kroger, H. Haebler (Z. Anorg. Allgem. Chem. **204** [1932] 81/96, 87/90).
[9] E.G. Brush, C.R. Rodd (KAPL-M-EGB-21 [1955]; N.S.A. **10** [1956] No. 10954).
[10] E.A. Gulbransen, K.F. Andrew (J. Electrochem. Soc. **97** [1950] 383/95).
[11] W. Bradshaw, E.S. Wright (LMSD-288232 [1960] 3.10/3.17; N.S.A. **14** [1960] No. 19374).
[12] J.L. Blumenthal, M.J. Santy (WSS-CI-65-5 [1965] 1/56, 5/6, 30/3; C.A. **63** [1965] No. 7687).
[13] F. Fichter, E. Brunner (Z. Anorg. Allgem. Chem. **93** [1915] 84/94, 86/8).
[14] I. Pechenyakov (Tekh. Misul. **11** No. 1 [1974] 89/94 from C.A. **81** [1974] No. 110034).
[15] A.S. Langsdorf, U.S. Atomic Energy Commission (U.S. 2567518 [1948/51]; N.S.A. **5** [1951] No. 6929).

6.1.5 With Fluorine, Chlorine, Bromine, and Iodine

The codeposition of Be atoms and F_2 or Cl_2 in an argon matrix at 12 K produces the radicals BeF and BeCl, which were studied by ESR investigation, Knight et al. [8]. The kinetics of the reaction of high purity sheet Be and gaseous fluorine were studied by means of the pressure drop method over the temperature range 125 to 775 °C and at pressures from 20 to 700 Torr. The reaction follows the parabolic law $v^2 = k \cdot t$, where v is the amount of consumed fluorine in mL/cm², k is the parabolic rate constant in (mL/cm²)²/min and t is the time in min (studied up to 560 min). Values of k, obtained under various conditions of temperature and pressure p:

temp. in °C . .	125	225	525	525	525	525	575
p in Torr . . .	200	200	60	200	630	700	200
$k \cdot 10^5$	1.05	0.94	0.20	0.69	3.77	5.70	3.30
temp. in °C . .	625	625	675	675	675	775	775
p in Torr . . .	200	630	20	60	200	60	200
$k \cdot 10^5$	17.0	41.0	4.03	16.30	69.20	130	280

A reaction similar to breakaway oxidation was not found with F_2; the BeF_2 layer formed is protective under the applied conditions. At 200 Torr pressure the rate constant shows a negative temperature dependence between T = 398 and 798 K with $k = 0.4 \times 10^{-5}$ exp (800/RT) and a positive temperature dependence between 798 and 1048 K with $k = 1.5 \times 10^{-2}$

exp (−8000/RT). The films formed at the lower temperatures are dull, while those formed at the higher temperatures resemble shiny mirrors. Their structure changes above 525 °C from hexagonal (α-quartz) to rhombic (tridymite). The pressure dependence of k is approximately first order with respect to the fluorine pressure p (in Torr); the values are $k = 7.404 \times 10^{-9} \cdot p^{1.3}$ at 525 °C and $k = 1.019 \times 10^{-6} \cdot p^{1.2}$ at 675 °C, O'Donnell [1]. Burning of Be powder in F_2 in a calorimeter bomb results in incomplete reaction and amorphous (glassy) BeF_2 forms. Better than 99% conversion of Be to BeF_2 was obtained by the device of burning pellets of a mixture of 0.1 g Be powder (≦12 µm particle size) with 1.7 g polytetrafluoroethylene instead of Be alone. In this way the enthalpy for the reaction $Be(s) + F_2(g) = BeF_2$ (amorphous) was found to be $\Delta H^{\circ}_{298} = -244.3 \pm 0.8$ kcal/mol, Churney, Armstrong [2]; see also Armstrong [3] and Gross et al. [4] for earlier attempts to determine the heat of combustion of Be in F_2. With the transition enthalpy −1.12 kcal/mol for BeF_2 (glass) → α-BeF_2 (hexagonal) the formation enthalpy of α-BeF_2 is $\Delta H^{\circ}_{298} = -245.4 \pm 0.8$ kcal/mol [2, p. 294]. BeF_2 is polymorphic, with structures corresponding to those of SiO_2, see e.g., Batsanov et al. [13]. Be powder (≈200 µm particle size) is apparently quite stable with respect to dry Cl_2 at room temperature. At temperatures up to 250 °C the amount of reaction during a 30 min exposure appears to be negligible. At about 350 °C, however, the reaction is spontaneous and rapid. The enthalpy of reaction $Be(s) + Cl_2(g) = BeCl_2(s)$ was found to be $\Delta H^{\circ}_{298} = -118.03 \pm 0.56$ kcal/mol ($= -493.85 \pm 2.35$ kJ/mol), Johnson, Gilliland [5], and as -117.1 ± 0.4 kcal/mol for the formation of α-$BeCl_2$, Gross et al. [6]. For α- and β-$BeCl_2$ the values $\Delta H^{\circ}_{298} = -117.2$ and −118.5 kcal/mol, respectiviely, are reported by Parker et al. [7].

For the reaction of Be powder with Br_2 in diethyl ether at low temperature to form $BeBr_2 \cdot 2(C_2H_5)_2O$ see p. 248. Be reacts with Br_2 in a closed glass system at 400 °C to form $BeBr_2$. By sublimation single crystals of the orthorhombic α-$BeBr_2$ (very hygroscopic) are obtained, isomorphous with α-$BeCl_2$, Lazarini [9]. By passing Br_2 vapor with a stream of argon over Be powder at 550 °C $BeBr_2$, which sublimes, is formed. In a similar manner BeI_2 is formed at 480 °C, Wood, Brenner [14].

Be reacts with iodine at 400 to 450 °C to form the hygroscopic BeI_2, which was purified by sublimation, Kopelman, Bender [10]. The reaction of Be chips in a stream of iodine and H_2, carried out in silica glass tubes between 750 and 1100 °C (the temperature is without influence on the yield), leads also to BeI_2. The sublimed BeI_2 was found to be tetragonal (single crystal X-ray diffraction), Johnson et al. [11], but it is isostructural with orthorhombic α-$BeBr_2$ according to Semenko, Naumova [12]. Three polymorphic forms of BeI_2 were found by [12] and two by [11]. Only one form was found for $BeBr_2$ [12]. The corrosion of Be by iodine vapor is accelerated by the presence of humidity, Gurovich [15].

References:

[1] P.M. O'Donnell (J. Electrochem. Soc. **114** [1967] 1206/9).
[2] K.L. Churney, G.T. Armstrong (J. Res. Natl. Bur. Std. A **73** [1969] 281/97; C.A. **71** [1969] No. 33998).
[3] G.T. Armstrong (AD-467028 [1965] 81/8; AD-451711 [1964] 73/7; C.A. **67** [1967] No. 15556, **63** [1965] No. 14146).
[4] P. Gross, C. Hayman, D.L. Levi, M.C. Stuart (PB-153445 [1960] 1/32, 3/4; C.A. **58** [1963] 7435).
[5] W.J. Johnson, A.A. Gilliland (J. Res. Natl. Bur. Std. A **65** [1961] 59/61).
[6] P. Gross, C. Hayman, P.D. Greene, J.T. Bingham (Trans. Faraday Soc. **62** [1966] 2719/24).
[7] V.B. Parker, D.D. Wagman, W.H. Evans (NBS-TN-270-6 [1971] 3).

[8] L.B. Knight, M.B. Wise, A.G. Childers, et al. (J. Chem. Phys. **73** [1980] 4198/202, **74** [1981] 4256/60).
[9] F. Lazarini (J. Appl. Cryst. **8** [1975] 568).
[10] B. Kopelman, H. Bender (J. Electrochem. Soc. **98** [1951] 89/93; AECU-1028 [1951] 1/17; N.S.A. **5** [1951] No. 2785).
[11] R.E. Johnson, E. Staritzky, R.M. Douglass (J. Am. Chem. Soc. **79** [1957] 2037/9).
[12] K.N. Semenko, T.N. Naumova (Zh. Strukt. Khim. **4** [1963] 67/72; J. Struct. Chem. [USSR] **4** [1963] 59/62).
[13] L.R. Batsanov, G.S. Yur'ev, V.P. Doromina (Zh. Strukt. Khim. **9** [1968] 79/85; J. Struct. Chem. [USSR] **9** [1968] 63/8).
[14] G.B. Wood, A. Brenner (J. Electrochem. Soc. **104** [1957] 29/37, 35).
[15] E.I. Gurovich (Tr. Soveshch. Vopr. Korrozii **1940** 135/7 from C.A. **1942** 6117).

6.1.6 With Sulfur, Selenium, Tellurium, and Polonium

The diffusion coefficient D of ^{35}S and ^{75}Se in polycrystalline Be was studied at 1055 to 1200 °C (for S) and 1100 to 1210 °C (for Se) by the method of layerwise radiometric analysis. At 1100 °C the values found are $D=(4.6\pm0.9)\times10^{-12}$ cm²/s for S and $D=(5.5\pm1.5)\times10^{-11}$ cm²/s for Se. The temperature dependences for the range studied are $D=(11^{+2.7}_{-8})$ exp $[(-77200\pm3400)/RT]$ for S and $D=3.2$ exp $[(-68000\pm6000)/RT]$ for Se with the activation energies 77.2 ± 3.4 and 68 ± 6 kcal/g-atom, respectively, Anan'in et al. [1].

Be reacts slowly with sulfur at 1350 °C with formation of BeS, v. Wartenberg [2]. Melting of Be with Se or Te produces beryllium selenide or telluride, respectively; the former bursts into flame, Illig [3]. BeTe is formed as a crystalline powder by passing Te vapor over Be metal at 1100 °C. The reaction product formed after about 3 h was removed, ground, and then exposed to the Te vapor at 1100 °C for an additional 3 h, Yim et al. [4].

The reaction of Be with Po vapor in a sealed capillary which contains the Be at one end at 600 °C and the Po at the other end at 575 °C results in a black powder of crystalline BePo (fcc) after 7 h. The formation has been followed by counting the γ-rays originating at the Be containing end of the capillary, Witteman et al. [5].

References:

[1] V.M. Anan'in [Ananyin], V.P. Gladkov, V.S. Zotov, et al. (Fiz. Metal. Metalloved. **52** [1981] 1318/20; Phys. Metals Metallog. [USSR] **52** No. 6 [1981] 167/70).
[2] H. v. Wartenberg (Z. Anorg. Allgem. Chem. **252** [1943] 136/43, 136).
[3] K. Illig (Wiss. Veröff. Siemens-Konzern **8** [1929] 74/82, 80/1).
[4] W.M. Yim, J.P. Dismukes, E.J. Stofko, R.J. Paff (J. Phys. Chem. Solids **33** [1972] 501/5).
[5] W.G. Witteman, A.L. Giorgi, D.T. Vier (J. Phys. Chem. **64** [1960] 434/40; AECD-4237 [1957] 1/23; N.S.A. **11** [1957] No. 3413).

6.1.7 With Boron

Several Be-B phases are observed by heating briquetted mixtures of Be and B powders with variable atomic ratio in an argon atmosphere up to 1600 °C: Be_4B (tetragonal), Becher, Schäfer [1], Be_2B (cubic), Markovsky et al. [2, 3], Sands et al. [4], Hoenig et al. [5], Markevich et al. [6], BeB_2 (hexagonal) [2, 4, 5] and solid solutions of Be in B (up to 13 wt%), Vekshina et al. [7], including the compositions BeB_6 [2, 7] and BeB_{12} [7], Becher [8]. The presence of dissolved Be favors various phase transformations of boron [7], see also [8], Becher, Schäfer [9]. See also Elliott [10] for a review of the Be-B phases, and also Stecher, Aldinger [11] who report the phase diagram Be-B up to 70 at% B.

Several experiments have been started in order to get adherent boride layers on a Be surface. After hot pressing of solid Be with amorphous boron powder for 24 h at 1000 °C in an ultraclean high-vacuum environment layers of Be, BeO, BeB_2 (10 μm thick), BeB_6 (≈200 μm thick), and B are observed with a clean separation at the BeO-BeB_2 interface. A bonded interface with a diffusion zone of about 10 μm thickness could be obtained as follows: a rod of β-boron was buried in high-purity Be powder and at first cold and then hot isostatic pressed at an Ar pressure of 1020 atm for 2 h at 950 °C. This technique appears to be successful for producing Be-B diffusion couples. Adherent hard boride films of ≦1 μm thickness were obtained by the chemical vapor deposition of a BCl_3-H_2 mixture on a polished Be disc at ≧850 °C. It is quite possible that the presence of oxygen may be responsible for the cracking and spalling of thicker films. By Auger spectra the presence of BeO, BeB_2, and Be_2B have been identified in the films, Das et al. [12].

References:

[1] H.J. Becher, A. Schäfer (Z. Anorg. Allgem. Chem. **318** [1962] 304/12).
[2] L.Ya. Markovskii, Yu.D. Kondrashev, G.V. Kaputovskaya (Zh. Obshch. Khim. **25** [1955] 1045/52; J. Gen. Chem. [USSR] **25** [1955] 1007/12).
[3] L.Ya. Markovskii, G.S. Markevich (Zh. Prikl. Khim. **33** [1960] 1667/9; J. Appl. Chem. [USSR] **33** [1960] 1647/8).
[4] D.E. Sands, C.F. Cline, A. Zalkin, C.L. Hoenig (Acta Cryst. **14** [1961] 309/10).
[5] C.L. Hoenig, C.F. Cline, D.E. Sands (J. Am. Ceram. Soc. **44** [1961] 385/9).
[6] G.S. Markevich, Yu.D. Kondrashev, L.Ya. Markovskii (Zh. Neorgan. Khim. **5** [1960] 1783/7; Russ. J. Inorg. Chem. **5** [1960] 865/7).
[7] N.V. Vekshina, L.Ya. Markovskii, Yu.D. Kondrashev, I.M. Stroganova (Zh. Prikl. Khim. **42** [1969] 1229/34; J. Appl. Chem. [USSR] **42** [1969] 1168/71).
[8] H.J. Becher (Z. Anorg. Allgem. Chem. **306** [1960] 266/72, **321** [1963] 217/23).
[9] H.J. Becher, A. Schäfer (Z. Anorg. Allgem. Chem. **306** [1960] 260/5).
[10] R.P. Elliott (Constitution of Binary Alloys, 1st Suppl., McGraw-Hill, New York 1965, p. 109).

[11] J. Stecher, F. Aldinger (Z. Metallk. **64** [1973] 684/9).
[12] D. Das, K. Kumar, E. Wettstein, J. Wollam (AD-A084 780 [1979] 1/70).

6.1.8 With Carbon

The heat for the reaction $Be(g)+2C(s)=BeC_2(g)$ has been studied by mass spectrometry, Chupka, Berkowitz [1].

The results of diffusion measurements [2 to 8] of C in Be show little agreement and the mechanism of diffusion is not completely clarified. The diffusion activation energy found varied between 16 and 90 kcal/mol, see e.g., Zotov [4]. There is a change in the diffusion mechanism between 600 and 870 K. At low temperatures, the C atoms are assumed to occupy interstitial positions and at higher temperatures positions of substitution. At about 720 K, C has the highest mobility, while at higher temperatures C becomes one of the least mobile impurities in Be [2, 3]. The change of diffusion mechanism is connected with activation energies between 40 and 50 kcal/mol according to [3]. Values for the diffusion coefficient D for the volume diffusion in Be at 620 K at diffusion times t=25 to 300 h [2] are:

t in h	25	50	125	300
$D \cdot 10^{13}$ in cm^2/s	2.8 ± 0.5	3.0 ± 0.6	2.6 ± 0.4	0.52 ± 0.08

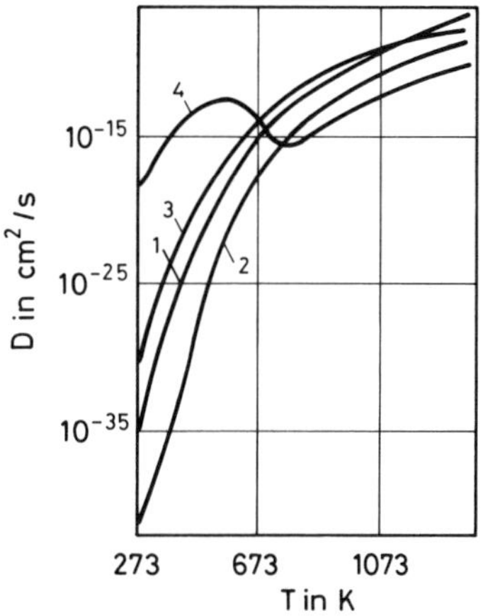

Fig. 6-12

Diffusion coefficient D vs. temperature T for the diffusion in Be of C (curve 4, calculated by [3]), of Cu (curve 1), of Fe (curve 2), and of Be (curve 3).

The experimental value at T = 620 K agrees well with the calculated curve D = f(T) by [3], shown in **Fig.** 6-**12** (curve 4), which also contains curves for the diffusion of Cu, Fe, and Be (self diffusion) in Be [2]. Studies for dependence of D on the orientation along c axis in single crystals at temperatures between 900 and 1240 °C see [5]. The results of diffusion measurements by means of the radiometric (^{14}C) method between about 1000 and 1500 K using arc-remelted and powder metallurgy Be are not sufficient to explain the diffusion mechanism completely. The relative low mobility of C (lower than that of other elements, as Fe, Ni, Cu, Ag, and of Be self diffusion) may be associated with the formation of dumbbell-shaped C-C pairs. No concentration of ^{14}C on the grain boundaries are found for both materials [4].

From metallographic studies the solubility of C in Be seems to be negligible, Kaufmann et al. [9]. The limit of solubility S_L of C in hexagonal α-Be (and other hexagonal metals) at the α-β transition temperature T_{tr} is expressed by $S_L = 64\ \exp(-9000/2\ T_{tr})$ with S_L in at% and T_{tr} in K. $S_L \approx 3$ at% is obtained for Be at $T_{tr} = 1473$ K, Anan'in et al. [10]. From the change of the lattice constant c in α-Be with increasing temperature, S_L is assumed to be about 1.5 at% C at T_{tr} by Amonenko et al. [11]. For equations to calculate the solubility of C in metals see the discussion of Burylev [12].

On heating two-layer Be-C films (consisting of a thin Be layer deposited on thin carbon films, 100 to 200 Å thick) in vacuum with a residual air pressure of 10^{-5} Torr, oxidation starts at 200 °C and continues to about 550 °C. At this temperature a mixture of BeO and Be_2C (cubic, a = 4.33 Å) has been formed, Trillat et al. [13].

Be sheets tightly packed with graphite powder in a steel capsule showed no carbon pick-up after 200 h at 500 or 600 °C, Baird et al. [14]. When Be powder in intimate mixture with carbon is heated at ≧900 °C in absence of air, hexagonal Be_2C is formed, Busch [15]. Be reacts with sugar charcoal in vacuum at 1000 °C to form Be_2C, Oishi, Hamano [16]. Molten Be reacts considerably with compact graphite, Markovskii et al. [17], to form Be_2C in an exothermic reaction. The reaction seems to start about 50 °C above the melting point of Be. The reaction could not always be brought about, apparently because, in some cases, a thin adherent layer of Be_2C prevented further reaction, Mallett et al. [18]. By heating pellets of Be to about 1400 °C for a few minutes, a layer of Be_2C was formed on the interior of graphite effusion cells, Pollock [19].

The kinetics of the reaction of Be with graphite has been studied by contact annealing of diffusion couples in vacuum between 750 and 1100 °C. The thickness δ of the cubic Be_2C

layer formed at 750 °C after 70, 191, and 300 h is 63, 108, and 133 μm, respectively. The rate of layer growth with time t, expressed by $\delta = k \cdot t^n$ with n = 0.517, is limited by the diffusion stage. The Be_2C layer grows mainly on graphite, which indicates that the mobility of Be in this layer greatly exceeds that of carbon. The temperature dependence of the rate constant k between 750 and 1100 °C can be expressed by $k = (5.47^{+1.69}_{-1.30}) \times 10^{-3}$ exp $[-(12290 \pm 600)/RT]$ with k in $cm \cdot s^{-1/2}$, T in K, and R in $cal \cdot mol^{-1} \cdot K^{-1}$, Zagryazkin et al. [20].

References:

[1] W.A. Chupka, J. Berkowitz (J. Phys. Chem. **62** [1958] 611/4).

[2] E.V. Deshkevich, V.S. Zotov, N.K. Lashchuk, et al. (Ukr. Fiz. Zh. **27** [1982] 1588/9; C.A. **98** [1983] No. 41016).

[3] N.K. Lashchuk, V.G. Tkachenko, V.I. Trefilov (Dokl. Akad. Nauk SSSR **258** [1981] 1103/6; Soviet Phys.-Dokl. **26** [1981] 608/10).

[4] V.S. Zotov (Fiz. Khim. Obrab. Mater. **4** [1979] 125/9; C.A. **91** [1979] No. 161723).

[5] V.P. Gladkov, V.S. Zotov, D.M. Skorov (Mater. At. Tekhn. No. 1 [1975] 28/32; Ref. Zh. Khim. **1976** No. 20B675).

[6] V.P. Gladkov, V.S. Zotov, I.I. Papirov, et al. (Poluch. Issled. Svoistv Chistykh Metallov II [Charkov] **1970** 56/61).

[7] V.P. Gladkov, V.S. Zotov, M.D. Skorov (At. Energ. [USSR] **32** [1972] 163/4; Soviet J. At. Energy **32** [1972] 179/81; C.A. **76** [1972] No. 118594).

[8] V.M. Anan'in, V.P. Gladkov, V.S. Zotov, D.M. Skorov (Diffuzionny Protsessy v Berillii, Energoizdat, Moscow 1981).

[9] A.R. Kaufmann, P. Gordon, D.W. Lillie (Trans. Am. Soc. Metals **42** [1950] 785/44).

[10] V.M. Anan'in, V.P. Gladkov, V.S. Zotov, D.M. Skorov (Obshch. Zakonomerm. Str. Diagramm Sostoyaniya Metal. Sist. Mater. 5th Vses. Soveshch., Moscow 1971 [1973], pp. 45/6; C.A. **81** [1974] No. 141562).

[11] V.M. Amonenko, V.E. Ivanov, G.F. Tikhinskii, V.A. Finkel' (Fiz. Metal. Metalloved. **14** No. 1 [1962] 128/9; Phys. Metals Metallogr. **14** No. 1 [1962] 114/5).

[12] B.P. Burylev (Zh. Fiz. Khim. **43** [1969] 1365/79; Russ. J. Phys. Chem. **43** [1969] 761/70).

[13] J.J. Trillat, L. Tertian, M. Bonnet-Gros (Compt. Rend. **251** [1960] 10/3; Mem. Sci. Rev. Met. **57** [1960] 845/51).

[14] J.D. Baird, G.A. Geach, A.G. Knapton, K.B.C. West (Proc. 2nd Intern, Conf. Peaceful Uses At. Energy, Geneva 1958, Vol. 5, pp. 328/33; C.A. **1960** 22066).

[15] L.S. Busch (in C.A. Hampel, Encyclopedia of Chemical Elements, Van Nostrand, New York 1968, pp. 49/56).

[16] Y. Oishi, Y. Hamano (Osaka Kogyo Gijutsu Shikenjo Kiho **8** [1957] 89/93 from C.A. **1959** 13855).

[17] L.Ya. Markovskii, N.V. Vekshina, R.A. Shtrikhman (Ogneupory **22** [1957] 42/6; C.A. **1957** 18526).

[18] M.W. Mallett, E.A. Durbin, M.C. Udy, et al. (J. Electrochem. Soc. **101** [1954] 298/305).

[19] B.D. Pollock (J. Phys. Chem. **63** [1959] 587/9; NAA-SR-2964 [1958]1/12; N.S.A. **13** [1959] No. 3579).

[20] V.N. Zagryazkin, A.S. Panov, M.M. Rysina (Izv. Akad. Nauk SSSR Neorgan. Materialy **12** [1976] 352/3; Inorg. Materials [USSR] **12** [1976] 304/5).

6.1.9 With Silicon, Phosphorus, and Arsenic

No compound is formed between Be and Si and the solubility of Si in Be is reported to be <0.97 wt%, Hindle, Slattery [1], or neglegible, respectively, Kaufmann et al. [2]. For the reaction of Be with SiO_2-coated Si wafers see p. 274. Deposition of Si onto a clean Be (0001) surface results in an ordered superstructure (up to about 300 °C). Epitaxial growth of crystalline Si films is hindered; at low temperature by the low mobility of the adsorbed atoms (the Si films are amorphous) and at high temperatures (400 or 500 °C) by the rapid diffusion of Si into the interior of the substrate, Jona [3]. For the diffusion coefficient and the solid solubility of electrically active Be in Si see Tomokage et al. [12].

The diffusion coefficient D of ^{32}P in polycrystalline Be was studied between 950 and 1240 °C by the method of layerwise radiometric analysis. D in cm^2/s in dependence of temperature t (two rows):

t in °C	950	1112	1150	1240
$D \cdot 10^8$	0.037; 0.041	2.6; 3.3	6.9; 8.2	45; 56

In the Arrhenius plot $D = D_0 \exp(-E_A/RT)$ the values $D_0 = (7.50^{+4.13}_{-2.66}) \times 10^6$ and $E_A = 91.2 \pm 1.2$ kcal/g-atom are obtained, Anan'in et al. [4]. Fine Be powder reacts at 750 °C with a stream of P vapor containing H_2 to form brown, hygroscopic Be_3P_2, von Stackelberg, Paulus [5].

Since the direct reaction of the elements in a sealed ampule is very violent, Be is heated at 700 °C successively with small quantities of P with intermediate crushing to obtain Be_3P_2, El Maslout et al. [6]. Heating 200 mg Be in several steps (Al_2O_3 or Ta crucible in an evacuated quartz ampule) with a total of 580 mg P (with intermediate homogenization) produces a mixture of Be_3P_2 and BeP_2 (red-brown, nonhygroscopic); further heating to 1000 °C with addition of P leads to BeP_2, Brice et al. [7]. In a similar manner mixtures of Be and P (<50 mol%P) are heated stepwise up to 1000 °C and black crystalline BeP_2 is obtained along with small amounts of (yellow) Be_3P_2, David, Lang [8]. Heating Be with P in the presence of traces of H_2 in sealed ampules under the same conditions as described above (under [7]) leads to the formation of nonstoichiometric ternary phases $BeP_{2-x}H_y$ (x = 1.70 to 1.85, y = 0.27 to 0.12) instead of BeP_2, Brice et al. [9, 10].

Heating pressed powdered mixtures of Be and As (atomic ratio 1:2) in sealed tubes at 570 °C (3 times for 24 h) and at 700 °C (2 times for 24 h), with intermediate homogenizing, produces black nonhygroscopic $BeAs_2$, Gerardin, Aubry [11].

References:

[1] E.D. Hindle, G.F. Slattery (Monogr. Rept. Ser. Inst. Metals [London] No. 28 [1963] 651/64, 660).
[2] A.R. Kaufmann, P. Gordon, D.W. Lillie (Trans. Am. Soc. Metals **42** [1950] 785/844, 801).
[3] F. Jona (J. Appl. Phys. **44** [1973] 4240/1).
[4] V.M. Anan'in [Ananyin], V.P. Gladkov, V.S. Zotov, et al. (Fiz. Metal. Metalloved. **52** [1981] 1318/20; Phys. Metals Metallog. [USSR] **52** No. 6 [1981] 167/70).
[5] M. von Stackelberg, R. Paulus (Z. Physik. Chem. B **22** [1933] 305/22, 308).
[6] A. El Maslout, J.P. Motte, A. Courtois, J. Protas, C. Gleitzer (J. Solid State Chem. **15** [1975] 223/8).
[7] J.F. Brice, R. Gerardin, M. Zanne, et al. (Mater. Res. Bull. **10** [1975] 1237/41).
[8] J. David, J. Lang (Compt. Rend C **282** [1976] 43/4).
[9] J.F. Brice, R. Gerardin, A. El Maslout, et al. (Mater. Res. Bull. **10** [1975] 1243/8).
[10] J.F. Brice, R. Gerardin, A. El Maslout, et al. (Fr. Demande 2316189 [1975/77]; C.A. **87** [1977] No. 186630).

[11] R. Gerardin, J. Aubry (J. Solid State Chem. **17** [1976] 239/44).
[12] H. Tomokage, M. Hagiwara, K. Hashimoto (Mem. Fac. Eng. Kyushu Univ. **42** [1982] 89/94 from C.A. **97** [1982] No. 173105).

6.2 Reactions with Metals

6.2.1 With Antimony

The Be–Sb phase diagram is not known. Be soaked in liquid antimony for 5 h at 700 °C was not wetted, no attack was apparent, and no Be was detected chemically in the antimony, Geach, Stubbs [1]. A similar result was also reported by Baird et al. [2], see also Miller, Baird [3].

In alloys prepared by the simultaneous condensation of the vaporized components on glass at room temperature, no intermetallic compound could be detected by electron diffraction studies, Palatnik, Kosevich [4]. At substrate temperatures below 400 °C, the condensed metals have the form of a mirror, Sb bright and Be mat. Between 450 and 480 °C, a narrow band, dark in reflected and transparent in transmitted light, appears on the substrate. Between 400 and 450 °C, this band is either not formed or it does not have such a sharply expressed character. Only one compound, Be_3Sb_2, was observed; it is air stable and has nonmetallic properties, Ugai et al. [5]. The composition of the compound is given as Be_2Sb_3 by Stonehouse [6] who detected this phase in hot pressed Be–Sb alloys with 0.1 to 16 wt% Sb. In alloys with a large excess of Be, the compound forms rapidly upon cooling. Be_2Sb_3 forms slowly from a mixture of powdered Be and powdered Sb at ~550 °C and within 2 to 5 h at 615 °C. Although Be_2Sb_3 will be formed using any proportions of the two metals, the best results are obtained with the Be being at least in excess of the stoichiometric amount required for Be_2Sb_3. The phase, which was characterized by X-ray diffraction (fcc lattice with a = 11.82 Å, order–disorder transformation near the melting point), melts between 620 and 630 °C and dissociates in Be and in molten Sb [6]. A reaction in the solid state between pressed Be and Sb mixtures with a Be:Sb atomic ratio $\geqq 1$ occurred between 550 and 600 °C over 100 h and the composition of the resulting compound was reported as $BeSb_2$. The compound was found to have pseudocubic symmetry (a′ = 5.88 Å) and to dissociate in Be and Sb at 630 °C, Gerardin, Aubry [7]. Another compound, $Be_{13}Sb$, was obtained from a molten Be–Sb mixture (atomic ratio 2:1) by cooling in two steps; from 1280 to 600 °C over 4 h and then slowly to room temperature over 14 h. A BeO crucible inside a closed Fe crucible was used as the container. Three different crystal species were found. Two have cubic symmetry (fcc) with a = 6.22(1) and 5.91(1) Å, respectively, and have not been identified. The third crystal species was identified as $Be_{13}Sb$ and the crystal structure was studied (space group Fm3c, a = 10.046(3) Å, Z = 8), Haase, Martinez–Ripoll [8].

References:

[1] G.A. Geach, M.S. Stubbs (in: G.E. Darwin, J.H. Buddery, Metallurgy of the Rarer Metals, No. 7: Beryllium, Academic, New York 1960, pp. 1/392, 261).

[2] J.D. Baird, G.A. Geach, A.G. Knapton, K.B.C. West (Proc. 2nd Intern. Conf. Peaceful Uses At. Energy, Geneva 1958, Vol. 5, pp. 328/33; C.A. **1960** 22066).

[3] P.D. Miller, W.K. Boyd (Mater. Eng. **68** No. 1 [1968] 33/6).

[4] L.S. Palatnik, V.M. Kosevich (Kristallografiya **4** [1959] 673/7; Soviet Phys.-Cryst. **4** [1959] 633/6).

[5] Ya.A. Ugai, V.L. Gordin, V.Z. Anokhin (Zh. Neorgan. Khim. **9** [1964] 218/20; Russ. J. Inorg. Chem. **9** [1964] 119/20).

[6] A.J. Stonehouse (U.S. 3574608 [1971]; C.A. **74** [1971] No. 145675).

[7] R. Gerardin, J. Aubry (J. Solid State Chem. **17** [1976] 239/44).

[8] A. Haase, M. Martinez-Ripoll (Acta Cryst. B **33** [1977] 555/7).

6.2.2 With Bismuth

No Be-Bi phase diagram has been reported. Be and Bi may be nearly immiscible in the liquid state, according to Kaufmann et al. [1]. However, a solubility of 0.1 wt% Bi in solid Be was reported in [2]. Good resistance of Be against liquid Bi at 300, 600, and 800 °C was reported by Koenig [3] and no attack was found after 24 h at 700 °C [2], Kelman et al. [4]. After soaking Be in molten Bi for 5 h at 1000 °C, the Bi contained <0.01 wt% Be [4]. In static tests in pure iron capsules, specimens of Be lost 15 to 25 mg per cm^2 after 24 h in Bi at 1000 °C, Koenig [10]. Using a Be crucible in an Ar atmosphere, the solubility S of Be in liquid Bi between 700 and 1000 °C was determined to be $\log S = 6.85 - 6140/T$ with T in K and S in ppm Be (by weight); $S \approx 5$ ppm at 1000 K and $S \approx 100$ ppm at 1273 K, Horsley, Maskrey [5]. At 600 to 800 K, the solubility of Be is approximately $\log S = 2.50 - 2750/T$. The data scatter widely, to $\sim +100\%$ and -50% from the log S-T curve. Quartz or graphite crucibles were used, Weeks [6].

In 1000 h capsule tests, Be shows good resistance in Bi fuel solutions (i.e., Bi containing U, Zr, and Mg additives) at 400 and 524 °C, Seifert [7], Seifert, Lowe [8], see also Lowe, Rozic [9].

References:

[1] A.R. Kaufmann, P. Gordon, D.W. Lillie (Trans. Am. Soc. Metals **42** [1950] 785/844, 801).

[2] J.F. Hogerton, R.C. Grass (Reactor Handbook, Vol. 3: Materials, Sect. 1: General Properties, AEC, Washington 1955, pp. 88, 92; AECD-3647 [1955] 88, 92; N.S.A. **9** [1955] No. 6476).

[3] R.F. Koenig (Iron Age **172** No. 8 [1953] 129/33).

[4] L.R. Kelman, W.D. Wilkinson, F.L. Yaggee (ANL-4417 [1950] 1/139, 85; N.S.A. **5** [1951] No. 400).

[5] G.W. Horsley, J.T. Maskrey (J. Inst. Metals **86** [1958] 401/2).

[6] J.R. Weeks (Am. Soc. Metals Trans. Quart. **58** [1965] 302/22, 303).

[7] J.W. Seifert (BAW-1067 [1959] 1/20; C.A. **1960** 2131).

[8] J.W. Seifert, A.L. Lowe (Corrosion **17** [1961] 475t/478t).

[9] A.L. Lowe, E.J. Rozic (Nucl. Sci. Eng. **2** Suppl. No. 1 [1959] 24/5; N.S.A. **13** [1959] No. 17333).

[10] R.F. Koenig (in: D.W. White Jr., J.E. Burke, The Metal Beryllium, ACS, Cleveland 1955, pp. 1/703, 549/54).

6.2.3 With Lithium

No Be-Li phase diagram is known and an X-ray study indicates that no compounds are formed, Yans [1], Klemm, Kunze [2]. It was shown experimentally [12] that ion implanted ^{6}Li ions in a Be single crystal (10^6 atoms/cm^2 are distributed over a depth range of $\sim$600 to 1800 Å) occupy substitutional sites. This result disagrees with expectations [13] based on considerations of chemical potentials and local electron densities, Kaufmann et al. [12].

By calculations of pseudopotential parameters, it is shown that there is a strikingly small interaction energy for Li in Be; with the result that Be is highly transparent to Li ions. The diffusion energy of Li in Be is shown to be much smaller than the self-diffusion energy of Be. The properties can lead to a quite plausible mechanism to explain the anomalous lattice sites for Li observed by [12] for ion-implanted Li in Be, Duesbery, Taylor [14].

Be is attacked by molten Li at 1000 °C, Miller, Boyd [3]. After a 200 h equilibration time, the approximate solubility of Be in Li at 1000 °C was found to be 0.22 wt% (0.17 at%); Bychkov et al. [4]. In 24 h capsule tests, the average solubility of Be in molten Li was found to be $\sim$3000 ppm (0.3 wt%) at 732 °C and 14000 ppm (1.4 wt%) at 1016 °C in the presence

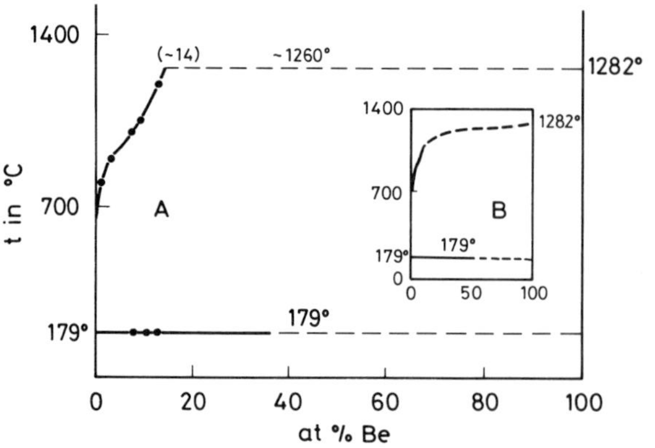

Fig. 6-13

Phase diagram Li-Be.

of Armco iron, Jesseman et al. [5]. The solubility of Be in molten Li (**Fig.** 6-**13**) could be determined in a reliable manner up to 1050 °C, but the values at higher temperatures are not certain due to vaporization of Li and the attack of the iron crucible. It is assumed that there is a restricted miscibility in the liquid state as indicated in Fig. 6-13, but considering the fact that the values of the miscibility at the highest temperature are uncertain, the presence of a eutectic system with a rather flat liquidus curve cannot be excluded (insert in Fig. 6-13) [2].

In static corrosion tests in Fe capsules of Be in molten Li at 1000 °C, Be is severely attacked. X-ray studies show the presence of Be_2Fe on the Armco iron ampule wall, but no Fe was found on the Be sample, Cunningham [6]. Similar static corrosion tests at 816 and 1000 °C were also carried out by Hoffman [7]. Poor resistance of Be against Li at 800 °C was reported by Koenig [8], however, good resistance (without presence of Fe) was reported by [7, p. 39]. In static tests at 600 °C, the solubility of Be in Li is extremely low and Be does not seem to react with Li, Wilkinson, Yaggee [9]. Good resistance of Be to Li at 600 °C was also reported by [3]. However, Be exposed to Li at 500 °C for a month is said to show a loss of 1.63 mg/cm^2 [10]. With mixtures of Li+As (atomic ratio 1:1:1), Be reacts above 500 °C to form the ternary compound BeLiAs, Schuster, Tiburtius [11].

References:

[1] F.M. Yans (NMI-1240 [1960] 1/41, 27; N.S.A. **15** [1961] No. 3084).

[2] W. Klemm, D. Kunze (Chem. Soc. [London] Spec. Publ. No. 22 [1967] 3/22, 11; C.A. **68** [1968] No. 16537).

[3] P.D. Miller, W.K. Boyd (Mater. Eng. **68** No. 1 [1968] 33/6).

[4] Yu.F. Bychkov, A.N. Rozanov, V.B. Yakovleva (At. Energiya [USSR] **7** [1959] 531/6; Soviet J. At. Energy **7** [1961] 987/92, 989).

[5] D.S. Jesseman, G.D. Roben, A.L. Grunewald, et al. (NEPA-1465 [1950] 1/15; N.S.A. **16** [1962] No. 6718).

[6] J.E. Cunningham (CF-51-7-135 [1957] 1/78, 27; N.S.A. **12** [1958] No. 5343).

[7] E.E. Hoffman (ORNL-2924 [1960] 1/150, 58; C.A. **1961** 6340).

[8] R.F. Koenig (Iron Age **172** No. 8 [1953] 129/33).

[9] W.D. Wilkinson, F.L. Yaggee (ANL-4990 [1950] 1/37, 18/9; N.S.A. **7** [1953] No. 705).

[10] J.F. Hogerton, R.C. Grass (Reactor Handbook, Vol. 3: Materials, Sect. 1: General Properties, AEC, Washington 1955, p. 88; AECD-3647 [1955] 88; N.S.A. **9** [1955] No. 6476).

[11] H.U. Schuster, C. Tiburtius (Z. Naturforsch. **31 b** [1976] 1536/7).

[12] E.N. Kaufmann, R. Vianden, T.E. Jackman, et al. (J. Phys. F **9** [1979] L23/L27).
[13] E.N. Kaufmann, R. Vianden, J.R. Chelikowsky, J.C. Phillips (Phys. Rev. Letters **39** [1977] 1671/5).
[14] M.S. Duesbery, R. Taylor (J. Phys. F **9** [1979] L19/L22).

6.2.4 With Sodium

No Be-Na phase diagram and no binary compound have been reported.

The thermotransport of Be in molten Na containing trace amounts of the ^7Be isotope (γ-emitter) was measured in a quartz capillary which was held at a known temperature gradient. The heat of Be transport was calculated as -6400 ± 2100 cal/g-atom from the slope of the plot ln c/c_0 vs. 1/T, where c/c_0 is the relative concentration of the solute (i.e., the resulting segregation) for a given temperature gradient within the range T=385 to 500 K. Assuming the heat of transport consists of an intrinsic and an extrinsic contribution, the intrinsic part was calculated to be -1400 cal/g-atom, Bhat et al. [1, 2].

No compound of Be and Na was detected by X-ray and chemical methods and melting point measurements. The solubility of Be in vacuum distilled Na (the oxygen content is not mentioned), using crucibles of soft iron in a purified Ar atmosphere, is about 1 at% at 400 °C and ~2.5 at% at 900 °C (values taken from the curve), Klemm, Kunze [3]. Contrary to [3], no solubility of Be in distilled Na was found in the loop (Na velocity 2.44 cm/s) in 1000 h dynamic tests at a temperature gradient of 700 → 575 °C. However, a fine gray powder of BeO was found at the surface of the Be sample and the bottom of the loop. No Be was found in the Na melt and no mass transfer of Be to the cold limb of the loop was detected, Bett, Draycott [4].

The corrosion of Be by liquid Na (or Na-K alloy) coolants has been studied and reported by several authors [e.g., 4 to 12, 20, 21] because of its interest to the nuclear reactor field. The corrosion of Be by molten Na (or Na-K alloys) is due to the formation of BeO from oxygen impurities in Na. The resulting BeO layer does not adhere to Be in flowing Na and so the reaction can proceed as long as oxygen is available [4 to 8]. Be can show good resistance to Na at 500 °C, if the oxygen content is lower than 0.01%, Beaver [13], Miller, Boyd [14]. According to [4, 5], Be is compatible with molten Na only when the Na has been thoroughly deoxidized. Cold trapping of Na does not remove the oxygen sufficiently and the attack was found to be catastrophic and very velocity dependent. Penetration rates of up to 0.0147 cm per month were found at 525 °C with a liquid metal (Na-K alloy) velocity of 213.36 cm/s (7 ft/s). The mechanism of corrosion was assumed to be the reduction of Na_2O by Be. The rate controlling factor is probably the diffusion of oxygen ions from the liquid metal stream to the Be surface. After removal of Na_2O by cold trapping, Ca can be used as a soluble deoxidant to reduce the rate of corrosion of Be. No mass transfer of the Ca or reaction with the N_2 atmosphere occurred and the reaction between Ca and Ni in the stainless steel pipework remained at a satisfactorily low level [4, 5]. However, because of the problem of removing CaO and other difficulties, the work on the liquid metal fuelled reactor system was not continued [5]. According to Kendall [9], the Be samples were nitrided if (~1%) Ca had been added as getter to the cold trapped sodium melt. After 520 h at 538 °C (1000 °F) under an Ar atmosphere, a black hard adherent film (thickness ≈50 µm) of Be_3N_2 was formed. This film showed no evidence of falling off under high velocity sodium and there was no evidence of erosion or material loss. The probable sources of nitrogen contamination were from Ca impurities or from gases carried into the system with Ca [9]. According to [15], the corrosion of Be in molten Na at 593 °C under an N_2 atmosphere occurred by the formation and removal of beryllium nitride. The rate of nitride formation varies with the amount of entrained nitrogen, the Ca content of the liquid metal, the temperature, and the distance of Be to the liquid metal nitrogen interface. However, no reaction of Be with the N_2 protecting atmosphere was reported by [4, 5].

Be can be used in Inconel systems (Ni alloys) containing molten Na if the temperature is held below about 650 °C, Adamson, Long [16]. In an Na melt, Be at 600 or 650 °C in contact or close proximity to hot Ni surfaces (Inconel) diffuses into the Ni to form a brittle Ni-Be alloy and causes surface voids in the Be, Bussard [17], see also [18].

With mixtures of Na and As (atomic ratio 1:1:1), Be reacts above 500 °C to form the ternary compound BeNaAs, Schuster, Tiburtius [19].

References:

[1] B.N. Bhat, S.P. Murarka, R.A. Swalin (COO-841-27 [1972] 1/6; C.A. **79** [1973] No. 139858).
[2] B.N. Bhat, S.P. Murarka, R.A. Swalin (Scr. Met. **7** [1973] 523/7; C.A. **80** [1974] No. 18366).
[3] W. Klemm, D. Kunze (Chem. Soc. [London] Spec. Publ. No. 22 [1967] 3/22, 11/2; C.A. **68** [1968] No. 16537).
[4] F.L. Bett, A. Draycott (Proc. 2nd Intern. Conf. Peaceful Uses At. Energy, Geneva 1958, Vol. 7, pp. 125/31, 127; C.A. **1961** 22022).
[5] A. Draycott (Chem. Process [Sydney] **13** No. 4 [1960] 27/35; C.A. **1960** 19199).
[6] L.F. Epstein (Proc. 1st Intern. Conf. Peaceful Uses At. Energy, Geneva 1956, Vol. 9, pp. 311/7; A-CONF-8-P-119 [1956]).
[7] A.J. Stonehouse, W.W. Weaver (Mater. Prot. **4** No. 1 [1965] 24/8).
[8] M. Davis, A. Draycott (Proc. 2nd Intern. Conf. Peaceful Uses At. Energy, Geneva 1958, Vol. 7, pp. 94/110; N.S.A. **13** [1959] No. 6764).
[9] W.W. Kendall (GEAP-3333 [1960] 1/15, 1, 12; C.A. **1960** 19418).
[10] A.B. McIntosh, K.Q. Bagley (J. Inst. Metals **84** [1956] 251/70, 254).

[11] G.C. Wheeler (AECU-3656 [1958] 1/23, 8; N.S.A. **12** [1958] 916).
[12] E.F. Batutis, C.A. Palladino, R. Gagne, J.W. Mausteller (NP-6178 [1956] 1/25 from N.S.A. **11** [1957] No. 3398).
[13] W.W. Beaver (Progr. Nucl. Energy V **1** [1956] 277/99, 294).
[14] P.D. Miller, W.K. Boyd (Mater. Eng. **68** No. 1 [1968] 33/6).
[15] E.F. Batutis, C.A. Palladino, R. Gagne, J.W. Mausteller (in: G.E. Darwin, J.H. Buddery, Beryllium: Metallurgy of the Rarer Metals, No. 7, Academic, New York 1960, p. 260; N.S.A. **14** [1960] No. 15917).
[16] G.M. Adamson, E. Long (CF-54-9-98 [1959] 1/12; N.S.A. **14** [1960] No. 2690).
[17] R.W. Bussard, R.E. MacPherson (CF-54-10-106 [1954] 1/36; N.S.A. **15** [1961] No. 17294).
[18] G.M. Adamson, E.E. Hoffman, C.R. Brooks (ORNL-2685 [1959] 88/107; N.S.A. **14** [1960] No. 24489).
[19] H.U. Schuster, C. Tiburtius (Z. Naturforsch. **31b** [1976] 1536/7).
[20] L.R. Kelman (CT-3726 [1955] 1/26, 6/7; N.S.A. **10** [1956] 869).

[21] R.F. Koenig (in: D.W. White Jr., J.E. Burke, The Metal Beryllium, ACS, Cleveland 1955, pp. 549/54).

6.2.5 With Potassium, Rubidium, and Caesium

No phase diagrams of Be with K, Rb, or Cs are known and no binary compounds have been found by means of X-ray and chemical analyses or melting point measurements, Klemm, Kunze [1]. However, the existence of KBe_2 was reported by Elliott [2].

No solubility of Be in the molten metals could be detected up to 400 °C. After 40 h annealing at 600 °C, the following solubilities of Be (in at%) were observed: 0.095 in K,

0.16 in Rb, and 0.18 in Cs [1]. According to Young [3], the solubility of Be in Rb between 538 and 760 °C is lower than 10^{-4} wt% (the lower limit of detection) and there are no indications of mass transfer effects or of Be corrosion by Rb liquid or vapor. In the 760 °C test, a small amount of black BeO was formed on the surface due to a trace amount of oxygen in the Rb [3]. The solubility of Be in Cs, which contained <0.01 wt% O_2, is reported as (in wt%) 1.2×10^{-4} at 50 °C, 1.1×10^{-4} at 150 °C, and 0.8×10^{-4} at 300 °C. A value of 7.1×10^{-4} is found at 300 °C with Cs containing 0.1 wt% O_2. For the experiments, Be was covered with Cs under vacuum, kept for 120 h in sealed ampules at the desired temperature, and then Cs was dissolved in H_2O. Metallographic studies of the Be surface show new phases indicating interaction between the Be and Cs, Godneva et al. [4].

Values for the desorption energy q of the desorption of Cs atoms and ions and for the electronic work function φ are calculated assuming a Cs coverage of nearly zero on the Be surface [5]: q (atom) = 2.50 eV, q (ion) = 2.69 eV, and $\varphi = 3.70$ eV. The frequency factor was calculated to be $3 \times 10^{13}\ s^{-1}$ for q (atom) = 2.50 eV. For the calculations, the experimental data of Wilson [6] for the electronic work function of polycrystalline Be in Cs vapor were used, Herion [5].

References:

[1] W. Klemm, D. Kunze (Chem. Soc. [London] Spec. Publ. No. 22 [1967] 3/22, 12; C.A. **68** [1968] No. 16537).
[2] R.P. Elliott (Constitution of Binary Alloys, 1st Suppl., McGraw-Hill, New York 1965/66, pp. 1/896, 165).
[3] P.F. Young (AGN-8063 [1962] 1/101, 34, 36; N.S.A. **17** [1963] No. 16634).
[4] M.M. Godneva, N.D. Sedel'nikova, E.S. Geizler (Zh. Prikl. Khim. **47** [1974] 2177/80; J. Appl. Chem. [USSR] **47** [1974] 2236/8).
[5] J. Herion (JUEL-1155 [1975] 1/67, 33, 36; C.A. **83** [1975] No. 49796).
[6] R.G. Wilson (J. Appl. Phys. **37** [1966] 3161/9).

6.2.6 With Alkaline Earth Metals

No phase diagrams Be-M with M = Mg, Ca, Sr, or Ba are known, however, the compounds MBe_{13}, with the fcc structure, are known. There is no doubt that MBe_{13} form eutectics with Be, but it cannot be decided whether the phase diagrams on the M-MBe_{13} side are eutectic or whether nonmiscibility in the melts exist. The methods of investigation were X-ray diffraction, chemical analysis, and melting point measurements, Klemm, Kunze [1].

Magnesium. An Mg-Be alloy with 0.2 wt% Be was used as the source to measure the Be diffusion in Mg. This alloy, which consisted of two phases, $MgBe_{13}$ and the Mg-Be solid solution, was prepared by simultaneous evaporation and condensation of Mg and Be under high vacuum. By annealing this alloy between two cylinders of Mg for a day in the temperature range 500 to 600 °C and then measuring the distribution of Be in Mg by means of local spectral analysis, the coefficient of diffusion D was measured to be $D = 8.06 \exp(-37490 \pm 2700)/RT$ with D in cm^2/s and T in K. (Measuring the diffusion under a hydrostatic pressure of 600 atm does not alter the results.) From the concentration distribution curves of Be in Mg at 500 to 600 °C, it is observed that the solid solubility of Be increases with increasing temperature from about 0.02 to 0.025 wt%, Yerko et al. [2]. A value of 0.005 wt% at 550 °C was established previously by Ivanov et al. [3]. A solid solubility of <0.05 wt% Be at 620 to 630 °C was found by Sinelnikov et al. [4]. At 750 to 800 °C, the maximum solubility of Be in molten Mg does not exceed 0.03 to 0.04 wt%, Ivanov et al. [4, 5]. Values for the solubility S of Be in liquid Mg between 650 and 1200 °C are given by [3]:

t in °C	650	750	850	950	1050	1100	1200
S in wt% . . .	0.011	0.028	0.062	0.16	0.2	0.28	0.39

In earlier publications, referred to by Hansen [6], no solubility of Be, even in boiling Mg, was reported and the solid solubility of Mg in Be is also negligible, Kaufmann et al. [7]. No reaction of Be with Mg was found after 8 d at 500 °C by Baird et al. [8]. In pressed mixtures of Be and Mg containing 5 to 20 wt% Mg (or in cold pressed Be powder) heated in molten Mg at 750 to 850 °C under argon, the compound $MgBe_{13}$ was formed which decomposes at 950 °C, Jones, Williams [9], also see Baker [10]. The decomposition at 950 °C was stated by [1]. Be also reacts with Mg by the powder metallurgical method to form $MgBe_{13}$, Baker, Williams [11], Ivanov et al. [5]. In Mg-Be alloys (studied up to 5 wt% Be), needles of $MgBe_{13}$ are observed after 40 h annealing at 600 °C [5]. The needles are only stable in vacuum up to 700 °C [3]. In the Mg-Be system, only one compound, $MgBr_{13}$, was found in equilibrium with Be [1] and Mg [4]. For the system Be-Mg, see also "Magnesium" A3, 1942, pp. 456/9.

Calcium. In the Be-Ca system, which was studied only at the Be rich side, a eutectic was found between Be and $CaBe_{13}$ at 0.056 wt% Ca (0.0126 at%), Potard et al. [16]. The solubility of Ca in Be is very low, <0.05 wt% [16]. $CaBe_{13}$ was present in alloys with 0.2 wt% Ca, Boisde, Schaub [12] and even in alloys with 0.1 wt% Ca, Hindle, Slattery [13]. Employing a spinel crucible, the solubility of Be in molten Ca was determined to be 5.1 at% at 1000 °C and 6.3 at% at 1100 °C [1]. Powder metallurgical methods can be used for the reaction of Be with Ca to form $CaBe_{13}$, Wolcott, Falge [14]. Alternatively, tinsels of Be can be heated for 60 h with Ca in a closed vessel at 1100 to 1200 °C [1] (but many authors [10, 11, 15] prepared $CaBe_{13}$ by reaction of BeO with Ca). The electrolytic deposition of Ca (from a eutectic CaF_2-$CaCl_2$ melt) on a Be cathode under Ar at 1000 °C forms in 2 h a surface layer containing $CaBe_{13}$ along with a second phase [12]. For Be-Ca alloys see also "Calcium" A2, 1957, p. 488.

Strontium and Barium. Solubility of Be in molten Sr and Ba [1]:

temperature in °C . .	800	900	1000	1100
at% Be in Sr	—	15.8	16.5	18.6
at% Be in Ba	16.7	18.9	21.7	25.5

Be tinsels with stoichiometric amounts of Sr were heated at 1000 °C for 20 h until the reaction to $SrBe_{13}$ was complete. With Ba, the formation of $BaBe_{13}$ is incomplete even after 60 h at 1000 °C [1].

References:

[1] W. Klemm, D. Kunze (Chem. Soc. [London] Spec. Publ. No. 22 [1967] 3/22, 12/4, 19/21; C.A. **68** [1968] No. 16537).

[2] V.F. Yerko, V.F. Zelenskii, V.S. Krasnorutskii (Fiz. Metal. Metalloved. **22** [1966] 112/4; Phys. Metals Metallog. [USSR] **22** [1967] 112/4).

[3] V.E. Ivanov, V.F. Zelenskii, V.K. Khorenko, et al. (Corrosion of Reactor Materials, Vol. 1, IAEA, Vienna 1962, pp. 343/67, 344).

[4] K.D. Sinelnikov, V.E. Ivanov, V.F. Zelensky (Proc. 2nd Intern. Conf. Peaceful Uses At. Energy, Geneva 1958, Vol. 5, pp. 234/40).

[5] V.E. Ivanov, V.F. Zelenskii, S.I. Failer, et al. (Poroshkovaya Met. **5** No. 5 [1965] 46/53; Soviet Powder Met. Metal Ceram. **1965** 385/90).

[6] M. Hansen (Constitution of Binary Alloys, 2nd Ed., McGraw-Hill, New York 1958, pp. 287/8).

[7] A.R. Kaufmann, P. Gordon, D.W. Lillie (Trans. Am. Soc. Metals **42** [1950] 785/844, 801).
[8] J.D. Baird, G.A. Geach, A.G. Knapton, K.B.C. West (Proc. 2nd Intern. Conf. Peaceful Uses At. Energy, Geneva 1958, Vol. 5, pp. 328/33; C.A. **1960** 22066).
[9] J.W.S. Jones, J. Williams (Powder Met. No. 8 [1961] 37/41).
[10] T.W. Baker (Acta Cryst. **15** [1962] 175/9).

[11] T.W. Baker, J. Williams (Acta Cryst. **8** [1955] 519).
[12] G. Boisde, B. Schaub (Fr. 1535585 [1967]; C.A. **71** [1969] No. 52953).
[13] E.D. Hindle, G.F. Slattery (Inst. Metals Monogr. Rept. Ser. No. 28 [1963] 651/64, 656).
[14] N.M. Wolcott, R.L. Falge (Phys. Rev. [2] **171** [1968] 591/5).
[15] J.H. Buddery, R.W. Thackray (J. Inorg. Nucl. Chem. **3** [1956] 190/3).
[16] C. Potard, G. Bienvenu, B. Schaub (Thermodyn. Nucl. Mater. Proc. Symp., Vienna 1967 [1968], pp. 795/807, 805/6).

6.2.7 With Zinc, Cadmium, and Mercury

No phase diagrams of Be with Zn, Cd, or Hg are known.

In the Be-Zn system, no binary compound exists and no solid solubility of Zn in Be was found, Yans [1]. At 700 to 900 °C, Be is wetted by liquid Zn and slightly dissolved, Nichkov, Smirnov [2]. The activity of Be in liquid Zn solutions, obtained from emf measurements at 594 to 790 °C, shows a positive deviation from Raoult's law. The dissolution of Be is accompanied by the absorption of heat and an increase in entropy. The temperature dependence of the activity coefficient γ fits satisfactorily the equation $\log \gamma_{Be}$ $(\pm 0.065) = -0.800 + 2945/T$ with T in K. The solubility of Be in liquid Zn can be expressed by $\log x_{Be}$ $(\pm 0.065) = 0.800 - 2945/T$ with x in atom fraction Be. The value obtained from emf measurements for the solubility at 700 °C, 0.082 wt%, is in good agreement with the directly measured value of 0.081 wt%, Dubinin et al. [3]. A value of 0.081 wt% at 696 °C was also found by Johnson, Anderson [4] who described the temperature dependence of the solubility S (in at% Be) between 429 and 696 °C with the equation $\log S = 3.211 - 3233/T$ with T in K.

The solubility of Be in liquid Hg in the absence of air at room temperature is <0.001 wt%, and no dissolved Be or weight loss was detected, Strachan, Harris [5]. With increasing temperature, the solubility increases uniformly from 0.01 ppm at 100 °C to 0.4×10^{-4} wt% at 800 °C. Be was not attacked or amalgamated after 330 h at 316 °C, but when the test vessel was open to air, a slight attack occurred and a white powdery layer (probably BeO) formed over the Be, Kelman et al. [6]. A pasty semisolid amalgam of Be with ~2 wt% Be was formed by the electrolysis of an equimolar salt mixture of $BeCl_2$ and NaCl at 300 to 350 °C. The electrolysis employed a stirred Hg cathode and a carbon anode under an $H_2 + Ar$ atmosphere. The resulting amalgam was not a solution, but a dispersion of Be or an intermetallic compound in Hg, Holden et al. [7], Kells et al. [8].

References:

[1] F.M. Yans (NMI-1240 [1960] 1/41, 12; N.S.A. **15** [1961] No. 3084).
[2] I.F. Nichkov, M.V. Smirnov (Izv. Vysshikh Uchebn. Zavedenii Tsvetn. Met. **1961** No. 3, pp. 105/7; C.A. **56** [1962] 1248).
[3] V.A. Dubinin, V.A. Lebedev, I.F. Nichkov, S.P. Raspopin (Zh. Fiz. Khim. **42** [1968] 678/81; Russ. J. Phys. Chem. **42** [1968] 356/8).
[4] I. Johnson, K.E. Anderson (ANL-6543 [1962] 92/3; N.S.A. **16** [1962] No. 28641).
[5] J.F. Strachan, N.L. Harris (J. Inst. Metals **85** [1956] 17/24).
[6] L.R. Kelman, W.D. Wilkinson, F.L. Yaggee (ANL-4417 [1950] 66/7; N.S.A. **5** [1951] No. 400).

[7] R.B. Holden, M.C. Kells, C.I. Whitman (Proc. 2nd Intern. Conf. Peaceful Uses At. Energy, Geneva 1958, Vol. 4, pp. 306/8).
[8] M.C. Kells, R.B. Holden, C.I. Whitman (J. Am. Chem. Soc. **79** [1957] 3925).

6.2.8 With Aluminium

There exists no compound between Be and Al and the mutual solid solubility is very small. After 8.5 h at 600 °C, no reaction was observed between Be and Al, Baird et al. [1]. The phase diagram is simple eutectic, Hansen [2], Aldinger, Petzow [3], Wright, Willey [9], and Gmelin Handbook "Aluminium" A4, 1936, pp. 637/8. Using thermal analysis and microscopic studies, a peritectic reaction on the Al rich side was found instead of a eutectic point at 646 °C, Nishi, Shinoda [4], but this was not confirmed by others [3]. The maximum solid solubility of Al in Be is ~0.02 at% at the eutectic temperature (646 °C), Jacobson, Hammond [5]. An arc melted alloy with 0.1 wt% Al (~0.03 at%) was two-phase, Hindle, Slattery [6]. A value for the solid solubility as low as $\leqq 7\times10^{-3}$ at% Al (the limit of detection at 600 and 800 °C) was reported by Myers, Smugeresky [7]. The solubility and the diffusion of Al in Be are studied between 795 and 1083 °C by means of the radioactive tracer method using the ^{26}Al isotope. The solubility S (in at%) is found to be (with T in K) $S=(13^{+18}_{-12})$ exp [(−12600±2500)/RT] and the diffusion coefficient D (in cm²/s) is $D=(1.0^{+2.1}_{-0.9})$ exp [(−40200±4300)/RT]. As shown in **Fig.** 6-**14**, the diffusion coefficient of Al is higher than those of Be, Ag, Cu, Fe, and Ni (taken from Papirov, Tikhinskii), Gladkov et al. [13].

For the solubility of Be in solid Al, no newer values exist. The older values, reported in [2], are: 0.05 to 0.06 wt% at 645 °C, 0.02 to 0.03 wt% at 600 °C, and 0.005 to 0.01 wt% at 500 °C; similar values are reported by Wright, Willey [9].

The solubility of Be in liquid Al between 680 and 800 °C was calculated from emf measurements to be $\log x(\pm0.051)=0.577-2056/T$ with T in K and x in atom fraction of Be [8]. For the concentration x of Be in the liquid phase between about 650 and 1100 °C, the equation $\ln x_{Be}=3.92-6580/T$ with x in atomic fraction was obtained by Potard et al. [10]. The activity of Be in the solutions shows a positive deviation from Raoult's law. The temperature depend-

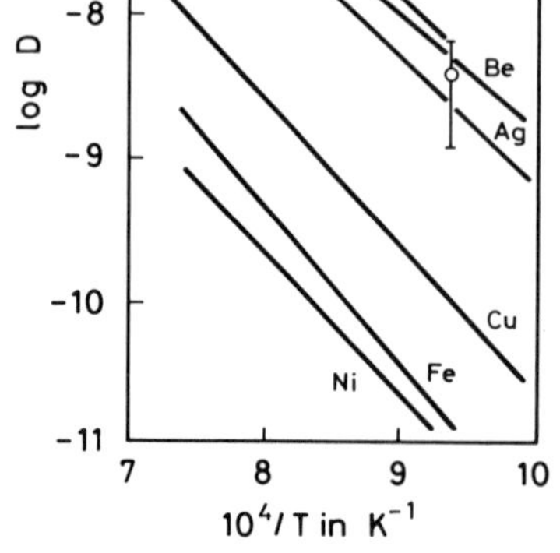

Fig. 6-14

Temperature dependence of diffusion coefficient of Al (and other metals) in Be with D in cm²/s.

ence of the activity coefficient fits satisfactorily a straight line which is described by the equation $\log \gamma_{Be}$ (± 0.051) $= -0.577 + 2056/T$ [8]. The activity coefficient of Al in an infinite dilute solution in liquid Be at 1600 K was determined to be $\gamma_{Al} = 4.64$, Bienvenu et al. [11]. The dissolution of Be in Al is accompanied by the adsorption of heat and an increase in excess entropy, Serebryakov et al. [8]. The thermodynamic properties of the liquid Be-Al system were studied at 1600 K. The thermodynamic activities and partial molar enthalpies were estimated from the partition data and verified by some measurements done with the Knudsen effusion method. The dependence of the values for the activity coefficients γ of Be and Al at 1600 K on the atom fraction x_{Al} in the melts:

x_{Al} . . .	0	0.1	0.2	0.3	0.4	0.5	0.6	0.7	0.8	0.9	1.0
γ_{Be} . .	1	0.9275	0.8858	0.8475	0.7927	0.7084	0.5910	0.4466	0.2894	0.1361	0
γ_{Al} . . .	0	0.2595	0.3431	0.3921	0.4435	0.5081	0.5886	0.6833	0.7880	0.8960	1

Fig. 6-**15** shows the integral values for the enthalpy ΔH, the Gibbs free energy ΔG, and the entropy ΔS for the Be-Al system at 1600 K. For the partial thermodynamic data see the tables in the paper, Schaub et al. [12].

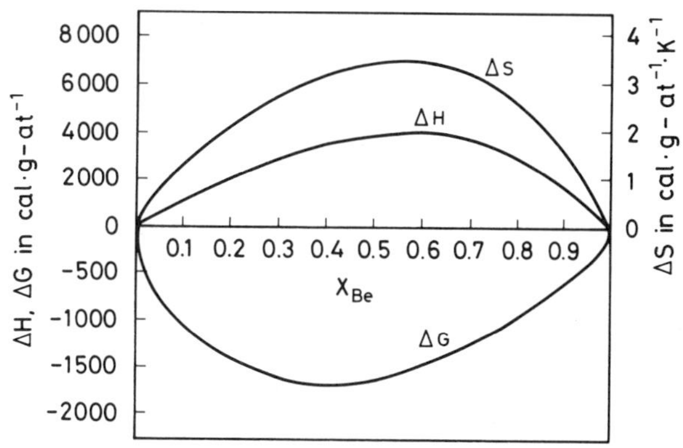

Fig. 6-15

The integral thermodynamic quantities ΔH, ΔG, and ΔS in the Be-Al system at 1600 K.

References:

[1] J.D. Baird, G.A. Geach, A.G. Knapton, K.B.C. West (Proc. 2nd Intern. Conf. Peaceful Uses At. Energy, Geneva 1958, Vol. 5, pp. 328/33; C.A. **1960** 22066).
[2] M. Hansen (Constitution of Binary Alloys, 2nd Ed., McGraw-Hill, New York 1958, pp. 73/4).
[3] F. Aldinger, G. Petzow (in: D. Webster, G.J. London, Beryllium Science and Technology, Vol. 1, Plenum, New York 1979, pp. 1/333, 235/305, 262/3).
[4] S. Nishi, T. Shinoda (Keikinzoku **16** [1966] 5/8; C.A. **69** [1968] No. 46508).
[5] M.I. Jacobson, M.L. Hammond (Trans. Met. Soc. AIME **242** [1968] 1385/91).
[6] E.D. Hindle, G.F. Slattery (Inst. Metals Monogr. Rept. Ser. No. **28** [1963] 651/64).
[7] S.M. Myers, J.E. Smugeresky (Met. Trans. A **7** [1976] 795/802, 799).
[8] G.A. Serebryakov, V.A. Lebedev, I.F. Nichkov, et al. (Zh. Fiz. Khim. **45** [1971] 2092/4; Russ. J. Phys. Chem. **45** [1971] 1186/7).
[9] E.H. Wright, E.A. Willey (ALCOA Res. Lab. Tech. Papers No. 15 [1960] 1/46, 5, 34; C.A. **61** [1964] 9243).
[10] C. Potard, G. Bienvenu, B. Schaub (Thermodyn. Nucl. Mater. Proc. Symp., Vienna 1967 [1968], pp. 809/25, 816; C.A. **69** [1968] No. 79620).

[11] G. Bienvenu, C. Potard, B. Schaub, P. Desré (Thermodyn. Nucl. Mater. Proc. Symp., Vienna 1967 [1968], pp. 777/87, 778).
[12] B. Schaub, C. Potard, P. Desré (Proc. 1st Intern. Conf. Calorim. Thermodyn., Warsaw 1969, pp. 1011/3; AED-CONF-69-261 [1971]; CONF-690834-2 [1969] 1/8; CEA-CONF-1502 [1969] 1/8; N.S.A. **24** [1970] No. 32436).
[13] V.P. Gladkov, A.V. Svetlov, D.M. Skorov, et al. (At. Energiya [USSR] **40** [1976] 257/8; Soviet J. At. Energy **40** [1976] 306/7).

6.2.9 With Gallium and Indium

In the Be-Ga system no binary compound was found, and immiscibility in the liquid state between 10 and 97.8 wt% Ga (~1.4 and 85.2 at%) was observed. There is no evidence for the mutual solid solubility of Ga and Be. The solubility S of Be in liquid Ga, taken from the straight curve log S = f(1/T) is about 0.03 wt% at 600 °C, 0.17 wt% at 800 °C, and 2.2 wt% at the melting point of Be (T in K), Elliott, Cramer [1]. The values $S = 4 \times 10^{-4}$ wt% at 600 °C and 0.9 wt% at 800 °C have been reported by Kelman et al. [2], Wilkinson [3]. In static corrosion tests, the resistance of Be to attack by Ga is good up to 500 °C, limited at 500 to 600 °C, and poor at 600 to 800 °C. However, when the Ga is saturated with Be in a static system, the resistance is good [2], Koenig [4]. At 600 °C, Ga diffuses very slowly into Be to form a reaction zone which retards further diffusion and which was believed to be a protective compound [3] (this was refuted by [1], see above). Since at 600 °C, fine-grained Be was more readily attacked by Ga than was coarse-grained Be, grain boundary attack may be the mechanism of the Be corrosion, Udy et al. [5]. Be is intergranularly attacked by Ga at 800 °C [1] and 815 °C, Jaffee et al. [6].

The Be-In phase diagram shows immiscibility in the liquid state, absence of intermediate phases, and insolubility in the solid state. The solubility of Be in liquid In is very low; the maximum being 0.02 wt% at the melting temperature of Be [1].

References:

[1] R.O. Elliott, E.M. Cramer (AECU-3022 [1952] 1/10; N.S.A. **9** [1955] No. 5954).
[2] L.R. Kelman, W.D. Wilkinson, F.L. Yaggee (ANL-4417 [1950] 111/3; N.S.A. **5** [1951] No. 400).
[3] W.D. Wilkinson (ANL-5027 [1953] 1/67, 9; N.S.A. **8** [1954] 24).
[4] R.F. Koenig (in: D.W. White Jr., J.E. Burke, The Metal Beryllium, ACS, Cleveland 1955, pp. 549/54).
[5] M.C. Udy, H.L. Shaw, F.W. Boulger (Nucleonics **11** No. 5 [1953] 52/9).
[6] R.I. Jaffee, R.M. Evans, E.A. Fromm, B.W. Gonser (AECD-3317 [1949/52] 1/44; BMI-T-17 [1952] 1/44 from [1]; N.S.A. **6** [1952] No. 2079).

6.2.10 With Rare Earth Metals

Be seems to form three compounds with Sc: $ScBe_{13}$, Laube, Nowotny [1], Sc_2Be_{17}, Gladyshevskii et al. [2], and $ScBe_5$ [2], Protosov, Gladyshevskii [3]. With the other rare earth elements M = Y, La, Ce, Pr, Nd, Sm, Eu, Tb, Dy, Ho, Er, Tm, Yb, and Lu, only the isotypic compounds, MBe_{13} with the fcc $NaZn_{13}$ structure are known, Gladyshevskii et al. [4]; for MBe_{13} with M = Sc, Y, La see also [1], with M = Y, La, Ce, and Er, Gschneidner [5], with M = Gd see Kruglykh et al. [10]. $ScBe_5$ forms by heating a Be-Sc mixture (84 at% Be) in a BeO crucible under Ar and subsequently annealing the resulting product in sealed tubes for 200 h at 800 °C [3]. The compounds MBe_{13} are formed by melting the elemental mixtures containing 92.3 at% Be in BeO or BeO lined Al_2O_3 crucibles under an Ar atmosphere [4] or by sintering the elemental mixtures at 1250 °C [1] or 1170 °C (M = Y), Matyushenko, Tikhinskii [6]. $GdBe_{13}$ was formed by diffusion sintering of powdered Be-Gd mixtures (slight Be excess) in vacuum in Ta crucibles for 2 h at 1200 °C [10]. The growth of the YBe_{13} layer, which has been measured up to t = 1000 h by the contact vapor method, follows the equation $\delta = kt^n$ with n close to 0.5 (n = 0.423 at 700 °C and 0.518 at 800 °C), where δ is the thickness of the interaction zone and k a constant determinated by the temperature. The formation of YBe_{13} occurs only on the Be side and is kinetically limited by the diffusion of Y through YBe_{13}, Vasina, Panov [7]. For the diffusion coefficient of Y in Be at 1260 °C calculated by [11] see p. 238.

Diffusion measurements have shown that the solid solution of Y and Ce in Be increases from 0.008 wt% Y at 1100 °C to 0.025 wt% at 1220 °C, and from 0.025 wt% Ce at 1050 °C to 0.7 wt% at 1250 °C, Anan'in et al. [8]. From lattice constant measurements up to 1200 °C, it was found that a Be alloy with 0.35 wt% Ce at 720 °C and above is mono-phase. With decreasing temperature, the solid solubility of Ce decreases (0.1 wt% at 200 °C) and $CeBe_{13}$ is present as second phase below 720 °C in an alloy with 0.35 wt% Ce, Papirov et al. [9].

References:

[1] E. Laube, H. Nowotny (Monatsh. Chem. **93** [1962] 681/3).

[2] E.I. Gladyshevskii, P.I. Kripyakevich, D.P. Frankevich, et al. (Vopr. Teor. Primen. Redkozem. Metal. Mater. Soveshch., Moscow 1963 [1964], pp. 153/4; JPRS-28849 [1964] 199/200; N.S.A. **19** [1965] No. 22873).

[3] V.S. Protosov, E.I. Gladyshevskii (Kristallografiya **9** [1964] 267/8; Soviet Phys.-Cryst. **9** [1964] 208/9).

[4] E.I. Gladyshevskii, P.I. Kripyakevich, D.P. Frankevich (Kristallografiya **8** [1963] 788/9; Soviet. Phys.-Cryst. **8** [1963] 628/9; Vopr. Teor. Primen. Redkozem. Metal. Mater. Soveshch., Moscow 1963 [1964], pp. 149/50; JPRS-28849 [1965] 195/6; C.A. **62** [1965] 2576).

[5] K.A. Gschneidner (Rare Earth Alloys, Van Nostrand, Princeton, N.J., 1961, pp. 112/4).

[6] N.N. Matyushenko, G.F. Tikhinskii (Kristallografiya **8** [1963] 451/3; Soviet Phys.-Cryst. **8** [1963] 351/3).

[7] E.A. Vasina, A.S. Panov (Zh. Fiz. Khim. **49** [1975] 735/6; Russ. J. Phys. Chem. **49** [1975] 427/8).

[8] V.M. Anan'in, V.P. Gladkov, V.S. Zotov, et al. (Obshch. Zakonomern. Str. Diagramm Sost. Met. Sist. **1973** 184/5; C.A. **81** [1974] No. 177330).

[9] I.I. Papirov, Yu.M. Smirnov, G.P. Tikhinskii, V.O. Finkel' (Ukr. Fiz. Zh. **11** [1966] 922/3; C.A. **65** [1966] No. 19796).

[10] A.A. Kruglykh, N.N. Matyushenko, V.S. Pavlov, G.F. Tikhinskii (Zh. Neorgan. Khim. USSR **10** [1965] 285/7; Russ. J. Inorg. Chem. **10** [1965] 149/50).

[11] S.G. Artysher, N.G. Volkov, V.A. Kashcheev (Fiz. Metal. Metalloved. **57** [1984] 930/5; Phys. Metal. Metallog. [USSR] **57** No. 5 [1984] 84/90).

6.2.11 With Titanium, Zirconium, Hafnium, and Thorium

A tentative phase diagram of the Be-Ti system was reported by Bedford [1] and is also shown in Elliott [2, pp. 172/3]; for earlier results see in "Titan" 1951, pp. 200/1. Be forms 4 compounds with Ti: $TiBe_{12}$, Ti_2Be_{17}, $TiBe_3$, and $TiBe_2$ [1], see also Aldinger, Wellner [3]. Also, a metastable compound, BeTi, has been reported by Tanner, Giessen [14]. In studies of the Be-Ti system in the range $\leqq$10 wt% Be, it was found that the solid solubility of Be in α-Ti does not exceed 0.05 wt% and that the maximum solid solubility in β-Ti is 1 wt% at about 1030 °C, Hunter [4]. Earlier values for the solid solubility of Be are <1 wt% in α-Ti and 1 to 2 wt% in β-Ti, Craighead et al. [5]. The solid solubility of Ti in Be is <1 wt%, Kaufmann et al. [6], Hindle, Slattery [7].

The diffusion of Be in β-Ti (body-centered cubic high temperature form) was studied between 915 and 1300 °C using the radioactive ^{7}Be isotope. The specimens were arc melted and annealed for 5 h at 1000 °C before the test. In the Arrhenius plot $D=D_0 \exp(-E_A/RT)$ the values are $D_0=0.80$ cm²/s and $E_A=40.2$ kcal/mol. At the melting point of Ti, D has the high value of (2 to 5) $\times 10^{-5}$ cm²/s, Pavlinov et al. [15]. The diffusion of Be in cast α- and β-Ti between 720 and 1040 °C was also studied by the radioactive tracer method with ^{7}Be. For the volume diffusion in α-Ti, the values $D_0=(14^{+31}_{-9}) \times 10^3$ cm²/s and $E_A=62.1 \pm 2.4$ kcal/mol were obtained. At the temperature of the $\alpha \rightleftharpoons \beta$ transition (~880 °C), the diffusion coefficient is $D=2.5 \times 10^{-8}$ cm²/s. The one value of D obtained for β-Ti at 1040 °C satisfied the diffusion equation found by [15], Shabalin et al. [16]. Be is compatible with Ti at 500 °C (31 d test). At 600 °C, a 3 μm thick reaction layer was formed after 30 d, Baird et al. [8], but no reaction between Be and Ti was observed after 2000 h at 600 °C in vacuum, Vickers [11]. After coextrusion, the reaction between a heated mixture of Be and Ti began at ~750 °C and $TiBe_2$ was formed. Be diffused much more rapidly into Ti than Ti into Be, Raymond, Kendall [13]. Be diffusion layers were formed on Ti in electrolytes of fused Be salts at 800 to 900 °C. The reaction was carried out in alundum crucibles placed in a sealed quartz cell. The thickness of the reaction layer was about 10 μm after 3 h, 20 μm after 10 h, and 30 μm after 15 h at 900 °C (values taken from the curve), Ilyushchenko, Kornilov [9]. For a theoretical treatment of the alloy formation in molten salts and the determination of the thermodynamic constants using the emf method, see Ilyushchenko et al. [10].

In the Be-Zr system, besides $ZrBe_{13}$, at least three compounds exist: Zr_2Be_{17}, $ZrBe_5$, and $ZrBe_2$. However, there are some discrepancies as to their exact compositions, see [2, pp. 175/6]. For earlier results see "Zirkonium" 1958, pp. 192/3. The solid solubility of Zr in Be is <0.1 wt% [7]. As for β-Ti, the diffusion of Be in β-Zr has been determined between 915 and 1300 °C. The resulting values for D_0 and E_A are 0.0833 cm²/s and 31.1 kcal/mol, respectively [15]. For the diffusion of ^{7}Be in α- and β-Zr, the values for D_0 are 0.33 and 0.083 cm²/s and for these E_A is 31.9 and 31.8 kcal/mol, respectively. At the $\alpha \rightleftharpoons \beta$ transition of Zr, the diffusion coefficient in α- and β-Zr is 2×10^{-7} and 5×10^{-8} cm²/s, respectively, Tendler et al. [17], see also [15]. Be is incompatible with Zr in vacuum at 600 °C. In 500 to 2000 h tests, extensive reaction occurs and $ZrBe_{13}$ was identified, Vickers [11]. At 500 °C, a reaction layer of 30 μm thickness was formed in 31 d [8]. In diffusion couples of Be and Zr pressed together and heated for 24 d at 700 °C, a layer forms on the Zr side of the interface with acicular grains projecting into the Zr from the interface. Tentative values of n and k in the equation $\delta^n=k \cdot t$, where δ is the thickness of the layer in cm, n the reaction index, k the rate constant in cm/d, and t the time in d, are: $n=1.7$, $k=4.32 \times 10^{-6}$ at 500 °C, $n=1.5$, $k=6.6 \times 10^{-5}$ at 650 °C, and $n=1.2$, $k=5.6 \times 10^{-4}$ at 700 °C. The tentative activation energy for the growth of the reaction layer between 500 and 700 °C is 11.3 kcal/mol, Kittel [12].

According to the tentative Be-Hf phase diagram reported by [1] (see also [2, pp. 162/3]), the compounds $HfBe_{13}$, Hf_2Be_{17}, $HfBe_5$, and $HfBe_2$ seem to exist.

In the Be-Th system only $ThBe_{13}$ was obtained and the mutual solid solubility is low, see [2, pp. 171/2].

References:

[1] R.G. Bedford (UCRL-5991-T [1960] 1/7; C.A. **57** [1962] 14480).

[2] R.P. Elliott (Constitution of Binary Alloys, 1st Suppl., McGraw-Hill, New York 1965/66, pp. 1/896).

[3] F. Aldinger, P. Wellner (Z. Metallk. **61** [1970] 344/9).

[4] D.B. Hunter (Trans. Met. Soc. AIME **236** [1966] 900/2).

[5] C.M. Craighead, O.W. Simmons, L.W. Eastwood (J. Metals **2** [1950] 485/513, 496/7).
[6] A.R. Kaufmann, P. Gordon, D.W. Lillie (Trans. Am. Soc. Metals **42** [1950] 785/844, 801).
[7] E.D. Hindle, G.F. Slattery (Inst. Metals Monogr. Rept. Ser. No. 28 [1963] 651/64, 660).
[8] J.D. Baird, G.A. Geach, A.G. Knapton, K.B.C. West (Proc. 2nd Intern. Conf. Peaceful Uses At. Energy, Geneva 1958, Vol. 5, pp. 328/33; C.A. **1960** 22066).
[9] N.G. Ilyushchenko, N.I. Kornilov (Tr. Inst. Elektrokhim. Akad. Nauk SSSR Ural'sk. Filial **1969** No. 12, pp. 85/91; C.A. **73** [1970] No. 80006).
[10] N.G. Ilyushchenko, N.I. Kornilov, A.I. Anfinogenov (Tr. Inst. Elektrokhim. Akad. Nauk SSSR Ural'sk. Filial **1966** No. 8, pp. 65/71; C.A. **65** [1966] No. 19797).
[11] W. Vickers (Inst. Metals Monogr. Rept. Ser. No. 28 [1963] 335/49, 339).
[12] J.H. Kittel (ANL-4937 [1949] 1/29, 9/12).
[13] L. Raymond, E.G. Kendall (TMS [The Metallurgical Soc.] Paper Select. No. F-70-3 [1970] 1/12 from N.S.A. **26** [1972] No. 18295).
[14] L.E. Tanner, B.C. Giessen (Metal. Trans. A **9** [1978] 67/9).
[15] L.V. Pavlinov, G.V. Grigor'ev, G.O. Gromyko (Izv. Akad. Nauk SSSR Metally **1969** No. 3, pp. 207/9; Russ. Met. **1969** No. 3, pp. 158/60).
[16] A.N. Shabalin, V.P. Gladkov, P.L. Gruzin, A.V. Svetlov (Fiz. Metal. Metalloved. **48** [1979] 663/5; Phys. Metal. Metallog. [USSR] **48** No. 3 [1979] 182/5).
[17] R. Tendler, J. Abriata, C.F. Varotto (J. Nucl. Mater. **59** [1976] 215/20).

6.2.12 With Germanium, Tin, and Lead

The diffusion coefficient of Be was measured by the diffusion annealing of a Ge rod, which was coated at both ends by Be, under vacuum at 720 to 920 °C. After 24 h to 150 h, the diffusion coefficient was found to be $D = 0.5\ \exp(-2.5/kT)$, where k is the Boltzmann constant and T is the temperature in K. The obtained values for the solubility S of Be in solid Ge rise from about 10^{16} atoms/cm³ at 720 °C to about 3×10^{16} atoms/cm³ at 920 °C (values taken from the curve), Belyaev, Zhidkov [1] (the Ge rod employed was "alloyed with Sb", but the possible effect of Sb on the values was not discussed). From electrical measurements, the solid solubility of Be in Ge was found to be limited to about 10^{19} atoms/cm³, Lashkarev et al. [2]. According to Kaufmann et al. [3], a Be alloy with 10 wt% Ge was biphasic with a light colored compound at the grain boundaries and within the grains. However, X-ray and metallographic studies of a cast alloy with 10 at% Ge show only the existence of the two phases, Be and Ge. If any intermediate compounds are formed between Be and Ge, they are stable only at high temperature, Yans [4].

No compounds are known between Be and Sn and the phase diagram shows a large miscibility gap in the liquid state, Elliott, Cramer [5]. Immiscibility in the liquid state has also been reported by [3]. No mutual solid solubility was obtained and only a small solubility of Be in liquid Sn was found: $S \approx 0.014$ at% at 600 °C and ≈ 2 at% Be at 1200 °C (values taken from the linear curve $\log S = f(1/T)$ shown in the paper). At the melting point of Be, $S = 0.3$ wt% [5]. From emf measurements of the cell Be(s) | KCl · LiCl + 2 wt% $BeCl_2$ | Sn-Be at 600 to 820 °C, the solubility of Be in liquid Sn is $\log x_{Be} = 0.93 - 3800/T$ with x in atom fraction and T in K; $\log x_{Be} = -\log \gamma_{Be}$, where γ is the activity coefficient of Be, Kober et al. [6]. The solubility of Be in liquid Sn (which contains about 5×10^{-4} atom fraction Fe) is $x = 0.0473$ at 1640 K and $x = 0.0492$ at 1740 K. The activity coefficients of Be in an infinitely dilute solution of liquid Sn at 1640 and 1740 K are $\ln \gamma^{\infty}_{Be(Sn)} = 3.4378$ and 3.2402, respectively, Bienvenu et al. [7]. The partial heat of solution of Be in liquid Sn at infinite dilution as experimentally measured by isoperibol solution calorimetry is $\Delta\bar{H}^{\circ}_{T} = 17.0$ kcal/g-atom at $T = 1200$ K, Boom [8], and the calculated value using a simple cellular model is 13.0 kcal/g-atom, Boom, de Boer [9].

No compound between Be and Pb and no binary phase diagram are known. The resistance of Be to Pb under static conditions is good at 600 °C, but limited at 800 °C, Koenig [10]. In quartz thermal convection loops with a temperature gradient of 800 to 500 °C, Be shows low resistance to mass transfer in liquid Pb, Cathcart, Manly [11]. Be shows fair resistance to attack by Pb after 40 h at 1000 °C, Koenig [12]. After a 5 h test at 1000 °C, the Pb of a Pb-Be mixture contained <0.01 wt% Be, Kelman et al. [13], cf. Gmelin Handbook "Blei" C3, 1970, p. 1159.

References:

[1] Yu.I. Belyaev, V.A. Zhidkov (Fiz. Tverd. Tela [Leningrad] **3** [1961] 182/3; Soviet Phys.-Solid State **3** [1961] 133/4).
[2] V.E. Lashkarev, R.M. Bondarenko, V.M. Dobrovol'skii, et al. (Ukr. Fiz. Zh. **4** [1959] 372/4).
[3] A.R. Kaufmann, P. Gordon, D.W. Lillie (Trans. Am. Soc. Metals **42** [1950] 785/844, 801).
[4] F.M. Yans (NMI-1240 [1960] 1/41, 21/3; N.S.A. **15** [1961] No. 3084).
[5] R.O. Elliott, E.M. Cramer (AECU-3022 [1952] 1/10; N.S.A. **9** [1955] No. 5954).
[6] V.I. Kober, V.A. Lebedev, I.F. Nichkov, et al. (Izv. Akad. Nauk SSSR Metally **1974** No. 3, pp. 104/6; Russ. Met. **1974** No. 3, pp. 60/2).
[7] G. Bienvenu, C. Potard, B. Schaub, P. Desre (Thermodyn. Nucl. Mater. Proc. Symp., Vienna 1967 [1968], pp. 777/87, 781, 784).
[8] R. Boom (Scr. Met. **8** [1974] 1277/81).
[9] R. Boom, F.R. de Boer (J. Less-Common Metals **46** [1976] 271/84, 280).
[10] R.F. Koenig (Iron Age **172** No. 8 [1953] 129/33).

[11] J.V. Cathcart, W.D. Manly (Corrosion **12** [1956] 87t/91t).
[12] R.F. Koenig (in: D.W. White Jr., J.E. Burke, The Metal Beryllium, ACS, Cleveland 1955, pp. 549/54).
[13] L.R. Kelman, W.D. Wilkinson, F.L. Yaggee (ANL-4417 [1950] 79; N.S.A. **5** [1951] No. 400).

6.2.13 With Vanadium, Niobium, and Tantalum

Be and V react to give the two intermetallic compounds VBe_2 and VBe_{12}, see e.g., Zalkin et al. [1]. The elements are reported to be miscible in the liquid state, Efimov [2]. Thickness δ of the reaction layer of VBe_2 and VBe_{12} between Be and V after heating at t=800 to 1200 °C for various times according to Vasina [14]:

t in °C	time in h	$\delta(VBe_2)$ in μm	$\delta(VBe_{12})$ in μm	t in °C	time in h	$\delta(VBe_2)$ in μm	$\delta(VBe_{12})$ in μm
800	500	80	35	1000	150	140	70
800	1000	154	56	1000	500	216	108
900	250	60	54	1000	1000	240	140
900	350	80	75	1000	3000	420	200
900	500	108	96	1050	200	185	84
900	1000	224	130	1200	10	175	35
950	50	50	44	1200	20	210	40
950	100	63	56	1200	25	240	60
950	200	77	70	1200	50	300	65
950	300	91	80	1200	100	316	100
1000	10	63	28	1200	200	332	130
1000	20	90	35	1200	500	520	140

Description of V-Be alloys in Gmelin Handbook "Vanadium" B2, 1967, pp. 578/9.

Nb reacts with Be forming several phases, which are described in Gmelin Handbook "Niob" B2, 1971, pp. 2/17. Zalkin et al. [1] supposed the existence of five compounds: $NbBe_{12}$, Nb_2Be_{17}, $NbBe_3$, $NbBe_2$, and Nb_3Be_2. For the phase diagram including these compounds, see Shunk [3]. In the phase diagram reported by Rayevskiy, Grigor'yev [4], the compound $NbBe_5$ is included instead of Nb_2Be_{17}. While the existence of Nb_2Be_{17} is confirmed by Wolcott, Falge [5] and Pugachev et al. [6], the existence of $NbBe_5$ was also reported by Panov, Rysina [7, 8] (see below).

By heating a diffusion couple of Be and Nb sheets (pressed together) in a reducing atmosphere at 650 °C, two layers of widely varying thickness are obtained. However, at 700 °C only one layer was observed to form. The growth rate at 700 °C was approximately 2.66 µm/d, Kittel [13].

Heating an Nb sheet in Be powder under vacuum (5 to 6 µ Torr) for 3 h at 1050 °C or 1 h at 1100 to 1200 °C leads to a reaction layer which contains the compounds Nb_3Be_2, $NbBe_2$, $NbBe_3$, Nb_2Be_{17}, and $NbBe_{12}$. After 1 to 6 h at 1000 °C, only Nb_3Be_2 and $NbBe_2$ are observed. The thickness of the layer grows parabolically with time over 6 h, see **Fig. 6-16**, p. 234. The saturation occurs by diffusion through the reaction layer [6]. Heating pressed diffusion couples (discs of 2 mm thickness) of Nb and "Be" ($ZrBe_{13}$ was used instead of Be) in vacuum or under an Ar atmosphere at 1100 °C, results in the development of a layer of $NbBe_2$ (17 µm thick after 50 h and 60 µm after 500 h) followed by the development of a layer of $NbBe_5$ (48 µm thick in 500 h) [7], see also [8]. According to Vasina [14], in the contact zone between Be and Nb, only Nb_3Be_2 is observed after 300 h at 700 °C. However, at higher temperatures, the phases $NbBe_{12}$, Nb_2Be_{17}, $NbBe_5$, and $NbBe_3$ can also occur. See the following table, in which the layer thickness of the phases after heating for various times and temperatures t in °C is tabulated, Al'tovskii, Panov [15]:

t in °C	time in h	layer thickness in µm			
		Nb_3Be_2	$NbBe_3$	$NbBe_5$(a) Nb_2Be_{17}(b)	$NbBe_{12}$
700	100	7	—	—	—
	200	12	—	—	—
	300	13	—	—	—
800	50	8	—	—	—
	100	12	7	15(b)	15
	200	23	2	—	20
	300	30	3	—	35
900	25	5	—	—	—
	50	15	1	—	27
	100	27	3	—	63
	300	49	2	25(b)	112
	500	100	5	17(b)	160
1000	25	10	1	26(b)	24
	50	21	2	25(b)	84
	100	44	2	10(a)	168
	150	67	3	10(b)	98
	200	70	3	10(b)	190
	200	94	3	15(b)	200
	300	105	3	20(a)	210
	500	150	2	24(b)	270
1050	10	40	4	—	56
	25	52	4	14(b)	84
	50	80	2	9(a)	110
	200	160	7	20(b)	245
1100	8	40	6	7(a)	88
	10	50	2	10(a)	110
	20	55	2	14(a)	96
1150	8	42	7	—	99
	10	60	5	—	110
	20	73	13	15(a)	140
1200	2	25	7	—	80
	5	42	7	—	85
	8	44	8	2(a)	130
	15	70	15	10(a)	150

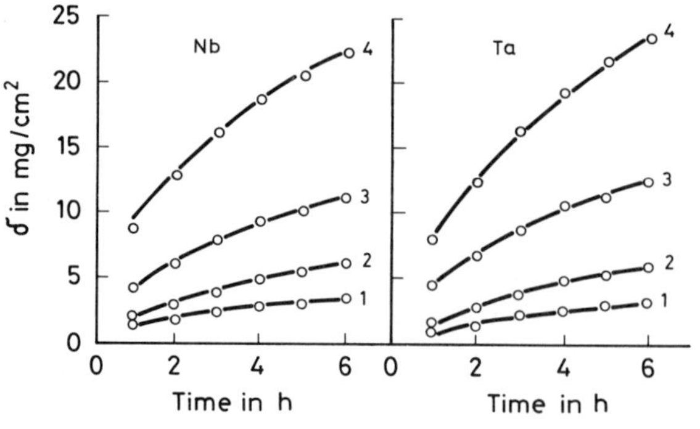

Fig. 6-16

Layer thickness δ vs. time on annealing Nb or Ta sheet in Be powder at various temperatures: 1) 1000 °C, 2) 1050 °C, 3) 1100 °C, 4) 1150 °C.

The beryllization of Nb also occurs in the electrolytes of molten Be salts (alundum crucibles in quartz cells) at 900 °C. The thickness of the resulting reaction layer is ~25 μm after 5 h, ~40 μm after 10 h, and ~50 μm after 15 h (values taken from the curve). The layer contains two intermetallic phases, Ilyushchenko, Kornilov [9]. For a theoretical treatment of alloy formation by this method, see Ilyushchenko et al. [10]. Be is incompatible with Nb at 500 °C. In 31 d, a reaction zone of 5 μm thickness was formed, Baird et al. [11]. No reaction at 600 °C in vacuum after 1000 and 2000 h was reported by Vickers [12].

No phase diagram has been reported for the Be-Ta system and probably the same five compounds exist as in the Be-Nb system: Ta_3Be_2, $TaBe_2$, $TaBe_3$, Ta_2Be_{17}, and $TaBe_{12}$ [1], cf. Gmelin Handbook "Tantal" B2, 1971, pp. 2/10. $TaBe_{12}$ and Ta_2Be_{17} were also obtained by [5]. By heating diffusion couples of Be and Ta at 650 °C one reaction layer was observed, with a growth rate of approximately 3.1 μm/d, Kittel [13]. As in the case of Nb (see above), the reaction layer between a Ta sheet and Be powder grows parabolically with time at 1000 to 1200 °C, see Fig. 6-16. After heating 3 h at 1000 °C, or 2 h at 1050 °C, or 1 h at 1100 to 1200 °C, the reaction layer contains the phases $TaBe_{12}$, Ta_2Be_{17}, $TaBe_3$, and Ta_3Be_2. After 1 h at 1000 or 1050 °C, only Ta_3Be_2 and $TaBe_3$ are observed [6]. Heating "Be" (as $ZrBe_{13}$) in contact with Ta (diffusion couples, see above) at 1100, 1200, or 1400 °C, produces layers of $TaBe_2$ and $TaBe_5$ whose thickness δ in μm is dependent on the heating time in h:

	at 1100 °C			at 1200 °C			at 1400 °C		
time in h . . .	50	100	500	25	100	250	25	50	100
$\delta(TaBe_2)$ . . .	44	84	152	54	144	190	203	219	380
$\delta(TaBe_5)$ . . .	—	12	24	32	48	76	—	200	300

The phases were identified by metallographic, microhardness, and X-ray studies [7], see also [8], and the review in [15, pp. 91/2]. Be is incompatible with Ta at 500 °C. In 31 d a reaction zone of 8 μm thickness was formed [11]. No reaction was observed after 2000 h at 600 °C (in vacuum) [12].

References:

[1] A. Zalkin, D.E. Sands, R.G. Bedford, O.H. Krikorian (Acta Cryst. **14** [1961] 63/5).

[2] Yu.V. Efimov (Izv. Akad. Nauk SSSR Neorgan. Materialy **1** [1965] 1298/305; Inorg. Materials [USSR] **1** [1965] 1186/92).

[3] F.A. Shunk (Constitution of Binary Alloys, 2nd Suppl., McGraw-Hill, New York 1970, pp. 108/9).

[4] I.I. Rayevskiy, A.T. Grigor'yev (Izv. Akad. Nauk SSSR Metally **1968** No. 5, pp. 198/202; Russ. Met. **1968** No. 5, pp. 134/6).
[5] N.M. Wolcott, R.L. Falge (Phys. Rev. [2] **171** [1968] 591/5).
[6] N.S. Pugachev, L.F. Verkhorobin, A.N. Derizemla, A.A. Matyash (Temperaturoustoich. Zashch. Pokrytiya Tr. 3rd Semin. Zharostoikim Pokrytiyam, Leningrad 1966 [1968], pp. 92/9; C.A. **71** [1969] No. 15439).
[7] A.S. Panov, M.M. Rysina (Izv. Akad. Nauk SSSR Metally **1972** No. 3, pp. 219/23; Russ. Met. **1972** No. 3, pp. 172/5).
[8] A.S. Panov, M.M. Rysina (Zh. Fiz. Khim. **51** [1977] 2580/3; Russ. J. Phys. Chem. **51** [1977] 1507/9).
[9] N.G. Ilyushchenko, N.I. Kornilov (Tr. Inst. Elektrokhim. Akad. Nauk SSSR Ural'sk. Filial No. 12 [1969] 85/91; Electrochemistry of Molten and Solid Electrolytes, Vol. 9, Plenum, New York 1972, pp. 68/72).
[10] N.G. Ilyushchenko, N.I. Kornilov, A.I. Anfinogenov (Tr. Inst. Elektrokhim. Akad. Nauk SSSR Ural'sk. Filial No. 8 [1966] 65/71; Electrochemistry of Molten and Solid Electrolytes, Vol. 5, Plenum, New York 1967, pp. 63/9).

[11] J.D. Baird, G.A. Geach, A.G. Knapton, K.B.C. West (Proc. 2nd Intern. Conf. Peaceful Uses At. Energy, Geneva 1958, Vol. 5, pp. 328/33; C.A. **1960** 22066).
[12] W. Vickers (Inst. Metals Monogr. Rept. Ser. No. 28 [1963] 335/49, 339).
[13] J.H. Kittel (ANL-4937 [1949] 1/29, 9/11).
[14] E.A. Vasina (in [15, pp. 86/7]).
[15] R.M. Al'tovskii, A.S. Panov (Korroziya i Sovmestimost Berilliya, Atomizdat, Moscow 1975, pp. 1/129, 88; C.A. **85** [1976] No. 84724).

6.2.14 With Chromium, Molybdenum, Tungsten, and Uranium

Data for the system Cr-Be before 1950 see Gmelin Handbook "Chrom" A1, 1963, pp. 715/7. At least two compounds are known: $CrBe_2$ [1 to 3] and $CrBe_{12}$ [2, 4, 5]. Edwards, Johnstone [1] studied the Be-Cr system up to 70 wt% Be and found that the solubility of Be in solid Cr is 1.5 at% at 900 °C and 9.2 at% at 1500 °C. The solubility of Cr in solid Be, which was measured between 1050 to 1220 °C, reached 0.076 at% (0.45 wt%) at 1220 °C, Jacobson, Hammond [6]. At lower temperatures the solubility of Cr is very low; an alloy with 0.1 wt% Cr contains two phases, Hindle, Slattery [5]. Be is compatible with Cr at 5000 °C (28 d test) and 600 °C (14 d test), Baird et al. [7]. Thickness δ in µm of the $CrBe_2$ and $CrBe_{12}$ layers in the reaction zone between Cr and Be after heating between t=700 and 1200 °C for various times, Vasina [23]:

t in °C	700	700	800	800	900	900	1000	1000	1200	1200
time in h	500	1000	100	1000	50	500	10	500	10	50
$\delta(CrBe_2)$	7	20	10	270	90	200	77	273	200	380
$\delta(CrBe_{12})$	—	—	—	10	47	100	40	106	50	76

In the Be-Mo system at least three compounds are formed: $MoBe_{22}$ [8, 9], $MoBe_{12}$ [2, 4, 9], and $MoBe_2$ [2, 3, 9, 10]. The existence of an Mo_3Be phase was reported by Paine, Carrabine [11]. Mo is compatible to Be up to 1000 °C, Braun [12]. Be did not react with Mo after heating for 2000 h in vacuum at 600 °C, Vickers [15]. For details of the reaction of Be with Mo see Gmelin Handbook "Molybdenum" Suppl. Vol. A3, 1983, pp. 96/7, see also the review in Al'tovskii, Panov [24, pp. 94/6]. The reactions of Be with W are extensively covered in Gmelin Handbook "Tungsten" Suppl. Vol. A4, 1986. In the Be-W system at least three compounds exist: WBe_{22}, WBe_{12}, and WBe_2 [9], Vasina, Panov [13], see also [8] for WBe_{22} and [10] for WBe_2. Studies of the W-Be system on the W side show WBe_2

to be the compound richest in W. The solubility of Be in solid W is on the order of 5 at% at the eutectic temperature (~2100 °C) and ~3 at% in the range 1300 to 1000 °C, Goldschmidt, Ham [14]. Other works [15, 19, 20, 22], see also the review in [24, pp. 96/8].

In the Be-U phase diagram, Buzzard [16] (see also [17]) found only one compound, UBe_{13}, whose existence has been confirmed by [4]. No solubility of U in solid Be and only a slight solubility of Be in γ-U was detected [16]. By means of a simple cellular model, the partial heat of solution of Be for infinite dilution in liquid U was calculated as $\Delta\bar{H}^{\circ} = -4.1$ kcal/g-atom, Boom, De Boer [21]. After heating in vacuum at 600 °C, polished Be was shown to be incompatible with U. The thickness δ of the UBe_{12} containing reaction layer between the clamped samples was 16 μm after 500 h (20.8 d) and 32 μm after 2000 h (83.3 d) [15]. The values for δ in μm for a layer formed between hot rolled Be sheet and compact U at 500 to 700 °C are shown in the following table:

time in d	14	28	56	112	224	254
δ at 500 °C	—	4	2	4	—	—
δ at 600 °C	5	10	10	14	—	20
δ at 700 °C	12	—	32	40	60	—

Knapton, West [18]. Older tests between 600 and 700 °C, see Kittel [22]. According to [7], cast Be is incompatible with U at 500 and 600 °C (δ=4 and 10 μm, respectively, after 28 d) and the reaction of rolled Be at 500 °C was visible only after 56 d. For the growth of reaction layers between Be and U, see also the review in [24, pp. 82/4].

References:

[1] A.R. Edwards, S.T.M. Johnstone (J. Inst. Metals **84** [1956] 313/7).
[2] A. Zalkin, D.E. Sands, R.G. Bedford, O.H. Krikorian (Acta Cryst. **14** [1961] 63/5).
[3] M. Stümke, G. Petzow (Z. Metallk. **66** [1975] 292/7).
[4] N.M. Wolcott, R.L. Falge (Phys. Rev. [2] **171** [1968] 591/5).
[5] E.D. Hindle, G.F. Slattery (Inst. Metals Monogr. Rept. Ser. No. **28** [1963] 651/64, 663).
[6] M.I. Jacobson, M. L. Hammond (Trans. Met. Soc. AIME **242** [1968] 1385/91).
[7] J.D. Baird, G.A. Geach, A.G. Knapton, K.B.C. West (Proc. 2nd Intern. Conf. Peaceful Uses At. Energy, Geneva 1958, Vol. 5, pp. 328/33; C.A. **1960** 22066).
[8] P.I. Kripyakevich, E.I. Gladyshevskii (Kristallografiya **8** [1963] 449/51; Soviet Phys.-Cryst. **8** [1963] 349/51).
[9] N.S. Pugachev, L.F. Verkhorobin, A.N. Derizemlya, A.A. Matyash (Temperaturoustoich. Zashch. Pokrytiya Tr. 3rd Semin. Zharostoikim Pokrytiyam, Leningrad 1966 [1968], pp. 92/9; C.A. **71** [1969] No. 15439).
[10] A.S. Panov, M.M. Rysina (Zh. Fiz. Khim. **51** [1977] 2580/3; Russ. J. Phys. Chem. **51** [1977] 1507/9).

[11] R.M. Paine, J.A. Carrabine (Acta Cryst. **13** [1960] 680/1).
[12] H. Braun (Metall **16** [1962] 646/54).
[13] E.A. Vasina, A.S. Panov (Izv. Akad. Nauk SSSR Metally **1974** No. 1, pp. 197/9; Russ. Met. **1974** No. 1, pp. 119/21).
[14] H.J. Goldschmidt, W.M. Ham (J. Less-Common Metals **10** [1966] 57/65).
[15] W. Vickers (Inst. Metals Monogr. Rept. Ser. No. 28 [1963] 335/49, 339).
[16] R.W. Buzzard (J. Res. Natl. Bur. Std. **50** [1953] 63/7).
[17] M. Hansen (Constitution of Binary Alloys, 2nd Ed., McGraw-Hill, New York 1958, pp. 299/300.

[18] A.G. Knapton, K.B.C. West (J. Nucl. Mater. **3** [1961] 239/40).
[19] C.R. Watts (Intern. J. Powder Met. **4** No. 3 [1968] 49/53; C.A. **70** [1969] No. 13874).
[20] P.M. Arzhanyi, R.M. Volkova, D.A. Prokoshkin (Izv. Akad. Nauk SSSR Met. Topl. **1962** No. 6, pp. 162/6; Russ. Met. Fuels **1962** No. 6, pp. 95/102).
[21] R. Boom, F.R. de Boer (J. Less-Common Metals **46** [1976] 271/84).
[22] J.K. Kittel (ANL-4937 [1949] 1/29, 9/12; N.S.A. **10** [1956] No. 5367).
[23] E.A. Vasina (from [24, p. 92]).
[24] R.M. Al'tovskii, A.S. Panov (Korroziya i Sovmestimost Berilliya, Atomizdat, Moscow 1975, p. 129; C.A. **85** [1976] No. 84724).

6.2.15 With Manganese

Besides $MnBe_2$ [1 to 3], at least one Be rich compound exists in the Be-Mn system. The composition $MnBe_{12}$ was reported by Batchelder, Raeuchle [4], Wolcott, Falge [5], and $MnBe_{13}$ by Anan'in et al. [6]. The phase $MnBe_8$, with a composition range $MnBe_3$ to $MnBe_{13}$ at 1100 °C, was reported by Cherkashin et al. [3]. The solubility of Mn in solid Be was determined by diffusion measurements employing the ^{52}Mn isotope [6], and was found to be 0.44 ± 0.1 wt% at 800 °C, 0.6 ± 0.1 wt% at 900 °C, and 3.8 ± 0.7 wt% at 1000 °C.

References

[1] L. Misch (Metallwirtschaft **15** [1936] 163/6).
[2] R.L. Berry, G.V. Raynor (Acta Cryst. **6** [1953] 178/86, 181).
[3] E.E. Cherkashin, E.I. Gladyshevskii, P.I. Kripyakevich, Yu.B. Kuzma (Zh. Neorgan. Khim. **3** [1958] 650/3; Russ. J. Inorg. Chem. **3** [1958] No. 3, pp.135/41).
[4] F.W. von Batchelder, R.F. Raeuchle (Acta Cryst. **10** [1957] 648/9).
[5] N.M. Wolcott, R.L. Falge (Phys. Rev. [2] **171** [1968] 591/5).
[6] V.M. Anan'in, V.P. Gladkov, V.S. Zotov, et al. (Obshch. Zakonomern. Str. Diagramm Sost. Metal. Sist. Mater. 5th Vses. Soveshch., Moscow 1971 [1973], pp. 184/5; C.A. **81** [1974] No. 177330).

6.2.16 With Nickel, Cobalt, and Iron

In the Be-Ni system treated in Gmelin Handbook "Nickel" B 1, 1965, pp. 15/29 two phases, NiBe and Ni_5Be_{21}, occur. The phases Ni_5Be_{21} and Ni_4Be_{22} are found after diffusion annealing of Ni foil in Be vapor at 1000 to 1300 °C, Verkhorobin et al. [1]. A thin alloy layer is formed by heating Ni plated Be in a vacuum for 18 h at 400 °C, Beach, Faust [16].

Be is incompatible with Ni at 500 or 600 °C. The thickness δ of the reaction layer is 40 µm after 31 d at 500 °C and 170 µm after 30 d at 600 °C, Baird et al. [2]. The values of δ in µm of the reaction layer between hot rolled Be sheet and compact Ni at 500 to 700 °C are shown below:

time in d	7	14	28	56	112
δ at 500 °C	64	120	206	210	180(?)
δ at 600 °C	44	60	84	124	280
δ at 700 °C	256	350	802	1120	1260

Knapton, West [3]. In the layer formed at 600 °C (160 µm after 1000 h), Ni_5Be_{21} was identified by X-ray diffraction, Vickers [4]. After heating diffusion couples of Be- and Ni-sheet for 6 to 48 d, two layers were observed at 500, 600, 650, and 700 °C. The layer nearest the Be was much thicker and at 600 °C frequently showed a precipitated phase. The thick layer formed at 700 °C contained columnar grains which were oriented perpendicular to the interface. The growth rate of the reaction layer was approximately 9 μm/d at 600 °C and 34 μm/d at 700 °C, Kittel [22]. When Ni was heated in a bath made up of 10 wt% powdered Be, 10 wt% NaF, and 10 wt% BeF_2 in a molten equimolar KCl-NaCl mixture, a reaction layer was formed consisting at an outer phase of Ni_5Be_{21}, an intermediate phase of NiBe and an Ni(Be) solid solution, Anfinogenov et al. [23]. The ^{7}Be isotope was used to determine the diffusion coefficient of Be in Ni between 1020 and 1400 °C and the resulting values fit the Arrhenius equation $D=0.019\ \exp(-46200/RT)\ cm^2/s$, Grigor'ev, Pavlinov [27]. The diffusion coefficient of Ni in cast and ceramic Be samples between about 900 and 1200 °C is $D=(0.20\pm0.12)\ \exp(-58000/RT)$ and $D=(0.22\pm0.33)\exp(-56000/RT)\ cm^2/s$, respectively. The results are compared in a graph with earlier results of other authors, Anan'in et al. [29]. For the diffusion of Ni in ^{63}Ni-covered Be at 1000 °C (from one side during 170 h) D was calculated regarding three types of boundary conditions (instantaneous source, constant concentration, and constant flux). D (in $10^{-11}\ cm^2/s$) $=4.75\pm20$, 6.80 ± 0.35, and 9.70 ± 0.90 were obtained for the three cases, Artyshev et al. [30]. The best agreement with the experiment [29] was observed for an instantaneous source at the boundary. This is also the case for the diffusion of ^{90}Y in Be at 1260 °C (calculated values in $10^{-9}\ cm^2/s$: 2.60 ± 0.15, 3.90 ± 0.15, and 5.25 ± 0.15 for the three assumptions) [30]. The distribution method was used to determine the activity coefficient of Ni in an infinitely dilute solution of Be: $\gamma^{\infty}_{Ni(Be)}=0.012$ at 1623 K, Bienvenu et al. [5, pp. 785/6].

In the Be-Co phase diagram reported by Aldinger, Jönsson [6], several intermetallic phases exist: $CoBe_9$, $CoBe_7$, $CoBe_3$, CoBe, Co_3Be, and solid solutions between $CoBe_7$ and $CoBe_3$ as well as between CoBe and Co_3Be. The older diagram reported by Hansen [7, pp. 278/80] shows only the phases Co_5Be_{21} and CoBe. The structure of a Be rich phase, $CoBe_{12}$, was studied by Batchelder, Raeuchle [8], see also Elliott [9, pp. 157/8], and the structure of $CoBe_x$ (and $FeBe_x$), isostructural with $RhBe_{6.6}$, was studied by Johnson et al. [20]. The mutual solid solubility of Be and Co is shown in the reported phase diagrams [6, 7].

The volume diffusion of Co in polycrystalline cast Be (twice distilled) between T = 1253 and 1483 K was studied by means of the radioactive isotope ^{57}Co (applied as $CoCl_2$) in an He atmosphere. The dependence of the diffusion coefficient D in cm^2/s on temperature T is shown below:

T in K	1253	1318	1363	1408	1468	1493
$D \cdot 10^{10}$	0.34 ± 0.5	0.87 ± 0.1	2.4 ± 0.4	4.8 ± 0.7	16 ± 2	23 ± 2

The values fit the Arrhenius equation: $D=27^{+25}_{-13}\ \exp(-68600\pm1800)/RT$. The activation energy of the diffusion of Co (68.6 kcal/mol) is higher than that of Ni (58.0 ± 2.2, see above) and of Fe (51.8 ± 1.1, see below), Gladkov et al. [26].

In the Be-Fe phase diagram reported by Hansen [7, p. 285] (see also Gmelin Handbook "Eisen" A8, 1936, pp. 1813/5), Elliott [9, p. 160], Shunk [10], the intermetallic phases $FeBe_{12}$, $FeBe_5$, and $FeBe_2$ occur and a mutual solid solubility is observed. The Be rich compound was identified as $FeBe_{12}$ by [8], see also [4], and Hindle, Slattery [11]. A phase $FeBe_x$, isostructural with $RhBe_{6.6}$, was found by [20]. There is a considerable solubility of Be in solid α-Fe [7, 10], see also Ko, Nishizawa [12], and Nishizawa et al. [21]. The maximum solid solubility of Fe in Be (0.86 at% at 1190 °C), Jacobson, Hammond [13], agrees quite well with the data in [9, 10]. The solubility S of Fe in single crystalline Be at t = 500 to 850 °C is shown below:

t in °C	500	600	700	800	850
S in at% . . .	0.017 ± 0.007	0.0320 ± 0.008	0.0620 ± 0.015	0.11 ± 0.02	0.16 ± 0.03

Fe was implanted into Be single crystals at room temperature by means of a high energy ion beam. The samples were then annealed isothermically and the Fe content determined by ion back scattering. The resulting data fit the equation $S = 14.0 \exp(-5246/T)$ with S in at% and T = 773 to 1123 K, Myers, Smugeresky [14]. The interaction of sheet Be with iron plasma bunches in vacuum was studied with Mössbauer spectroscopy, Lisichenko et al. [25]. By means of a simple cellular model, the partial heat of solution of Be in an infinitely dilute solution of liquid Fe was calculated as $\Delta\bar{H}^{\circ}_{Be} = -7.3$ kcal/mol (−30.6 kJ/mol), Boom, de Boer [18]. The calculated values $\Delta\bar{H}^{\circ}_{Be} = -31$ kJ/mol and $\Delta\bar{H}^{\circ}_{Fe} = -44$ kJ/mol have been reported by Miedema et al. [19]. The activity coefficient of Fe in an infinitely dilute solution of liquid Be was determined using the distribution method to be $\gamma^{\infty}_{Fe(Be)} = 0.085$ at 1640 K and 0.093 at 1740 K, Bienvenu et al. [5, p. 784]. An iron plated Be rod was heated in a vacuum for 18 h at 400 or 500 °C and no diffusion of Fe or Be was observed. However, at 550, 600, and 800 °C, an alloy layer was formed, Beach, Faust [16]. For diffusion of Fe in Be in dependence on temperature, see Fig. 6-12, p. 215. Be was shown to be incompatible with Fe in 29 d tests at 500 °C, in which an 8 μm thick reaction layer was formed [2]. The thickness δ in μm of the reaction layer between a hot rolled Be sheet and compact Fe at 500 to 700 °C [3] is shown below:

time in d	7	14	28	56	112
δ at 500 °C	none	8	12	24	24
δ at 600 °C	10	10	30	40	80
δ at 700 °C	48	80	228	240	312

Diffusion couples of Be and Fe sheet which were heated at 650 and 700 °C produced two layers of varying thickness. The layers formed mostly on the Be side of the interface and the growth rate of the reaction layer was approximately 6.6 μm/d at 650 °C and 10.2 μm/d at 700 °C. The activation energy is approximately 12.0 kcal/mol, Kittel [22]. In the layer formed at 600 °C on mild steel (15 μm thickness after 500 h), $FeBe_{12}$ was identified by X-ray diffraction [4]. When molten Ca, into which Be had been introduced, was used as the transport medium, Be diffusion layers of 80 to 100 μm thickness were obtained on steel at 900 °C after 10 h. The microstructure of the Be layer consisted of columnar grains oriented perpendicular to the surface of the specimen. The grains consisted of the $FeBe_2$ phase and an α-solution of Be in Fe, Shatinskii, Zbozhnaya [17]. For the growth of reaction layers between Be and Fe see also the review in Al'tovskii, Panov [24]. The diffusion coefficients D of Fe diffusion in Be at 600 and 800 °C [14] agree satisfactorily with the data reported by Naik et al. [15] who found the following values by means of the radioactive tracer method with ^{59}Fe:

t in °C	700	750	800	836	900	950	1026	1076
$10^{10} \cdot D$ in cm²/s	0.0106	0.0593	0.145	0.361	1.12	2.58	12.5	20.3

The values fit the equation $D = (0.53 \pm 0.20) \exp(-51800 \pm 1100)/RT$ with T in K [15].

The diffusion coefficient D of Be in Fe (fcc phase) between 1100 and 1350 °C was determined by means of the ^{7}Be isotope. The resulting values fit the Arrhenius equation $D = 0.1 \exp(-57600/RT)$ cm²/s [27]. For the Be diffusion in the low-temperature bcc Fe phase (studied with the Fe/1 wt% Be alloy, which is stable as the bcc phase to high temperatures) the equation $D = 5.34 \exp(-52100/RT)$ cm²/s was obtained, Grigor'ev, Pavlinov [28]. See also Fig. 6-14, p. 226, for the diffusion coefficient of Fe and Ni in Be between about 800 and 1100 °C.

References:

[1] L.F. Verkhorobin, N.N. Matyushenko, N.S. Pugachev, Yu.G. Titov (Tezisy Dokl. 2nd Vses. Konf. Kristallokhim. Intermetal Soedin., Lvov 1974, pp. 84/5 from C.A. **87** [1977] No. 9720).

[2] J.D. Baird, G.A. Geach, A.G. Knapton, K.B.C. West (Proc. 2nd Intern. Conf. Peaceful Uses At. Energy, Geneva 1958, Vol. 5, pp. 328/33; C.A. **1960** 22066).

[3] A.G. Knapton, K.B.C. West (J. Nucl. Mater. **3** [1961] 239/40).

[4] W. Vickers (Inst. Metals Monogr. Rept. Ser. No. 28 [1963] 335/49, 339).

[5] G. Bienvenu, C. Potard, B. Schaub, P. Desre (Thermodyn. Nucl. Mater. Proc. Symp., Vienna 1967 [1968], pp. 777/87).

[6] F. Aldinger, S. Jönsson (Z. Metallk. **68** [1977] 362/7).

[7] M. Hansen (Constitution of Binary Alloys, 2nd Ed., McGraw-Hill, New York 1958).

[8] F.W. von Batchelder, R.F. Raeuchle (Acta Cryst. **10** [1957] 648/9).

[9] R.P. Elliott (Constitution of Binary Alloys, 1st Suppl., McGraw-Hill, New York 1966).

[10] F. Shunk (Constitution of Binary Alloys, 2nd Suppl., McGraw-Hill, New York 1970, pp. 111/2).

[11] E.D. Hindle, G.F. Slattery (Inst. Metals Monogr. Rept. Ser. No. 28 [1963] 651/64, 658).

[12] M. Ko, T. Nishizawa (Trans. Japan. Inst. Metals **16** [1975] 369/71; Nippon Kinzoku Gakkaishi **43** [1979] 118/26).

[13] M.I. Jacobson, M.L. Hammond (Trans. Met. Soc. AIME **242** [1968] 1385/91).

[14] S.M. Myers, J.E. Smugeresky (Met. Trans. A **7** [1976] 795/802, 799/800, A **8** [1977] 609/16, 614).

[15] M.C. Naik, J.M. Dupouy, Y. Adda (Mem. Sci. Rev. Met. **63** [1966] 488/94).

[16] J.G. Beach, C.L. Faust (BMI-732 [1952] 1/46, 39/46; C.A. **1956** 10568).

[17] V.F. Shatinskii, O.M. Zbozhnaya (Fiz. Khim. Mekhan. Mater. **8** [1972] 59/61; C.A. **79** [1973] No. 34135).

[18] R. Boom, F.R. de Boer (J. Less-Common Metals **46** [1976] 271/84, 280).

[19] A.R. Miedema, F.R. de Boer, R. Boom (CALPHAD Comput. Coupling Phase Diagrams Thermochem. **1** [1977] 341/59, 347; C.A. **89** [1978] No. 95926).

[20] O. Johnson, J.S. Smith, O.H. Krikorian (Acta Cryst. B **26** [1970] 109/13).

[21] T. Nishizawa, M. Hasebe, M. Ko (Acta Met. **27** [1979] 817/28).

[22] J.H. Kittel (ANL-4937 [1949] 1/29, 9/11; N.S.A. **10** [1956] No. 5367).

[23] A.I. Anfinogenov, G.I. Belyaeva, N.G. Ilyushchenko, et al. (Neorgan. Organosilik. Pokryt. Tr. 6th Vses. Soveshch. Zharostoikim Pokrytiyam, Leningrad 1973 [1975], pp. 229/33; C.A. **85** [1976] No. 36341).

[24] R.M. Al'tovskii, A.S. Panov (Korroziya i Sovmestimost Berilliya, Atomizdat, Moscow 1975, pp. 1/129, 98/102; C.A. **85** [1976] No. 84724).

[25] V.I. Lisichenko, S.I. Grinyuk, N.N. Petrichenko (Fiz. Khim. Obrab. Mater. **1975** No. 4, pp. 23/6; C.A. **84** [1976] No. 78102).

[26] V.P. Gladkov, A.V. Svetlov, D.M. Skorov, A.N. Shabalin (Fiz. Metal. Metalloved. **48** [1979] 871/2; Phys. Metal. Metallog. [USSR] **48** No. 4 [1979] 170/2).

[27] G.V. Grigor'ev, L.V. Pavlinov (Fiz. Metal. Metalloved. **25** [1968] 836/9; Phys. Metal. Metallog. [USSR] **25** No. 5 [1968] 79/82).

[28] G.V. Grigor'ev, L.V. Pavlinov (Fiz. Metal. Metalloved. **26** [1968] 946/7; Phys. Metal. Metallog. **26** No. 5 [1968] 179/80).

[29] V.M. Anan'in, V.P. Gladkov, V.S. Zotov, D.M. Skorov (At. Energiya SSSR **29** No. 3 [1970] 220/1; Soviet J. At. Energy **29** [1970] 941/3).

[30] S.G. Artyshev, N.G. Volkov, V.A. Kashcheev (Fiz. Metal. Metalloved. **57** [1984] 930/5; Phys. Metal. Metallog. [USSR] **57** No. 5 [1984] 84/90).

6.2.17 With Copper, Silver, and Gold

In the Cu–Be system, three intermediate phases are observed: the δ-phase which includes $CuBe_3$ and $CuBe_2$, the narrow β′-phase at ~48 at% Be (12 wt%), and the β-phase with a composition around 31 at% Be which is stable only above 605 °C. There is a considerable mutual solid solubility (α and ε phase); see [1], [18, p. 159], [19, p. 110] for the literature before 1970. For studies of the Cu–Be system in the composition range of the β-phase, see also Auvray et al. [2], Golikov, Tyapkin [3], Tyapkin, Golikov [4], Tadaki et al. [5]. The phenomenon of contact melting of Cu–Be phases or Cu, several hundred degrees below their melting points, was observed in reaction layers formed between Cu and powdered Be heated in H_2+HCl or H_2+BeCl_2 atmospheres at 800 to 950 °C, Patskhverova [6].

The solubility S of Cu in Be can be expressed by the equation $S = 12.6 \exp(-842/T)$ with S in at% Cu and T = 593 to 1373 K, Myers, Smugeresky [7]. For the calculation of the solubility limit of Cu (Ag and Au) in dilute Be alloys using the pseudopotential theory of metals, see Ausloos, Lalevic [8]. Be is incompatible with copper at 600 °C in vacuum and the thickness of the reaction layer is 104 μm after 500 h, 139 μm after 1000 h, and 279 μm after 2000 h, Vickers [9]. Be is also incompatible with Cu at 500 °C and after 29 d (696 h) the thickness of the reaction layer is 18 μm, Baird et al. [10].

In order to determine the diffusion in the Cu–Be system, plates of Be pressed between two Cu plates were heated in a vacuum for 5 h at the desired diffusion temperature (550 to 840 °C). They were then heated without pressure in an H_2 atmosphere for a longer period followed by quenching in H_2O. The diffusion zones were studied by microhardness and microscopic measurements. In the interdiffusion of Be and Cu, all possible phases of the Cu–Be system (see above) were formed and their growth rate was proportional to the square root of the time. The values of the common diffusion coefficients D in cm²/s, of D_0 and E_A in the equation $D = D_0 \exp(-E_A/RT)$ (assuming a linear concentration gradient in the phases) are shown below [11]:

phase	α (Cu side)	β	β′	δ	ε (Be side)
D at 840 °C*)	1.5×10^{-9}	6.7×10^{-8}	4.7×10^{-8}	4.5×10^{-10}	5.1×10^{-10}
D at 750 °C	2.8×10^{-10}	2.3×10^{-8}	1.4×10^{-8}	1.2×10^{-10}	—
D at 650 °C	2.9×10^{-11}	5.1×10^{-9}	1.9×10^{-9}	1.9×10^{-11}	—
D at 550 °C	3.2×10^{-12}	—	3.7×10^{-10}	2.6×10^{-12}	—
D_0 in cm²/s	0.19	0.084	0.054	0.0012	—
E_A in kcal/mol	41.5	27.5	31	33	—

*) The value 884 °C in the paper is probably an error.

Values for the partial diffusion constants of Be and Cu related to the contact zone in the β-phase:

$D_{Be}(\beta) = 7.2 \times 10^{-8}$ at 840 °C and 4.9×10^{-9} at 650 °C with $D_0 = 3.5 \times 10^{-2}$ and $E_A = 29$ kcal/mol; $D_{Cu}(\beta) = 5.8 \times 10^{-8}$ at 840 °C and 5.7×10^{-9} at 650 °C with $D_0 = 4.5 \times 10^{-3}$ and $E_A = 25$ kcal/mol, Reimbach, Krietsch [11]. Older results of the diffusion of Be in Cu (α solid solution) by Rhines, Mehl [12], Nowick [13], are discussed in [11].

The diffusion of Cu in single crystals of Be along the a and c axes ($D_\perp$ and $D_\parallel$, respectively) between 420 and 640 °C was studied by Myers et al. [14]. The Cu was introduced by ion implantation at 100 keV or by vapor deposition and the concentration-versus-depth profiles were obtained by ion backscattering methods. There is a significant anisotropy with $D_\parallel/D_\perp = 0.5 \pm 0.1$ at 600 °C. Dupouy et al. [15] obtained similar results over the range 700 to 1000 °C, using the radioactive tracer method with ⁶⁴Cu (see p. 242). The combined results for the temperature dependence of D_{Cu} between 420 and 1000 °C can be expressed by the equations $D_{\perp}=0.416 \exp(-23230/T)$ and $D_{\parallel}=0.381 \exp(-23870/T)$ with D in cm^2/s and T in K. The corresponding activation energies are 2.00 ± 0.10 eV and 2.05 ± 0.10 eV for the diffusion along the a axis or c axis, respectively [14]. The values obtained for $D_{\perp}$ and $D_{\parallel}$ between 700 and 1000 °C by [15] are: $D_{\perp}=0.35 \exp(-45900/RT)$ and $D_{\parallel}=0.90 \exp(-49500/RT)$ cm^2/s. The ratio $D_{\perp}/D_{\parallel}$ is 2.59 at 699 °C, 1.85 at 883 °C, and 1.60 at 1000 °C.

In the Ag-Be system (cf. Gmelin Handbook "Silber" C, 1972, pp. 9/12), an intermetallic phase (δ) at ≈73 at% Be was observed. There are conflicting data on the existence of a phase $AgBe_{12}$, a γ-phase at ~63 at% Be and on the mutual solid solubilities [1, pp. 9/10], [18, p. 4], [19, p. 1]. The diffusion coefficients D of ^{110}Ag in single crystalline Be parallel and perpendicular to the c axis were measured between 650 and 910 °C by the radioactive tracer method and can be described by the equations $D_{(\parallel c)}=0.43 \exp(-39300/RT)$ and $D_{(\perp c)}=1.76 \exp(-43200/RT)$ with D in cm^2/s and T in K. The anisotropy decreases with increasing temperature: $D_{\parallel}/D_{\perp}=2.24$ at 656 °C to 1.24 at 897 °C. The equation governing the diffusion of ^{110}Ag in polycrystalline Be, takes the following form: $D=0.9 \exp(-46200/RT)$, Naik et al. [16]. Previously, the following data had been reported: $D=6.2 \exp(-46100/RT)$, $D_{\parallel}=(0.41 \pm 0.15) \exp[(-39100 \pm 1600/RT)]$ and $D_{\perp}=(1.98 \pm 0.15) \exp[(-45700 \pm 1800)/RT]$, Naik [17]. See also Fig. 6-14, p. 226 for the diffusion coefficient of Cu and Ag in Be between about 700 and 1100 °C. For the diffusion coefficient of Cu in Be in dependence on temperature, see Fig. 6-12, p. 215.

In the Au-Be system five intermetallic phases are observed at room temperature: $AuBe_5$, $AuBe_3$, AuBe, Au_2Be, and Au_3Be. The solubility of Be in solid Au is very small and the solubility of Au in solid Be is 2 to 3 at% [1, pp. 187/8], [19, p. 64], cf. Gmelin Handbook "Gold" 3, 1954, pp. 782/4.

References:

[1] M. Hansen (Constitution of Binary Alloys, 2nd Ed., McGraw-Hill, New York 1958).
[2] X. Auvray, R. Graf, A. Guinier (Scr. Met. **8** [1974] 995/1004).
[3] V.A. Golikov, Yu.D. Tyapkin (Fiz. Metal. Metalloved. **37** [1974] 322/7; Phys. Metals Metallog. [USSR] **37** No. 2 [1974] 89/93).
[4] Yu.D. Tyapkin, V.A. Golikov (Fiz. Metal. Metalloved. **36** [1973] 1058/70]; Phys. Metals Metallog. [USSR] **36** No. 5 [1973] 144/54).
[5] T. Tadaki, T. Sahara, K. Shimuzi (Trans. Japan. Inst. Metals **14** [1973] 401/7).
[6] L.S. Patskhverova (Fiz. Metal. Metalloved. **27** [1969] 1128/9; Phys. Metals Metallog. [USSR] **27** No. 6 [1969] 186/7; Fiz. Khim. Poverkh. Yavlenii Vys. Temp. **1971** 205/8; C.A. **77** [1972] No. 167663).
[7] S.M. Myers, J.E. Smugeresky (Met. Trans. A **8** [1977] 609/16, 614).
[8] M. Ausloos, B. Lalevic (Phys. Status Solidi A **7** [1971] K51/K53; Appl. Phys. **6** [1975] 229/32).
[9] W.Vickers (Inst. Metals Monogr. Rept. Ser. No. 28 [1963] 335/49, 339).
[10] J.D. Baird, G.A. Geach, A.G. Knapton, K.B.C. West (Proc. 2nd Intern. Conf. Peaceful Uses At. Energy, Geneva 1958, Vol. 5, pp. 328/33; C.A. **1960** 22066).

[11] R. Reimbach, F. Krietsch (Z. Metallk. **54** [1963] 173/9).
[12] F.N. Rhines, R.F. Mehl (Trans. AIME **128** [1938] 185/222).
[13] A.S. Nowick (J. Appl. Phys. **22** [1951] 1182/6).
[14] S.M. Myers, S.T. Picraux, T.S. Prevender (Phys. Rev. [3] B **9** [1974] 3953/64).
[15] J.M. Dupouy, J. Mathie, Y. Adda (3rd Conf. Intern. Met. Beryllium Commun., Grenoble and Paris 1965 [1966], pp. 159/61).
[16] M.C. Naik, J.M. Dupouy, Y. Adda (Mem. Sci. Rev. Met. **63** [1966] 488/94).

[17] M.C. Naik (CEA-R-2621 [1965] 1/50, 28, 37; Diss. Univ. Paris 1964).
[18] R.P. Elliott (Constitution of Binary Alloys, 1st Suppl., McGraw-Hill, New York 1966).
[19] F.A. Shunk (Constitution of Binary Alloys, 2nd Suppl., McGraw-Hill, New York 1970).

6.2.18 With Ruthenium, Rhodium, and Palladium

Ru-Be phase diagram and alloys see Gmelin Handbook "Ruthenium" Erg.-Bd., 1970, pp. 177/8. In the tentative Ru-Be phase diagram reported by Shunk [6, p. 117], several intermetallic phases occur in the range 0 to 40 at% Ru (0 to ~90 wt% Ru): $RuBe_{13}$, Ru_3Be_{17}, Ru_3Be_{10}, $RuBe_2$, Ru_2Be_3. The hexagonal phases, Ru_2Be_{17} (and Rh_2Be_{17}), were obtained from the metal mixtures, melted in vacuum in a BeO crucible at 1450 to 1500 °C, Verkhorobin et al. [2].

X-ray diffraction and metallographic studies in the Rh-Be system in the range of 0 to 14 wt% Be show the existence of two phases, $RhBe_{\approx 2}$ and RhBe, Kruglykh et al. [3], see also Hansen [1, p. 295]. A phase with the composition $RhBe_{6.6}$ was established by Johnson et al. [4] by crystal structure studies; the reported lattice constants [4] are the same as those which have been found by [2] for the Rh_2Be_{17} phase.

Besides $PdBe_{12}$, $PdBe_5$, and PdBe, several intermediate phases containing >50 at% Pd have been observed in the Pd-Be system, see Hansen [1, p. 293] and Elliott [5, p. 168]. For the Pd-Be phase diagram and alloys see also Gmelin Handbook "Platin" 1951, pp. 579/80.

References:

[1] M. Hansen (Constitution of Binary Alloys, 2nd Ed., McGraw-Hill, New York 1958).
[2] L.F. Verkhorobin, G.P. Kovtun, A.A. Kruglykh, et al. (Izv. Akad. Nauk SSSR Metally **1971** No. 6, pp. 168/71; Russ. Met. **1971** No. 6, pp. 121/2).
[3] A.A. Kruglykh, M.M. Matyushenko, G.P. Tiklinskii (Ukr. Fiz. Zh. **13** [1968] 1107/10).
[4] Q. Johnson, G.S. Smith, O.H. Krikorian (Acta Cryst. B **26** [1970] 109/13).
[5] R.P. Elliott (Constitution of Binary Alloys, 1st Suppl., McGraw-Hill, New York 1966).
[6] F.A. Shunk (Constitution of Binary Alloys, 2nd Suppl., McGraw-Hill, New York 1970).

6.2.19 With Osmium, Iridium, and Platinum

In the Os-Be system, a hexagonal phase Os_2Be_{17} was described by Verkhorobin et al. [1] and a cubic phase Os_3Be_{17} by Sands et al. [2]. In addition to these two phases, the phase $OsBe_2$ also exists, see Hansen [3, p. 292]; see also Gmelin Handbook "Osmium" Suppl. Vol. 1, 1980, p. 22. In the Ir-Be system, the hexagonal phase is postulated to have the composition Ir_2Be_{17} by [1] and the composition $IrBe_{6.6}$ by Johnson et al. [4]. The phase $IrBe_2$ is also thought to exist, Hansen [3, p. 287]. For alloys of Be with Ir, see Gmelin Handbook "Iridium" Erg. Bd. 1, 1978, p. 48. In the Pt-Be system, some phases seem to exist, but the reported compositions are often contradictory, see Hansen [3, p. 294], Elliott [5, p. 168], Shunk [6, p. 116].

References:

[1] L.F. Verkhorobin, G.P. Kovtun, A.A. Kruglykh, et al. (Izv. Akad. Nauk SSSR Metally **1971** No. 6, pp. 168/71; Russ. Met. **1971** No. 6, pp. 121/2).
[2] D.E. Sands, Q.C. Johnson, O.H. Krikorian, K.L. Kromholtz (Acta Cryst. **15** [1962] 1191/5).
[3] M. Hansen (Constitution of Binary Alloys, 2nd Ed., McGraw-Hill, New York 1958).
[4] Q. Johnson, G.S. Smith, O.H. Krikorian (Acta Cryst. B **26** [1970] 109/13).
[5] R.P. Elliott (Constitution of Binary Alloys, 1st Suppl., McGraw-Hill, New York 1966).
[6] F.A. Shunk (Constitution of Binary Alloys, 2nd Suppl., McGraw-Hill, New York 1970).

6.2.20 With Technetium, Rhenium, and Transuranium Elements

Only one phase, $TcBe_{22}$, is described as yet in the Tc-Be system, Bucher, Palmy [1].

In the Re-Be system, at least two phases, $ReBe_{22}$, Sands et al. [2], and $ReBe_2$ exist, see Hansen [3, p. 295], Elliott [4, p. 170], and Shunk [5, p. 116], for $ReBe_2$, see also Gmelin Handbook "Rhenium" 1941, p. 84.

In the M-Be systems with M = Np, Pu, Am, and Cm the compounds MBe_{13} are obtained, see Gmelin Handbook "Transurane" B2, 1976, p. 12 (Np), p. 23 (Pu), B3, 1977, pp. 111/9 (Pu), and p. 270 (Am). For $AmBe_{13}$ and $CmBe_{13}$, see also p. 281.

References:

[1] E. Bucher, C. Palmy (Acta Cryst. A **24** [1967] 340/1).
[2] D.E. Sands, Q.C. Johnson, A. Zalkin, O.H. Krikorian, K.L. Kromholtz (Acta Cryst. **15** [1962] 832/4).
[3] M. Hansen (Constitution of Binary Alloys, 2nd Ed., McGraw-Hill, New York 1958).
[4] R.P. Elliott (Constitution of Binary Alloys, 1st Suppl., McGraw-Hill, New York 1966).
[5] F.A. Shunk (Constitution of Binary Alloys, 2nd Suppl., McGraw-Hill, New York 1970).

6.3 Reactions with Nonmetal Compounds

6.3.1 With Liquid Water and Aqueous Hydrogen Peroxide

As the aqueous corrosion of Be depends on many factors (e.g., purity, metallurgical history and surface treatment of Be, purity of the H_2O, gas atmosphere) and is extremely complex, the reported results show little agreement with respect to reproduction and comparison, especially at high temperatures.

Reviews on the aqueous corrosion of Be are given, e.g., by [1 to 4, 34]. Compact Be shows good resistance to pure H_2O at room temperature. In a static 427 h test in demineralized H_2O at 30 °C, for example, the corrosion rate of Be is <0.00254 mm per year, English [1, p. 537]. Be has performed for ten years without corrosion problems in the cooling water of a nuclear test reactor; the water used was demineralized by cation and anion resin beds and corrected to a pH of 5.5 to 6.5 with small additions of HNO_3, Stonehouse, Beaver [3]. By placing Be pieces in the cooling water of a nuclear reactor, no corrosion occurred within 3 months, Ganzha et al. [5].

The resistance of Be is better in aerated H_2O than in air-free demineralized H_2O, as shown by tests at 70 to 90 °C [1]. Chloride and sulfate ions in very low concentrations as well as Cu^{2+} and Fe^{3+} ions accelerate the corrosion markedly; for this reason beryllium's performance in tap water is poor and the metal has low resistance to seawater (see p. 287) [1 to 4]. But the addition of fluoride salts (NH_4F, LiF, or $NaBeF_4$) or $NaNO_2$ to the tap water acts as corrosion inhibitors and improves the resistance of Be, Schmitt [6].

At Room Temperature

In the wet milling of Be to <25 μm particle size the amount of surface oxidation of Be is directly dependent on the H_2O content of the organic mill solution (naphtha, benzene, hexane, etc.). After milling, e.g., 1500 g Be chips in a mixture of 1500 mL naphtha and 45 mL distilled H_2O for 20 h, the freshly fractured powder surface had reacted practically stoichiometrically with the water according to $Be + H_2O = BeO + H_2$, Morana [7].

Compact Be specimens showed good resistance to demineralized H_2O of pH 7.0, tested for 6 months, Burnham, Barts [8]. Cross-rolled, surface ground Be sheet which had been flash pickled with a HF-HNO_3-H_2O solution is resistant to distilled H_2O at 15 °C for about 14 days. After 18 days a weight loss of about 0.16 mg/cm² and a maximum pit depth of 0.02 mm were observed (see Fig. 6-44, p. 287, for a comparison with seawater), Prochko et al. [9]. Static corrosion tests of similarly pickled Be specimens at 25 °C for 105 d in distilled, aerated H_2O ($2\,\Omega^{-1}\cdot cm^{-1}$) and tap H_2O ($140\,\Omega^{-1}\cdot cm^{-1}$) showed the latter to be considerable more corrosive. By analyzing the white gelatinous corrosion-product sludge for Be (after dissolving in H_2SO_4), the relative corrosion rates are 0.006 and 0.015 $mg\cdot cm^{-2}\cdot d^{-1}$ (0.012 and 0.0279 mm/a), respectively. The corresponding values for anodized Be are for both sorts of H_2O $<5\times10^{-4}$ $mg\cdot cm^{-2}\cdot d^{-1}$ (<0.00127 mm/a). Addition of corrosion inhibitors to the tap water improves the resistance of Be; in tap H_2O containing 2.5 g $NaNO_2$/L no corrosion was observed after a 2 month exposure, Schmitt [6]. The weight gain of Be strips in distilled and tap H_2O in static tests up to 7000 h is shown in figures by O'Donnell [10], but no relation to the surface area is given. The corrosion resistance can be improved by coatings (0.2 μm thick) of rhombic BeF_2 [10, 35].

At 30 to 100 °C

The corrosion resistance of Be to H_2O at atmospheric pressure at 87 °C is quite good. No attack was noted in dynamic long-time tests at a H_2O flowing rate of 914 cm/s, Evans [11]. The results of static and dynamic corrosion tests with extruded Be in air-free and aerated demineralized water at t=30 to 90 °C are shown in the following table from English [1, p. 537]:

t in °C	test type	aeration	time in h	max. pit depth in mm	corrosion rate in mm/a
30	static	no	427	negligible	<0.00254
70	static	no	427	0.0737	0.0406
80	static	no	427	0.104	0.0813
90	static	no	427	0.119	0.0813
70	static	air	406	0.0406	0.00254
80	static	air	406	0.0279	0.00254
90	static	air	406	negligible	0.00254
70	dynamic [a)]	no	473	negligible	0.00254
80	dynamic [a)]	no	473	negligible	0.0127
90	dynamic [a)]	no	473	negligible	0.0152

[a)] Specimens rotated (~150 to 244 cm/s).

Change of pH between about 4 and 8 has little effect. Dynamic tests in nearly air-free H_2O for 645 h at 80 °C at pH 4.1, 5.9, and 7.6 result in corrosion rates of 0.00762 mm/a at pH 4.1, and 0.00254 mm/a at pH 5.9 and 7.6. The maximum pit depths are 0.0889 at pH 4.1, 0.0508 at pH 7.6, and negligible at pH 5.9 [1]. The corrosion rate of Be in 27.6 L deionized H_2O at 93.3 °C (200 °F), flowing at 13.41 m/s was shown to be dependent on the surface area of Be exposed to the volume of H_2O in the system. Corrosion rates of 0.028 and 0.071 mm/a were found at surface areas of 125 and 12.5 cm², respectively. The corrosion rate at a low H_2O flow rate of 1 cm/s is very low, between 0.00254 and 0.0041 mm/a, Griess, English [18].

Tests at 85 °C for 1 year with and without the presence of 0.005 M H_2O_2 in the demineralized H_2O_2 were carried out to simulate cooling water conditions under irradiation in the MTR (Materials Testing Reactor). The pH was adjusted daily to 5.5 to 6.5 with dilute HNO_3

or NaOH. The test samples, extruded Be (containing 0.07% Be_2C and ~0.25% BeO) and remelted Be (containing 1.86% BeO), were polished and etched with 25% HNO_3 for 20 min. The presence of 0.005 M H_2O_2 stimulates the corrosion rates for both test materials, but the acceleration is smaller on the extruded material. The corrosion rate in mm/a for both materials at 85 °C was 0.00178 in H_2O. In the presence of 0.005 M H_2O_2 the value is 0.00432 for the extruded and 0.00838 for the remelted material with maximum pit depths of 0.518 and 0.406 mm, respectively. The maximum pit depths in the pure H_2O are 0.416 and 0.183 mm, respectively, English [12]. Static two-month tests at 85 °C in 0.005 M H_2O_2 (pH 5.5 to 6.5) with hot pressed and vacuum cast extruded Be show that the corrosion damage was comparatively low. Ground surfaces were more corrosion resistant than milled surfaces. The most noticeable damage was caused by shallow pits, Reed [13], see also Reed [14]. It was found that for vacuum cast and extruded Be subsequent annealing increased the susceptibility of Be to corrosion in demineralized H_2O at 85 °C containing 0.0005 and 0.005 M H_2O_2. An increase of the H_2O_2 concentration from 0.0005 M to 0.005 M will markedly decrease pit frequency and increase overall corrosion attack, Olsen [15], see also Olsen [16] for earlier tests. The bond at the interfaces, formed by the corrosion products in the tests described above [12], was very tenacious and separation of interfaces was accomplished with great difficulty. The appearance of large blisters on the Be surfaces after long exposure was believed due to the hydrolysis of segregated beryllium carbide inclusions to amorphous beryllium hydroxide. This transition is accompanied by a large volume increase which may be sufficient to create internal pressures great enough to cause the blistering effect [12]. Stress corrosion tests of extruded and hot pressed Be in circulating 0.002 to 0.006 M H_2O_2 solution at 85 to 93 °C show no evidence of stress-corrosion cracking in any of the specimens tested, Logan, Hessing [17].

In dynamic loop tests at 100 °C for 733 d with H_2O containing 5 ppm oxygen at pH 5, flowing at 381 and 701 cm/s, the corrosion rates were constant and averaged 0.0482 mm/a for hot pressed reactor grade Be. Only very thin films formed on all specimens during the exposure. All specimens were subject to pitting attack, which gradually increased to a maximum depth between 0.178 and 0.229 mm, Griess et al. [19], see also [20, 21]. While the corrosion rate of untreated Be discs in static 30 d tests in H_2O at 100 °C is 1.254 $mg \cdot cm^{-2} \cdot a^{-1}$, the rate for passivated Be is 0.0155 $mg \cdot cm^{-2} \cdot a^{-1}$ under the same conditions. The passivating occurred with an aqueous solution containing $K_2Cr_2O_7$ and H_3PO_4, Morana [22].

Above 100 °C

Sheet Be (containing 1.5% BeO, 0.35% Be_2C, 0.65% Al+Fe+Mg) is not resistant to deionized H_2O under a 2:1 H_2/O_2 atmosphere at 149 °C in a stainless steel autoclave. The weight gain after 1128 h was 30 mg/dm^2 and a flaky white corrosion product had formed, Boyd, Peoples [23]. In pure H_2O the corrosion rate of Be is on the order of 0.5 to 1 $mg \cdot cm^{-2} \cdot month^{-1}$ at 150 °C and 2 to 4 $mg \cdot cm^{-2} \cdot month^{-1}$ at 260 °C. The best resistance is observed in H_2O degassed or saturated with H_2, Coriou [24].

The corrosion behavior of single crystals and polycrystalline Be was studied in a steel autoclave at 200 and 250 °C. When single crystals of Be, exposing a basal plane and two or more prism faces, were immersed in H_2O at 200 °C, the attack occurred initially and with considerable rapidity on the basal plane (0001) by the hydrolysis of the Be_2C inclusions (and to a lesser extent Al and Si), leaving holes surrounded by crater-like deposits of hydrated beryllium oxide. Plates in the basal plane adjacent to the holes began to lift and the crystal began to break up by cleavage from the prism face along the basal planes. The final destruction is believed to be due to stress failure rather than to corrosion. It was found that the corrosion resistance of polycrystalline Be with relatively low purity to

H_2O is good at 250 °C (tests up to 860 h). Vacuum cast ingots of relatively pure Be but containing 50 to 70 ppm Cl^- have poor resistance to H_2O at 250 °C. The resistance can be improved by addition of about 3000 ppm Fe to the Be in tests at 250 °C, but not at 325 °C, Pearsall [33].

Corrosion tests in distilled H_2O at 274 °C (350 h) and 316 °C (16 to 54 h) show that sintered grade Be is more resistant to disintegration and localized attack than vacuum cast, extruded material, Peoples, Bulkowski [25]. Tests at 288 °C up to 350 h on sintered and vacuum cast, extruded Be in distilled H_2O show that 288 °C is beyond the safe limit for the material tested, Peoples et al. [26]. Polished Be samples placed on stainless steel screen trays in the autoclave exposed to high purity distilled H_2O ($\varrho = (0.5 \text{ to } 1) \times 10^6\ \Omega \cdot cm$ at 25 °C) at 316 °C had a maximum life time of 140 to 160 days before accelerated local attack took place. Weight gains from 0.3 to 1.5 mg/cm^2 were observed in 140 to 160 days. Kneppel [27]; (for more details concerning Be purity see below [31]). The corrosion rates rapidly decrease with time during the first 28 days at 316 °C as shown by tests of reactor grade Be in deionized H_2O ($\varrho \geq 18 \times 10^5\ \Omega \cdot cm$). While the rate after 1 day is about 195 to 280 mg $\cdot$ dm^{-2} $\cdot$ month^{-1}, the rate after 7 and 28 days is 31 to 44 or 11 to 20.8 mg $\cdot$ dm^{-2} $\cdot$ month^{-1}. No pitting attack was observed. Introduction of 20 ppm Cl^- ions results in severe pitting of Be and even disintegration of the sample after only 7 days (see p. 286). The density of the Be sample is very important; densities below 98% of theoretical giving apparent corrosion rate values 10 to 100 times higher than the corresponding full-density material [28].

When stainless steel wire was used as supports for the tested Be samples in H_2O at 316 °C, a weight gain was observed after 20 days, but with an insulation of Al_2O_3 between Be and the steel a remarkable decrease in corrosion rate became evident; after 57 d exposure no measurable change in weight of the specimens and no evident change in the surface appearance occurred, Bennett [29]. An increase of corrosion by coupling Be to stainless steel was not found by Demant et al. [30] and Kneppel [31] in H_2O at 316 or 343 °C, respectively; (see also below the corrosion tests in heavy water by Griess et al. [32]).

The corrosion resistance of pickled Be in H_2O at 325 °C is low; pressed and sintered samples ($\approx$98% theoretical density) are inferior to sheet material. Alloying with Ni improves the corrosion resistance [30]. It was found that high-purity Be prepared by distillation and/or zone-refining as well as commercial high-purity Be invariably corrodes catastrophically in exposures of two days or less in H_2O at 343 °C. This poor resistance was exhibited even in samples fabricated by powder metallurgical techniques. However, the material can be improved by alloying. Heat-treated alloys containing (in ppm wt) 4000 Ni, 2000 Ni +2000 Fe, or 5000 Ni+5000 Fe showed good resistance in H_2O at 343 °C for as long as 43 days with no sign of attack. A control sample prepared from high purity Be powder with no additions, processed in the same manner as the alloys, was completely converted to oxide after just 3 days in H_2O at 343 °C. Unalloyed Be fabricated by powder metallurgy from less pure commercial powders is capable of good corrosion resistance at 343 °C for at least up to about 160 days, but sometimes only for shorter times. The presence of minute amounts of Cu in the water accelerated corrosion of Be samples in an unpredictable manner. The corrosion of Be is also influenced by its heat treatment and surface preparation. It appeared that change of pH of the water between 6 and 8 has no major effect on the corrosion, Kneppel [31]. Autoclave tests of hot-pressed reactor grade Be in degassed and in oxygenated heavy water at 250 and 300 °C showed no significant attack after 120 days and no apparent differences in degassed H_2O at 250 and 300 °C. Specimens exposed to both temperatures developed a thin, tenacious film, which could not be completely removed by standard descaling techniques and exhibited small weight gains. Some pitting was observed; pit depths ranged up to 0.04 mm at 250 °C and up to 0.09 mm at 300 °C. In the presence of O_2, thicker films were found at both temperatures with a maximum pit depth of 0.06 mm and "defilmed" corrosion rates (not all scale removed) for 120 days were ≦0.005 mm/a. Be coupled with stainless steel in degassed H_2O at both temperatures underwent little attack after 60 days; a slight weight gain and mild pitting were observed on contact surfaces. But the presence of O_2 in the water resulted in near-catastrophic attack on the Be members of the couples, particularly at 250 °C. Massive pits, approximately 2.54 to 3.18 mm in depth, formed over all surfaces. Heavy tubercles of white corrosion product were very evident, Griess et al. [32].

References:

[1] J.L. English (in: D.W. White, J.E. Burke, The Metal Beryllium, ASM, Cleveland 1955, pp. 1/703, 533/48).

[2] G.E. Darwin, J.H. Buddery (Beryllium: Metallurgy of Rarer Metals, No. 7, Academic, New York 1960, pp. 1/392, 253/8; N.S.A. **14** [1960] No. 15917).

[3] A.J. Stonehouse, W.W. Beaver (Mater. Protect. **4** No. 1 [1965] 24/8; CONF-492-9 [1964] 1/31, 2/6; N.S.A. **19** [1965] No. 20495).

[4] J.J. Mueller, D.R. Adolphson (Beryllium Sci. Technol. **2** [1979] 417/33, 422/4).

[5] V.D. Ganzha, K.A. Konoplev, R.N. Rodionova, et al. (Rev. Phys. Acad. Rep. Populaire Roumaine Suppl. **6** [1961] 629/33; C.A. **60** [1964] 5026).

[6] C.R. Schmitt (Y-1397-Rev. 1 [1975] 1/24, 6/8; C.A. **83** [1975] No. 123011).

[7] S.J. Morana (U.S. 3322582 [1964/67]; C.A. **67** [1967] No. 56845).

[8] J.B. Burnham, M.H. Bartz (AECD-3923 [1956] 1/12; N.S.A. **10** [1956] No. 10830).

[9] R.J. Prochko, J.R. Myers, R.K. Saxer (Mater. Protect. **5** No. 12 [1966] 39/42).

[10] P.M. O'Donnell (Corros. Sci. **7** [1967] 717/8).

[11] G.E. Evans (CF-52-8-148-Chapter 11-Del. [1957] 1/34, 29/32; N.S.A. **11** [1957] No. 12474).

[12] J.L. English (ORNL-772 [1956] 1/56; N.S.A. **10** [1956] No. 6273).

[13] J. Reed (CF-50-5-81 [1956] 1/15; N.S.A. **10** [1956] No. 8862).

[14] J. Reed (ORNL-942 [1957] 1/90; N.S.A. **11** [1957] No. 7619).

[15] A.R. Olsen (ORNL-1146 [1956] 1/59; N.S.A. **10** [1956] No. 7253).

[16] A.R. Olsen (ORNL-733 [1950] 1/33; N.S.A. **10** [1956] No. 4281).

[17] H.L. Logan, H. Hessing (NBS-6 [1955] 1/12; N.S.A. **10** [1956] No. 6272).

[18] J.C. Griess, J.L. English (ORNL-4034 [1967] 1/42, 29/40; C.A. **67** [1967] No. 38982).

[19] J.C. Griess, J.L. English, L.L. Fairchild, P.D. Neumann (ORNL-3789 [1965] 122/5; N.S.A. **19** [1965] No. 30671).

[20] J.C. Griess, H.C. Savage, J.L. English, et al. (ORNL-3591 [1964] 70/3; N.S.A. **18** [1964] No. 22013).

[21] J.C. Griess, H.C. Savage, J.L. English (ORNL-3417 [1963] 64/8; N.S.A. **17** [1963] No. 25065).

[22] S.J. Morana (U.S. 3301718 [1967]; C.A. **66** [1967] No. 68542).

[23] W.K. Boyd, R.S. Peoples (BMI-1047 [1955] 1/20; N.S.A. **10** [1956] No. 3006).

[24] H. Coriou (Bull. Soc. Franc. Electr. [7] **7** [1957] 490/510, 496/7).

[25] R.S. Peoples, H. Bulkowski (BMI-HAP-101 [1957] 1/17; N.S.A. **11** [1957] No. 3807).

[26] R.S. Peoples, H. Bulkowski, F.W. Fink (BMI-HAP-102 [1957] 1/7; N.S.A. **11** [1957] No. 3808).

[27] D.S. Kneppel (NMI-1190 [1957] 1/21; C.A. **1958** 1015).

[28] AMES Laboratory (TID-7526-Pt. 7 [1956/59] 1/18; N.S.A. **14** [1960] No. 6681).

[29] W.D. Bennett (AECL-1236 [1961] 1/56, 15/6; N.S.A. **15** [1961] No. 23933).

[30] J.T. Demant, R.E. Wiggins, P. Gibson (AERE-R-4422 [1963] 1/29; N.S.A. **18** [1964] No. 4192).

[31] D.S. Kneppel (NMI-1911 [1963] 1/95; N.S.A. **18** [1964] No. 4198).
[32] J.C. Griess, J.L. English, L.L. Fairchild, et al. (ORNL-3262 [1962] 90/8; N.S.A. **16** [1962] No. 17563).
[33] C.S. Pearsall (MIT-1109 [1957] 1/19; N.S.A. **11** [1957] No. 7680).
[34] W.P. Jepson (in: H.P. Godard, W.B. Jepson, M.R. Bothwell, R.L. Kane, The Corrosion of Light Metals, Wiley, New York 1967, pp. 1/372, 221/56, 246/56).
[35] P. O'Donnell (U.S. 3591426 [1968/71]; C.A. **75** [1971] No. 90837).

6.3.2 With Water Vapor

For the reaction with moist air, with moist O_2, and moist CO_2 see p. 200, 197, and 268, respectively.

At Low Pressures ($\leq$760 Torr)

The weight-gain vs. time curves for oxidation of hot rolled chemically polished Be sheet (prepared from electrolytic flake Be) in H_2O vapor (12 Torr), studied with a vacuum microbalance, from 550 to 750 °C are shown in **Fig.** 6-**17** and **Fig.** 6-**18**. At 550 °C the oxidation effectively ceased after about 100 h, since after a further 200 h the additional weight gain was only 0.7 μg/cm². At 600 °C the rate of oxidation was again comparatively large at first but it progressively decreased to a constant small value of 0.06 $\mu g \cdot cm^{-2} \cdot h^{-1}$ after about 80 h. At 650 °C the oxidation is no longer protective. After an initial decrease, the rate increases and breakaway occurs after about 35 h (weight gain 65 μg/cm²). There is no signifi-

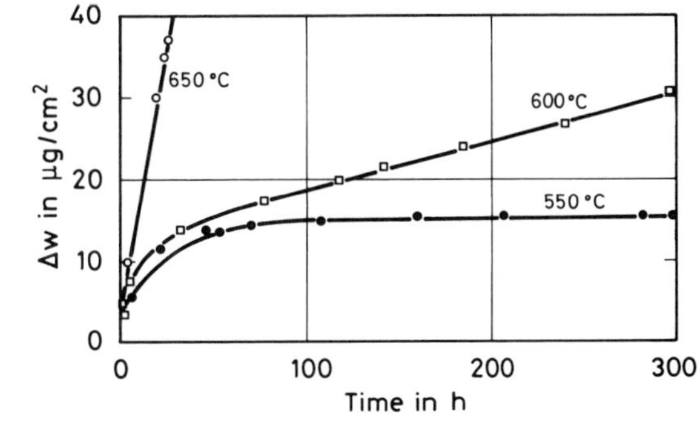

Fig. 6-17

Weight gain Δw vs. time for Be in H_2O vapor (12 Torr) at various temperatures.

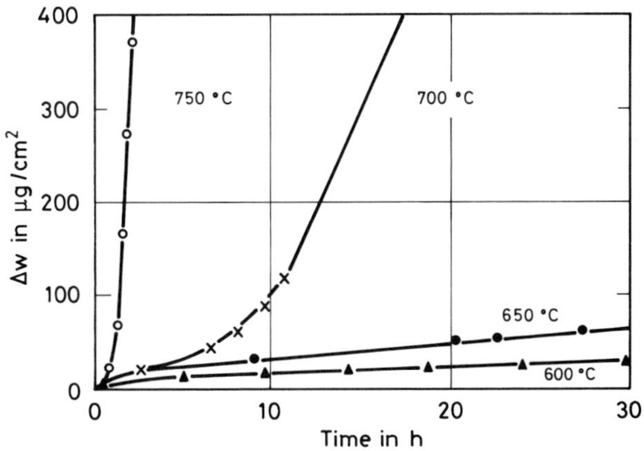

Fig. 6-18

Weight gain Δw vs. time for Be in H_2O vapor (12 Torr) at various temperatures.

cant difference between the oxidation in H_2O vapor (12 Torr) and in moist oxygen (100 Torr total pressure, 12 Torr partial pressure of H_2O) at 650 °C, see **Fig.** 6-**19**, but breakaway occurred later in the moist oxygen at 700 and 750 °C. The weight gain per cm^2 is of formal significance only, since breakaway takes the form of localized intergranular penetration of Be. This can explain the discontinuities in the curves at 650 °C (see Fig. 6-19). At 750 °C breakaway occurred after only 1 h (weight gain 5 $\mu g/cm^2$). At 650 and 700 °C the rate of reaction was apparently still increasing when the runs were stopped, whereas at 750 °C a constant rate (700 $\mu g \cdot cm^{-2} \cdot h^{-1}$) was reached after about 2 h. The pressure dependence of the oxidation rate at 700 °C and $p(H_2O) = 5 \times 10^{-4}$, 0.58 and 12 Torr is shown in **Fig.** 6-**20**, Aylmore et al. [1]. The short time oxidation behavior in H_2O vapor at 20 Torr of Be sheet (prepared from reactor grade Be by arc melting, pickling, and hot rolling), containing 0.14 wt% ^{14}C-labelled carbon, is shown in **Fig.** 6-**21** in comparison with that in O_2 and CO_2 (at 76 Torr). Breakaway occurs in H_2O and O_2. (Inclusions of Be_2C are attacked at 700 °C preferentially by O_2 and to a lesser extent by H_2O vapor according to $Be_2C + 2H_2O \rightarrow 2BeO + CH_4$), Jepson et al. [12]. The oxidation kinetics of sintered Be were studied in H_2O vapor (5 and 30 Torr pressure) at 600 to 800 °C with the use of a Gulbransen type microbalance. At 600 °C the oxidation rate initially follows the parabolic rate law and then it conforms to a new logarithmic law. At 650 and 700 °C the oxidation proceeds in two successive steps: at first according to the new logarithmic and then changing to follow a rectilinear rate law. The latter step is accompanied by breakaway of the oxide film. At 800 °C the oxidation obeyed the rectilinear rate law during the whole period of the run. Photographic and electron micrographic studies

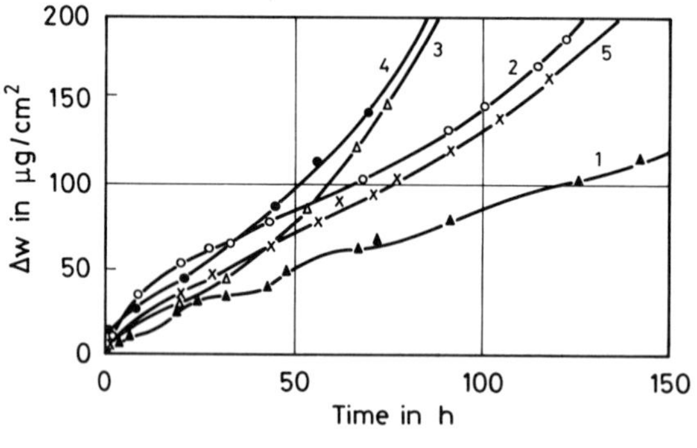

Fig. 6-19

Weight gain vs. time for Be at 650 °C in H_2O vapor at 12 Torr (curves 1 to 3) and in moist O_2 at 100 Torr (12 Torr H_2O) (curves 4 and 5).

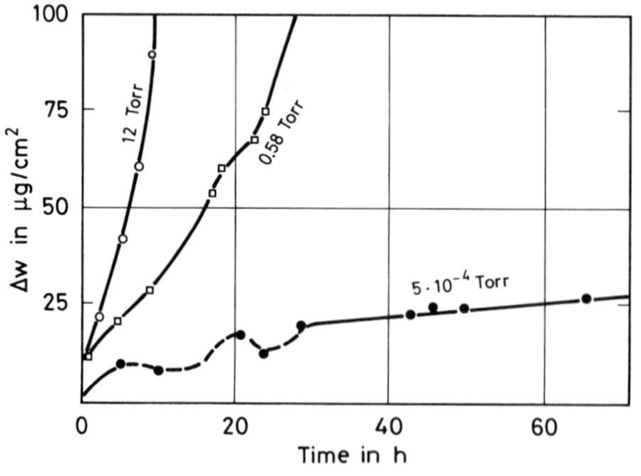

Fig. 6-20

Weight gain Δw vs. time for Be at 700 °C in H_2O vapor at various pressures.

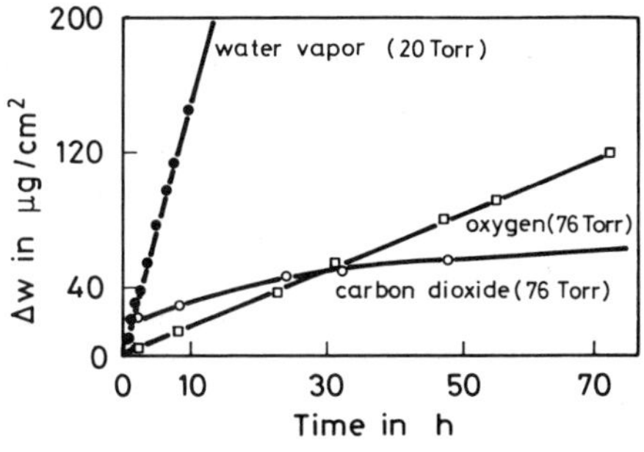

Fig. 6-21

Weight gain Δw vs. time of Be sheet in H_2O vapor (20 Torr), O_2, and CO_2 (76 Torr) at 700 °C.

of the oxidized Be samples obtained at each stage, indicated that the logarithmic law can be attributed to the generation of blisters at the oxide-metal-interface, while the stage of breakaway is initiated along the grain boundaries through the failure of the blisters, Sasabe, Nomura [2]. Corrosion tests between 660 and 905 °C at $p(H_2O) = 5$ to 19 Torr showed that Be samples fabricated by powder metallurgical routes suffered extensive intergranular attack, whereas samples of cast and of extruded metal showed an improved oxidation resistance; a longer time elapsed before breakaway and the rate of oxidation after breakaway was less than with the powder-route samples. The oxidation of the last was pressure-dependent, Phennah, Davies according to Jepson [10].

Some short time corrosion tests of Be samples with varying metallurgical history at 650 °C in flowing steam at atmospheric pressure showed that in samples prepared from powder, the initiation of attack in the form of blistering over the whole surface occurred after a 2 h exposure. The higher purity cast samples showed relatively little blistering (one sample was not visibly attacked after 500 h exposure in 650 °C steam). This result was in direct opposition to the corrosion behavior in high-temperature liquid water, where the less pure materials fabricated from powder exhibited the best corrosion resistance, Kneppel [8].

The weight gain of commercial grade Brush powders (surface area ~0.3 m²/g, impurities: 1.74% BeO, 1100 ppm C, 1400 ppm Fe, ~1000 ppm Mg+Al+Si) and of high purity Pechiney flake (surface area ~0.1 m²/g, impurities <500 ppm) in slowly flowing (~20 mL/min) He containing 4 vol% H_2O vapor at 550 to 700 °C and atmospheric pressure up to 250 h is shown in **Fig. 6-22**. In contrast to [1] no breakaway was observed. The less pure material shows initially a period of rapid weight gain, but the weight gain between about 5 and 250 h is lower than for the purer material at comparable temperatures. The weight gain

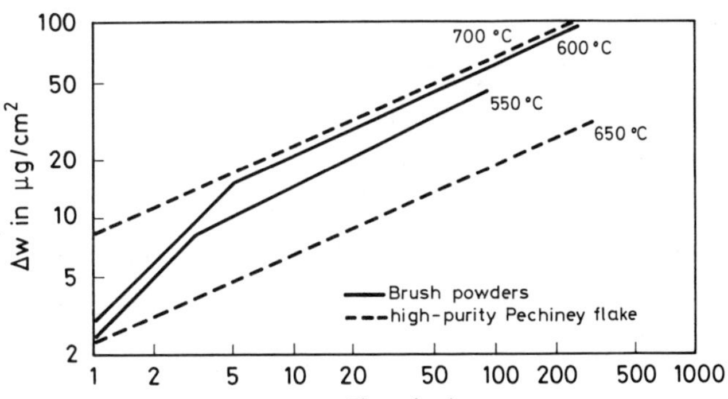

Fig. 6-22

Weight gain Δw vs. time of Be powder in moist He (~4 vol% H_2O) at atmospheric pressure.

of the powders (in Pt containers) was measured by the extension of a spring with a reproducibility of ±5% in duplicate runs, Werner, Inouye [9]; the reaction rate for the high-purity material was parabolic, while the less pure material exhibited a short period of linear oxidation followed by the parabolic rate, Werner [11]. The kinetics of the oxidation of reactor grade Be rods (98.3 wt% Be, major impurity BeO), which were contained in alumina reaction tubes, was studied between 777 and 1007 °C (1050 to 1280 K) in a steady-flow system at H_2O vapor pressures of 150, 300, and 650 Torr. The total sample weight change from the beginning to the end of an experiment corresponded very well with the weight change calculated from the total volume of H_2, liberated according to the reaction $Be+H_2O \rightarrow BeO+H_2$. The reproducibility of replicate experiments was approximately ±10%. It was found that the "breakaway reaction" occurred at about 1050 K (777 °C), and that above 1050 K a rapid, nonprotective linear surface oxidation took place with a rate of reaction which was proportional to the H_2O vapor concentration and which increased exponentially with increasing temperature. From the Arrhenius plot in **Fig. 6-23**, the apparent activation energies 23.6, 25.8, and 27.8 kcal/mol were obtained for $p(H_2O)=150$, 300, and 650 Torr. Above 1280 K (1007 °C) the reaction rate was too rapid to be followed quantitatively in the steady-flow system. The Be-H_2O reaction at 1280 K was approximately 2.5 times faster than the Be-O_2 reaction at 1320 K. From the logarithmic plot of the constant reaction rates vs. log-$p(H_2O)$ for each for the 4 temperatures (1050, 1100, 1200, and 1280 K), the rate vs. $p(H_2O)$ dependence varied from an order of 1.0 at 1280 K to an order of 0.6 at 1050 K. A variation of the H_2O vapor flow rate in the range 50 to 200 cm/s had no effect on the overall reaction kinetics at temperatures up to 1200 K, but at 1280 K ($p(H_2O)=650$ Torr) increasing flow from 50 to 200 cm/s caused about a 25% increase in the overall reaction rate. It would appear likely that at ≧1280 K and low flow rates, the diffusion of H_2O vapor to the Be surface through the H_2 gas generated (as reaction product) would play an important role in determining the reaction kinetics. The diffusional effects are avoided in a steady-flow system up to 1200 K, Blumenthal, Santy [3].

From earlier studies in a static system in the high temperature range at 900 to 1500 °C and $p(H_2O)=20$ Torr, which showed small effect of temperature on the reaction kinetics,

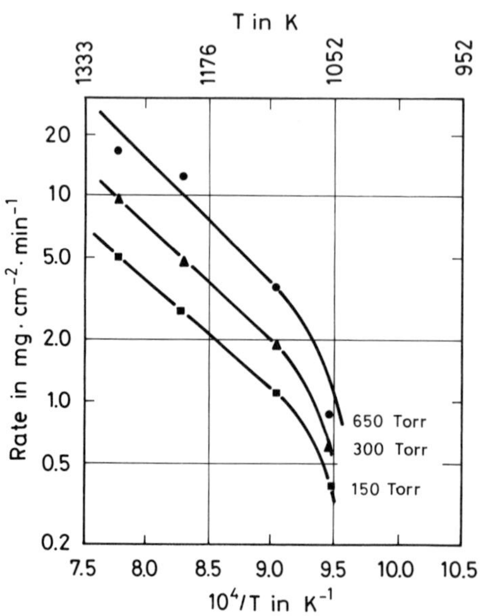

Fig. 6-23

Reaction rate of Be vs. reciprocal temperature in H_2O vapor at various pressures.

it was concluded that the overall reaction rate was controlled by diffusional effects, at least at the higher temperatures, and that the actual surface reaction between Be and H_2O vapor may be considerably faster and a great deal more temperature-dependent than indicated by the experimental results, Blumenthal, Santy [4].

By melting electrolytic Be, either by inductive heating at $\sim 10^{-4}$ Torr in a BeO crucible (capacity 3 L), or by electron bombardment at 4×10^{-7} Torr of Be bars lying on a water-cooled metal plate (Cr plated Cu), Be reacts with the H_2O vapor of the residual furnace atmosphere. The reaction, studied by mass spectrometry of the gas atmosphere, occurs above 600 °C resulting in an increase of the H_2 partial pressure above Be. The hydrogen content of the cooled Be after melting is four times higher in the inductively melted sample than in the electron beam melted Be in high vacuum, Schaub [5].

At High Pressures

Corrosion tests of Be with varying metallurgical history show that wet saturated steam at 325 °C (vapor pressure of water at 325 °C $= 90.45 \times 10^3$ Torr) is more aggressive than liquid H_2O at the same temperature. There are large and irregular differences of corrosion behavior between specimens of Be prepared by various fabrication routes. Most of the specimens were disintegrated after 1 day. The resistance of Be to the saturated steam can be somewhat improved by alloying with Ni (0.5 to 5 wt%) or by preexposure to air at 500 °C. The corrosion of Be in dry steam at 500 °C and 51.75×10^3 Torr pressure seems to be less severe than in wet steam at 325 °C, Demant [6]. The corrosion of chemically etched Be sheet (produced from electrolytically bake Be) in dry steam at 500 °C and 51.75×10^3 Torr can be inhibited by the presence of 2.4 g boric acid per 7.5 g steam in the autoclave. Thus, weight gains of about 0.50 mg/cm^2 are obtained after 24 h, whereas the samples in the absence of boric acid were completely oxidized in 24 h, Wanklyn, Britton [7]. The corrosion behavior of various Be samples in dry steam at 400 °C and 77.6×10^3 Torr pressure, in tests up to 48 days, is comparable to the corrosion behavior in water at 343 °C (see p. 247). Samples which behaved poorly in water at 343 °C also failed in steam at 400 °C. Weight gains in the steam were generally somewhat lower than in the water, but localized attack in steam was somewhat more severe and appeared to progress with increased exposure time. Breakdown of corrosion resistance occurred in steam after 60 to 100 days, in the form of small broken oxide blisters. The corrosion resistance of different Be samples in steam at 400 °C does not look promising. No alloys were tested, Kneppel [8].

Ignition and Combustion

The theoretical burning rate of Be droplets in H_2O vapor and H_2O-Ar mixtures with 50 and 70 mol% H_2O at 20 atm pressure and 440 K are reported by Kuehl [13]; see p. 204 for the ignition and combustion of Be in H_2O-containing gas mixtures or flames.

References:

[1] D.W. Aylmore, S.J. Gregg, W.B. Jepson (J. Nucl. Mater. **3** [1961] 190/200).
[2] M. Sasabe, S. Nomura (Nippon Genshiryoku Gakkaishi **10** [1968] 319/24 from C.A. **70** [1969] No. 14797).
[3] J.L. Blumenthal, M.J. Santy (Symp. Intern. Combust. Proc. **11** [1966/67] 417/25; C.A. **67** [1967] No. 110222).
[4] J.L. Blumenthal, M.J. Santy (Western States Sect. Combust. Inst. Paper [WSS-CI] No. 65-5 [1965] 1/56, 9, 37/8; C.A. **63** [1965] 7687).
[5] B. Schaub (Compt. Rend. Colloq. Intern. Met. Vide, Brussells 1965 [1966], pp. 103/7).

[6] J.T. Demant, R.E. Wiggins, P. Gibson (AERE-R-4422 [1963] 1/29; N.S.A. **18** [1964] No. 4192).
[7] J.N. Wanklyn, C.F. Britton (Brit. 922707 [1963]; N.S.A. **17** [1963] No. 20428).
[8] D.S. Kneppel (NMI-1911 [1963] 1/95, 12/3, 43; N.S.A. **18** [1964] No. 4198).
[9] W.J. Werner, H. Inouye (Inst. Metals Monogr. Rept. Ser. No. 28 [1963] 283/93, 289/90; C.A. **60** [1964] 10325).
[10] P.J. Phennah, M.W. Davies (in: H.P. Godard, W.B. Jepson, M.R. Bothwell, R.L. Kane, The Corrosion of Light Metals, Wiley, New York 1967, pp. 1/372, 221/56, 252).

[11] W.J. Werner (ORNL-3160 [1961] 143/4; N.S.A. **15** [1961] No. 29716).
[12] W.P. Jepson, J.B. Warburton, B.L. Myatt (J. Nucl. Mater. **10** [1963] 127/33).
[13] K. Kuehl (AIAA [Am. Inst. Aeronaut. Astronaut.] J. **3** [1965] 2239/47, 2243).

6.3.3 With NH_3 and Ammonium Salts

Be is not soluble in liquid NH_3, Ostertag et al. [1]. With NH_3 gas at 800 °C the compound Be_3N_2 is formed. For the reaction $3Be(s)+2NH_3(g) \rightarrow \alpha\text{-}Be_3N_2(s)+3H_2(g)$ the reaction enthalpy $\Delta H^\circ_{298} = -118.4 \pm 0.3$ kcal/mol was determined calorimetrically, Gross et al. [2].

Be (1 g) reacts with molten NH_4Cl (6 g) at 500 °C in 60 min to form Be_2N_2. Additional heating at 700 °C for 30 min improves the purity of the reaction product, Taratunin, Sobol'eva [3]. Be (0.2 mol) reacts with NH_4Cl (0.4 mol) or $(NH_4)_2SO_4$ (0.2 mol) in liquid NH_3 to give 73% or 15%, respectively, of the theoretical volume of H_2, Krug, Tocker [4]. The ammonium halides NH_4X with X=Cl, Br, I, dissolved in liquid NH_3, are reduced by metallic Be, which dissolves in the solutions. At first solutions of BeX_2 are formed, from which compounds such as $BeCl_2 \cdot 4NH_3$ or $BeBr_2 \cdot 4NH_3$ can be isolated. With an excess of Be (or by dissolution of Be in liquid NH_3 solutions of BeX_2) complex compounds such as $3Be(NH_2)_2 \cdot BeX_2 \cdot 4NH_3$ (with X=Br or I) or others are formed, Bergstrom [5]. By electrolysis of liquid NH_3 with addition of some NH_4I using a Be anode, blue streaks of dissolved Be are observed at the cathode after an initial decomposition of NH_4^+ ions [1].

References:

[1] W. Ostertag, W. Ruedorff, H. Suhr (Z. Naturforsch. **20b** [1965] 599/601).
[2] P. Gross, C. Hayman, P.D. Greene, J.T. Bingham (Trans. Faraday Soc. **62** [1966] 2719/24).
[3] P.I. Taratunin, N.A. Sobol'eva (U.S.S.R. 535217 [1974/1976]; C.A. **86** [1977] No. 92641).
[4] R.C. Krug, S. Tocker (J. Org. Chem. **20** [1955] 1/8, 2).
[5] F.W. Bergstrom (J. Am. Chem. Soc. **50** [1928] 657/62).

6.3.4 With NO, N_2O_4, NO_2F, $NOF \cdot 3HF$, and N_2O_4+NOCl

The reaction of Be rods (purity 98.3 wt% Be) with dry NO is very slow below the melting point of Be and rather slow with molten Be at 1500 °C. The weight gain with time of the sample in NO (330 to 358 Torr pressure) at 1500 °C in comparison with that in O_2, N_2, CO, CO_2, and H_2 is shown in **Fig.** 6-**24**, Blumenthal, Santy [1].

High-strength forged Be specimens (BeO content about 5 to 8.6%) show no attack by N_2O_4 (or by monomethyl hydrazine) at room temperature after 7 months exposure, Soffa, Basl [2]. No reaction between Be and NO_2F is observed up to 300 °C, Aynsley et al. [3].

Be reacts quickly with liquid $NOF \cdot 3HF$ to form $BeF_2 \cdot 2NOF$, Seel et al. [4].

The reaction of Be (2 to 3 g) with liquid N_2O_4-NOCl mixtures (15 to 20 g) at −4 to +6 °C for a period of 4 to 16 h leads to different reaction products depending on the NOCl content

Fig. 6-24

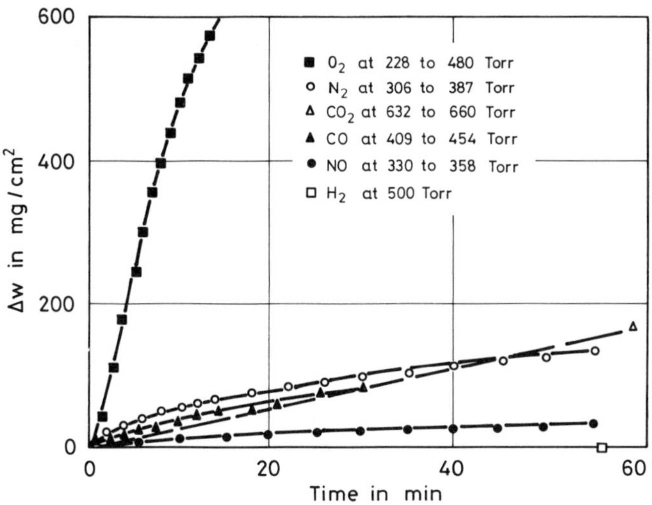

Weight gain Δw of Be in NO at 1500 °C in comparison with other gases.

of the mixture. With 38 to 75 wt% NOCl, $Be(NO_3)_2$ is produced at first and at longer reaction time additional $Be(NO_3)_2 \cdot 2N_2O_4$ is formed. With 75 to 95% NOCl, a mixture of $Be(NO_3)_2$ and $BeCl_4(NO)_2$ is formed and with 95 to 100% NOCl, only $BeCl_4(NO)_2$ is formed, Serezhkina et al. [5].

References:

[1] J.L. Blumenthal, M.J. Santy (Western States Sect. Combust. Inst. Paper [WSS-CI] No. 65-5 [1965] 1/56, 6, 34; C.A. **63** [1965] 7684).

[2] L.L. Soffa, G.J. Basl (Met. Soc. Conf. Proc. **33** [1964/66] 1047/78, 1064/6; CONF-729-21 [1964] 1/37; N.S.A. **19** [1965] No. 23107).

[3] E.E. Aynsley, G. Hetherington, P.L. Robinson (J. Chem. Soc. **1954** 1119/24).

[4] F. Seel, W. Birnkraut, D. Werner (Angew. Chem. **73** [1961] 806).

[5] L.B. Serezhkina, M.P. Lobacheva, V.N. Serezhkin, et al. (Zh. Neorgan. Khim. **17** [1972] 3191/4; Russ. J. Inorg. Chem. **17** [1972] 1678/80).

6.3.5 With Nonaqueous HF, HCl, HBr, and HI

Be reacts with the gaseous hydrogen halides HX at elevated temperatures according to $Be + 2HX \rightarrow BeX_2 + H_2$, Vivet [1] (X=Cl, Br, I), Besson [2] (X=Cl), Besson, Vivet [3] (X=Br, I). According to Gschneidner [4] Be reacts also with HF at elevated temperature. The reaction becomes accelerated only above ~410 °C with HCl and above ~440 °C with HBr and HI. At lower temperatures protective layers of BeX_2 are formed which evaporate at the higher temperatures [1 to 3].

Be reacts with dimethyl ether (saturated at −70 °C with HCl) under reflux (at −23 °C) and stirring for 20 to 30 min according to $Be + 2HCl + 2(CH_3)_2O = BeCl_2 \cdot 2(CH_3)_2O + H_2$, Turova et al. [5]. In a similar manner the reactions of Be with diethyl ether saturated with HCl, HBr, or Br_2 at dry ice temperature lead to $BeCl_2 \cdot 2(C_2H_5)_2O$ or $BeBr_2 \cdot 2(C_2H_5)O$, respectively, Turova et al. [6], see also [11]. For calculations of the potential energy surface of the gas reaction Be+HF → BeF+H, see Kuntz, Roach [7, 8], Schor et al. [9], Chapman et al. [13], see also [14, 15]. For a calculation of the activation energy and the high-temperature rate constant for the gas reaction Be+HCl=BeCl+H, see Mayer et al. [10], and for the H-transfer reaction Be+HX=BeH+X with X=F, Cl, Br, and I, see Mayer, Schieler [12].

References:

[1] G. Vivet (Diss. Univ. Saarbrücken 1958, pp. 11/8, 25/30, 36/44).
[2] J. Besson (Bull. Soc. Chim. France **1950** 1175/9).
[3] J. Besson, G. Vivet (Compt. Rend. **236** [1953] 1788/90).
[4] K.A. Gschneidner (IS-1757 [1968] 1/158, 5/6; C.A. **69** [1968] No. 38219).
[5] N.Ya. Turova, A.V. Novoselova, K.N. Semenenko (Zh. Neorgan. Khim. **5** [1960] 1705/9; Russ. J. Inorg. Chem. **5** [1960] 828/30).
[6] N.Ya. Turova, A.V. Novoselova, K.N. Semenenko (Zh. Neorgan. Khim. **4** [1959] 1215; Russ. J. Inorg. Chem. **4** [1959] 550).
[7] P.J. Kuntz, A.C. Roach (J. Chem. Phys. **74** [1981] 3420/34).
[8] A.C. Roach, P.J. Kuntz (J. Chem. Phys. **74** [1981] 3435/43).
[9] H. Schor, S. Chapman, S. Green, R.N. Zare (J. Chem. Phys. **69** [1978] 3790/806, **83** [1979] 920/2).
[10] S.W. Mayer, L. Schieler, H.S. Johnston (11th Symp. Combust., Berkeley, Calif., 1966 [1967], pp. 837/44, 840; C.A. **69** [1968] No. 5606).
[11] N.Ya. Turova, A.V. Novoselova (Zh. Neorgan. Khim. **8** [1963] 525/8; Russ. J. Inorg. Chem. **8** [1963] 273/5).
[12] S.W. Mayer, L. Schieler (J. Phys. Chem. **72** [1968] 236/40).
[13] S. Chapman, M. Dupuis, S. Green (Chem. Phys. **78** [1983] 93/105).
[14] J.M. Lucas, A. Aguilar, A. Sole, J. Virgili (THEOCHEM **21** [1985] 351/6 from C.A. **102** [1985] No. 191507).
[15] S. Chapman (J. Chem. Phys. **81** [1984] 262/79).

6.3.6 With H_2S, H_2Se, S_2Cl_2, B_2O_3, BF_3, and BCl_3

Condensed films of Be are resistant to a 50% H_2S-air mixture (50% relative humidity) at 20 °C in a 30 d test, Polatnik, Gorban [1]. The reaction of Be with H_2S at 900 °C produces BeS, Staritzky [2]. Crystalline powders of BeS or BeSe are formed by passing H_2S or H_2Se over Be metal twice for 3 h at 1100 °C with intermediate grinding of the reaction product, Yim et al. [3].

0.2 g Be powder (2 mm particle size) reacts with 3 mL S_2Cl_2 in a sealed tube at 400 °C in 10 min to form $BeCl_2$, Funk et al. [4]. Heating of Be-B_2O_3 mixtures with various ratios shows that the reduction reaction to form beryllium boride occurs slowly and incompletely and requires a high initial temperature of >1000 °C, Mikheeva et al. [5].

The fluorination of Be(s) with BF_3(g) in a beryllia-lined Knudsen cell between 1300 and 1600 K was studied by a mass spectrometer. The species BeF_2^+, BeF^+, and Be^+ were detected (for the equilibrium Be(g)+BeF_2(g) $\rightleftharpoons$ 2BF(g), see p. 284), Hildenbrand, Murad [6]. The reaction of fine Be powder with BCl_3 at 850 °C for 1 h yields unstable beryllium boride and amorphous boron, Deiss, Andrieux [7].

References:

[1] L.S. Polatnik, N.D. Gorban (Fiz. Metal. Metalloved. **18** [1964] 220/5; Phys. Metals Metallog. [USSR] **18** No. 2 [1964] 61/6).
[2] E. Staritzky (Anal. Chem. **28** [1956] 915).

[3] W.M. Yim, J.P. Dismukes, E.J. Stofko, R.J. Paff (J. Phys. Chem. Solids **33** [1972] 501/5).
[4] H. Funk, K.H. Berndt, G. Henze (Wiss. Z. Martin-Luther-Univ. Halle-Wittenberg **6** [1957] 815/22; C.A. **1960** 12860).
[5] V.I. Mikheeva, F.I. Shamrai, E.Ya. Krylova (Zh. Neorgan. Khim. **2** [1957] 1223/31; Russ. J. Inorg. Chem. **2** No. 6 [1957] 12/26, 20/1).
[6] D.L. Hildenbrand, E. Murad (J. Chem. Phys. **44** [1966] 1524/9).
[7] W.J. Deiss, J.L. Andrieux (Bull. Soc. Chim. France **1959** 178/82).

6.3.7 With CO

As shown by LEED (low energy electron diffraction) studies CO is adsorbed on Be single crystals (0001 face) but does not change the surface structure of Be, Adams [1]. The total coverage and the sticking coefficient of CO on thick Be films (glass substrate) are measured at room temperature. The amount of adsorbed CO molecules per cm² is 4×10^{13} at 7×10^{-8} Torr and 2.4×10^{14} at 7×10^{-7} Torr with initial sticking coefficients of 0.5 and 0.1, respectively. The initial sticking coefficient is roughly proportional to the square root of the pressure, Hurd, Adams [2].

The oxidation of hot rolled chemically polished Be sheets (prepared from electrolytic French Flake Be) with CO (100 Torr) was studied between 500 and 750 °C by means of a radioactive tracer technique using ^{14}C. At 550 °C and above, the reaction product spalls from the sample; a quantitative chemical analysis shows the presence of BeO, Be_2C, and C. It is argued that the reaction $Be + CO = Be_2O + C$ occurs and that the reaction $2Be + C = Be_2C$ is more probable than the reaction $3Be + CO = BeO + Be_2C$, Gregg et al. [3]. According to optical and electron microscopic and electron diffraction studies, BeO and Be_2O are present in the corrosion product of Be in CO at 700 °C, Scott [4].

The kinetics of the reaction of Be (the same material as used in [3]) with ^{14}C-labelled CO between 500 and 750 °C was studied up to 150 h; the runs were terminated before the initial CO pressure of 100 Torr had decreased to 50 Torr. The weight gain of the samples are shown in **Fig. 6-25**. At 550 °C and above the oxidation is nonprotective. All the curves show a rapid initial weight gain which varied from 0.1 mg/cm² at 500 °C to 0.5 mg/cm²

Fig. 6-25

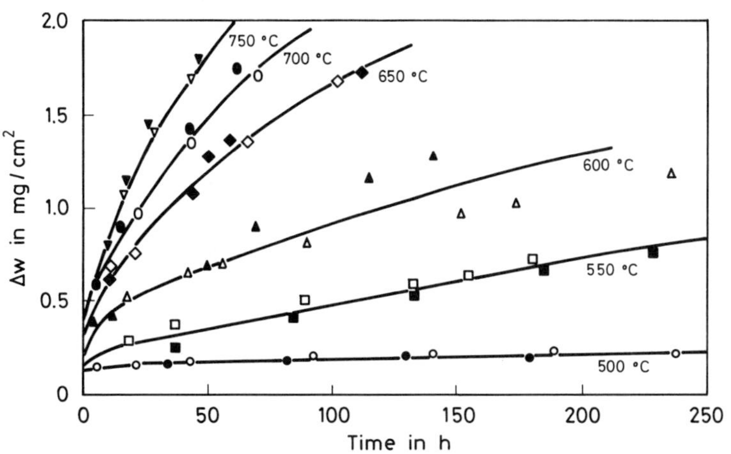

Weight gain Δw of Be in CO vs. time at various temperatures. Black and white symbols are results of duplicate runs.

at 700 °C. The initial rapid weight gain is absent when the samples, instead of being preheated, are heated in CO from room temperature to the reaction temperatures, shown in runs at 500 and 600 °C. A comparison of the results for CO with those for CO_2 shows that both the total weight gain and the rate of oxidation are considerably greater in CO than in CO_2 for a given temperature, Gregg et al. [5]. At 600 to 700 °C with the same material the kinetics of the oxidation in moist CO are studied (total pressure 100 Torr, H_2O partial pressure 12 Torr). The reaction is essentially one with H_2O vapor, the reaction with CO only contributing a few percent to the total weight gain (measured with a vacuum microbalance), Gregg et al. [6].

The rate of the reaction of Be rods (purity 98.3 wt% Be) with CO at 1000 to 1500 °C was studied in the pressure range 0.5 to 500 Torr; the weight gain of Be with time at low CO pressures is shown in **Fig. 6-26** at 1000, 1100, and 1200 °C, and in **Fig. 6-27** at 1350 and 1500 °C. Fig. 6-24, p. 255, compares the reaction rates of Be with CO with those of H_2, O_2, N_2, NO, and CO_2 at higher pressures at 1500 °C. The reaction product in CO up to 1350 °C consists of a tightly adherent inner yellowish white coating (probably a mixture

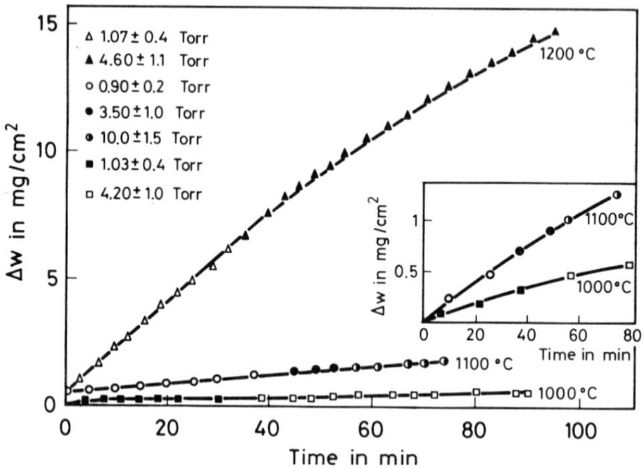

Fig. 6-26

Weight gain Δw of Be vs. time in low pressure CO at 1000, 1100, and 1200 °C.

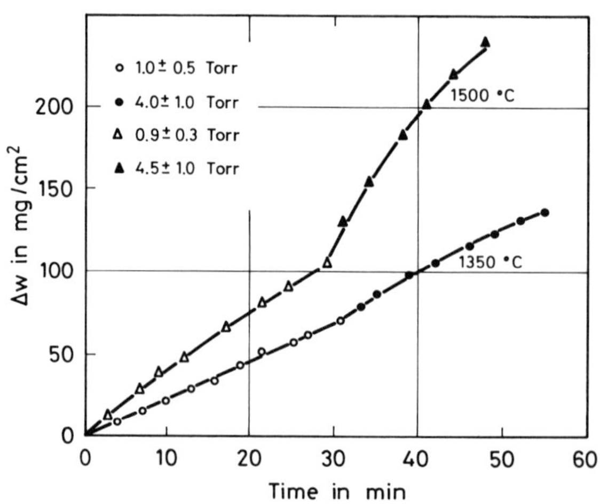

Fig. 6-27

Weight gain Δw of Be vs. time in low pressure CO at 1350 and 1500 °C.

of BeO and Be_2C) and a black outer layer of C which could be easily removed. At 1500 °C the black outer layer had disappeared and Be was coated only with one strongly adherent orange-yellow layer, Blumenthal, Santy [7].

References:

[1] R.O. Adams (REP-1148 [1969] 1/9; Struct. Chem. Solid Surfaces No. 35 [1969] 70.1/70.9; C.A. **75** [1971] No. 134140; N.S.A. **25** [1971] No. 29894).
[2] J.T. Hurd, R.O. Adams (J. Vacuum Sci. Technol. **6** [1969] 229/33).
[3] S.J. Gregg, R.J. Hussey, W.B. Jepson (J. Nucl. Mater. **2** [1960] 225/33; Nature **186** [1960] 468/9).
[4] V.D. Scott (Nature **186** [1960] 466/7).
[5] S.J. Gregg, R.J. Hussey, W.P. Jepson (J. Nucl. Mater. **3** [1961] 175/89).
[6] S.J. Gregg, R.J. Hussey, W.B. Jepson (J. Nucl. Mater. **4** [1961] 46/58).
[7] J.L. Blumenthal, M.J. Santy (Western States Sect. Combust. Inst. Paper [WSS-CI] No. 65-5 [1965] 1/56, 7, 35/6, 42; C.A. **63** [1965] 7684).

6.3.8 With CO_2

Introduction. The corrosion resistance of Be to dry and moist CO_2 has found much interest because of its possible use as a canning material and as a moderator in gas-cooled nuclear reactors. The following reactions occur when Be is oxidized in dry CO_2:

(1) $Be+CO_2=BeO+CO$ (2) $2Be+CO_2=2BeO+C$
(3) $Be+CO=BeO+C$ (4) $2Be+C=Be_2C$

According to reactions (1) and (3) Be is oxidized by CO_2 as well as by the reaction product CO and furthermore a deposition of carbon and formation of Be_2C occur. The reaction rate is much higher in moist than in dry CO_2. For reviews on the reaction of Be with CO_2 see, e.g., Jepson [1], Dawson, Sowden [2], Papirov [3]. Electron microscopy and electron diffraction studies of the reaction of thin Be films (200 to 500 Å thick) with dry flowing CO_2 between 300 and 800 °C show that Be_2C forms rapidly in the initial stages of the reaction, where reaction (2) seems to be faster than reaction (1). In the later stages reaction (1) predominates and Be_2C disappeared, probably according to the reaction $Be_2C+3CO_2=2BeO+4CO$ in dry CO_2 and according to the reaction $Be_2C+2H_2O=2BeO+CH_4$ in wet CO_2, Khorenko et al. [4].

References:

[1] W.B. Jepson (in: H.P. Godard, W.B. Jepson, M.R. Bothwell, R.L. Kane, The Corrosion of Light Metals, ASM, New York 1967, pp. 219/52, 232/40).
[2] J.K. Dawson, R.G. Sowden (Chemical Aspects of Nuclear Reactors, Vol. 1: Gas-Cooled Reactors, Butterworth, London 1963, pp. 91/5; N.S.A. **18** [1964] No. 11372).
[3] I.I. Papirov (Okislenie i Zashchita Berilliya [Oxidation and Protection of Beryllium], Metallurgia, Moscow 1968, pp. 1/121, 44/53; C.A. **70** [1969] No. 14035).
[4] V.K. Khorenko, L.A. Kornienko, B.V. Matvienko (Zh. Fiz. Khim. **40** [1966] 1903/8; Russ. J. Phys. Chem. **40** [1966] 1020/3).

6.3.8.1 With Dry CO_2 below Atmospheric Pressure

The reaction rate of fine Be powder (~0.7 m^2/g surface area) with CO_2 at pressures of 1×10^{-3} Torr in a dynamic system between 600 and 850 °C was found to be nearly parabolic. The weight gain Δw in mg/cm^2 with time is expressed by the equation $\Delta w=kt^{1/n}$, where t is the time in h and k and n are constants. Experimental values for ln k and 1/n in the equation $\ln \Delta w = \ln k + 1/n \ln t$:

temp. in °C . . .	600	720	(720) [a]	838	846	850
time in h	350	1000	(400)	350	350	1000
−ln k	7.354	5.461	(5.167)	4.963	5.778	5.869
1/n	0.6132	0.3224	(0.4347)	0.3838	0.3427	0.3414

[a] At 760 Torr.

The rate constants obtained at 10^{-3} and 760 Torr exhibit only a slight difference (see also p. 263). This shows that the rate-controlling step is the diffusion of the oxidizing species through the oxide layer, Werner [1]. X-ray and chemical analyses of the powder samples (Brush powder with ~0.3 m^2/g surface area, see p. 263) after testing in dry CO_2 at 1×10^{-3} Torr pressure showed the presence of BeO and Be_2C. The amount of C after testing, although exhibiting some deviation, appears to increase with time and temperature:

temp. in °C . .	600	600	600 [a]	650	650 [a]	720	720	720 [a]	838 [b]	838 [b]
time in h . . .	334	477	528	433	431	222	1006	409	335	1014
wt% C	0.57	0.14	0.64	1.3	0.7	0.13	2.64	2.0	0.63	1.82

[a] Runs at 760 Torr pressure (see p. 263). — [b] Specimens sintered.

Since electron diffraction studies performed on oxide scraped from tested sheet specimens showed no evidence of Be_2C or C in the BeO film, it was postulated that the Be_2C is present in the metal substrate, Werner, Inouye [2]. Electron diffraction studies of the surface of Be sheets subjected to CO_2 at 700 °C showed that the Be_2C particles are dispersed within the oxide layer, Scott [3]. To study the mechanism of the reaction of Be sheet (chemically polished, hot rolled, prepared from electrolytic flake Be) in static CO_2 (labelled with ^{14}C) at 10 and 20 Torr pressure and 500 to 750 °C up to 300 h, weight gain measurements at the end of the runs as well as a volumetric radioactive tracer technique was applied. To distinguish between the carbon deposition by reaction (2) or (3), see p. 259, a series of runs were made in which CO was removed from the gas phase as fast as it was formed by oxidation to CO_2 with CuO. It was shown that reaction (1) occurs and predominates at all temperatures (500 to 750 °C), while at ≧550 °C also carbon deposition occurs, either by reaction (2) or (3). After the tests 40% of the carbon was present as Be_2C. The formation reaction (4) is believed to be more probable than the reactions $4Be + CO_2 = 2BeO + Be_2C$ or $3Be + CO = BeO + Be_2C$. The reaction $2C + O_2 = 2CO$ does not occur, Gregg et al. [4]. (The attack on Be_2C inclusions in Be at 700 °C by CO_2 at 10 Torr pressure according to $Be_2C + 3CO_2 = 2BeO + 4CO$ was found to be negligible, Jepson et al. [5].) In contrast to the runs in CO_2 with increasing content of the reaction product CO, in the runs with back-oxidized CO, the carbon deposition was much lower and was confined to the initial period of the run. Weight gain Δw_c and Δw_o (as carbon and oxygen) after the test time of the Be samples in ^{14}C labelled CO_2 (20 Torr) with back-oxidized CO:

temp. in °C . . .	650	650	700	700	700	750
time in h	260	315	90	125	240	190
Δw_c in µg/cm² . .	15.7	4.9	16.3	18.2	23.9	22.5
Δw_o in µg/cm² . .	74.3	66.4	165.2	148.5	183.9	241.0
$\Delta w_c/\Delta w_o$	0.21 [a]	0.07	0.10	0.12	0.13	0.09

[a] This sample seems to be anomalous.

In the runs, the CO_2 was added to the outgassed, preheated (to reaction temperature) samples, Gregg et al. [4]. In similar tests with "CO free" $^{14}CO_2$ at 650 °C and 5 Torr pressure up to a total weight gain of 120 μg/cm² after 120 h, the deposited carbon increased linearly from about 3 to 22 μg/cm² between about 2 and 120 h (values taken from the curves; it was not mentioned whether the samples were preheated or not), Jepson [6]. If the CO_2 is admitted to the preheated sample, the total weight gain ($\Delta w_o + \Delta w_c$) is very rapid at first and most of the carbon deposition occurs over this initial period. In a test where the Be sample was heated to 650 °C in $^{14}CO_2$ (100 Torr, CO back-oxidized) the amount of deposited C steadily increased over the first 150 h, levelling off at a value about half that obtained when the gas was admitted to the preheated sample. Most of the tests are carried out up to 300 h with preheated samples (prepared from the same batch as that used before, see above under [4]) in unlabelled CO_2 at 100 Torr (without oxidizing the formed CO) using a vacuum microbalance to determine the total weight gain Δw (O+C) during the runs. The oxidation is protective between 500 and 700 °C, with the rate of reaction continuously decreasing with time t to reach a very small value (e.g., 0.07 $\mu g \cdot cm^{-2} \cdot h^{-1}$ after 300 h at 700 °C). But the reproducibility of the curves Δw = f(t) was rather poor, whether the sample was "as-received" or was chemically polished; so Δw varied between 17 and 28 μg/cm² after 70 h at 600 °C in the first case and between 24 and 52 μg/cm² for the polished samples. At 750 °C breakaway occurs, and the rate increases with time after an initial decrease. Curves Δw = f(t) for the polished samples are shown in **Fig. 6-28**. All the curves show the unusual feature of a rapid weight gain over the first few minutes. Along the flatter portions of the graphs the data fitted (within experimental error) the empirical equation $a \cdot \Delta w^2 + b \cdot \Delta w = t$, where Δw is the weight gain at time t and a and b are constants. In at least one run at each temperature ^{14}C-labelled CO_2 was used and the amount Δw_c of carbon deposited on the samples after test time t in h was estimated by directly counting the radiation over succesive areas of the specimen surface, Gregg et al. [7].

Results [7]:

temp. in °C	500	550	600	600	650	700	750	750
time in h	305	306	99	502	325	307	0.5	312
Δw_c in μg/cm² . .	not detd.	1.4	0.8	6.0	1.2	5.6	4.5	6.6
Δw_o in μg/cm² . .	14.7	28.9	31.2	87.5	84.1	100.1	33.2	150.6

Fig. 6-28

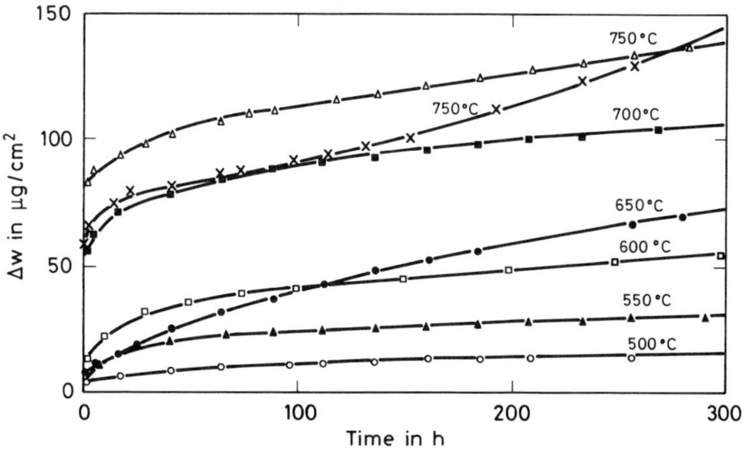

Weight gain Δw of Be vs. time in CO_2 (100 Torr) at various temperatures. The curve for 600 °C is the average of two runs.

The reaction of Be rods (2.5 cm long, 1.27 cm in diameter, prepared from high purity flake Be by compaction and hot extrusion, not etched) with flowing ^{14}C-labelled CO_2 (~5 cm/s) at 200 Torr pressure between 593 and 815 °C up to 500 h was studied at the end of each run by weight gain, surface counting, autoradiography, metallography, and chemical analysis. The CO formed was reoxidized by CuO. The results, summarized in the following table, are not in accordance with what one would expect on the basis of the time and temperature of each test:

temp. in °C	593	704	704 [b)]	704 [b)c)]	760 [b)]	815
time in h	500	330	355	356	285	164
Δw in mg/cm²	4.10	44.2	10.9	spalling	1.02	2.37
wt% C after test [a)]	0.077	0.084	0.066	0.26	—	0.050

a) C-content of as-received material 0.05 wt%.

b) Drying agent present in gas stream.

c) Specimen annealed 1 h at 1200 °C in Ar prior to testing.

The reaction layers consisted of Be, BeO, and Be_2C, shown by X-ray diffraction. The carbon was found to remain associated with the reaction product and did not diffuse into the Be. The reaction of Be with CO_2 proceeded along preferred longitudinal directions which seemed to be influenced by the voids present in the as-received material. Three mechanisms have been proposed to account for the greater rate in the longitudinal direction (for details see the paper). Since the weight changes of the specimens were not followed continuously during the tests, it is difficult to say definitely which specimens exhibit nonprotective oxidation (breakaway). The oxide formed appeared quite porous and metallic particles were trapped in the reaction product, McCoy [8]; in earlier reports no Be_2C was found, McCoy [9, 10]. The surfaces of short-time oxidized Be rods in CO_2 (76 Torr) at 700 °C were studied by electron microscopy. After 1 and 2 h the surface scale was almost crack-free but blisters were evident. After 4 h larger blisters and circumferentially cracked regions were present. After 150 h exposure to CO_2 the circumferentially cracked regions have apparently healed and the surface is, on the whole, uniform. But there was significant outward growth of needle-like crystallites and the surface still contained blisters and cracks, which were longitudinal, Jepson et al. [11].

The high temperature reaction of Be-rods (reactor grade purity) in CO_2 at different pressures, studied between 1000 and 1500 °C by weight gain measurements after 1 h reaction time, shows the following results (Δw in mg/cm²):

temp. in °C	1000	1100	1200	1200	1350	1350	1500	1500
p in Torr . .	1.1 to 7	0.9 to 2.1	≈1	495	0.4 to 2	≈485	0.7 to 2.5	632 to 660
Δw	1.1	5.8	9.7	3.7	260	61	375	171

There is a negative pressure effect at 1200 to 1500 °C, Blumenthal, Santy [12].

References:

[1] W.J. Werner (ORNL-2988 [1960] 424/5; N.S.A. **15** [1961] No. 569).

[2] W.J. Werner, H. Inouye (Inst. Metals Monogr. Rept. Ser. No. 28 [1963] 283/93, 288; C.A. **60** [1964] 10325).

[3] V.D. Scott (Nature **186** [1960] 466/7).

[4] S.J. Gregg, R.J. Hussey, W.B. Jepson (J. Nucl. Mater. **2** [1960] 225/33).

[5] W.B. Jepson, J.B. Warburton, B.L. Myatt (J. Nucl. Mater. **10** [1963] 127/33).

[6] M.B. Jepson (J. Appl. Chem. **14** [1964] 309/15).
[7] S.J. Gregg, R.J. Hussey, W.B. Jepson (J. Nucl. Mater. **3** [1961] 175/89; Nature **186** [1960] 468/9).
[8] H.E. McCoy (J. Nucl. Mater. **15** [1965] 249/62; ORNL-TM-622 [1963] 1/36; C.A. **60** [1964] 15389).
[9] H.E. McCoy (ORNL-3302 [1962] 234/9; N.S.A. **16** [1962] No. 28379).
[10] H.E. McCoy (ORNL-3313 [1962] 38/9; N.S.A. **16** [1962] No. 30167).

[11] W.B. Jepson, B.L. Myatt, J.B. Warburton, J.E. Antill (J. Nucl. Mater. **10** [1963] 224/32).
[12] J.L. Blumenthal, M.J. Santy (Western States Sect. Combust. Inst. Paper [WSS-CI] No. 65-5 [1965] 1/56, 8/9; C.A. **63** [1965] 7684).

6.3.8.2 With Dry CO_2 at Atmospheric Pressure

For the ignition in O_2-containing gas mixtures, see p. 204.

Ultrafine Be powder (particle size <0.1 μm) ignites in CO_2 at room temperature. The obtainable energy for the reaction to form BeO is 11 kcal/g Be, Rhein [1]. Ignition of the powder at room temperature occurs in CO_2 as well as in a simulated venusian atmosphere (83.73% CO_2, 9.17% N_2, 4.10% Ar), Rhein [2]. The reaction enthalpy of the reaction $2Be + CO_2 = 2BeO + C$ obtained from emf measurements between 682 and 1040 °C was found to be $\Delta H = -10.41$ kcal per g Be, with little change in this temperature range, Smirnov, Chukreev [16].

The reaction of thin Be films (200 to 500 Å thick) with flowing dry CO_2 (5×10^{-4} wt% H_2O) studied by electron microscopy or electron diffraction is very fast at 700 and 800 °C. After 3 to 5 min at 700 °C or after 30 s at 800 °C the Be lines in the diffraction pattern disappear. At 500 °C the oxidation is incomplete even after 5 h (for the reaction mechanism and the initial formation of Be_2C, see p. 259), Khorenko et al. [3].

Weight gain studies on Brush powder samples (particle size 60 to 70 μm, surface area ~0.3 m^2/g, impurities: 1400 ppm Fe, 1100 ppm C, 460 ppm Mg, 400 ppm Si, 500 ppm Al, 1.74 wt% BeO) in dry flowing CO_2 (~20 mL/min) at 600 to 720 °C up to t=1000 h show that duplicate runs were reproducible to ±5%. The reaction layer is protective and the weight gain Δw with time t can be described by $\Delta w = k \cdot t^{1/n}$ with $n \approx 2$. Values for k in $\mu g \cdot cm^{-2} \cdot h^{-1}$ and 1/n at various temperatures:

temp. in °C . . .	550	600	650	720
k	0.48 [a)]	2.025	3.844	15.21
1/n	0.508 [a)]	0.492	0.483	0.517

a) The same values are obtained also at 10^{-3} Torr (see also p. 260).

The obtained Δw vs. t curves are shown in **Fig.** 6-**29**, p. 264 (in logarithmic scales), which also include data for high purity Pechiney flake in wet CO_2 (~4 vol% H_2O, dashed curves). The reaction products after the tests are BeO and Be_2C (see the table on p. 260), as shown by X-ray diffraction, Werner, Inouye [4]. The oxidation temperature of Be powder (particle size about 200 μm) in CO_2 is decreased from about 621 to 140 °C by prior activation. For the activation the Be powder was exposed in H_2 to 85 keV glow discharge for 5 min, Inoue [15].

As the corrosion behavior of compact Be specimens in dry CO_2 seems to depend on many factors (e.g., the source and the metallic history of the samples, impurities, the test time, traces of H_2O vapor), the weight gain vs. time curves show low reproducibility and

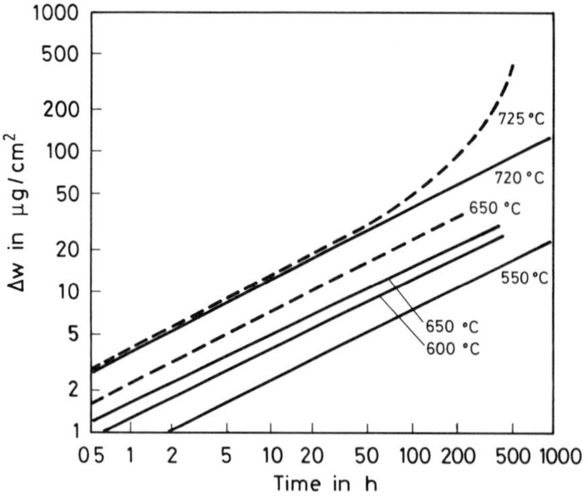

Fig. 6-29

Weight gain vs. time of Be powder in dry CO_2 at 700 °C (dashed curves wet CO_2, see text).

the results obtained by different authors do not agree. So the reported temperatures, where the oxidation is still protective, vary between 600 °C (3000 h test), Munro, Williams [5], 850 °C (10000 h test), or even 1000 °C (up to 1500 h), Higgins, Antill [6, 7]. Temperatures of 650 °C (2000 h test), Davies [8], 700 °C (1000 h test), Ivanov et al. [9], 720 °C (1000 h test, powder) [4], and 900 °C (100 h test), Draycott et al. [10], are also reported.

The oxidation of extruded, pickled Be (prepared from electrolytic flake Be) was studied in flowing CO_2 (flowing rate ~20 mL/min) between 500 and 700 °C up to 3000 h by weight gain (Δw) measurements. At 500 and 600 °C the corrosion rate k decreases with time t and the curves closely follow the parabolic law $\Delta w^n = kt$ with $n \approx 2$. The values of k are approximately $5 \times 10^{-16}\ g^2 \cdot cm^{-4} \cdot s^{-1}$ at 500 °C and $7 \times 10^{-16}\ g^2 \cdot cm^{-4} \cdot s^{-1}$ at 600 °C. The weight gain is 60 μg/cm² after 1835 and 2070 h at 500 °C and 80 μg/cm² after 2150 and 2990 h at 600 °C. At 650 and 700 °C the surface film is no longer protective; it is assumed that at 650 °C the formation of Be_2C may become significant and make the oxide film nonprotective. The Δw-time curves are shown in **Fig. 6-30**, in which also the curves for wet CO_2 (see p. 268) are included, Munro, Williams [5], Livey, Williams [11]. The oxidation kinetics of the reaction of pure Be with flowing CO_2 at 550 to 700 °C can be initially described by the cubic law, followed by the parabolic law in the second stage. This feature was related to a change of the oxide morphology, Nishigaki et al. [12]. A small Ca content in Be markedly improved the corrosion resistance, studied at 600 to 700 °C [13]. Sintered Be samples (with varying BeO content) are corrosion resistant to flowing dry CO_2 ($\leqq 5 \times 10^{-4}$ wt% H_2O) at 700 °C during the whole testing time of 1000 h. After an initial rapid weight gain the reaction rate decreases considerably with the time (t) and can be expressed by $\Delta w = kt^{1/n} + b$ (with $n > 1$). After 1000 h Δw varied between 0.3 and 0.6 mg/cm² for different samples. Electron microscopy and electron diffraction studies of the sample surfaces show, especially during the initial period of oxidation, the formation of a large number of small, discrete Be_2C crystals in the BeO films, Ivanov et al. [9].

Nine compact Be specimens (three different sources and fabricated by a variety of different routes) were studied between 500 and 1000 °C in dry CO_2. Protective films were formed on most of the specimens from 500 to 850 °C for times up to 10000 h, but a high purity sample (which was cast in an ingot) showed breakaway at 700 °C (between 200 and 4500 h). At 1000 °C the oxide formed was initially protective, but breakaway occurred at the latest after 1500 h. The value of n in the relation $\Delta w^n = kt$ is >1 in the protective range and

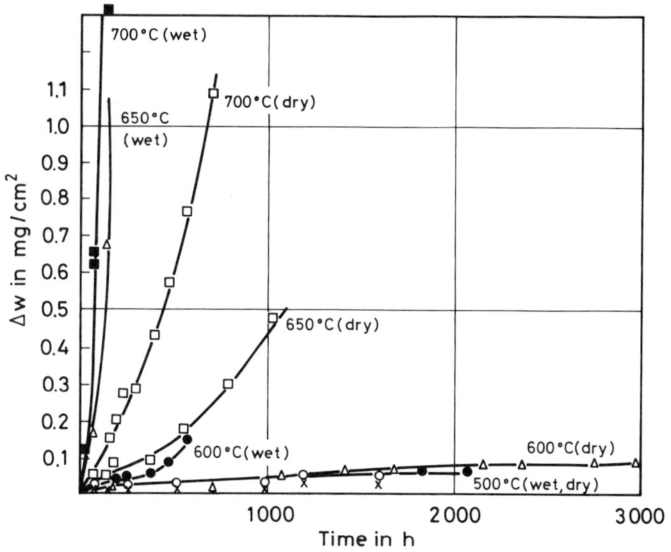

Fig. 6-30

Weight gain Δw of Be vs. time in flowing dry and wet (~2 wt% H_2O) CO_2 at various temperatures.

was found to vary between 2 and 9.5; n approaches 2 at 850 and 1000 °C (between ~200 and 1500 h), and in some cases at 700 °C. In the range of nonprotective oxidation the rate of reaction is constant with time (n = 1), or increases continuously (n < 1). Some values for Δw in mg/cm² after an exposure of some samples for 5000 h in dry CO_2: ~0.05 at 500 °C, ~0.13 at 600 °C, ~0.1 to 0.22 at 700 °C, and ~0.5 to 1 at 850 °C. After 500, 2000, or 5000 h at 1000 °C Δw is approximately 5, 10, or 40 mg/cm², respectively. At 1000 °C the samples were covered with a hard white scale and some of the samples were bent and the oxide blistered. At lower temperatures the films were gray or merely produced interference colors [6, 7]. Two materials, hot pressed Brush powder and rolled Be sheet prepared from the purer Pechiney flake, were studied in flowing CO_2 (100 mL/min) between 600 and 1000 °C. In the range 600 to 850 °C the samples were preheated to reaction temperature before exposing to CO_2. The rate of oxidation (studied up to 2500 h) decreased with time after initial rapid oxidation at 600 to 1000 °C for the rolled sheet, while for the hot pressed samples breakaway occurred at 850 °C (after ~120 or 600 h at weight gains of 0.09 and 0.37 mg/cm², respectively, in two cases). The kinetics in the range of protective oxidation do not conform to any known rate law. Although, after breakaway, the rate of oxidation initially increased, it subsequently decreased with time and only an average of 0.9% and 5.7% of the specimen had been oxidized after 2500 h at 850 and 1000 °C, respectively. The reproducibility of the runs at 850 and 1000 °C is very bad. No significant effect of prior fast neutron irradiation of the samples upon the oxidation in CO_2 was found, when the swelling of the samples (caused by He) was less than 5%, but there was an appreciable increase in the magnitude of oxidation coinciding with maximum (~30%) swelling, i.e. for the hot pressed sample between 300 and 1100 h at 850 °C or after 2 h at 1000 °C, Bennett et al. [14].

References:

[1] R.A. Rhein (Western States Sect. Combust. Inst. Paper [WSS-CI] No. 64-25 [1964] 1/8; C.A. **62** [1965] 6331; NASA-CR-60125 [1964] 1/29, 2; C.A. **63** [1965] 14628).

[2] R.A. Rhein (Astronaut. Acta **11** [1965] 322/7).

[3] V.K. Khorenko, L.A. Kornienko, B.V. Matvienko (Zh. Fiz. Khim. **40** [1966] 1903/8; Russ. J. Phys. Chem. **40** [1966] 1020/3).

[4] W.J. Werner, H. Inouye (Inst. Metals Monogr. Rept. Ser. No. 28 [1963] 283/93; C.A. **60** [1964] 10325).
[5] W. Munro, J. Williams (AERE-M-M-108 [1956] 1/16; N.S.A. **10** [1956] No. 4622).
[6] J.K. Higgins, J.E. Antill (J. Nucl. Mater. **5** [1962] 67/80, 70/1, 75; C.A. **57** [1962] 1952).
[7] J.E. Antill, J.K. Higgins (TID-7597 [1960] 796/805; N.S.A. **15** [1961] No. 19779).
[8] M.W. Davies (At. Energy Rev. **2** No. 1 [1959] 11/7).
[9] V.E. Ivanov, V.F. Zelenskii, V.K. Khorenko, et al. (Poroshkovaya Met. **8** No. 8 [1968] 73/7; Soviet Powder Met. Metal Ceram. **1968** 647/50).
[10] A. Draycott, F.D. Nicholson, G.H. Price, W.I. Stuart (AAEC-E-83 [1961] 1/34, 5; N.S.A. **16** [1962] No. 12049).
[11] D.T. Livey, J. Williams (Proc. 2nd Intern. Conf. Peaceful Uses At. Energy, Geneva 1958, Vol. 5, pp. 311/8; C.A. **1960** 22249).
[12] S. Nishigaki, Y. Kondo, R. Nagasaki (Nippon Kinzoku Gakkaishi **33** [1969] 134/40 from C.A. **70** [1969] No. 80441).
[13] S. Nishigaki, Y. Kondo, R. Nagasaki (Nippon Kinzoku Gakkaishi **33** [1969] 455/60 from C.A. **72** [1970] No. 6189).
[14] M.J. Bennett, N.W. Crick, P.C. Blythe, J.E. Antill (J. Nucl. Mater. **17** [1965] 60/72).
[15] K. Inoue (U.S. 3598566 [1971]; 3738828 [1973]; C.A. **75** [1971] No. 112289, **79** [1973] No. 139304).
[16] M.V. Smirnov, N.Ya. Chukreev (Zh. Neorgan. Khim. **3** [1958] 2445/9; Russ. J. Inorg. Chem. **3** No. 11 [1958] 18/24).

6.3.8.3 With Dry CO_2 above Atmospheric Pressures

Lengthy tests (up to 12000 h) were carried out in dry CO_2 (<10 ppm H_2O vapor) at 11.2 atm pressure between 600 and 750 °C with two Be specimens prepared from Brush powder, either by hot pressing or by cold pressing, sintering, and extrusion at 450 °C. The deviation in results is shown in **Fig.** 6-**31** a, while Fig. 6-31 b shows that the hot pressed metal has a higher resistance than the warm extruded metal. In the early stages of the oxidation both samples showed protective oxide films showing interference colors, but the weight gain vs. time curves did not obey the parabolic law. At 750 °C the weight gain

Fig. 6-31

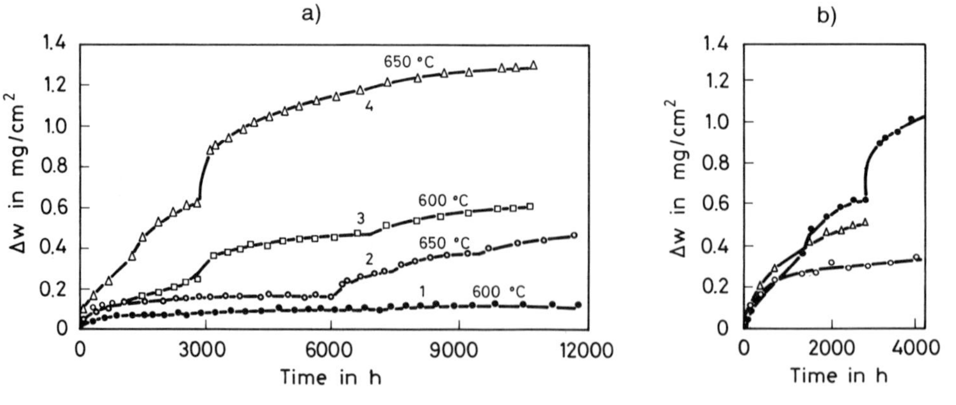

Weight gain Δw vs. time of Be in dry CO_2 at 11.2 atm pressure: a) hot pressed Be (curves 1, 2) and warm extruded Be (curves 3, 4); b) scatter of values on warm extruded Be at 650 °C.

Fig. 6-32

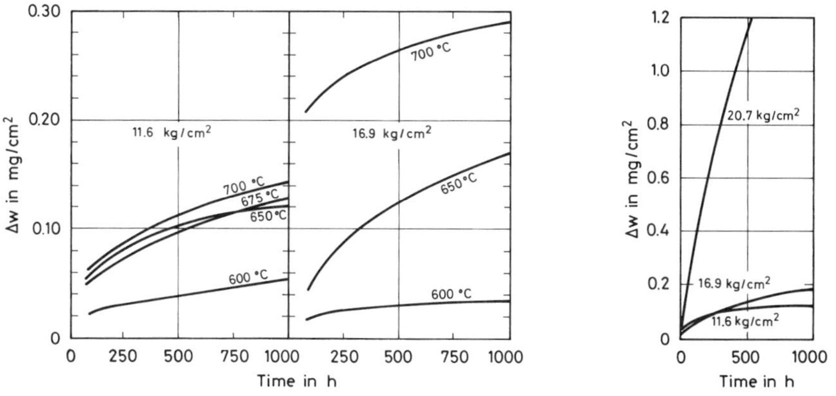

Weight gain Δw vs. time of Be in dry CO_2 at various pressures at 600 to 700 °C (left) and at 650 °C (right).

of the hot pressed material reached 0.15 mg/cm² in 400 h and the warm extruded material 3.5 mg/cm² in 1000 h, Phennah et al. [1]. The results of 1000 h tests of extruded Be (impurities about 550 ppm Al, 400 ppm Si, 360 ppm Fe, 200 ppm Ca, 200 ppm Cl, 0.5% O) in commercially pure CO_2 (<10 ppm H_2O content) between 600 and 700 °C and pressures of 11.6, 16.9, and 20.7 kg/cm² are shown in **Fig. 6-32**. No breakaway occurs at 11.6 and 16.9 kg/cm² up to 700 °C, but at 20.7 kg/cm² breakaway is observed even at 650 °C. Etched specimens show breakaway at 16.9 atm CO_2 pressure at 700 °C, Draycott [2]. Extruded Be (prepared from Pechiney powder) is resistant up to 20 atm pressure of commercially pure CO_2 (10 to 20 vpm H_2O content) up to 650 °C; no breakaway occurs up to 1500 h, Draycott et al. [3]. Be sheet (prepared from induction-melted ingots) exposed to "dry" CO_2 (containing 30 ppm H_2O) at 7.8 and 21.5 atm pressure at 700 and 800 °C shows protective oxidation only at 700 °C and 7.8 atm for the period of test, 6650 h. The weight gain after 3000 h is ~0.5 mg/cm². At 7.8 atm and 800 °C breakaway occurs after 1100 h. At the higher pressure, the protective oxide film broke down after only a few hours and dark gray oxide blisters were formed, Raine, Robinson [4]. Be, which had been melted in vacuum, showed better corrosion resistance at 700 °C than other samples in CO_2 at 20.4 atm pressure: The weight gain of arc-melted Be after 2000 h exposure was 0.28 mg/cm²; the weight gains for Be single crystals and hot rolled Be (prepared from hot pressed powder) after 1000 h exposure were 0.48 and 38.0 mg/cm², respectively, Raine, Robinson [5], see also [6]. Hot extruded Be discs prepared from either cold compacted Brush powder or electrolytic Be were tested in "dry" CO_2 (containing 30 ppm H_2O) at 21.4 atm pressure at 650 °C. Breakaway occurred after about 7000 h, Menzies [7]. A Be alloy containing 0.4 wt% Ca (produced by arc melting and direct fabrication) is resistant to CO_2 at 700 °C and 20.5 atm pressure. The weight gain is 0.184 mg/cm² after 5740 h exposure, Raine, Robinson [8]. Commercial pure Be shows good corrosion resistance in dry CO_2 (<5 vpm H_2O) at 60 atm up to 650 °C (tests up to 10000 h). The reaction rate found at 625 °C, higher than that at 650 °C, is possibly due to a preferred deposition of carbon at 625 °C (see the equations on p. 259), Dewanckel et al. [9], see also Darras et al. [10, 11]. An equipment for ellipsometric studies of film growth on Be exposed to flowing dry (and wet) CO_2 at pressures up to 20.4 atm at up to 800 °C is described by Hayfield [12, 13].

References:

[1] P.J. Phennah, M.W. Davies, B.C. Woodfine (Inst. Metals Monogr. Rept. Ser. No. 28 [1963] 294/313; C.A. **60** [1964] 10338).
[2] A. Draycott (Dechema Monogr. **45** [1962] 87/103; C.A. **59** [1963] 3594).
[3] A. Draycott, F.D. Nicholson, G.H. Price, W.I. Stuart (AAEC-E-83 [1961] 1/34, 8/9; N.S.A. **16** [1962] No. 12049).
[4] T. Raine, J.A. Robinson (J. Nucl. Mater. **7** [1962] 263/78).
[5] T. Raine, J.A. Robinson (Brit. 875539 [1959]; C.A. **56** [1962] 2265).
[6] T. Raine, J.A. Robinson (Ger. Offen. 1164671 [1960/64]; C. **1965** No. 9-2512).
[7] I.A. Menzies (Corrosion Sci. **3** [1963] 35/49).
[8] T. Raine, J.A. Robinson (J. Nucl. Mater. **5** [1962] 341/3).
[9] B. Dewanckel, D. Leclerc, R. Darras (3rd Conf. Intern. Met. Beryllium Commun., Grenoble, Fr., 1965 [1966], pp. 171/91; C.A. **67** [1967] No. 119756).
[10] R. Darras, B. Dewanckel, D. Leclercq (Energie Nucl. [Paris] **8** [1963] 588/96; C.A. **60** [1964] 6577).

[11] R. Darras (Bull. Inform. Sci. Tech. CEA [Paris] No. 62 [1962] 43/55; C.A. **58** [1963] 7669).
[12] P.C.S. Hayfield (Surface Sci. **16** [1969] 370/81).
[13] P.C.S. Hayfield, G.W.T. White (NBS-MP-256 [1963/64] 157/99; C.A. **67** [1967] No. 47989).

6.3.8.4 With Moist CO_2

Below Atmospheric Pressure. The surface of Pechiney rods (double extruded, polished) short time ($\leqq$35 h) oxidized in moist CO_2 (total pressure 76 Torr, $p_{H_2O}=15$ Torr) at 700 °C was studied with electron microscopy by the replica technique. As the oxidation proceeds, numerous blisters developed on the oxide surface and were transformed into areas bounded by a crack. This is observed after short exposure times and well before breakaway occurs. The cracked areas are judged to be the principal sites of entry of reactant to the interior of the Be. One of the surprising results was the virtual absence of longitudinal cracks in the oxide along the grain boundaries of the underlying metal, Jepson et al. [1]. As in the tests in dry CO_2 (see p. 260) for the short time tests in moist ^{14}C-labelled CO_2 (100 Torr total pressure, $p_{H_2O}=12$ Torr), hot rolled, chemically polished Be sheet, prepared from electrolytic flake, was used. In the range 500 to 600 °C the oxidation is protective within the test time of 300 h. At 700 and 750 °C breakaway occurred after 8 and 2 h, respectively. At 650 °C breakaway occurred in two of three runs. The rate of reaction after breakaway is considerable, that at 750 °C being 200 $\mu g \cdot cm^{-2} \cdot h^{-1}$. The rate after breakaway takes the form of penetration between the grains of the metal. The kinetics are remarkably similar to those in H_2O vapor. The reaction of Be with H_2O ($Be+H_2O=BeO+H_2$) predominates over the reaction with CO_2 in the proportions of approximately 4:1, the reaction with CO_2 according to $Be+CO_2=BeO+CO$ predominates over the reaction $2Be+CO_2=2BeO+C$ in the ratio of about 6:1, as shown for the runs at 650 °C, Gregg et al. [2].

At Atmospheric Pressure. The reaction rate of thin Be films with flowing moist CO_2 studied by electron microscopy or electron diffraction is faster than in dry CO_2 (see p. 263). With CO_2 containing 2 to 4 wt% H_2O no Be_2C is detected. With CO_2 with lower humidity (0.05 to 1 wt% H_2O) Be_2C is formed (besides BeO), but decomposes fairly rapidly according to the hydrolytic reaction $Be_2C+2H_2O=2BeO+CH_4$, Khorenko et al. [3].

The reaction rate of high-purity Be powder (electrolytic flake, surface area 0.1 m^2/g) in flowing CO_2, containing about 4 vol% H_2O, at 650 and 725 °C is shown in Fig. 6-29,

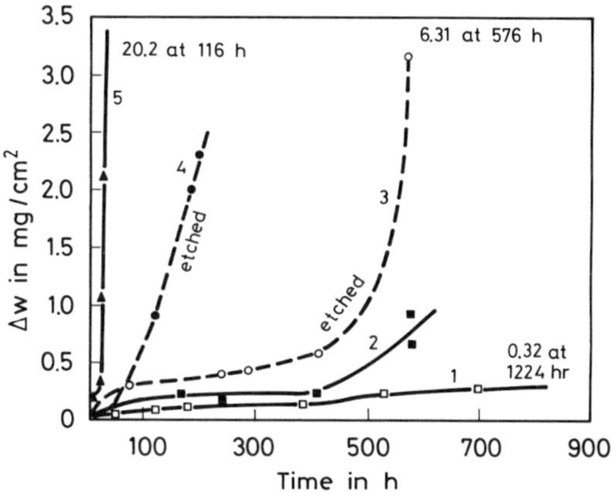

Fig. 6-33

Weight gain Δw vs. time of extruded Be (curves 1 to 4) and vacuum cast Be (curve 5) in moist CO_2 (2 wt% H_2O) at 700 °C and atmospheric pressure.

p. 264. At 650 °C the oxidation is protective for the test time of ~250 h. At 725 °C breakaway occurred after 60 to 70 h. The C-content of the samples after the runs is 0.031 wt% at 650 °C and 0.10 wt% at 725 °C (90 ppm before the runs), Werner, Inouye [4].

The reaction rate of compact Be (extruded and etched) in flowing moist (and dry) CO_2 containing ~2 wt% H_2O between 500 and 700 °C for times up to 3000 h is shown in Fig. 6-30, p. 265. In the moist CO_2, breakaway occurred at 600 °C. At 500 °C the corrosion rate is assumed to be parabolic with time t ($\Delta w^2 = k \cdot t$) with the rate constant $k \approx 5 \times 10^{-16}$ $g^2 \cdot cm^{-4} \cdot s^{-1}$, Munro, Williams [5], Livey, Williams [6]. Extruded Be showed breakaway oxidation at all temperatures between 640 and 900 °C in CO_2 containing about 2 vol% H_2O, Draycott et al. [7]. Ten Be specimens (different sources and different fabrication methods) were tested in 3 vol% H_2O containing CO_2 at 700 °C. All show breakaway, at once or later. The time after which the linear rate law was observed varied between <10 h (for 3 samples), 20 to 300 h (4 samples), and 500 to 600 h (3 samples), Higgins, Antill [8], see also Antill, Higgins [9].

The breakaway reaction of two types of Be samples at 700 °C in moist CO_2 (2 wt% H_2O) was studied metallographically and by electron and X-ray diffraction. Hot extruded Be (prepared from electrolytic powder) and vacuum-cast material (prepared from some of the extruded rods) were used for the tests. The extruded rods were not uniform in structure as the central core showed a much higher concentration of grain-boundary oxide than the outer rim. The results of tests are shown in **Fig.** 6-**33**. The vacuum-cast material was less resistant than the extruded metal, and chemically etched samples (curves 3, 4) were less resistant than samples with machined surfaces (curves 1, 2). The outer rim of the extruded discs oxidized at a faster rate than the central core, i.e., the material which contained less intergranular oxide was less oxidation resistant than the material containing greater amounts of oxide, Smith [10]. To study the inhibiting effect of intergranular BeO content more thoroughly, the Be powder was partially preoxidized in O_2 at 800 °C (0.5 to 1.5 h) or fine BeO powder (2 to 30 μm particles) was added before hot extruding or hot pressing and rolling. The materials made from preoxidized powder (BeO content 1.4 and 2.2 wt%) showed no evidence of breakaway at 700 °C in moist CO_2 (2 wt% H_2O) at exposure times up to 3000 h, but the materials fabricated from as-received powder (0.3 to 0.5 wt% BeO content) showed breakaway within the first few hours of exposure, as did the material to which BeO powder had been added before fabricating. When exposed samples were heated in vacuum, appreciable quantities of H_2 and CH_4 were evolved; these quantities were greater, the greater the extent of the breakaway reaction. X-ray diffraction from samples which showed severe breakaway effects revealed that the principal reaction product was BeO, that there was a slight increase in the intensity of Be_2C reflections after breakaway, and that there were a number of reflections that could not be identified, Smith et al. [11]; unidentified reflections were also reported earlier in [7].

Electron diffraction studies of hot pressed extruded discs after heating in moist CO_2 (2 wt% H_2O) at 700 °C showed the growth of small BeO crystals (after 10 min) and BeO needles (after 1 h), which grew into large BeO flakes (after several hours). No Be_2C was detected on the sample surface (due to surface hydrolysis), but its presence was made evident by X-ray diffraction, van Peer [12]. The corrosion resistance of sintered Be samples in moist CO_2 at 700 °C is improved by preoxidation of the starting Be powder as well as by alloying with Ca. In CO_2 containing 12 vol% H_2O, only $CaBe_{13}$, Be-Ca alloys with 5 wt% Ca, and a Be sample with 10 to 12 wt% BeO (prepared from a Be powder of $\leqq 1$ µm particle size, which was obtained by arc atomization in an inert atmosphere) withstood testing times of 1000 h without signs of attack or increased oxidation. In CO_2 containing only 0.2 vol% H_2O, the weight gain of the three samples after 1000 h was still lower, and a specimen prepared from preoxidized powder with 3.5 wt% BeO (by heating in a rotated furnace in an O_2-Ar atmosphere) showed also no breakaway under these conditions. It was stated that the low resistance of Be in humid gaseous atmospheres is mainly due to H_2 absorption. The hydrogen pickup by the corrosion-resistant specimens was shown to be lower than by the nonresistant samples (see also above under [11]), Ivanov et al. [13].

The oxidation of a Be-Ca alloy with 0.4 wt% Ca in moist CO_2 (14 to 25 Torr H_2O partial pressure) between 700 and 800 °C occurred through three reaction stages. The rates of reaction of the two initial stages were parabolic. The H_2O pressure affected the activation energy of the reaction, thus, the rate-controlling step is different from that in dry CO_2. The third stage followed a linear rate law, resulting in a rapid disintegration, Kondo et al. [14].

Above Atmospheric Pressure. As with dry CO_2 (see p. 266 under [1]), with CO_2 containing 230 to 1100 ppm H_2O, corrosion tests were carried out at 11.2 atm pressure (absolute) at 600 to 750 °C on hot pressed or warm extruded Be. The deviation of results at 600 °C are shown in **Fig.** 6-**34** and the effect of H_2O vapor concentration is shown in **Fig.** 6-**35**. In all cases breakaway occurred as low as 600 °C, but the time to the onset of breakaway oxidation decreased with increasing H_2O content (see Fig. 6-35) and with increasing temperature for a fixed H_2O content, see **Fig.** 6-**36**. The accelerated reaction rates after breakaway are associated with severe intergranular oxidation under blisters on the metal surface, Phennah et al. [15]. The weight gain of a vacuum hot pressed Be specimen after 1062 h exposure at 650 °C to CO_2 containing 600 ppm H_2O at 11.2 atm pressure, was found to

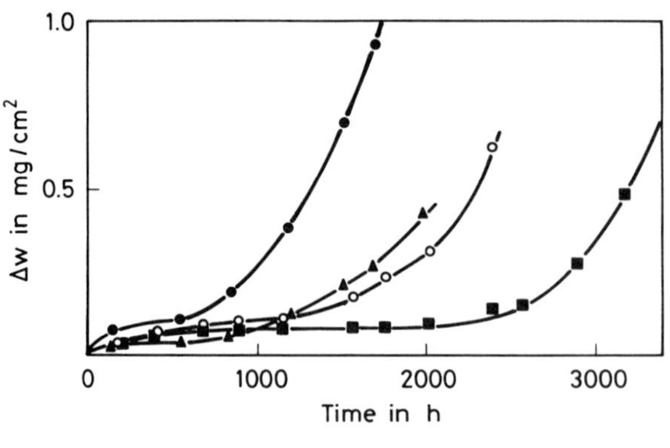

Fig. 6-34

Weight gain Δw of hot-pressed Be in moist CO_2 (600 ppm H_2O) at 11.2 atm and 600 °C. The curves illustrate scatter in results.

Fig. 6-35

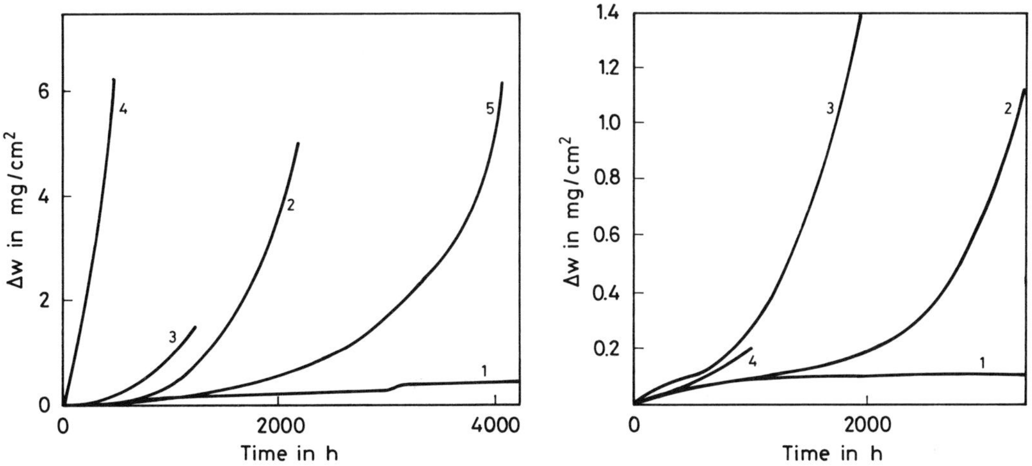

Weight gain Δw vs. time of warm extruded Be (left) and hot-pressed Be (right) in moist CO_2 at 11.2 atm and 600 °C: 1) <10 ppm H_2O, 2) 230 ppm H_2O, 3) 600 ppm H_2O, 4) 1100 ppm H_2O, 5) 600 ppm H_2O + 2% Co.

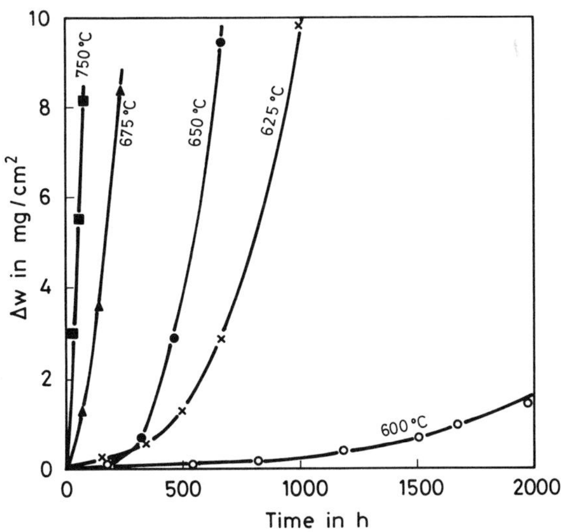

Fig. 6-36

Weight gain Δw of hot-pressed Be in moist CO_2 (600 ppm H_2O) at 11.2 atm at 600 to 750 °C in dependence of exposure time.

be 38.1 mg/cm² (in comparison <0.5 mg/cm² in dry CO_2), Davies [16]. For extruded Be, breakaway readily takes place at 625 °C in CO_2 containing 0.01 vol% H_2O at 18 atm pressure, Draycott et al. [7], see also Draycott [17].

Three Be materials are tested in moist CO_2 (300 ppm H_2O) at 21.4 atm pressure: sheet (I) prepared from "Brush" powder by forging and hot cross-rolling and hot extruded discs prepared from "Brush" powder (II) or electrolytic flake (III) after cold compacting. The sheet has the least resistance to oxidation and shows the greatest specimen-to-specimen variation. Times to breakaway for the materials I, II, III at t=600 to 700 °C:

material	I	I	I	II	II	II	III	III	III
t in °C . .	600	650	700	600	650	700	600	650	700
time in h	<2000	<1000	<300	>7500	<1000 to ~3000	600 to 1000	1500 to 5000	>3000	>7000

For comparison the times to breakaway at 650 °C and 21.4 atm in CO_2 containing only 30 ppm H_2O: ~4000 h for material I and >7000 h for II and III, Menzies [18]. The effect of gas pressure on the corrosion behavior of Be (arc-melted and then hot forged and rolled) in CO_2 containing 300 or 30 ppm H_2O at 700 °C is shown in **Fig.** 6-**37**. In CO_2 with 300 ppm H_2O severe corrosion takes place at 700 °C at 1 atm as well as at 21.5 atm pressure. No breakaway oxidation is observed in the moist CO_2 at 21.5 atm when Be is alloyed with 0.29 wt% Ca; the weight gain for the alloy after 5000 h exposure at 600 and 700 °C is about 0.1 or 0.2 mg/cm², respectively. Alloys with 0.2 wt% Ca show satisfactory lifetimes at 700 °C for periods of at least 3 or 5 years, Raine, Robinson [19]. While the weight gain of unalloyed (arc-melted and fabricated) Be at 700 °C in moist CO_2 (300 ppm H_2O) at 20.5 atm pressure is 7.76 mg/cm² after 300 h, alloys with 0.35 wt% Ca show a weight gain of only 0.24 mg/cm² after 8000 h, Raine, Robinson [20].

Similar, unalloyed hot pressed, hot extruded ("reactor grade") Be discs show varying weight gains Δw between 13.7 and 53.2 mg/cm² after 350 h exposure to moist CO_2 (300 ppm H_2O) at 21 atm and 700 °C, Raine, Robinson [21 to 23]; in alloys with the preferred composition 99.5 wt% Be, 0.5 wt% Ca, Δw is only about 0.21 mg/cm² after 2800 h exposure to the moist CO_2 [23, 24]. Alloying of Be with 0.5 wt% Zr or Ti causes a corrosion-inhibiting effect similar to that due to Ca [21]. Be is not resistant to moist CO_2 at 60 atm above 600 °C at H_2O contents of 0.01 vol%, Dewanckel et al. [24], or 0.03 vol%, Darras [25], Darras et al. [26]. Alloying of Be with 0.2 wt% Ca improved the corrosion resistance up to 700 °C [24]. 5000 h tests at 650 °C and 60 atm pressure in CO_2 (0.01 vol% H_2O) show that the resistance of Mg-coated Be is even better than that of Be alloys with 0.4 wt% Ca or 0.3 wt% Mg, Darras et al. [27].

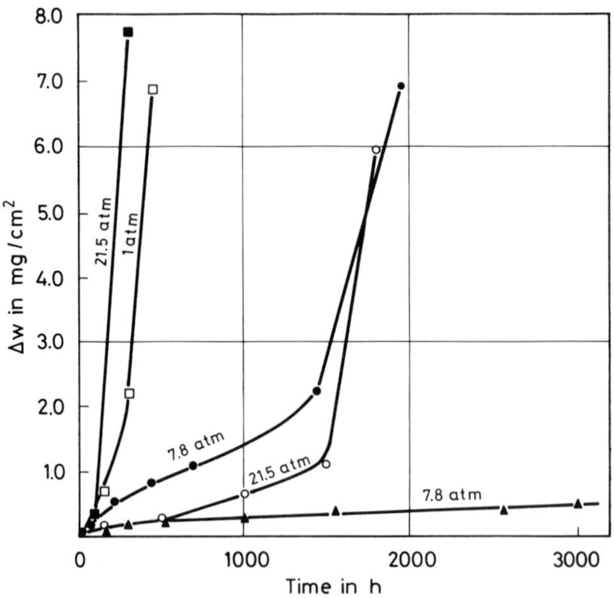

Fig. 6-37

Weight gain of Be vs. time in moist CO_2 (300 ppm H_2O, curves ■, □) and in "dry" CO_2 (30 ppm H_2O, curves ●, ○, ▲) at 700 °C (curve ● at 800 °C).

References:

[1] W.B. Jepson, B.L. Myatt, J.B. Warburton, J.E. Antill (J. Nucl. Mater. **10** [1963] 224/32).
[2] S.J. Gregg, R.J. Hussey, W.B. Jepson (J. Nucl. Mater. **4** [1961] 46/58).
[3] V.K. Khorenko, L.A. Kornienko, B.V. Matvienko (Zh. Fiz. Khim. **40** [1966] 1903/8; Russ. J. Phys. Chem. **40** [1966] 1020/3).
[4] W.J. Werner, H. Inouye (Inst. Metals Monogr. Rept. Ser. No. 28 [1963] 283/93; C.A. **60** [1964] 10325).
[5] W. Munro, J. Williams (AERE-M-M-108 [1956] 1/16; N.S.A. **10** [1956] No. 4622).
[6] D.T. Livey, J. Williams (Proc. 2nd Intern. Conf. Peaceful Uses At. Energy, Geneva 1958, Vol. 5, pp. 311/8; C.A. **1960** 22249).
[7] A. Draycott, F.D. Nicholson, G.H. Price, W.I. Stuart (AAEC-E-83 [1961] 1/34; N.S.A. **16** [1962] No. 12049).
[8] J.K. Higgins, J.E. Antill (J. Nucl. Mater. **5** [1962] 67/80).
[9] J.E. Antill, J.K. Higgins (TID-7597 [1961] 796/805; N.S.A. **15** [1961] No. 19779).
[10] R. Smith (Fuel Elem. Fabric. Spec. Emphasis Cladding Mater. Proc. Symp., Vienna 1960 [1961], Vol. 2, pp. 147/56; N.S.A. **15** [1961] No. 25160).

[11] R. Smith, W.I. Stuart, W.J. van Peer, G. Price (Inst. Metals Monogr. Rept. Ser. No. 28 [1963] 325/34; C.A. **60** [1964] 10338).
[12] W.J. van Peer (Australian J. Phys. **14** [1961] 191/2).
[13] V.E. Ivanov, V.F. Zelenskii, V.K. Khorenko, et al. (Poroshkovaya Met. **1968** No. 8, pp. 73/7; Soviet Powder Met. Metal Ceram. **1968** 647/50).
[14] Y. Kondo, R. Nagasaki, T. Sano (Nippon Kinzoku Gakkaishi **33** [1969] 750/4 from C.A. **71** [1969] No. 41650).
[15] P.J. Phennah, M.W. Davies, B.C. Woodfine (Inst. Metals Monogr. Rept. Ser. No. 28 [1963] 294/313; C.A. **60** [1964] 10338).
[16] M.W. Davies (At. Energy Rev. **2** No. 1 [1959] 11/7).
[17] A. Draycott (Dechema Monogr. **45** [1962] 87/103; C.A. **59** [1963] 3594).
[18] I.A. Menzies (Corrosion Sci. **3** No. 1 [1963] 35/49).
[19] T. Raine, J.A. Robinson (J. Nucl. Mater. **7** [1962] 263/78).
[20] T. Raine, J.A. Robinson (J. Nucl. Mater. **5** [1962] 341/3).

[21] T. Raine, J.A. Robinson (Ger. Offen. 1188295 [1962/65]).
[22] T. Raine, J.A. Robinson (Brit. 911381 [1960/62]; C.A. **58** [1963] 3181).
[23] Associated Electrical Ind., Ltd. (Fr. 1307236 [1961/63]; C. **1964** No. 40-2272).
[24] B. Dewanckel, D. Leclerc, R. Darras (3rd Conf. Intern. Met. Beryllium Commun., Grenoble, Fr., 1965 [1966], pp. 171/92; C.A. **67** [1967] No. 111756).
[25] R. Darras (Bull. Inform. Sci. Tech. CEA [Paris] No. 62 [1962] 43/55; C.A. **58** [1963] 7669).
[26] R. Darras, B. Dewanckel, D. Leclercq (Energie Nucl. [Paris] **8** [1963] 588/96; C.A. **60** [1964] 6577).
[27] R. Darras, B. Dewanckel, D. Leclercq (Fr. 1457463 [1965/66]; C.A. **67** [1967] No. 17370).

6.3.8.5 With CO Containing Dry and Moist CO_2

During the reaction of Be with CO_2 the amount of CO (if not back-oxidized with CuO) increases according to the reaction $Be + CO_2 = BeO + CO$, see p. 259. When Be is oxidized at 650 °C in dry CO_2-CO mixtures containing up to 7.5% CO (total pressure 100 Torr), the kinetics are the same as in pure CO_2 (see p. 261 under [7]), Gregg et al. [1]. This was also stated for the oxidation in moist gas mixtures (total pressure 100 Torr, $P_{CO} = 5$ Torr, $p_{H_2O} = 12$ Torr) at 700 °C. Here the reaction with CO only contributes 0.2% to the total weight gain (short time tests), Gregg et al. [2]. The reaction rate of Be in dry CO_2 containing 5%

CO at 11.2 atm pressure and 650 °C is the same as in pure CO_2 (2000 h test, see p. 270), Davies [3]. Corrosion tests of hot pressed Be at 600 and 650 °C and 11.2 atm pressure in dry CO_2 containing 2 or 5 vol% CO show that CO has no significant effect on the oxidation behavior for the test time of about 2000 h. In moist CO_2 (600 ppm H_2O) the content of 2 or 5 vol% CO delays the onset of breakaway at 600 and 650 °C (see e.g., Fig. 6-35, p. 271); at 675 °C contents of 2 to 10 vol% CO also delay the onset of breakaway, but there is no correlation between time to breakaway and CO content, Phennah et al. [4]. At 60 atm pressure between 600 and 700 °C no distinct effect of the addition of 5% CO to CO_2 was found either in dry gas, Darras [5], Dewanckel et al. [6], or in the presence of 0.03 vol% H_2O, Darras et al. [7].

The effect of alloying with 0.7 wt% Ca on the oxidation of Be (up to 490 h) was studied in moist CO_2 containing 2% CO and 110 ppm H_2O at 650 °C and 20.4 atm pressure by electron microscopy, electron diffraction, and electron probe microanalysis. The most significant effect of adding Ca to Be was to reduce the extent of intergranular oxidation of the metal. The surface oxide scales formed on alloyed and unalloyed Be were found to be quite similar, consisting of BeO platelets (several 1000 Å diameter) together with localized regions of enhanced attack which contained some BeO whiskers. However, sub-surface regions of the corroded alloy were found to contain a layer enriched with Ca (assumed to be $CaBe_{13}$). It is proposed that this layer contributes to the mechanical stability of the BeO scale and may also act as a diffusion barrier between the gas and the unreacted alloy. A Ca-containing corrosion product (such as $CaCO_3$) was not detected, Scott, Ranzetta [8].

References:

[1] S.J. Gregg, R.J. Hussey, W.B. Jepson (J. Nucl. Mater. **3** [1961] 175/89, 184/5).
[2] S.J. Gregg, R.J. Hussey, W.B. Jepson (J. Nucl. Mater. **4** [1961] 46/58, 58).
[3] M.W. Davies (At. Energy Res. **1** [1959] 11/7).
[4] P.J. Phennah, M.W. Davies, B.C. Woodfine (Inst. Metals Monogr. Rept. Ser. No. 28 [1963] 294/313, 306; C.A. **60** [1964] 10338).
[5] R. Darras (Bull. Inform. Sci. Tech. CEA [Paris] No. 62 [1962] 43/55, 51; C.A. **58** [1963] 7669).
[6] B. Dewanckel, D. Leclerc, R. Darras (3rd Conf. Met. Beryllium Commun., Grenoble, Fr., 1965 [1966], pp. 171/91, 176; C.A. **67** [1967] No. 119756).
[7] R. Darras, B. Dewanckel, D. Leclercq (Energie Nucl. [Paris] **8** [1963] 588/96, 593; C.A. **60** [1964] 6577).
[8] V.D. Scott, G.V.T. Ranzetta (J. Nucl. Mater. **9** [1963] 277/89).

6.3.9 With SiO_2 and SiC

Be is incompatible with fused silica in vacuum at 600 °C. In the 2000 h test a brown reaction product was formed and BeO and Si were identified by X-ray diffraction, Vickers [1]. By heating of Be-metallized integrated circuit Si wafers having a 4000 Å thermal oxide layer in Ar for 5 to 20 min, a reaction between the Be layer (2500 or 5000 Å thick) and SiO_2 occurs above 440 °C. The SiO_2 was partially converted to an HF-insoluble residue with green or gray color at 535 to 765 °C. The possible reactions are $2SiO_2+2Be \rightarrow Be_2SiO_4+Si$ and $SiO_2+2Be \rightarrow 2BeO+Si$. The Be layer adheres to both the Si and SiO_2, and does not crack or lift the oxide layer upon being quenched in water from temperatures as high as 765 °C. There is no visible attack upon the Si substrate up to 765 °C for 5 min, Mooney, McCaldin [2]. By heating Be flake in a silica furnace and boat for 16 h at 1500 °C in a H_2 atmosphere which contain traces of H_2O vapor, BeO whiskers and platelets have been

Fig. 6-38

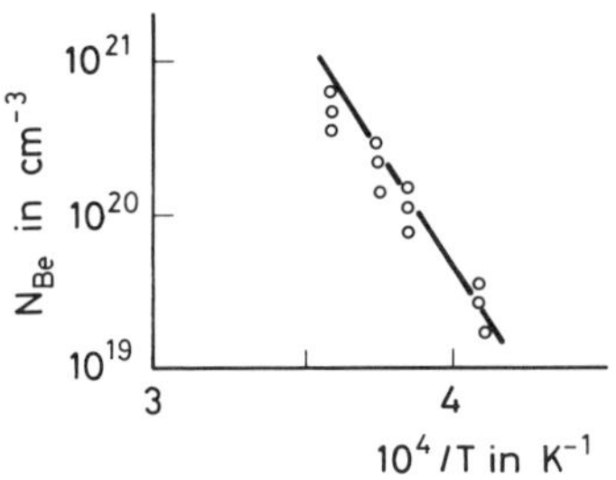

Solubility of Be in SiC vs. temperature.

Fig. 6-39

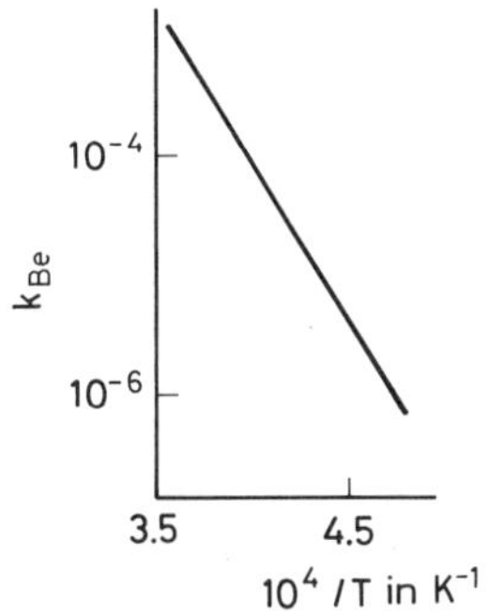

Partition coefficient k_{Be} of Be in the system SiC_g-SiC_s vs. temperature.

grown. As the initial H_2O content is several orders of magnitude less than the amount of BeO formed, the source for oxygen must be SiO_2. SiO_2 reacts directly with the molten Be, or with H_2 to form H_2O, which reacts with Be or can serve as transport medium for BeO, Edwards, Happel [5].

The diffusion of Be in SiC is treated in "Silicium" B2, Pt. 1, 1984, pp. 117/8. Annealing of SiC in Be vapor for 3 h produces a surface layer, which contains Be_2C and whose thickness increases from 27 μm at 1050 °C to 350 μm at 1150 °C. The reaction $SiC + 2Be \rightleftharpoons Be_2C + Si$ takes place, Matyushenko et al. [3]. The solubility of Be in solid SiC was determined at 2400 to 2800 K by recrystallization of very pure Be_2C-doped SiC via the gas phase and analyzing the crystals obtained using emission spectral analysis. The temperature dependence of the solubility is shown in **Fig.** 6-**38** and the partition coefficient $k_{Be} = (N_{Be}/N_{SiC})/(p_{Be}/p_{SiC})$ of Be in the system SiC (gas)-SiC (solid) in **Fig.** 6-**39**. N_{Be} and N_{SiC} are the Be and SiC concentration in the solid solution SiC(Be), p_{Be} is the vapor pressure of Be during the dissociative evaporation of Be_2C, and p_{SiC} is the vapor pressure over SiC. The enthalpy of solution of a neutral Be impurity in SiC was found to be 85 kcal/mol, Safaraliev et al. [4].

References:

[1] W. Vickers (Inst. Metals Monogr. Rept. Ser. No. 28 [1963] 335/49, 343).
[2] J.B. Mooney, J.O. McCaldin (J. Electrochem. Soc. **124** [1977] 625/7).
[3] N.N. Matyushenko, A.A. Rozen, N.S. Pugachev (Poroshkovaya Met. **6** No. 4 [1966] 61/4; Soviet Powder Met. Metal Ceram. **1966** 310/2).
[4] G.K. Safaraliev, Yu.M. Tairov, V.F. Tsvetkov (Izv. Akad. Nauk SSSR Neorgan. Materialy **13** [1977] 1763/6; Inorg. Materials [USSR] **13** [1977] 1423/6).
[5] P.L. Edwards, R.J. Happel (J. Appl. Phys. **33** [1962] 943/8).

6.3.10 With Organic Liquids, Coolants, and Explosives

Compact Be specimens exhibit a slight gain in weight after exposure to CH_3OH, C_2H_5OH, or its mixtures with H_2O at room temperature for 96 h. The weight gain of ~0.5 mg of Be pieces ($12.7 \times 38.1 \times 4.8$ mm^3) may be attributed to the formation of a protective film on the metal surface, Schmitt [1]. Crushed Be (0.5 g) reacts with 25 mL absolute ethyl alcohol which contains 0.5 g $BeCl_2$, with evolution of H_2 within 1 week to form $Be(OC_2H_5)_2$. Instead of $BeCl_2$, iodine or $HgCl_2$ can also be used as catalysts for the reaction, Turova et al. [2].

Be dissolves in the nonaqueous mixture of $(CH_3)_2SO$ (DMSO) and SO_2, but not in DMSO or SO_2 separately, Gill et al. [3], Harrison et al. [4]. Be is attacked by mixtures of CH_3OH with Freon or perchloroethylene and by mixtures of methyl ethyl ketone with Freon or H_2O, Steele [5].

Be is not attacked by a mixture of terphenyls ("Santowax"), used as a coolant in autoclave tests at ~400 °C for 720 h, Neymark [6], Parkins [7], Davies [8]. Besides a slight brown discoloration, the samples showed no sign of attack in 300 h tests at 450 °C or 1000 h tests at 400 °C; the weight gains are of the order of mg/dm^2. H_2O additions resulted in higher weight gains. The vapor pressure of the terphenyl mixture was about 2 atm at 400 °C, Schleicher [9]. Be showed excellent resistance to "pentalene 290" (polyamyl 1-naphthalene) in 1000 h tests at 274 °C (under H_2 pressure of about 2.4 atm), Boyd, Peoples [10], Boxall, Daniel [11].

Mixtures of finely divided Be powder with 73.1 wt% tetranitromethane or 75.5 wt% bis(trinitroethyl)nitramine are highly explosive with a heat of explosion which is 4 times higher than that of TNT, Brunauer [12].

References:

[1] C.R. Schmitt (Y-1397-Rev. 1 [1975] 1/24, 12/3; C.A. **83** [1975] No. 123011).
[2] N.Y. Turova, A.V. Novoselova, K.N. Semenenko (Zh. Neorgan. Khim. **4** [1959] 997/1001; Russ. J. Inorg. Chem. **4** [1959] 453/5).
[3] J.B. Gill, D.C. Goodall, W.D. Harrison (Hydrometallurgy **6** [1981] 347/51).
[4] W.D. Harrison, J.B. Gill, D.C. Goodall (J. Chem. Soc. Commun. **1976** 540/1).
[5] J.R. Steele (Mater. Prot. **1** No. 7 [1962] 59/62).
[6] R.S. Neymark (NAA-SR-7755 [1963] 1/36, 24; N.S.A. **17** [1963] No. 23891).
[7] W.E. Parkins (Corrosion Reactor Mater. Proc. Conf., Salzburg 1962, Vol. 2, pp. 503/34, 508, 528; C.A. **60** [1964] 5027).
[8] H.E. Davies (AERE-M-M-153 [1957] 1/12, 5; N.S.A. **11** [1957] No. 13321).
[9] H.W. Schleicher (EUR-379.E [1963] 1/46, 24/5; N.S.A. **17** [1963] No. 37478; 2nd Intern. Congr. Metal. Corrosion, New York City 1963, pp. 93/4).
[10] W.K. Boyd, R.S. Peoples (BMI-1046 [1955] 1/9; C.A. **1956** 7706).
[11] D.G. Boxall, A.R. Daniel (R-58CAP25 [1958] 1/91, 10/11; N.S.A. **14** [1960] No. 14065).
[12] S. Brunauer (U.S. 3111439 [1949/63]; C.A. **60** [1964] 2720).

6.3.11 With Acids

Be reacts with aqueous hydrogen halides in all concentrations at room temperature. It reacts with dilute H_2SO_4 readily and with concentrated H_2SO_4 slowly. It is attacked by dilute HNO_3. It is resistant to concentrated HNO_3 at room temperature, but not at higher temperatures, Miller, Boyd [1]. The relative rate of dissolution of a given compact sample of Be in various mineral acids of equal concentration decreases in the order $HF > H_2SO_4 \approx HCl \gg HNO_3$, Hardy, Scargill [2]. Be also dissolves in peroxy trifluoroacetic acid (mixture of $CF_3COOH + H_2O_2$), Scott, Shell [3], and in hot concentrated H_3PO_4, Inglis, Cotton [4]. Be is attacked by acetic acid at room temperature, but not by glacial acetic acid [1]. Some organic acids, such as acetic, citric, and tartaric acid, attack Be initially, but the reaction stops after a period of time owing to the formation of a protective film, Pettibone [18], see also Illig [19]. A solution of the composition 65 to 80 mL H_3PO_4 (density 1.75), 0 to 5 mL HNO_3 (density 1.42), and 15 to 30 mL H_2O is suitable for the chemical polishing of Be, Bottom, Ince [5]. For an electron diffraction study of the Be surface after etching with 12 N acids of HF, HCl, HBr, HI, HNO_3, or H_2SO_4 see Yamaguchi [6]. Be is insoluble in $HAsF_6$ and HF solutions of BF_3 and SnF_4, Clifford et al. [20].

With HNO_3

Dissolution rates k in $mg \cdot cm^{-2} \cdot min^{-1}$ of massive extruded Be in 1.67 and 16.22 M HNO_3 (10 and 72 wt%, respectively) at room and boiling temperature:

HNO_3 concentration	temperature	test time	k
1.67 M	room	60 h	4.67×10^{-2}
1.67 M	boiling	6 min	0.1333
16.22 M	room	60 h	5×10^{-4}
16.22 M	boiling	6 min	0.333

The cold concentrated acid passivates the Be. The reaction with boiling acid is far from violent with massive Be, but might be so if powdered metal was used, Darwin, Buddery [7]. The rate of dissolution k varies with the source and method of fabrication of the Be and the initial rate can be higher than the final rate after 30 to 60 min. **Fig.** 6-**40** shows k for untreated and descaled extruded Be (prepared from Brush powder by extrusion of the loose sintered billets into tubes of 7.6 mm internal diameter) at 97 °C in 14 and 6 M HNO_3. Values for the final constant dissolution rate k at 97 °C of a Be sheet with a rough dark gray surface (prepared by powdering of Pechiney "flakes", purity ~99.5%, followed by sintering and hot rolling) in relation to the HNO_3 concentration c:

c in mol/L	1	2	4	6	8	6[a)]
k in $mg \cdot cm^{-2} \cdot min^{-1}$ . .	0.0167	0.0167	0.0414	0.052	0.05	0.026
wt% dissolved/h	1.75	1.75	4.33	5.5	6.0	1.7

[a)] Extruded Be.

Values of k for the sheet material at 25, 50, and 75 °C: 6×10^{-4}, 30.7×10^{-4}, and 2.28×10^{-2}, respectively, in 8 M HNO_3 and 11.8×10^{-4}, 75.2×10^{-4}, and 2.8×10^{-2} in 14 M HNO_3. As shown in a figure in the paper, the dissolution rate in 14 M HNO_3 at 50 °C increases strongly in presence of HF. At contents of >0.1 M HF the reaction became very vigorous and caused the temperature to rise almost to the boiling point [2]. In boiling 4 M HNO_3 containing 0.05 M NaF, the initial dissolution rate for the first 10 min is higher than in the absence of NaF by a factor of 40, Warren, Ferris [8].

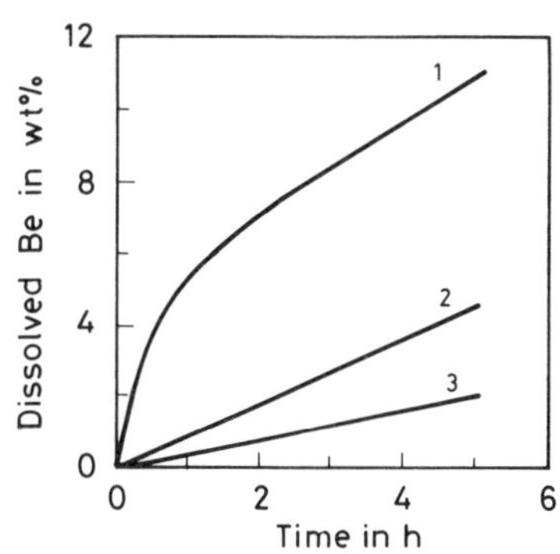

Fig. 6-40

Amount of dissolved Be in dependence of time at 97 °C: 1) untreated Be, 14 M HNO_3; 2) descaled Be (2 M HNO_3+0.05 M HF), 14 M HNO_3; 3) chemically polished Be, 6 M HNO_3.

With HF, HCl, $HClO_4$, HBr, H_2SO_4, and H_3PO_4

The dissolution rate of Be discs (cut from vacuum melted, cast, and extruded material of about 99.5% purity) in dilute HF, HCl, and H_2SO_4 at 30 to 33 °C is shown in **Fig. 6-41**. The initial rate in the 1 N acids (in $mg \cdot cm^{-2} \cdot min^{-1}$) is 0.8 in HF, 0.28 in HCl, and 0.31 in H_2SO_4 [2]. A value of 63 $mg \cdot cm^{-2} \cdot min^{-1}$ in boiling 4 M H_2SO_4 for the first 10 min is reported by [8]. The dissolution of about 30 mg Be (vacuum cast metal lump of ~99.5% purity) in 3 N HF, or 2 N HCl, or 2 N H_2SO_4 at 36 °C is complete in about 30 min. The measured volume of H_2 evolved from complete dissolution agrees satisfactorily with the equations $Be+2HX \rightarrow BeX_2+H_2$ (X=F, Cl) or $Be+H_2SO_4 \rightarrow BeSO_4+H_2$, Straumanis, Mathis [9]. At 80 °C the dissolution of Be in 40% aqueous HF is very rapid giving a product having an F content of 77.4% after drying and 73.5% after 6 h heating at 500 °C (F content in BeF_2 is 80.82 wt%), Hyde et al. [10].

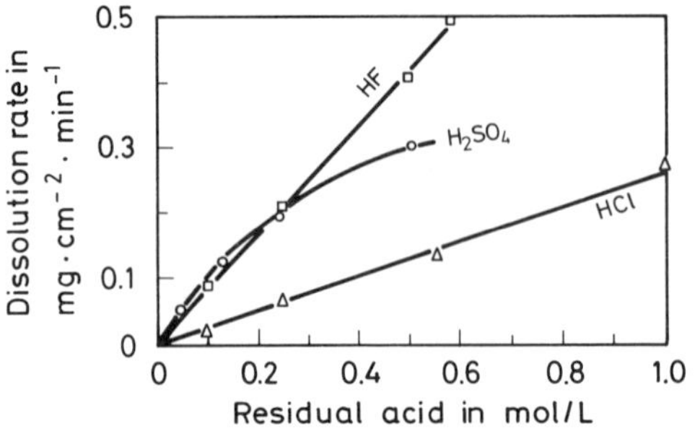

Fig. 6-41

Dissolution rate of Be discs in diluted mineral acids at 30 to 33 °C in dependence of the acid concentration.

While vacuum cast Be is dissolving in dilute HCl (≦0.5 or 0.3 N depending of the purity of Be), a black deposit is observed, which separates from the Be and settles slowly. The deposit consists of needles or twinned crystals of metallic Be. The appearance of a black dispersion during the dissolution of Be in HCl (or during anodic dissolution in NaCl solution) is due to the disintegration of the metal during the dissolution reaction [9, 11]. HBr [9] and $HClO_4$ react in a similar manner with Be, but no disintegration effect has been observed in HF or H_2SO_4 [11]. This effect is also due to the negative difference effect which is observed using a Be anode and a Pt cathode [9, 12]. For the dissolution potentials of Be in dilute HF, HCl, $HClO_4$, and H_2SO_4 at 30 °C see Straumanis, Gnanamuthu [13]. The enthalpy of solution $|\Delta H|$ of ultrapure Be in HF decreases slightly with increasing HF concentration:

wt% HF	12	22.6	30	40
$-\Delta H$ in kcal/mol	101.45±0.6	101.0±0.3	100.5±0.6	100.5±0.9

Bear, Turnbull [14]. In H_2SO_4 (1.163 and 1.019 M) $\Delta H^{\circ}_{298} = -91.814 \pm 0.085$ was obtained by Navratil, Oetting [15]; $\Delta H^{\circ}_{298} = -92.83 \pm 0.11$ was measured by Taylor et al. [16]. A value of $\Delta H = -81.4 \pm 0.2$ kcal/mol at 80 °C was obtained by Wartenberg [17]. The enthalpy of solution of Be in a chemical milling solution (an aqueous solution which contains 750 mL H_3PO_4, 31 mL H_2SO_4, and 71 g CrO_3 per L) was determined to be -119.8 ± 6.8 kcal/mol at 25 °C, Navratil, Oetting [21].

References:

[1] P.D. Miller, W.K. Boyd (AD-824446 [1967] 1/37, 5; C.A. **70** [1969] No. 60126).
[2] C.J. Hardy, D. Scargill (J. Chem. Soc. **1961** 2658/63).
[3] A.F. Scott, J.G. Shell (J. Am. Chem. Soc. **81** [1959] 2278/9).

[4] N.P. Inglis, J.B. Cotton (Corrosion Prevent. Control **5** No. 11 [1958] 59/63, 73; C.A. **1959** 9980).
[5] A.E. Bottom, A. Ince (Brit. 928454 [1958/63]; N.S.A. **17** [1963] No. 27745).
[6] S. Yamaguchi (Bull. Chem. Soc. Japan **18** [1943] 53/91, 67/8, 59).
[7] G.E. Darwin, J.H. Buddery (Metallurgy of Rarer Metals, Academic, New York 1960, pp. 243/4; N.S.A. **14** [1960] No. 15917).
[8] K.S. Warren, L.M. Ferris (CF-61-2-3 [1961] 1/10, 3; N.S.A. **15** [1961] No. 11049).
[9] M.E. Straumanis, D.L. Mathis (J. Electrochem. Soc. **109** [1962] 434/6).
[10] K.R. Hyde, D.J. O'Connor, E. Wait (J. Inorg. Nucl. Chem. **6** [1958] 14/8).

[11] M.E. Straumanis, D.L. Mathis (J. Less-Common Metals **4** [1962] 213/5).
[12] M.E. Straumanis (Metall **16** [1962] 102/7).
[13] M.E. Straumanis, D.S. Gnanamuthu (AD-299972 [1963] 1/20; N.S.A. **17** [1963] No. 19989).
[14] I.J. Bear, A.G. Turnbull (J. Phys. Chem. **69** [1965] 2828/33).
[15] J.D. Navratil, F.L. Oetting (J. Inorg. Nucl. Chem. **35** [1973] 3943/7).
[16] A.R. Taylor, B.B. Letson, D.R. Smith (BM-RI-6724 [1966] 1/8, 1).
[17] H. v. Wartenberg (Z. Anorg. Allgem. Chem. **252** [1943] 136/43, 138).
[18] J.S. Pettibone (in: F.L. LaQue, H.R. Copson, Corrosion Resistance of Metals and Alloys, 2nd Ed., Reinhold, New York 1963, pp. 1/736, 623/31, 623; N.S.A. **18** [1964] No. 22451).
[19] K. Illig (Wiss. Veröffentl. Siemens-Werken **8** [1929] 74/82, 82).
[20] A.F. Clifford, H.C. Beachell, W.M. Jack (J. Inorg. Nucl. Chem. **5** [1957] 57/70, 16, 63, 67).

[21] J.D. Navratil, F.L. Oetting (REP-1985 [1973] 1/6; C.A. **79** [1973] No. 108785).

6.4 Reactions with Metal Compounds

6.4.1 With NaH, $NaNH_2$, KNH_2, and $Pb(N_3)_2$

The rate constant k for the reaction Be(g) + NaH(g) = BeH(g) + Na(g) at 2500 K was calculated to be $5.7 \times 10^{13}\ cm^3 \cdot mol^{-1} \cdot s^{-1}$ and between 1000 to 4000 K as $k = 21.6 \times 10^{11} \cdot T^{0.66} \cdot \exp(-9300/RT)$, Mayer et al. [1]. Be dissolves in molten $NaNH_2$ to form a mixed amide having the approximate composition $Be(NH_2)NHNa \cdot NH_3$, Fernelius, Bergstrom [2]. A solution of Na or K in liquid NH_3 reacts with an excess of Be to give $Be(NH_2)NHNa \cdot NH_3$ or $Be(NH_2)NHK \cdot NH_3$ (also formulated as $Be(NH_2)_2 \cdot KNH_2$ or $K[Be(NH_2)_3]$). Be reacts with KNH_2 in liquid NH_3 to give the intermediate $Be(NH_2)_2$ which reacts further with KNH_2 to give $Be(NH_2)NHKNH_3$. Also, in the course of the reaction, a blue solution of metallic K is formed: $Be + 2KNH_2 \rightleftharpoons Be(NH_2)_2 + 2K$; $2K + 2NH_3 \rightleftharpoons 2KNH_2 + H_2$, $Be(NH_2)_2 + KNH_2 \rightarrow Be(NH_2)NHK \cdot NH_3$, Bergstrom [3]. The heat of explosion of $Pb(N_3)_2$ in mixtures with powdered Be is higher than that without the addition of Be because of the exothermic formation of Be_3N_2 (see p. 209), Apin et al. [4].

References:

[1] S.W. Mayer, J.B. Szabo, L. Schieler, H.S. Johnston (WSCI-65-24 [1965] 1/19, 8; C.A. **64** [1966] 4315).
[2] W.C. Fernelius, F.W. Bergstrom (J. Phys. Chem. **35** [1931] 740/55, 749/50).
[3] F.W. Bergstrom (J. Am. Chem. Soc. **50** [1928] 652/6).
[4] A.Ya. Apin, Yu.A. Lebedev, O.I. Nefedova (Zh. Fiz. Khim. **32** [1958] 819/23; C.A. **1958** 21108).

6.4.2 With Metal Oxides

Molten Be reacts with most of the metal oxides. Only Ca and Th have thermodynamical more stable oxides than Be (but the compounds $CaBe_{13}$ and $ThBe_{13}$ are formed by reaction of Be with CaO or ThO_2), Darwin, Buddery [1]. For the Gibbs free energy ΔG of BeO and other oxides up to about 1200 K, see [2] and [3]. For the calculation of ΔG°_T for the exothermic reduction reactions between Be and several metal oxides at 300 to 1500 K, see Mason, Walton [26]. Because of its strong deoxydizing action, Be can be used as a deoxydant for molten Cu and other heavy metals, see, e.g., Illig [4], see also Gmelin-Handbuch "Kupfer" A 1, 1955, p. 354.

With Na_2O

For the reaction $Be + Na_2O \rightarrow BeO + 2Na$ the Gibbs free energy was calculated to be $\Delta G^\circ_{800} = -52$ kcal/mol at 800 K, Clough [5].

With BeO, MgO, CaO, SrO, and BaO

BeO is the only satisfactory crucible material for the melting of Be, Kaufmann [6]. A crucible of BeO was not attacked by molten Be after 15 min at 1500 °C in an H_2 atmosphere, Winzer [7]. When Be is evaporated from a BeO crucible in a high vacuum ($\sim 10^{-6}$ Torr), the reaction $Be + BeO \rightarrow Be_2O$ takes place at ≧1400 °C. Be_2O is more volatile than BeO and has been found in the mass spectrum of the vapor, Amonenko et al. [8]. A slight reaction (black surface discoloration of BeO and penetration) between Be and BeO at 1600 °C in an inert atmosphere was also observed by Economos [9], Economos, Kingery [10], and Jaeger [11].

Powdered MgO was completely reduced by powdered Be after 135 min at 1275 °C with evaporation of the Mg, Matignon, Marchal [12]. Sintered MgO is compatible with Be in vacuum at 550 °C and incompatible at 600 °C. A heavy brown reaction product was formed and penetration into the Be occurred after 2000 h at 600 °C. BeO was identified by X-ray diffraction, Vickers [13]. For the reaction $Be(l) + MgO(s) \rightarrow BeO(s) + Mg(g)$, the reaction constant log K = 4.9 and $\Delta G^\circ_{2073} = -47.0$ kcal/mol were calculated at 1800 °C, Norton, Kingery [15]. An excess of Be reacts with CaO to form the compound $CaBe_{13}$ [1].

The reaction $14Be + SrO = SrBe_{13} + BeO$ was carried out at 1200 to 1250 °C in a Ta crucible in vacuum (10^{-3} Torr), Matyushenko et al. [16], see also Karev, Reshetova [17].

Powdered BaO was reduced by powdered Be at 1250 °C with the evaporation of Ba [12]. At 1000 °C the reaction constant K for the reaction $Be(s) + BaO(s) \rightleftharpoons BeO(s) + Ba(g)$ was calculated to be $K = 2.5 \times 10^{-3}$ and the vapor tension of Ba was calculated to be 2 Torr, Zaitseva [18].

With Al_2O_3

Powdered Al_2O_3 was reduced to Al by powdered Be at 1280 °C [12] and was incompatible with Be (in vacuum) at temperatures as low as 550 to 600 °C. After 2000 h at 600 °C, a sample of Al_2O_3 was disintegrated by Be, and BeO was identified. Sintered Al_2O_3 was compatible with Be at 550 and 600 °C, Vickers [13] and for short times at high temperatures. A crucible of sintered Al_2O_3 was not attacked by molten Be after 15 min at 1400 and 1600 °C (He atmosphere), however, a slight reaction and formation of $BeO \cdot Al_2O_3$ was observed at 1800 °C [9 to 11]. For the reaction $Be(l) + {}^1/_3\,Al_2O_3(s) = BeO(s) + {}^2/_3\,Al(l)$ at 1800 °C the reaction constant and Gibbs free energy were calculated to be: log K = 1.9 and $\Delta G^\circ_{2073} = -18$ kcal/mol [15]. When molten Be (at its boiling point) reacts with Al_2O_3, the Be lies like a pan cake on the surface of the Al_2O_3 and reacts rapidly to form Al and BeO, Grosse, Stokes [25].

With Rare Earth Oxides

The reaction $M_2O_3+29Be=2MBe_{13}+3BeO$ occurred when M_2O_3 (with M=Tm or Lu) was heated with an excess of Be in vacuum at 1200 to 1300 °C [17].

With TiO_2, ZrO_2, and ThO_2

Be is incompatible with powdered ZrO_2 (in vacuum) at 550 and 600 °C. At 600 °C, the specimen was disintegrated after 2000 h, Vickers [13]. The resistance of ceramic crucibles (TiO_2, ZrO_2, or ThO_2) against molten Be was tested in an He atmosphere at 1400 to 1800 °C. Slight reaction at 1400 °C and violent reaction at 1600 °C was observed with TiO_2, no reaction at 1400 °C and slight reaction at 1600 °C with ZrO_2, and slight reaction at 1600 °C with ThO_2 [14], [9 to 11]. Equilibrium constants K at 1800 °C and the Gibbs free energies ΔG°_{2073} in kcal/mol were calculated for the following reduction reactions [15]:

$Be(l)+3\,TiO_2=BeO+Ti_3O_5$;	log K = +8.6,	$\Delta G^\circ_{2073}=-82$
$Be(l)+{}^1/_2\,ZrO_2=BeO+{}^1/_2\,Zr(s)$;	log K = +1.7,	$\Delta G^\circ_{2073}=-16$
$Be(l)+{}^1/_2\,ThO_2=BeO+{}^1/_2\,Th(s)$;	log K = −1.2,	$\Delta G^\circ_{2073}=+11$

With Pa_2O_5, UO_2, and U_3O_8

The reaction $Pa_2O_5+31\,Be \rightarrow 2\,PaBe_{13}+5\,BeO$ takes place rapidly in BeO crucibles at the melting point of Be (~1277 °C), Benedict et al. [19].

Mixed powders of UO_2 and excess Be reacted completely to form UBe_{13} after 28 d at 700 °C and partially after 28 d at 600 °C, Knapton, West [20]. When stoichiometric UO_2-Be mixtures where heated at a rate of 20 °C/min, ignition occurred at 218 °C, Fleming, Johnson [21]. The following table gives the thickness δ in μm of the reaction layer formed by heating hot rolled Be sheet clamped in contact with UO_2 pellets (95% of theoretical density) between 600 and 800 °C for up to 224 d:

time in d . . .	28	56	112	224
δ at 600 °C . .	2	2	—	36
δ at 700 °C . .	8	10	80	110
δ at 800 °C . .	20	80	90	140

After 7 d at 1000 °C, δ = 160 μm and a localized UBe_{13} formation was observed [20]. Contrary to [20] and [21], Murray and Williams [22] found no reaction between Be and UO_2 specimens over prolonged heating periods at 600 to 700 °C.

For calculations of the enthalpy ΔH and ΔG for the reduction reactions of Be with UO_2 and U_3O_8 up to about 1300 °C, see curves in [21].

With Oxides of the Transuranium Elements

PuO_2 [23], AmO_2, and CmO_2 [19] are reduced by Be at high temperatures. The alloys MBe_{13} (M=Pu, Am, Cm) are formed with excess Be. The reactions $MO_2+15\,Be \rightarrow MBe_{13}+2\,BeO$ (with M=Am or Cm) were carried out under He. The reaction mixture was slowly heated to 1500 °C and maintained at this temperature for 5 min. A BeO crucible was more suitable than a Ta crucible [19]. The reaction of Be with AmO_2 was carried out at 1300 °C. The resulting alloy, $AmBe_{13}$, has a high specific neutron emmission of $3.57\times10^6\ n\cdot s^{-1}\cdot g^{-1}$ resulting from the nuclear reaction $^9Be(\alpha, n) \rightarrow {}^{12}C$, Brachet, Vasseur [24]. PuO_2 is reduced with Be in vacuum in 30 min at 850 °C, Runnals [23].

References:

[1] G.E. Darwin, J.H. Buddery (Metallurgy of the Rarer Metals, No. 7: Beryllium, Academic, New York 1960, pp. 1/392, 244; N.S.A. **14** [1960] No. 15917).
[2] H.H. Hausner (Beryllium Its Metallurgy and Properties, Univ. Calif. Press, Berkeley 1965, pp. 18/9).
[3] R. Darras (Ind. At. **3** No. 9/10 [1959] 41/61, 44).
[4] K. Illig (Wiss. Veröffentl. Siemens-Werken **8** [1929] 74/82; C. **1931** I 245).
[5] W.S. Clough (J. Nucl. Energy **21** [1967] 225/32).
[6] A.R. Kaufmann (in: T. Lyman, Metals Handbook, 8th Ed., Vol. 1, ASM, Metals Park, Ohio, 1961, pp. 1198/9).
[7] R. Winzer (Angew. Chem. **45** [1932] 429/31).
[8] V.M. Amonenko, L.N. Riabchikov, G.F. Tikhinskii, V.A. Finkel (Dokl. Akad. Nauk SSSR **128** [1959] 977/8; Proc. Acad. Sci. USSR Phys. Chem. Sect. **124/129** [1959] 825/6).
[9] G. Economos (Ind. Eng. Chem. **45** [1953] 458/9).
[10] G. Economos, W.D. Kingery (J. Am. Ceram. Soc. **36** [1953] 403/9).

[11] G. Jaeger (Metall **9** [1955] 358/66).
[12] C. Matignon, G. Marchal (Compt. Rend. **184** [1927] 715/7).
[13] W. Vickers (Inst. Metals Monogr. Rept. Ser. No. 28 [1963] 335/49, 343).
[14] F.H. Norton, W.D. Kingery (NYO-3136 [1951] 1/14, 4; N.S.A. **6** [1952] No. 27).
[15] F.H. Norton, W.D. Kingery (NYO-3137 [1952] 1/23, 6; N.S.A. **6** [1952] No. 185).
[16] N.N. Matyushenko, L.F. Verkhorobin, V.N. Karev (Kristallografiya **9** [1964] 273/5; Soviet Phys.-Cryst. **9** [1964] 213/4).
[17] V.N. Karev, L.I. Reshetova (Zavodsk. Lab. **31** [1965] 440/1; Ind. Lab. [USSR] **31** [1965] 534/5).
[18] L.S. Zaitseva (Izv. Akad. Nauk SSSR Ser. Fiz. **20** [1956] 1123/6).
[19] U. Benedict, K. Buijs, C. Dufour, J.C. Toussaint (J. Less-Common Metals **42** [1975] 345/54).
[20] A.G. Knapton, K.B.C. West (J. Nucl. Mater. **3** [1961] 239/40).

[21] J.D. Fleming, J.W. Johnson (ORO-325-Pt. 2 [1959] 1/30, 9, 11, 18; N.S.A. **16** [1962] No. 8718).
[22] P. Murray, J. Williams (Proc. 2nd Intern. Conf. Peaceful Uses At. Energy, Geneva 1958, Vol. 6, pp. 538/49, 542).
[23] O.J.C. Runnals (AECL-543 [1958] 1/34, 20/1; N.S.A. **12** [1958] No. 1234).
[24] G. Brachet, C. Vasseur (CEA-R-3875 [1969] 1/14; C.A. **72** [1970] No. 50068).
[25] A.V. Grosse, C.S. Stokes (PB-161460 [1960] 1/26, 16/7; C.A. **56** [1962] 9432).
[26] C.R. Mason, J.D. Walton (NP-7117 [1958] 1/113, 48/9; N.S.A. **13** [1959] No. 4714).

6.4.3 With Alkali Hydroxides

Be reacts completely with an excess of molten NaOH in an Ni crucible at about 650 °C. The reaction is as follows: $Be + 2NaOH \rightarrow BeO + Na_2O + H_2$, Williams et al. [1], Walker et al. [2]. The exothermic reaction $Be + KOH \rightarrow BeO + K + {}^1/_2\, H_2$ was said to occur in vacuum at 700 °C with the evaporation of K, Matignon, Marchal [3].

References:

[1] D.D. Williams, J.A. Grand, R.R. Miller (J. Am. Chem. Soc. **78** [1956] 5150/5).
[2] B.E. Walker, C.T. Ewing, D.D. Williams (NRL-MR-393 [1954] 1/6; N.S.A. **9** [1955] No. 190).
[3] C. Matignon, G. Marchal (Compt. Rend. **184** [1927] 715/7).

6.4.4 With Alkaline Aqueous Solutions

Be dissolves slowly in dilute aqueous solutions of NaOH or KOH, but rapidly in more concentrated solutions, liberating H_2 and forming a water-soluble beryllate. It does not react with aqueous solutions of NH_3, Darwin, Buddery [1]. In boiling 6 M NaOH an initial dissolution rate of 2 $mg \cdot cm^{-2} \cdot min^{-1}$ for the first 10 min is reported by Warren, Ferris [2].

References:

[1] G.E. Darwin, J.H. Buddery (Metallurgy of Rarer Metals, Academic, New York 1960, p. 244; N.S.A. **14** [1960] No. 15917).
[2] K.S. Warren, L.M. Ferris (CF-61-2-3 [1961] 1/10, 3; N.S.A. **15** [1961] No. 11049).

6.4.5 With Metal Halides

With Solid or Liquid Halides

Be reacts with NaF according to the reaction: $Be+2NaF \rightarrow BeF_2+2Na$. With molten NaF, the reaction was complete under nonequilibrium conditions within 30 min at 1000 °C. In the presence of an excess of NaF, sodium fluoroberyllate was formed, Novoselov et al. [1]. The reaction $Be+PbF_2 \rightarrow BeF_2+Pb$ was carried out by igniting a pressed powder mixture of Be and PbF_2 with an electrically heated W-wire in an H_2 atmosphere. The calorimetrically measured reaction enthalpy is $\Delta H_{298} = -84.12 \pm 0.28$ kcal/mol, Gross et al. [2], Gross, Hayman [3]. FeF_2 (12 mmol) dissolved in 1 kg of a molten LiF-BeF_2 mixture (34 mol% BeF_2) was reduced by 0.25 g metallic Be, Shaffer et al. [4].

The trifluorides of Ac [5] and of the transuranium elements Np [6], or Pu, Am, Cm [5], are reduced by powdered Be (BeO crucible) at 1100 to 1200 °C in vacuum. In addition to BeF_2 (which readily evaporates), transuranium metals M, alloys with Be, or compounds MBe_{13} are formed in quantities dependent on the amount of Be initially present.

Two examples are: $2AmF_3+xBe \rightarrow 2AmBe_{(x-3)}+3BeF_2$, Runnals, Boucher [7] or $2PuF_3(s)+3Be(s) \rightarrow 2Pu(l)+3BeF_2(g)$. The calculated vapor pressure of BeF_2 at 1400 K is ~20 Torr, Runnals, Boucher [5]. Pu-Be alloys with atomic ratios from 1:10 to 1:300 were prepared by Be reduction of PuF_3. Increases in the neutron yield (due to the Be(α, n) reaction) at 900 and 1125 °C indicate that the reaction proceeds at a measurable rate at 900 °C and is rapid at 1125 °C; 400 mg PuF_3 were completely reduced in 25 min at 1125 °C, Runnals [19]. The neutron emission rate for the reaction of 279.8 mg Be with 25.9 mg Am (as the trifluoride) indicated that the reaction was also rapid at 1125 °C (finished in 8 min), but the neutron count increased again rapidly on melting the alloy (~1375 °C) [7].

Be is slightly soluble in molten $BeCl_2$, Markow, Delimarskii [8]. The reduction of $BeCl_2$ by metallic Be in molten salts (eutectic LiCl-KCl mixture containing 10 wt% $BeCl_2$) between 350 and 600 °C was studied by redox potential changes (Mo electrode) which reached values of 1.4 V. The equilibrium constant $K=[Be^+]^2/[Be^{2+}]$ of the reaction $Be(s)+Be^{2+}(fused) = 2Be^+(fused)$ in the molten salt mixture can be expressed by $\log K = (2.36-4904/T) \pm 0.10$. Addition of KF to the melt displaces the equilibrium towards the Be^{2+} side because of the formation of the more stable BeF^{4-} anions, Smirnov, Chukreev [9]. The equilibrium $Be^{2+}+Be \rightleftharpoons 2Be^+$, in melts of LiCl, LiCl-KCl, KCl, and CsCl containing Be^{2+}, was measured by the emf of a cell Be| melt containing Be^{2+} and $Be^+ | Cl_2$, C. The equilibrium constants, shown in **Fig. 6-42**, p. 284, decrease as the radius of the alkali metal increases, Smirnov, Chukreev [10]; for the electrode potentials and the anodic dissolution of Be in the molten salts, see a following volume of "Beryllium". The reaction $Be+2NaCl \rightarrow BeCl_2+Na$ goes to completion at 1000 °C, if the Na vapor is allowed to condense, Rohmer [20].

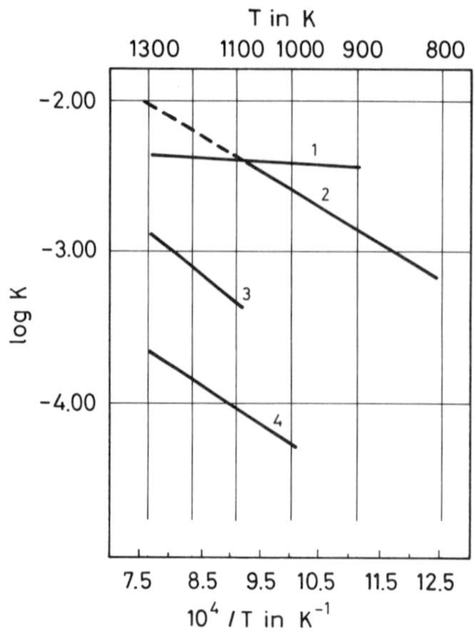

Fig. 6-42

Equilibrium constant K of the reaction $Be^{2+}(melt) + Be(s) \rightleftharpoons 2Be^{+}(melt)$ in melts of LiCl(1), LiCl+KCl(2), KCl(3), CsCl(4).

The Gibbs free energy for the reaction $^1/_2$ Be + NaI → $^1/_2$ BeI_2 + Na was calculated to be $\Delta G^{\circ}_{500} = 47$ kcal/mol at 500 K, Clough [11].

With Gaseous Halides

The reaction $Be(s,l) + BeF_2(g) = 2BeF(g)$ was studied in the temperature range T = 1425 to 1675 K using the molecular flow-effusion method. The reaction enthalpy $\Delta H = 91.5 \pm 3.8$ kcal/mol and the entropy $\Delta S = 44.3 \pm 2.4$ cal · mol^{-1} · K^{-1} were obtained over this temperature range, from a least squares analysis of the experimental data [13, 14]. The equilibrium constant K, in the form of log K, is a linear function of 1/T and changes from log K = −4.354 at 1420 K to log K = −2.396 at 1675 K. The Gibbs free energy ΔG decreases linearly with increasing T from 28.2 kcal/mol at 1420 K to 18.4 kcal/mol at 1675 K [13]. The reaction $Be(s,l) + BeCl_2(g) = 2BeCl(g)$ was also studied by the molecular flow-effusion method between 1573 and 1723 K and gave $\Delta H = 89.1 \pm 7.6$ kcal/mol and $\Delta S = 39.4 \pm 4.6$ cal · mol^{-1} · K^{-1} [14, 15]; log K for this reaction changes from −3.7844 at 1575 K to −2.9287 at 1724 K and ΔG changes from 27.275 kcal/mol at 1575 K to 23.105 kcal/mol at 1724 K [15]. The equilibrium constant for the reaction $Be(s) + BeCl_2(g) = 2BeCl(g)$, in the temperature range 980 to 1235 °C, fits the equation log K = −20480/T + 9.19, Gross, Hayman [3]; see also Gross, Lewin [16] for a discussion of the results of [3, 15, 18] with respect to differences in the obtained enthalpy of formation of BeCl(g). The transport reaction $Be(s) + 2NaCl(g) = BeCl_2(g) + 2Na(g)$ occurs over a temperature gradient 1000 → 900 °C, Schäfer [12].

For the gas reactions $Be(g) + BeF_2(g) = 2BeF(g)$ and $Be(g) + CaF(g) = BeF(g) + Ca(g)$, the enthalpies $\Delta H_{298} = +28.7 \pm 2$ and -8.0 ± 2 kcal/mol, respectively, have been obtained. The vaporous species are produced by the fluorination of elemental Be with CaF_2 and BF_3 in a Knudsen effusion cell and a mass spectrometer has been used to study the high temperature equilibria, Hildenbrand, Murad [17]. Mass-spectrometric investigation of the reactions $Be(g) + BeCl_2(g) = 2BeCl(g)$ and $Be(g) + AlCl(g) = BeCl(g) + Al(g)$ give $\Delta H_{298} = 36.1 \pm 2$ and $\Delta H_{298} = 26.2 \pm 2$ kcal/mol, respectively, for the reactions. The gaseous species were produced by passing HCl gas over a Be-Al mixture in a Knudsen cell molecular source, Hildenbrand, Theard [18].

References:

[1] G.F. Novoselov, I.N. Kashcheev, A.B. Zolotarev (At. Energiya SSSR **30** [1971] 383; Soviet J. At. Energy **30** [1970] 468/9).
[2] P. Gross, C. Hayman, J.T. Bingham (AD-721476 [1971] 1/17, 9; C.A. **75** [1971] No. 81000).
[3] P. Gross, C. Hayman (AD-704139 [1970] 1/14, 1/2, 5/8; C.A. **73** [1970] No. 92306).
[4] J.H. Shaffer, F.A. Doss, W.K.R. Finnel, W. Jennings, W.P. Teichert (ORNL-3591 [1964] 59/64; C.A. **62** [1965] 1354).
[5] O.J.C. Runnals, R.R. Boucher (Can. J. Phys. **34** [1956] 949/58).
[6] O.J.C. Runnals (Acta Cryst. **7** [1954] 222/3).
[7] O.J.C. Runnals, R.R. Boucher (Nature **176** [1955] 1019/20).
[8] B.F. Markov, Yu.K. Delimarskii (Ukr. Khim. Zh. **19** [1953] 255/63, 262; RAE-TRANS-526 [1953] 1/10; AD-72451 [1953] 1/10; N.S.A. **8** [1954] No. 87, **10** [1956] No. 1076).
[9] M.V. Smirnov, N.Ya. Chukreev (Zh. Neorgan. Khim. **4** [1959] 2536/43; Russ. J. Inorg. Chem. **4** [1959] 1168/72).
[10] M.V. Smirnov, N.Ya. Chukreev (Tr. Inst. Elektrokhim. Akad. Nauk SSSR Ural'sk. Filial **1962** No. 3, pp. 3/15, 4, 12; C.A. **59** [1963] 9398).

[11] W.S. Clough (J. Nucl. Energy **21** [1967] 225/32).
[12] H. Schäfer (Naturwissenschaften **50** [1963] 53/5).
[13] M.A. Greenbaum, R.E. Yates, M.L. Arin, et al. (J. Phys. Chem. **67** [1963] 703/7).
[14] M.A. Greenbaum, M. Farber, et al. (Proc. 1st Meeting Interagency Chem. Rocket Propulsion Group Thermochem., New York 1963 [1964], Vol. 1, pp. 101/4; C.A. **62** [1965] 71).
[15] M.A. Greenbaum, M.L. Arin, M. Wong, M. Farber (J. Phys. Chem. **68** [1964] 791/5).
[16] P. Gross, R.H. Lewin (AD-728679 [1971] 1/25; C.A. **76** [1972] No. 7206).
[17] D.L. Hildenbrand, E. Murad (J. Chem. Phys. **44** [1966] 1524/9).
[18] D.L. Hildenbrand, L.P. Theard (J. Chem. Phys. **50** [1969] 5350/5).
[19] O.J.C. Runnals (AECL-543 [1958] 1/34, 20/1; N.S.A. **12** [1958] No. 1234).
[20] R. Rohmer (Compt. Rend. **214** [1942] 744/6).

6.4.6 Reactions with Salt Solutions

Be is attacked by neutral salt solutions, Inglis, Cotton [1]. It reduces salts of many elements in solution, e.g., Te, As, Bi, Sb, Zn, Cd, Hg, Pb, Sn, U, Cu, Co, Ni, Au, Ag, Illig [2].

With Solutions of Ammonium Fluoride

The dissolution of Be tubes (prepared from "Brush" Be powder by loose sintering and extruding) in aqueous 5 M NH_4F solution is fairly rapid, with evolution of H_2 at initial rates of 0.6 and 2.4 $mg \cdot cm^{-2} \cdot min^{-1}$ at 50 and 97 °C, respectively. The rate of dissolution at 50 °C decreases exponentially with time and approaches zero as the molar ratio of fluoride to dissolved Be approaches 4; the reaction can be described by the equation $Be + 4NH_4F + 2H_2O \rightarrow (NH_4)_2BeF_4 + H_2 + 2NH_4OH$. However, at 97 °C a precipitate, probably $Be(OH)_2$, is formed and the final molar ratio tended towards 3; possibly the reaction $(NH_4)_2BeF_4 + 2NH_4OH \rightarrow Be(OH)_2 + 4NH_4F$ takes place, Hardy, Scargill [3]. The rate of dissolution of a Be disc in boiling 3.6 and 5.6 M NH_4F solution exhibits pseudo first-order dependence on the "free" fluoride concentration (i.e., the total fluoride concentration minus fluoride assumed to be present as BeF_4^{2-}). Even when the "free" fluoride concentration is zero, the metal continues to dissolve, with the formation of a slight precipitate (in accordance with [1]), Whitefield [4].

Be rapidly dissolves in ammonium hydrogen fluoride solutions according to the equation $Be + NH_4HF_2 \rightarrow (NH_4)_2BeF_4 + H_2$, Darwin, Buddery [5].

With Solutions of NaCl and Seawater

Be is subject to pitting attack in aqueous environments containing chloride ions, Miller, Boyd [6]. The attack of forged Be (see below) by 3 wt% NaCl solution at 37.8 °C (100 °F) in 2, 10, and 1080 min has been studied microscopically; pitting is initiated at a few local anodic sites and then propagates quite rapidly over the surface. The chemical attack does not follow along grain boundaries but propagates through the bulk of the grains. Furthermore, the most rapid attack occurs in the longitudinal direction of forging. In high purity single-crystal Be the most rapid dissolution rate is parallel to the basal plane; a longer time is required to initiate local pitting than with the forged Be, Mackay, Gilpin [7].

The weight loss of extruded Be in aerated distilled H_2O containing 1, 5, and 10 ppm Cl^- ions (as NaCl) in dependence of time at 90 °C is shown in **Fig.** 6-**43**, from N. Goldowski, reported by English [8]. The weight loss of the Be strips in 150 ppm Cl^- containing H_2O in dynamic 1500 h tests at room temperature increases slowly up to about 750 h and is enhanced at longer times, O'Donnell [14]. The weight loss of a cross-rolled, surface-ground, 1.52 mm thick Be sheet (purity 98.3 wt%), which was at first pickled with aqueous $HF-HNO_3$ solution, in aqueous NaCl solutions at 15 °C has been measured up to 30 d. The obtained corrosion rates per year, based on the 30 d test, are 0.546 mm in 3 wt% NaCl solution and 0.848 mm in 3.5 wt% NaCl solution. The maximum pit depth is 0.173 mm after 8 d in the solution with 3.5% NaCl. The corrosion rate is higher than in sea water, see **Fig.** 6-**44**, Prochko et al. [9]. The corrosion of pickled Be sheet in synthetic sea water (pH 8.2, chloride content 19500 ppm) at 25 °C is shown in **Fig.** 6-**45**. The corrosion rate decreases asymptotically from about 0.54 mm/a after 5 days exposure to approximately 0.051 mm/a after 150 days. This suggested that a build-up of the white, flocculent corrosion product (assumed to be hydrated beryllium oxide) provided some form of pseudo-protective layer, but the cleaned surface after 150 days exposure revealed a significant amount of pitting attack. While the maximum pit depth increased asymptotically from about 0.04 to 0.08 mm, the percentage of the surface pitted increased linearly from about 2 to 45% as the exposure period increased from 5 to 150 days. The interior surfaces of all pits were covered with black amorphous Be, Miller et al. [15].

When bare forged Be coupons (prepared from a vacuum hot pressed block by a single forward extrusion) are exposed to a fog spray of 5 wt% aqueous NaCl at 37.8 °C (100 °F), pitting starts to occur on polished samples at localized spots after 1 d and continued to increase with additional exposure time. The corrosion rate from the weight loss after 30 d is 0.0559 mm/a and pits as large as 1.143 mm across and 0.635 mm deep are observed. The weight loss after 1 d exposure is about 0.012 mg/cm^2 (the coupons were rinsed with distilled H_2O after exposure) [7]. To simulate operating conditions in an engine compressor,

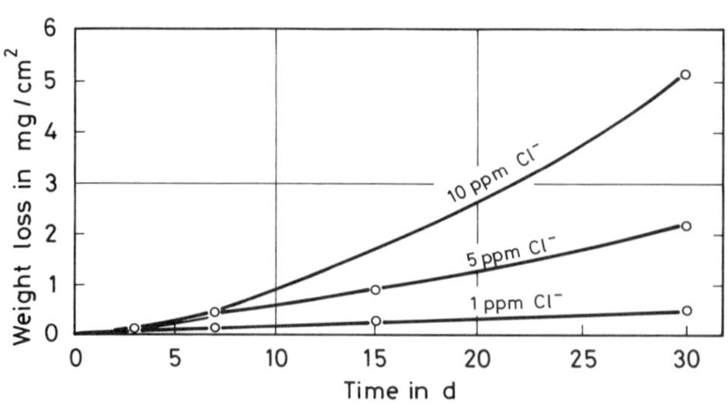

Fig. 6-43

Effect of chloride content in distilled water at 90 °C on corrosion of Be.

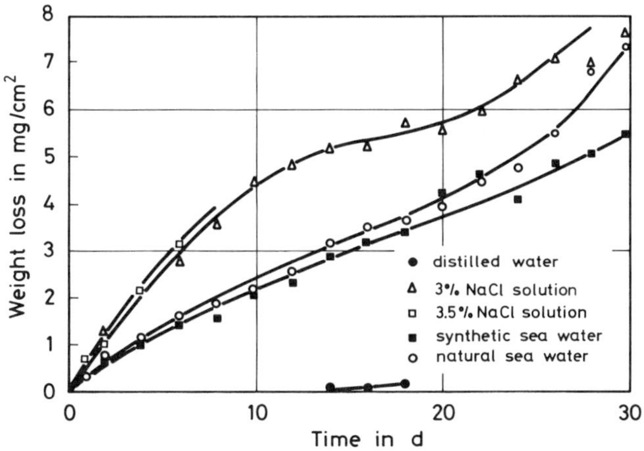

Fig. 6-44

Weight loss of Be sheet vs. time in five environments at 15 °C.

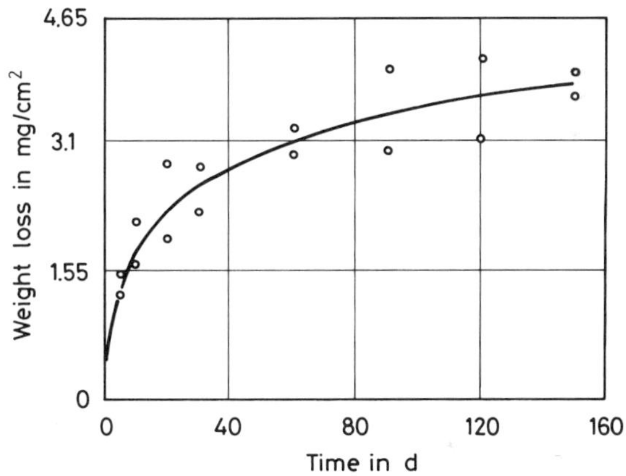

Fig. 6-45

Weight loss of Be sheet in synthetic sea water at 25 °C as a function of time.

the coupons are alternately exposed to the salt fog (16 h) at 37.8 °C and heated in air at elevated temperatures (8 h at 93.3, 204.4, 315.6, or 426.7 °C) for a total of 14 cycles. At 204.4 and 315.6 °C after the first few cycles, a thin protective oxide film was formed which retarded the chemical attack by the salt environment. Bare Be shows its maximum corrosion resistance at 315.6 °C (600 °F). At higher temperatures "catastrophic" oxidation begins to occur. Anodized coatings and aluminized coatings offer good corrosion protection for Be in the presence of aqueous NaCl environments and elevated temperature of 315.6 °C [7, 10], see also [13]. Corrosion tests of pickled and anodized Be sheet in natural sea water at 15 °C during application of stress show that its use is limited by pitting and stress corrosion failure [9]. Stress corrosion test of pickled Be sheet in synthetic sea water at 25 °C confirmed that Be is subject to premature failure when exposed to the combined effect of salt water and applied tensile stress. The average time-to-failure decreased from approximately 2350 to 40 h as the applied stress was increased from 85.8 to 2812 kg/cm², respectively. Failure in all cases was in the direction normal to the applied stress. Electron fractography studies suggested that stress-corrosion failure occurred transgranularly [15].

The effect of galvanostatic anodic and cathodic currents on the stress corrosion of Be sheet material in aerated, synthetic sea water at 22 °C and an applied stress of 2109 kg/cm² was studied by King, Myers [16]. Applied anodic currents reduced the time-to-failure while cathodic currents prevented pitting and subsequent stress corrosion. Mechanism for premature failure was associated with rapid anodic dissolution along an active path nearly normal to the applied tensile stress. Hydrogen cracking was excluded as a primary factor in the failure mechanism [16]. The corrosion of cross-rolled, ground, and etched 1.016 mm thick Be sheet (purity 99.3% Be) in flowing 3.5 wt% NaCl solution (pH 6.8 to 7.2) at 21.1 °C (70 °F) was studied during exposure to sustained tensile stress (10 to 90% of yield strength, 100 to 300 MN/m^2). In 90 days 2160 test cycles consisting of a 10 min immersion phase followed by a 50 min drying phase in still air were made. The time required for the Be specimens to crack decreased linearly with increasing tensile strength. The absence of a threshold stress (a stress below which cracking does not occur) is not only unusual, but suggests that the Be specimens under these test conditions cracked from a mechanism other than stress corrosion. There appears to be a greater susceptibility to stress-accelerated surface corrosion, Moore [11]. Be is also immune to stress corrosion cracking in a salt spray under the following conditions: 3 wt% NaCl solution, 35 °C, 90% relative humidity, testing period 100 h, Be stressed up to 90% of the yield strength, Packer [12].

References:

[1] N.P. Inglis, J.B. Cotton (Corrosion Prevent. Control **5** No. 11 [1958] 63/73; C.A. **1959** 9980).
[2] K. Illig (Wiss. Veröffentl. Siemens-Werken **8** [1929] 74/82, 79; C **1931** I 245).
[3] C.J. Hardy, D. Scargill (J. Chem. Soc. **1961** 2658/63).
[4] H.J. Whitefield (AAEC-TM-137 [1962] 1/9; N.S.A. **16** [1962] No. 18848).
[5] G.E. Darwin, J.H. Buddery (Metallurgy of Rarer Metals, Vol. 7: Beryllium, Academic, New York 1960, p. 244; N.S.A. **14** [1960] No. 15917).
[6] P.D. Miller, W.K. Boyd (AD-824446 [1967] 1/37, 3/5; C.A. **70** [1969] No. 60126).
[7] T.L. Mackay, C.B. Gilpin (Electrochem. Technol. **6** [1968] 235/40).
[8] J.L. English (in: D.W. White Jr., J.E. Burke, The Metal Beryllium, ACS, Cleveland 1955, p. 539).
[9] R.J. Prochko, J.R. Myers, R.K. Saxer (Mater. Prot. **5** No. 12 [1966] 39/42).
[10] C.P. Gilpin, T.L. Mackay (AD-646508 [1967] 1/67 from C.A. **68** [1968] No. 32628).
[11] H.D. Moore (Natl. SAMPE Symp. Exhib. Proc. **15** [1969] 853/9; C.A. **73** [1970] No. 90643).
[12] C.M. Packer (LMSD-49735 [1959] 1/23, 16; N.S.A. **14** [1960] No. 20555).
[13] T.L. McKay, C.B. Gilpin, C. Barclay, N.A. Tiner (Proc. 1st Jt. Aerosp. Mar. Corros. Technol. Semin., Los Angeles 1968 [1969], pp. 48/53; C.A. **73** [1970] No. 123071).
[14] P.M. O'Donnell (Corros. Sci. **7** [1967] 717/8).
[15] R.A. Miller, J.R. Myers, R.K. Saxer (Corrosion **23** [1967] 11/4).
[16] T.T. King, J.R. Myers (Corrosion **25** [1969] 349/51).

6.4.7 With CeS, FeS, PbTe, TiB_2, and Metal Carbides

Melts of Be do not attack CeS crucibles below 1800 °C (vacuum or inert atmosphere), Eastman et al. [1]. A dense adhering ingot was formed when Be was melted in a CeS crucible at 1400 °C for 5 min in vacuum, Jaeger [2]. Crucibles of CeS have also been tested for Be resistance by Lyu Fu-Yui [3]. Be reacts violently with molten FeS under layer separation forming BeS (upper layer) and Fe. Be can desulfurize iron melts, Kroll [18]. Diffusion couples composed of Be and PbTe showed no reaction after 200 h at 649 °C (1200 °F), Elkins [4]. Be was compatible with PbTe in 200 h tests at 650 °C and in 100 h tests at 700 °C, Goodspeed [17]. A TiB_2 crucible was resistant against molten Be at 1500 and 1650 °C in vacuum, while crucibles made of TiC and ZrC were not, Lunde, Meier [5].

The reaction of Be with $ZrC_{0.98}$ and $NbC_{0.86}$ [6] as well as with $NbC_{0.78}$ and $NbC_{0.915}$ [7] has been studied by contact annealing of diffusion couples in vacuum between 750 and 1000 °C [6] or at 1100 °C [7], respectively. X-ray and microhardness measurements showed that, with $ZrC_{0.98}$, $NbC_{0.86}$, and $NbC_{0.915}$, a layer of Be_2C was formed in the contact zone. The growth rate of the Be_2C layer is limited by the diffusion stage. For the reaction between Be and $NbC_{0.86}$ at 750 to 1000 °C, the rate constant k can be expressed by the equation $k=(1.23^{+1.47}_{-0.67})\times10^{-3}\exp[-(10750\pm1560)/RT]$ with k in $cm\cdot s^{-1/2}$, R in $cal\cdot mol^{-1}\cdot K^{-1}$ and T in K. The rate constant for the reaction with $ZrC_{0.98}$ is about two orders of magnitude higher [6]. For the reaction with $NbC_{0.915}$ at 1100 °C, the thickness of the Be_2C layer reached 300±50 µm after 100 h with $k=(5.0\pm0.5)\times10^{-5}\ cm\cdot s^{-1/2}$. The extrapolated value of k for the reaction with $NbC_{0.86}$ at 1100 °C is $2.4\times10^{-5}\ cm\cdot s^{-1/2}$. When Be was heated with $NbC_{0.78}$ (1100 °C) the phase $NbBe_{12}$ was formed instead of Be_2C, with $k=(3.3\pm0.2)\times10^{-5}\ cm\cdot s^{-1/2}$ and a layer thickness of 200±10 µm after 100 h [7].

The reaction of Be with the C deficient monocarbides $TiC_{0.95}$, $TiC_{0.65}$, $VC_{0.8}$, $NbC_{0.77}$, $NbC_{0.72}$, and $TaC_{0.76}$ was studied by heating the pressed mixtures in vacuum for 10 h at 1100 °C and then further for 40 h at 1500 °C. X-ray and density investigation of the powdered samples revealed that after 10 h at 1100 °C no dissolution or reaction had occurred. After 40 h at 1500 °C, ~11 at% Be was dissolved in $NbC_{0.77}$ and $NbC_{0.72}$ as was evidenced by the increase of the cubic lattice constant from NbC_x to the solid solution NbC_xBe_y. The reaction with $TaC_{0.76}$ was similar and the solubility of Be reached 12 at% at 1500 °C, Zainulin et al. [8, 9]; however, an additional phase was also formed at the composition $TaC_{0.76}Be_{0.43}$, which has not been identified [9]. Be was insoluble in the Ti carbides. The solution of Be in $VC_{0.8}$ requires at least 30 h at 1500 °C, but even after 40 h, no homogeneous samples were obtained [8, 9].

Be and UC did not react below 600 °C, Murray, Williams [10], but at 600 °C [11, 12] and above, they reacted to form UBe_{13} [13 to 16]. The thickness of the UB_{13} reaction layer between a hot rolled Be sheet and sintered UC after heating in vacuum for 14 d (336 h) was 3 µm at 600 °C and 43 µm at 700 °C [11]. Be-UC diffusion couples heated in an Ar atmosphere between 700 and 1000 °C showed a marked increase in the rate of reaction between 750 and 800 °C. The following table gives the thickness δ of the UBe_{13} reaction layer formed in couples of hot pressed Be and arc-cast UC:

temp. in °C . .	700	750	800	850	900	950	1000
time t in h . .	185	187	162	144	121	103	90
δ in µm . . .	7	27	115	106	150	131	176

Murdock [13]. Similar studies using UC with 4.8 and 5.1 wt% C show that δ increases approximately linearly with $t^{1/2}$ at 1000 °C. The reaction is controlled by the diffusion of Be in UBe_{13} with reaction occurring at the UBe_{13}-UC interface. An activation energy of 33 kcal/mol was determined for the process. An uniformly dispersed precipitate, which could not be identified, was found in the reaction layer. In addition to these unidentified particles, needles of UC_2 were found in the UBe_{13} layer indicating that Be reacts at a somewhat lower rate with UC_2 than with UC, Murdock [14].

Considerable reaction between Be and UC in the contact zone of couples occurred at 820 °C after 4000 h. UBe_{13} was identified by X-ray diffraction and a molten metallic phase was observed visually, Strasser, Stahl [15], Bolta, Strasser [16].

References:

[1] E.D. Eastman, L. Brewer, L.A. Bromley, et al. (J. Am. Ceram. Soc. **34** [1951] 128/34).
[2] G. Jaeger (Metall **9** [1955] 358/66).

[3] Lyu Fu-Yui (Hua Hsueh Tung Pao **1962** No. 4, pp. 237, 249 from Ref. Zh. Khim. **1963** No. 14M52).
[4] P.E. Elkins (NAA-SR-MEMO-7172 [1962] 1/31, 10/1; N.S.A. **16** [1962] No. 24161).
[5] M.C. Lunde, W.P. Meier (REP-882 [1968] 1/13, 5; C.A. **69** [1968] No. 109176).
[6] V.N. Zagryazkin, A.S. Panov, M.M. Rysina (Izv. Akad. Nauk SSSR Neorgan. Materialy **12** [1976] 352/3; Inorg. Materials [USSR] **12** [1976] 304/5).
[7] V.N. Zagryazkin, A.S. Panov (Zh. Fiz. Khim. **49** [1975] 550/1; Russ. J. Phys. Chem. **49** [1975] 322/3).
[8] Yu.G. Zainulin, S.I. Alyamovskii, G.P. Shveikin, E.N. Shchetnikov, P.V. Gel'd (Zh. Prikl. Khim. **42** [1969] 693/5; J. Appl. Chem. [USSR] **42** [1969] 657/8).
[9] Yu.G. Zainulin, S.I. Alyamovskii, G.P. Shveikin, E.N. Shchetnikov, P.V. Gel'd (Tr. Inst. Khim. Akad. Nauk SSSR Ural'sk. Filial **1970** No. 20, pp. 68/70; C.A. **75** [1971] No. 133865).
[10] P. Murray, J. Williams (Proc. 2nd Intern. Conf. Peaceful Uses At. Energy, Geneva 1958, Vol. 6, pp. 538/49).
[11] J.D. Baird, G.A. Geach, A.G. Knapton, K.B.C. West (Proc. 2nd Intern. Conf. Peaceful Uses At. Energy, Geneva 1958, Vol. 5, pp. 328/33; C.A. **1960** 22066).
[12] A. Accary, R. Delmas (CEA-R-2674 [1964] 1/16, 6; C.A. **62** [1965] 10031; Proc. 3rd Intern. Conf. Peaceful Uses At. Energy, Geneva 1964, Paper 59, pp. 1/15; A-CONF-28-P-59 [1964] 1/15; N.S.A. **18** [1964] No. 37602).
[13] J.F. Murdock (ORNL-3160 [1961] 45/7; N.S.A. **15** [1961] No. 29716).
[14] J.F. Murdock (J. Nucl. Mater. **7** [1962] 192/6).
[15] A.A. Strasser, D. Stahl (UNC-5134-Vol. 1 [1965] 1/117, 80; C.A. **65** [1966] 19603; N.S.A. **20** [1966] No. 27416).
[16] C. Bolta, A. Strasser (NDA-2145-6 [1960] 1/84, 22; N.S.A. **15** [1961] No. 12469).
[17] R.C. Goodspeed (WCAP-1868 [1961] 1/35, 7; N.S.A. **17** [1963] No. 23971).
[18] W. Kroll (Metallwirtschaft **13** [1934] 21/3).

6.4.8 With $Hg(CH_3)_2$

The reaction $Be + Hg(CH_3)_2 \rightarrow Be(CH_3)_2 + Hg$ requires 24 h at about 115 to 130 °C (above the boiling point of $Hg(CH_3)_2$) and is carried out in a sealed evacuated tube [1]. Alternatively, a reaction mixture of 1 g powdered Be +4.3 mL $Hg(CH_3)_2$ is refluxed for 36 h in a N_2 filled reaction tube, Goates et al. [2]. The reaction with microquantities (2 mg Be + 42 mm^3 $Hg(CH_3)_2$) is carried out in vacuum at 125 °C for 72 h, Muxart et al. [3].

References:

[1] I. Sherman, W. Rabideau, M. Alei, C.E. Holley (LA-1687 [1954] 1/11, 5/6; C.A. **1956** 3992).
[2] G.E. Goates, F. Glockling, N.D. Huck (J. Chem. Soc. **1952** 4496/501).
[3] R. Muxart, R. Mellet, R. Jaworsky (Bull. Soc. Chim. France **1956** 445/6).

6.4.9 With GaP

The solubility isotherm of Be in GaP at the temperature of equilibrium between liquid and solid GaP is shown in **Fig.** 6-**46**. The coefficient k of segregation, which characterizes the distribution of Be between the liquid and the solid phase, is constant only in the linear section of the isotherm and is calculated from the slope to be $k = 1.6$, Yu.L. Il'in, V.S. Sorokin, A.V. Ustinov, D.A. Yas'kov (Izv. Akad. Nauk SSSR Neorgan. Materialy **12** [1976] 1936/8; Inorg. Materials [USSR] **12** [1976] 1590/2).

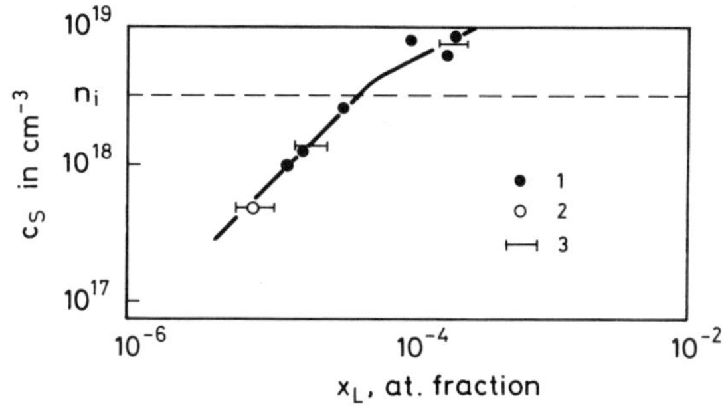

Fig. 6-46

Solubility isotherme of Be in GaP. c_S = solubility of Be atoms in solid GaP, x_L = atomic fraction of Be in liquid GaP, n_i = concentration of intrinsic carriers at the melting point of GaP.

6.4.10 With GaAs

The diffusion coefficient of Be into GaAs between 700 and 900 °C is $D = D_0 \exp(-E/kT)$ with $D_0 = 0.655\ cm^2/s$ and $E = 2.43$ eV, Masu et al. [1]. For the diffusion of Be implanted in GaAs and the influence of the semiconducting properties of GaAs, see Kanber et al. [2]. Implanted Be in n type semiconducting $Ga_{0.47}In_{0.53}As$ acts as p type dopant, Vescan et al. [3].

References:

[1] K. Masu, M. Konagai, K. Takahashi (J. Appl. Phys. **54** [1983] 1574/8).
[2] H. Kanber, M. Feng, J.M. Whelan (Proc. SPIE-Intern. Soc. Opt. Eng. No. 463 [1984] 67/71 from C.A. **101** [1984] No. 47096).
[3] L. Vescan, J. Selders, M. Maier, H. Kräutle, H. Beneking (J. Cryst. Growth **67** [1984] 353/7).

7 The Chemical Behavior of the Be^{2+} Ion in Solution

The behavior of the Be^{2+} ion is governed by its very high charge to radius ratio. In the following only the action of the ions with H_2O, i.e., hydration, exchange, and hydrolysis are briefly treated and the possibilities of precipitation from aqueous solutions are mentioned.

The main facts are to be reported in subsequent volumes of the Gmelin Handbook series concerning beryllium compounds and coordination compounds, including their solutions.

The Be^{2+} ion is colorless and has a remarkably sweet taste. It coordinates very readily giving tetracoordinated cations or anions. The salts are hydrolized to a considerable extent.

7.1 Hydration

For a comprehensive survey on hydration, see the general review in Gmelin Handbook "Sauerstoff" 5, 1963, pp. 1655/68.

7.1.1 **Hydration Number H** in mol H_2O/g ion Be^{2+}

A hydration number of about H=4 in the inner coordination sphere of Be^{2+} in acid salt solutions was found by different methods: electrical conductivity, Gusey [1]; ^{17}O NMR spectra in presence of the paramagnetic ions Co^{2+}, Jackson et al. [2], Connick, Fiat [3], or Dy^{3+}, Alei, Jackson [4]; proton magnetic resonance spectra, Fratiello, Douglass [5], Akitt, Duncan [6]; densimetrically (molar volume), Jedinakova [7], also see Akitt [8], Eucken [9], and Journet, Vadon [10] for the determination of the ion hydration from the apparent molar volume, which shows a negative temperature coefficient. $H=3.4\pm20\%$ was calculated from the heat of hydration and the Be^{2+}-H_2O distance, Gorski, Koch [11]. For the decrease of the hydration number below 4 due to the hydrolysis of the $Be(H_2O)_4^{2+}$ ion, see p. 295. From measurements of the viscosity, freezing point, and electrical conductivity it was found that Be^{2+} has the largest hydration number in aqueous solutions compared with other divalent cations, Fricke, Schützdeller [12]. From diffusion measurements $H\approx10$ was obtained, Spandau, Spandau [13], and 13.5 was calculated by the method of Robinson and Stokes from extrapolated values of the mobility for infinite solution, Haase [14]. The charge of the hydrated beryllium ion is distributed over the H atoms of the hydration sphere. The number of the charged H atoms in the hydration sphere is calculated as 12 from value assigned to ionization potential and hydration energy [16].

The thickness of the Be^{2+} ion-free layer in aqueous beryllium salt solutions of the surface (gas-solution interface) was calculated as 1.98 Å, Khentov, Vlasov [15].

References:

[1] N.I. Gusey (Zh. Fiz. Khim. **45** [1971] 2243/6; Russ. J. Phys. Chem. **45** [1971] 1268/71).
[2] J.A. Jackson, J.F. Lemons, H. Taube (J. Chem. Phys. **32** [1960] 553/5).
[3] R.E. Connick, D.N. Fiat (J. Chem. Phys. **39** [1963] 1349/51).
[4] M. Alei, J.A. Jackson (J. Chem. Phys. **41** [1964] 3402/4).
[5] A. Fratiello, D.C. Douglass (J. Chem. Phys. **39** [1963] 2017/22).
[6] J.W. Akitt, R.H. Duncan (J. Chem. Soc. Faraday Trans. 1 **76** [1980] 2212/20, 2216).
[7] V. Jedinakova (Sb. Vys. Sk. Chem. Technol. Praze Anorg. Chem. Technol. B **18** [1974] 145/56; C.A. **83** [1975] No. 66397).
[8] J.W. Akitt (J. Chem. Soc. A **14** [1971] 2347/50).
[9] A. Eucken (Z. Elektrochem. **52** [1948] 6/24).
[10] G. Journet, J. Vadon (Bull. Soc. Chim. France **1955** 593/607).

[11] N. Gorski, H. Koch (Z. Naturforsch. **23a** [1968] 629/30).
[12] R. Fricke, H. Schützdeller (Z. Anorg. Allgem. Chem. **131** [1923] 130/9).
[13] H. Spandau, G. Spandau (Z. Physik. Chem. **192** [1943] 211/28).
[14] R. Haase (Z. Elektrochem. **62** [1958] 279/81).
[15] V.Ya. Khentov, Yu.V. Vlasov (Zh. Fiz. Khim. **52** [1978] 3170/2; Russ. J. Phys. Chem. **52** [1978] 1809/11).
[16] N.D. Biryukov (Izv. Sibirsk. Otd. Akad. Nauk SSSR **1962** No. 8, pp. 40/52; C.A. **58** [1963] 2900).

7.1.2 Thermodynamic Data of Hydration and Formation

Enthalpy of hydration ΔH_h° (in kcal/mol), free enthalpy of hydration ΔG_h° (in kcal/mol), and entropy of hydration ΔS_h° (in cal · mol^{-1} · K^{-1}) for the gaseous Be^{2+} ion, $Be^{2+}(g) \rightarrow Be^{2+}(aq)$ at 25 °C and infinite solution: $\Delta H_h^\circ = -601$, $\Delta G_h^\circ = -577$, and $\Delta S_h^\circ = -81$ are obtained as the most probable values; as ΔG and ΔS are related to a standard entropy of the proton in solution of $S_{H^+}^\circ(aq) = 0$, these values have to be corrected. With $S_{H^+}^\circ(aq) = -3.4$, e.g., the values $\Delta S_h^\circ = -87.8$ and $\Delta G_h^\circ = -575$ are obtained for Be^{2+}. The enthalpy of hydration is an average of two calculation methods.

a) From the standard enthalpies ΔH_f° of the formation (from solid Be, see below) of $Be^{2+}(g)$ and $Be^{2+}(aq)$ ($\Delta H_f^\circ(aq) - \Delta H_f^\circ(g)$ is the so-called conventional hydration enthalpy) and that of the hydrated proton in infinite solution $\Delta H_f^\circ(H^+aq)$, calculated as -102.5 kcal/mol (with the assumption of equal hydration enthalpy for Cs^+ and I^- ions): $\Delta H_h^\circ(Be^{2+}) = \Delta H_h^\circ$(convent.) $- (2 \times 102.5)$.

b) From the lattice energy and the integral heat of solution of several Be^{2+} salts, Vasil'ev et al. [1] (the thermodynamic standard quantities are taken from Rossini et al. [2] and Latimer [3]).

$\Delta H_h^\circ = -608$, $\Delta G_h^\circ = -582.3$, and $\Delta S_h^\circ = -86$ are calculated with the assumption of zero energy for the hydration of neutral atoms by method (a) with a calculated value ΔG_h°(convent.) $= -790.6$ kcal/mol (with values from [2]) and $\Delta G_f^\circ(H^+aq) = -104.8$ kcal/mol, and $S_{H^+}^\circ(aq) = -3.3$ cal · mol^{-1} · K^{-1}, Noyes [4], see also Noyes [5]. The conventional value ΔG_h°(convent.) for Be^{2+} was calculated as -790.14 (from data in [2]) by Plambeck [6]. $\Delta H_h^\circ = -601$ and $\Delta G_h^\circ = -571$ are obtained by Mikhailov, Drakin [7] from conventional data, and $\Delta H_h^\circ = -591$ and $\Delta G_h^\circ = -565$ by Latimer [8]. From experimental electrochemical measurements $\Delta G_h^\circ = -584$ and -580 are calculated, De Bethune [9], see also Jakuszewski [10], and Matsuda [11]. A calculation of the free enthalpy of hydration by the theoretical Born equation $\Delta G = -n^2/2r\,(1 - 1/\varepsilon)$, where n is the charge and r is the radius of Be^{2+}, and ε is the dielectric constant, gives $\Delta G_h^\circ = -560$ using an effective ion radius r = 1.15 Å for Be^{2+}. The value is most consistent with $S_{H^+}^\circ(aq) = -2.1$, giving $\Delta S_h^\circ = -87$ for the hydration of Be^{2+}, Latimer [8]. Using r = 0.62 Å for Be^{2+} and $S_{H^+}^\circ(aq) = -2.1$ the theoretical calculation gave $\Delta S_h^\circ = -104.5$ for Be^{2+}, Drakin, Mikhailov [12], see also Krestov [13] for the calculation of ΔS_h° and Krestov [14] for the change of the heat capacity during hydration.

Thermodynamic data for the formation of hydrated Be^{2+} from solid Be in hypothetical ideal solution of unit molality at 25 °C: $\Delta H_f^\circ = -91.5$ kcal/mol, and $\Delta G_f^\circ = -90.75$ kcal/mol. For $Be(s) \rightarrow Be^{2+}(g)$: $\Delta H_f^\circ = +715.398$ kcal/mol, Parker et al. [15].

References:

[1] V.P. Vasil'ev, E.K. Zolotarev, A.F. Kapustinskii, et al. (Zh. Fiz. Khim. **34** [1960] 1763/7; Russ. J. Phys. Chem. **34** [1960] 840/2).

[2] F.D. Rossini, D.D. Wagman, W.H. Evans, S. Levine, I. Jaffe (NBS-C-500 [1952] 368).
[3] W.M. Latimer (The Oxidation States of the Elements and Their Potentials in Aqueous Solutions, 2nd Ed. Prentice-Hall, Englewood Cliffs, N.J., 1952, pp. 313/5).
[4] R.M. Noyes (J. Am. Chem. Soc. **84** [1962] 512/22).
[5] R.M. Noyes (J. Am. Chem. Soc. **86** [1964] 971/9).
[6] J.A. Plambeck (Can. J. Chem. **47** [1969] 1401/10).
[7] V.A. Mikhailov, S.I. Drakin (Zh. Fiz. Khim. **29** [1955] 2133/44; C.A. **1956** 13570).
[8] W.M. Latimer (J. Chem. Phys. **23** [1955] 90/2).
[9] A.J. De Bethune (J. Chem. Phys. **31** [1959] 847/8).
[10] B. Jakuszewski (J. Chem. Phys. **31** [1959] 846/7).

[11] A. Matsuda (J. Res. Inst. Catal. Hokhaido Univ. **27** No. 3 [1979] 167/9; C.A. **93** [1980] No. 32592).
[12] S.I. Drakin, V.A. Mikhailov (Zh. Fiz. Khim. **33** [1959] 1544/50; Russ. J. Phys. Chem. **33** [1959] 45/8).
[13] G.A. Krestov (Theor. Eksperim. Khim. **1** [1965] 479/87; Theor. Exptl. Chem. [USSR] **1** [1965] 313/8).
[14] G.A. Krestov (Izv. Vysshikh Uchebn. Zavedenii Khim. Khim. Tekhnol. **6** [1963] 228/32 from C.A. **59** [1963] 12247).
[15] V.P. Parker, D.D. Wagman, W.H. Evans (NBS-TN-270-6 [1971] 1).

7.1.3 Water Exchange and Substitution

The rate of water exchange between hydrated Be^{2+} ions and solvent H_2O at 25 °C has been measured by observation of the ^{17}O NMR signal of the H_2O in the first coordination sphere. Values for the rate constant k for one H_2O molecule as well as for the enthalpy ΔH^{*} and entropy ΔS^{*} of activation for the H_2O exchange: $k = 2.1 \times 10^3\ s^{-1}$, $\Delta H^{*} = 8.3$ kcal/mol, and $\Delta S^{*} = -15\ cal \cdot mol^{-1} \cdot K^{-1}$. The solution contained 0.4 g $BeCl_2$, 0.913 g H_2O (46.5% $H^{18}O$, 20% $H_2{}^{17}O$), and 0.2 g $Mn(ClO_4)_2 \cdot 2H_2O$, Neely [1]. Also by ^{17}O NMR spectra the lifetime of H_2O molecules in the first coordination sphere on Be^{2+} at room temperature was found to be about 3×10^{-4} s (corresponding to $k \approx 3.3 \times 10^3\ s^{-1}$) or somewhat greater. The solution contained (in 10^{-3} mol): 4.868 $BeCl_2$, 86.47 H_2O, 1.72 $HClO_4$, 0.520 $Co(ClO_4)_2$, Connick, Fiat [2].

The value obtained by [2] is only an approximate estimation. Combination of two methods, ^{17}O NMR and ^{1}H spin-echo NMR permitted study of the individual contribution of water exchange and protolysis, $[Be(H_2O)_4]^{2+} + H_2O \rightarrow [Be(H_2O)_3OH]^{+} + H_3O^{+}$ (see p. 295 for the hydrolysis reactions or deprotonation, respectively). For the exchange of 1 H_2O molecule $k = (1.8 \pm 0.1) \times 10^3\ s^{-1}$, $\Delta H^{*} = (41.5 \pm 2)$ kJ/mol, and $\Delta S^{*} = -(44 \pm 15)\ J \cdot mol^{-1} \cdot K^{-1}$ were obtained (3 molal $Be(NO_3)_2$ solutions at pH < 1 containing Eu^{3+} as shift reagent for the bulk ^{17}O water line). The protolytic dissociation of the $Be(H_2O)_4^{2+}$ ion at 25 °C is faster than the H_2O exchange: $k = (8 \pm 2) \times 10^4\ s^{-1}$, $\Delta H^{*} = (31.2 \pm 4)$ kJ/mol, and $\Delta S^{*} = -(45 \pm 15)\ J \cdot mol^{-1} \cdot K^{-1}$ were obtained for Be concentrations up to 0.2 molal, Frahm, Füldner [3]. By an isotopic dilution method (using $H_2^{18}O$), $k = 2 \times 10^2\ s^{-1}$ was found in aqueous acid solutions at room temperature, Anderson [4]. The rate constant of H_2O substitution in the first coordination sphere by sulfate ions, $Be^{2+}(H_2O)_x \rightarrow Be^{2+}(H_2O)_{x-1}SO_4$, was found as $k = 1 \times 10^2\ s^{-1}$ at 25 °C (ionic strength 0.1 M) by sound absorption measurements (relaxation technique), Eigen [5], Eigen, Tamm [6]. From the rate constant given by Eigen [5, 6], the free enthalpy of activation of H_2O exchange was calculated as $\Delta G^{*} = 14.7$ kcal/mol. Relating this value to the binding energy of the H_2O molecules (calculated for the first and second solvation sphere by quantum chemical MO calculations) leads to strong support for an exchange mechanism involving the second hydration shell and a higher coordinated transition state, Rode et al. [7].

A comparison of the rates of solvent exchange in solvated Be ions with those of H_2O as solvent gives the following order with decreasing rate (at 25 °C): $H_2O > DMF > DMSO > HMPT > TMPA$ (DMF = dimethylformamide, DMSO = dimethyl sulfoxide, HMPT = hexamethylphosphorotriamide, TMPA = trimethylphosphate), Füldner et al. [8].

References:

[1] J.W. Neely (UCRL-20580 [1971] 1/72, 16; C.A. **76** [1972] No. 132258).
[2] R.E. Connick, D.N. Fiat (J. Chem. Phys. **39** [1963] 1349/51).
[3] J. Frahm, H.H. Füldner (Ber. Bunsenges. Physik. Chem. **84** [1980] 173/6).
[4] L.B. Anderson (Diss. Univ. Ann. Arbor 1962 from Diss. Abstr. **22** [1962] 3002).
[5] M. Eigen (Pure Appl. Chem. **6** [1963] 97/115, 103; Proc. 6th Intern. Conf. Coord. Chem., Detroit, Mich., 1961, pp. 371/7).
[6] M. Eigen, K. Tamm (Z. Elektrochem. **66** [1962] 107/21, 116).
[7] B.M. Rode, G.J. Reibnegger, S. Fujiwara (J. Chem. Soc. Faraday Trans. II **76** [1980] 1268/74).
[8] H.H. Füldner, D.H. Devia, H. Strehlow (Ber. Bunsenges. Physik. Chem. **82** [1978] 499/503).

7.2 Hydrolysis of the $[Be(H_2O)_4]^{2+}$ Ion

Weakly Acid Medium

In beryllium salt solutions, which show acid reaction, the main Be species is $[Be(H_2O)_4]^{2+}$. With increase of pH (addition of NaOH or $NaHCO_3$ solution), H_2O is substituted by OH, but the solutions remain clear up to an [OH]/[Be] ratio of $Z \approx 1$, see, e.g., Kakihana, Sillén [1], or only slightly higher, see, e.g., Akitt, Duncan [2], Mesmer, Baes [3]. The precipitation begins at $Z > 1$ and is complete at $Z \approx 1.8$ to 2.0 depending on the anion [1]. At $Z < 1$ a variety of soluble aquo-hydroxo polycations may be formed with maintenance of the tetrahedral environment of Be. The decrease of the average hydration number with increase of Z is shown in **Fig.** 7-**1** [2]. The overall (acid) hydrolysis reaction is

$$m\,Be^{2+}aq + n\,H_2O = [Be_m(OH)_n\,aq]^{(2m-n)^+} + n\,H^+ \qquad (1)$$

The following cationic Be species, formulated as $[Be_m(OH)_n]^{(2m-n)^+}$ without consideration of hydrated water, have been reported in the literature (for the older literature see, e.g., reviews in [1, 2]):

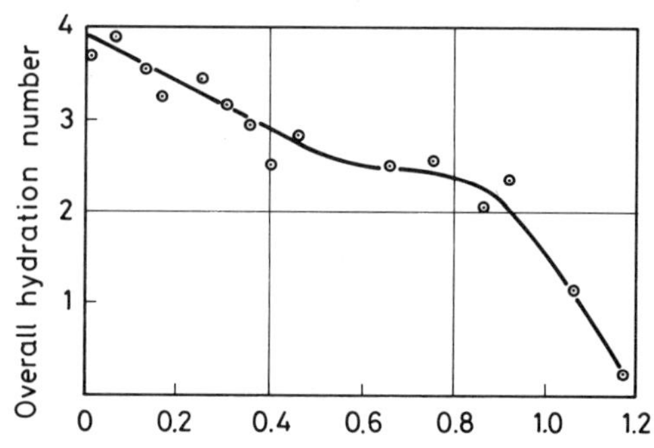

Fig. 7-1

Overall hydration number of Be in hydrolized $BeCl_2$ solutions as a function of degree of hydrolysis Z.

$[Be_2(OH)^{3+}$ [1 to 19, 41]
$[Be_2(OH)_2]^{2+}$ [5, 20, 24]
$[Be_3(OH)_4]^{2+}$ [4, 5, 12, 13]
$[Be_6(OH)_8]^{4+}$ [4, 12, 18]
$[Be(OH)_2]^0$ [1, 6, 7, 13, 14, 16, 17, 19, 32, 40, 41]
$[Be_3(OH)_3]^{3+}$ [1 to 12, 14 to 17, 19 to 23, 29, 41]
$Be(OH)^+$ [13, 14, 18, 31, 40]
$[Be_5(OH)_7]^{3+}$ [3]
$[Be_6(OH)_9]^{3+}$ [4, 12]

The methods of investigation are potentiometric titration at 25 °C [1, 4, 6 to 8, 11 to 15, 17 to 21, 23, 41], at 60 °C [5], or at 0, 25, and 60 °C [3], polarography [16], ^1H and ^9Be NMR spectroscopy [2], Raman spectroscopy [9], ion exchange [22, 28], coagulation methods [29], and distribution between two layers [32]. Most of the authors agree that at early stages of hydrolysis at room temperature at some extent the dimer species $[Be_2(OH)]^{3+}$ appears besides $[Be(H_2O)_4]^{2+}$ and that the trimer species $[Be_3(OH_3)]^{3+}$ becomes predominant (besides $Be(OH)_2$) as further alkali is added, see, e.g., [1, 2]. Near the precipitation point ($Z>1$) [4] or in hot solutions [5] (where the reaction is faster and the solubility of the species is higher) further species may occur [2].

By pH measurements during the first deciseconds after mixing beryllium salt solution and NaOH at 20 °C (streaming chamber) it was found that the equilibrium $Be^{2+}+H_2O \rightleftharpoons Be(OH)^+ + H^+$ is reached within 5×10^{-3} s (in acid free salt solution) and that at $I=0.1$, initially, the main species (at $Z=1$) is $Be(OH)^+$. The deprotonation of two protons produces in the first instance an unstable $Be(OH)_2^*$ species which transforms into the stable isomer $Be(OH)_2$ (solvatation isomerism) in a first order reaction with a half-life at 7 ms. This isomerization causes almost complete disappearance of $BeOH^+$ from the equilibrium $Be^{2+} \rightleftharpoons BeOH^+ \rightleftharpoons Be(OH)_2$. The formation of the polynuclear species $Be_3(OH)_3^{3+}$ needs about 30 s to go to completion, Schwarzenbach, Wenger [40].

It was assumed that the tetrahedrally coordinated Be atoms in the polymeric species are linked by sharing OH, e.g., $(H_2O)_3Be(OH)Be(OH_2)_3$ in Be_2OH^{3+}. For $Be_3(OH)_3^{3+}$ a six-membered trinuclear ring structure

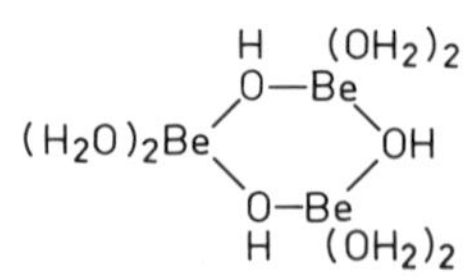

was proposed by Kakihana, Sillén [1] and, in fact, a planar ring structure was confirmed by Raman spectroscopy [9].

For the equilibrium $[Be_m(OH)_n{}^{(2m-n)^+}]\cdot[H^+]^n/[Be^{2+}]^m$ (according to equation (1), p. 295), the following overall stability (formation) constants $\beta_{n,m}$ of the species $[Be_m(OH)_n]^{(2m-n)^+}$ at 25 °C and ionic strength I have been reported:

I	$[Be_2(OH)]^{3+}$ $-\log \beta_{1,2}$	$[Be_3(OH_3)]^{3+}$ $-\log \beta_{3,3}$	$[Be_3(OH)_4]^{2+}$ $-\log \beta_{4,3}$	$[Be(OH_2)]^0$ $-\log \beta_{2,1}$	Ref.
3 M $NaClO_4$	3.24 (3.28) [a)]	8.66 (9.40) [a)]	—	10.9 (11.89) [a)]	[1, 11] [a)]
3 M $NaClO_4$	3.22	8.66	—	—	[8, 27]
3 M $NaClO_4$	2.9	7.7	13.22	—	[5] [b)]
1 M $NaClO_4$	3.51	—	—	—	[13]
0.5 M $NaClO_4$	3.20	8.81	—	11.0	[6, 7]
3 M $LiClO_4$	3.04	8.671	—	—	[18]
3 M $LiClO_4$	3.27	8.74	—	—	[14]
0.1 M $LiClO_4$	3.32	8.807	—	11.35	[18]

I	$[Be_2(OH)]^{3+}$ $-\log \beta_{1,2}$	$[Be_3(OH_3)]^{3+}$ $-\log \beta_{3,3}$	$[Be_3(OH)_4]^{2+}$ $-\log \beta_{4,3}$	$[Be(OH_2)]^0$ $-\log \beta_{2,1}$	Ref.
3 M KCl	3.18	8.88	—	—	[15]
2 M KCl	3.65	8.03	15.6	—	[4]
1 M NaCl	3.43	8.91	—	—	[3]
2 M KNO_3	3.28	8.90	16	—	[12]
1 M KNO_3	3.22	8.87	—	11.26	[19]
0.1 M KNO_3	2.955	8.804	—	11.320	[41]
0.1 M K_2SO_4	2.66	7.44	14	—	[4]
dilut. solution	—	—	—	13.65	[32]

a) Heavy water. — b) 60 °C.

Further values:

$-\log \beta_{1,1} \approx >6.3$ (0.1 M $LiClO_4$) [18], 6.5 (1 M $NaClO_4$) [13], 5.4 (3 M $LiClO_4$) [14], 5.7 at 20 °C (0.1 M $NaClO_4$) [40], and a value of 6.4 is evolved from the data of [1] (see the table above) for I=3 M $NaClO_4$ [40];
$-\log \beta_{2,1} \approx 11.2$ at 20 °C (0.1 M $NaClO_4$) [40];
$-\log \beta_{2,2} = 6.25$ (3 M $NaClO_4$) at 60 °C [5], also see Prytz [24];
$-\log \beta_{3,3} = 8.9 \pm 0.1$ at 25 °C and ionic strength zero calculated from the solubility of $Be(OH)_2$, Schindler, Garett [39]; $-\log \beta_{7,5} = 25.33$ (1 M NaCl) [3];
$-\log \beta_{8,6} = 28.1$ (2 M KCl), 23.4 (0.1 M K_2SO_4) [4], 27.5 (2 M KNO_3) [12], and 27.337 (3 M $LiClO_4$) [18];
$-\log \beta_{9,6} = 34.5$ (2 M KNO_3) [12], 29.2 (0.1 M K_2SO_4) [4].

For a study of the begin of polymerization of Be in very diluted solutions by ion exchange, see Feldman, Havill [28].

For the stability constants of the cationic species in mixtures of H_2O with methanol, ethanol, acetone, or dioxane, see [18], see also [25, 26].

Basic Medium

In basic solutions prepared by dissolution of freshly precipitated gelatinous $Be(OH)_2$ in NaOH, anionic beryllate species exist. Studies of solutions with Na/Be mole ratios of 4 to 20 (by the Tyndall effect) suggest that at low Na/Be ratios polymeric beryllate anions are present which range from ionic to colloidal dimensions and that at high Na/Be ratios probably simple beryllate anions are formed. With increasing Na/Be ratio, from 4 to 16, the amount of $Be(OH)_2$, which is precipitated after heating the solutions for 2 h at 95 °C, decreased from nearly 100% to zero for an initial Be concentration of 3.125 g BeO/L (see figure in the papers), Everest et al. [34], Everest [35]. The simple anions $Be(OH)_4^{2-}$ and $Be(OH)_3^-$ occur in the solid compounds $Na_2Be(OH)_4$ and $NaBe(OH)_3$ prepared by Scholder et al. [36]. The stability constant $\log \beta_{3,1} = -24.11$ for the formation $Be^{2+} + 3H_2O \rightleftharpoons Be(OH)_3^- + 3H^+$ (or $[Be(H_2O)_4]^{2+} \rightleftharpoons 3H^+ + [Be(H_2O)(OH)_3]^-$, respectively) is obtained in diluted solution at 25 °C (distribution between two layers) [32].

In aqueous KOH solutions (1 to 9 M) the species $[BeH_2O(OH)_3]^-$ are present according to densitimetric studies of Jedinakova [37]. From solubility studies of the dissolution of BeO in molten KOH-NaOH mixtures it is concluded that the major species may be the $Be(OH)_4^{2-}$ ion, Shying et al. [38].

Thermodynamic Data

Thermodynamic data for the hydrolysis reaction (1), p. 295, at 25 °C are obtained calorimetrically by Carell, Olin [30], Ishiguro, Ohtaki [10], and from the temperature dependence of the stability constants by [3], based on measurements in 1 M NaCl [3], 3 M $LiClO_4$ [10], or 3 M $NaClO_4$ [30] (see the table on p. 296 for the stability constants):

species	ΔG° in kcal/mol		ΔH° in kcal/mol			ΔS° in $cal \cdot mol^{-1} \cdot K^{-1}$		
$Be_2(OH)^{3+}$	4.5	4.4	5.0	4.4	4.4	1.4	0.2	1.04
$Be_3(OH)_3^{3+}$	12.2	11.8	16.0	15.2	14.8	15.3	11.3	9.8
$Be_5(OH)_7^{3+}$	34.8	—	45.3	—	—	35.2	—	—
Ref.	[3]	[30]	[3]	[30]	[10]	[3]	[30]	[10]

According to the tabulated data, there is an increase of the thermodynamic values with increasing polymerization [3]. The values in mixtures of H_2O with 20 mol% dioxane are lower than those in H_2O [10].

For a review of the thermodynamic parameters of Be species in aqueous solutions, see Soboleva et al. [33].

References:

[1] H. Kakihana, L.G. Sillén (Acta Chem. Scand. **10** [1956] 985/1005).
[2] J.W. Akitt, R.H. Duncan (J. Chem. Soc. Faraday Trans. I **76** [1980] 2212/20).
[3] R.E. Mesmer, C.F. Baes (Inorg. Chem. **6** [1967] 1951/60).
[4] E. Lanza (Rev. Chim. Minerale **6** [1969] 653/85).
[5] L. Ciavatta, M. Grimaldi (Gazz. Chim. Ital. **103** [1973] 731/46).
[6] F. Bertin, G. Thomas, J.C. Merlin (Compt. Rend. **260** [1965] 1670/3).
[7] F. Bertin, G. Thomas, J.C. Merlin (Bull. Soc. Chim. France **1967** 2393/404).
[8] B. Carell, Å. Olin (Acta Chem. Scand. **15** [1961] 1875/84).
[9] S. Ishiguro, M. Maeda, S. Ono, H. Kakihana (Denki Kagaku **46** [1978] 553/9).
[10] S. Ishiguro, H. Ohtaki (Bull. Chem. Soc. Japan **52** [1979] 3198/203).

[11] H. Kakihana, M. Maeda (Bull. Chem. Soc. Japan **43** [1970] 109/13).
[12] E. Lanza, G. Carpéni (Electrochim. Acta **13** [1968] 519/33).
[13] G. Mattock (J. Am. Chem. Soc. **76** [1954] 4835/8).
[14] H. Ohtaki, H. Kato (Inorg. Chem. **6** [1967] 1935/7).
[15] M.R. Paris, C. Gregoire (Anal. Chim. Acta **42** [1968] 431/7).
[16] P.J. Shirvington, T.M. Florence, A.J. Harle (Australian J. Chem. **17** [1964] 1072/8).
[17] L.G. Sillén (Quart. Rev. [London] **13** [1959] 146/68).
[18] H. Tsukuda, T. Kawai, M. Maeda, H. Ohtaki (Bull. Chem. Soc. Japan **48** [1975] 691/5).
[19] A. Vanni, M.C. Gennaro, G. Ostacoli (J. Inorg. Nucl. Chem. **37** [1975] 1443/51).
[20] L.P. Adamovich, G.S. Shupenko (Ukr. Khim. Zh. **25** [1959] 155/61).

[21] M. Bartusek, J. Zelinka (Collection Czech. Chem. Commun. **32** [1967] 992/1005).
[22] M.K. Cooper, D.E.J. Garman, D.W. Yaniuk (J. Chem. Soc. Dalton Trans. **1974** 1282/5).
[23] H. Kakihana, M. Maeda (Bull. Chem. Soc. Japan **42** [1969] 1458/60).
[24] M. Prytz (Z. Anorg. Allgem. Chem. **231** [1937] 238/48, **197** [1931] 103/12, **180** [1929] 355/69).
[25] H. Ohtaki, H. Tsukuda (Proc. 16th Intern. Conf. Coord. Chem., Dublin 1974, Abstr. No. R 73, pp. 1/3; C.A. **85** [1976] No. 37829).
[26] H. Ohtaki (Inorg. Chem. **6** [1967] 808/13).

[27] S. Hietanen, L.G. Sillén (Acta Chem. Scand. **18** [1964] 1015/6).
[28] I. Feldman, J.R. Havill (J. Am. Chem. Soc. **74** [1952] 2337/40).
[29] E. Matijević (J. Colloid. Sci. **20** [1965] 322/9).
[30] B. Carell, Å. Olin (Acta Chem. Scand. **16** [1962] 2357/62).

[31] A.I. Zhukov, G.P. Baranov, P.V. Plyasunov (Zh. Neorgan. Khim. **7** [1962] 1452/7; Russ. J. Inorg. Chem. **7** [1962] 745/8).
[32] R.W. Green, P.W. Alexander (Australian J. Chem. **18** [1965] 651/8).
[33] G.L. Soboleva, I.A. Tugarinov, V.F. Kalinina (Geokhimiya **1977** 1013/24; C.A. **87** [1977] No. 91576).
[34] D.A. Everest, R.A. Mercer, R.P. Miller, G.L. Milward (J. Inorg. Nucl. Chem. **24** [1962] 525/34).
[35] D.A. Everest (The Chemistry of Beryllium, Elsevier, London 1964, pp. 7/22).
[36] R. Scholder, H. Hund, H. Schwarz (Z. Anorg. Allgem. Chem. **361** [1968] 284/95).
[37] V. Jedinakova (Sb. Vys. Sk. Chem. Technol. Praze Anorg. Chem. Technol. B **18** [1974] 145/56; C.A. **83** [1975] No. 66397).
[38] M.E. Shying, J. Aggett, R.B. Temple (Australian J. Chem. **18** [1965] 1719/29).
[39] P. Schindler, A.B. Garett (Helv. Chim. Acta **43** [1960] 2176/8).
[40] G. Schwarzenbach, H. Wenger (Helv. Chim. Acta **52** [1969] 644/65).

[41] P.L. Brown, J. Ellis, R.N. Sylva (J. Chem. Soc. Dalton Trans. **1983** 2001/4).

7.3 Precipitation

As most beryllium compounds are soluble in H_2O there are only a few possibilities to precipitate Be^{2+} from solutions. The precipitation as $Be(OH)_2$, e.g., with NH_3, or as $Be(NH_4)PO_4$ with $(NH_4)_2HPO_4$ are the usual methods for quantitative precipitation. The direct precipitation with NaOH, $NaHCO_3$, or Na_2CO_3 is not suitable because Be forms soluble beryllate and carbonate complexes, but by boiling these complex solutions, insoluble crystalline β-$Be(OH)_2$ is formed, see the special literature of analytical chemistry and "Production" (Chapter 1).

From soluble carbonate complexes, Be can be quantitatively precipitated with $Co(NH_3)_6Cl_3$ as yellow $[Co(NH_3)_6][(H_2O)_2Be_2(Co_3)_2(OH)_3] \cdot 3H_2O$, Pirtea, Mihail [1]. Be can further be precipitated quantitatively by substituted arsinic acid, R_2AsO_2H, with R = ethyl, n-propyl, or n-butyl, Pietsch [2], or with R = phenyl, Pietsch [3], and with butylarsonic acid, $C_4H_9AsO_3H_2$, Pietsch [4]. The precipitation with carboxyphenyl arsinic acids (o-, m-, or p-) is not quantitative, Pietsch [5]. All precipitates are white [2 to 5].

References:

[1] T.I. Pirtea, G. Mihail (Z. Anal. Chem. **159** [1958] 205/8).
[2] R. Pietsch (Mikrochim. Acta **1962** 37/47).
[3] R. Pietsch (Mikrochim. Acta **1957** 161/6).
[4] R. Pietsch (Mikrochim. Acta **1960** 539/52).
[5] R. Pietsch (Mikrochim. Acta **1959** 861/9).

8 Toxicology of Beryllium

A. Seidel
Institut für Genetik und für Toxikologie von Spaltstoffen
Kernforschungszentrum Karlsruhe
Karlsruhe, Federal Republic of Germany

The practical importance of acute and chronic beryllium induced diseases in occupationally exposed persons and for the general public has decreased during the last three decades due to improved industrial hygiene standards. Nevertheless these diseases are very interesting for the experimental toxicologist as well as for the pulmologist. This is due to the fact that studies on the pathogenesis of chronic non-malign beryllium-induced lung disease became of increasing importance for understanding the etiology of other, nonoccupational chronic lung diseases (see Section 8.3.3, p. 308). The various aspects of beryllium toxicology have been summarized by several authors [4, 33, 34, 66, 70]. In view of the scope of this handbook and considering the fact, that especially acute intoxications with Be are presently very rare, the toxicological animal experiments with Be will not be discussed in this volume (see [4]). The paragraphs on chronic beryllium-induced lung disease and on carcinogenesis will mainly deal with the broad informations on these subjects for humans which are already available directly.

8.1 Historical Survey

The first reports on effects of Be on animals date back to the end of the last century, when comparative pharmacological and toxicological studies with the element were published [1, 2]. Between 1930 and 1960 a large series of animal studies with Be was performed, dealing with acute and chronic non-malign effects, effects on enzymes and carcinogenesis. Due to the increasing medical interest in Be, a great number of these studies were focussed on the beryllium induced changes in the lungs. A review on these papers was given by Vorwald et al. [3]. A general review about these experimental studies, including experimental carcinogenesis, is also contained in several monographs, e.g. [4, 5]. Reports on pulmonary and dermatological effects of Be on humans started in 1933 [6], a paper which was followed by a large number of publications describing effects attributed to Be in occupationally exposed persons but also in those living in the neighborhood of beryllium plants (references quoted in [7]). It became evident that Be can cause acute and chronic non-malign effects and the term "Berylliosis" was born. In 1952 the "Beryllium Case Registry" was established at the Massachusetts General Hospital, being in 1978 transferred to the National Institute for Occupational Safety and Health in Cincinnati. More than 900 cases of beryllium disease have been collected as shown in a recent report from Eisenbud, Lisson [8]. While there is no controversy concerning these cases and beryllium induced carcinogenesis in animals, there is still some debate about possible carcinogenic effects of Be in man. Bearing all this information in mind it is, on the other hand very important to note the drastic reduction in the number of beryllium induced diseases since 1950, when limits of exposure were introduced [8, 9]. For example, only 15 cases of acute chemical pneumonitis due to accidental overexposure have occurred in USA since 1950, and the number of chronic berylliosis cases has also become very low. No more cases concerning the general public have occurred. Be has become another example of an element which can safely be handled provided that effective measures of control exist.

8.2 Biological Behavior of Be in Mammals

Ingested compounds of Be are poorly absorbed from the gastrointestinal tract, regardless of the chemical form. For example, only 0.2% of an oral $^{7}BeCl_2$ dose was absorbed by rats after administration [10]. The absorption from intramuscular, intraperitoneal and subcutaneous injection sites is much better, at least as far as relatively soluble compounds are concerned [10]. The injected mass will have a decisive influence since higher mass doses will impair the solubilization. The resorption from the lung varies from almost complete mobilization within a few days in the case of the citrate complex [11] to only negligable transfer to other organs in the case of high-fired BeO [12]. Beryllium is a bone-seeking radionuclide, provided that it has gained access to and is transported in the blood stream in essentially monomeric form. This is the case when it is injected intravenously as a citrate [10, 13], or when it has been resorbed from an injection site in the form of low-molecular weight complexes [14]. Results from intravenous studies are presented in Table 8/1 as an example. It can be assumed that under these conditions between 30 and 50% of the injected Be dose is deposited in the skeleton. Of the rest, a few per cent is found in the liver. Within bone, Be is a so-called surface seeker, incorporation into the calcified portions of bone being unimportant [15]. The biological half-life of Be in bone is very long (this is common to many bone-seeking radionuclides). Any condition favoring the formation of Be colloids in the blood stream leads to a drastic change in its organ distribution (Table 8/2). This may be induced by addition of a nonradioactive carrier or by raising the pH of the injection solution. Under these circumstances, the typical "colloid-type" distribution is found, with a low skeleton and high liver and spleen burden. It is also common to many radiocolloids that the element is later transferred from the liver to the skeleton (Table 8/2, p. 302). Also the intraorgan distribution is different from that after injection in monomeric form since the cells of the reticuloendothelial system, capable of phagocytosis of foreign

Table 8/1
Distribution of Intravenously Injected Beryllium Compounds within 24 h.
(The values represent the average per cent of the total recovered radioactivity per organ; average deviation in parentheses [4].)

material injected	excretion in %	bone + marrow[a)] in %	liver in %	spleen in %	number of rats
$^{7}BeCl_2$, carrier-free, pH 2	47 (±4)	43 (±6)	4 (±0.4)	0.1 (±0.1)	10
$^{7}BeCl_2$+0.15 μmol $^{9}BeCl_2$, pH 2	39 (±2)	53 (±8)	3 (±0.5)	0.05 (±0.05)	2
$^{7}BeCl_2$+1 μmol $^{9}BeCl_2$, pH 2	33 (±3)	37 (±2)	25 (±3)	1 (±0)	2
$^{7}BeCl_2$, carrier-free, pH 6	18 (±2)	17 (±4)	59 (±5)	1.7 (±0.7)	9
$^{7}BeCl_2$+1 μmol $^{9}BeCl_2$, pH 6[b)]	11 (±1)	13 (±0)	44 (±1)	6 (±2)	2
$^{7}BeCl_2$+0.15 μmol $^{9}BeCl_2$+ 3 μmol citrate, pH 6	35 (±1)	50 (±6)	2 (±1)	0.15 (±0.05)	2
$^{7}Be(OH)_2$+0.3 μmol $^{9}Be(OH)_2$[c)]	8 (±3)	15[a)] (±3)	61 (±8)	8 (±3)	5

a) Femoral marrow, counted separately, had minimal activity except following the injection of $Be(OH)_2$. In this case the activity corresponded to 7%/g tissue.

b) The acid solution was neutralized and injected immediately before any visible precipitation occurred.

c) Precipitated with NH_3, coagulated by heating, washed by high speed centrifugation, and suspended in saline.

References on pp. 313/5

Table 8/2
Redistribution and Excretion of Beryllium in Per Cent (for explanation see legend of Table 8/1) [4].

state of ^{7}Be injected	bone+marrow			liver			excretion			No. of rats	
	1d	21d	difference	1d	21d	difference	1d	21d	difference	1d	21d
carrier-free, pH 2	46	48	+2	4	0.4	−3.6	39	49	+11	4	4
$^{7}Be(OH)_2$ + 3 μmol $^{9}Be(HO)_2$	15	28	+13	61	23	−38	8	31	+23	5	5
carrier-free, pH 6	12	22 [a]	+12	66	36 [a]	−30	17	35 [a]	+18	4	3 [a]

[a] Animals sacrificed after 7 d.

particles, are the primary deposition site and not the bone surfaces or the hepatocytes. Monomerically injected Be is mainly excreted with the urine [14, 16]. The cumulative urinary excretion can amount to 50% of the administered dose, depending on the experimental conditions. When colloids have been formed the fecal excretion is relatively higher, probably due to increased biliary elimination.

The distribution of Be subsequent to inhalation or intratracheal injection strictly depends on the biological solubility of the administered compound in the lung. When citrate is involved, the element is resorbed within a few days, excreted via the urine and the fraction remaining in the body is predominantly deposited in the skeleton [11]. The other extreme is represented by high-fired BeO which is not transferred at all to other organs except for some translocation to the pulmonary lymph nodes [12]. The biological half-life of such types of Be in the lungs will be very long, probably >1 year in larger animal species. It will, in addition, also be influenced by the Be dose, since the element has deleterious effects on pulmonary macrophages, thus, reducing its own clearance rate from the lung. The differences in the in vivo solubility between high-fired and low-fired BeO and compounds like the sulfate, hydroxide or others are of importance for the type of toxic effects which may develop later on. After in vivo exposure BeO has been found in phagosomes and phagolysosomes of alveolar macrophages.

In order to explain the mechanisms of hepatotoxicity after experimental injection of Be, some studies on the cellular and subcellular uptake of the element in liver have been performed. When injected in colloidal state, whatever chemical forms may be involved, Be will be taken up by the macrophages of the liver (Kupffer cells) [17], the relative distribution between parenchymal and non-parenchymal liver cells depending on the size and properties of the colloids. By techniques of subcellular fractionation it has been shown that injected colloidal Be probably becomes stored ultimately in lysosomes [18], in analogy to ^{239}Pu injected into rats in colloidal form. The gross subcellular distribution as analyzed by differential centrifugation is strongly dependent on the injected Be dose. With increasing dose the fraction sedimenting with the nuclei drastically increases whereas the Be concentration in the final supernatant (cytosol) decreases. This is in agreement with studies with other hepatotoxic metals [19] (Table 8/3). The chemical form in which the element exists in the cytosol has not yet been determined. Ferritin, to which Be as well as many other metals is bound, may play some role [20]. In a biochemical binding and transport study with lung, it has been shown that Be electrophoretically migrates with the proteins [21]. Again ferritin, which has been identified as one of the binding sites for soluble ^{241}Am in

Table 8/3
Beryllium in Subcellular Fractions from Rat Liver after Various Doses.

dose of $BeSO_4$ in µmol/kg	specific activity of beryllium (in % of that of homogenate)				
	nuclear	heavy mitochondrial	light mitochondrial	microsomal	supernatant
0.083	44	110	260	70	160
0.83	141	98	295	93	125
1.8	98	—	—	—	—
8.3	98	—	315	—	—
28	280	200	310	57	35
83	340	175	204	63	—
110	410	—	—	—	20

Rats were injected with various amounts of $[^7Be]BeSO_4$ from 0.083 to 110 µmol/kg body weight and killed 24 h later. Specific activity is expressed as nanomoles of beryllium/mg of protein. Nuclei were isolated in different experiments by using the high-density sucrose method [18].

rat lung [22] is mentioned as one of the possible candidates. A very interesting and very important finding is the occurrence of Be in the nuclei cell fraction [18].

A true association with isolated and purified liver cell nuclei could be demonstrated. Also other studies using different techniques seem to confirm that Be has a true affinity to nuclei [23 to 27], for example those of fibroblasts, tissue, and alveolar macrophages.

Fig. 8-1

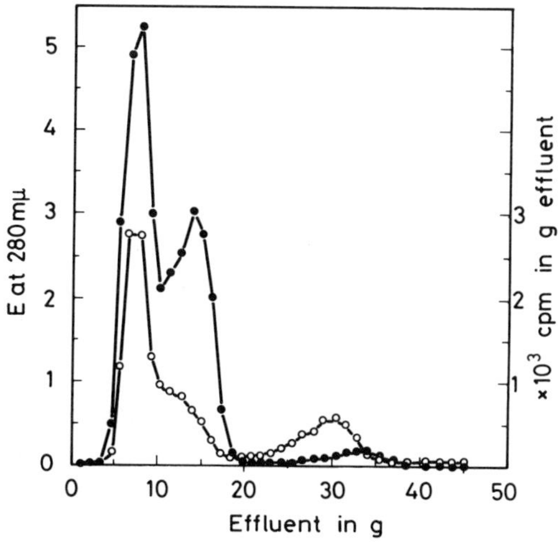

Radioactivity (o——o) and UV extinction ($E_{280\,m\mu}$; •——•) profiles of the effluent from a Sephadex column developed with 0.01 M Tris[1])-HCl buffer (pH 7.5) after application of 1.0 mL plasma from a rat 5 min after an i.v. injection of labeled $BeSO_4$ (0.03 mg Be/kg body weight). Gel bed volume: 34 mL. Flow rate: 13.3 g buffer/h [29].

[1]) Tris = tris(hydroxymethanyl)-amino methane.

References on pp. 313/5

The binding of Be within the nuclei is not yet fully understood, competition with Mg may play some role [27], references also in [28]. The tetravalent Pu is another example of a foreign metal for which nuclear binding could directly be demonstrated (by electron microscopic autoradiography with ^{241}Pu [22]).

The chemical binding of Be in blood plasma is of great importance for its organ distribution and excretion. When injected with higher mass doses phosphate colloids will be formed, the chemical composition of which is not yet known. The colloids are taken up by the cells of the reticuloendothelial system as already mentioned. According to gel filtration and ion exchange chromatography studies [29] these phosphates are associated with plasma proteins (**Fig.** 8-**1**, p. 303), probably globulins. These plasma proteins probably play some role in the formation of protective colloids. However, the results from different authors concerning binding to plasma proteins are not fully consistent [29 to 32]. Certainly the dose of Be injected into the blood stream is of central importance. In addition to the colloidal and/or protein bound Be fraction the element also exists in a low-molecular, diffusible form [13, 29] in the blood, for example as bound to citrate [30]. In this form it can be eliminated rapidly via the urine and exhibits its affinity to bone. Generally the fact that Be can inhibit various enzymes, especially alkaline phosphatase (a detailed summary is contained in [4]) indicates that Be binding to proteins can occur.

8.3 Beryllium Toxicity in Man

8.3.1 Acute Beryllium Disease

As already mentioned above the acute beryllium disease has become an extremely rare event, with only 15 cases reported between 1950 and 1983 and no acute case in USA since 1968 [8, 9]. The lung reactions are usually caused by soluble Be compounds but also by less soluble ones and also the oxide. When the threshold limits for Be concentration in air are maintained, no acute case of a beryllium disease occurs. The disease can begin within a few days after an exposure to high concentrations of Be, or more slowly during some months of exposure to lower concentrations. It is characterized by the signs of toxic, chemical irritation of the airways with all symptoms of respiratory distress (cough, breathlessness, anorexia, cyanosis). Also the skin can be afflicted. The roentgenological changes will be those of acute chemical pneumonitis: the microscopic appearance of the lung also corresponds to this reaction (edema in the alveolar space with occurrence of various types of cells). The disease may be fatal but complete recovery or transition into a chronic form is more probable. The pathogenetic mechanism is a direct toxic effect on the cells and their components. Since there are no typical symptoms for the acute beryllium disease the diagnosis essentially depends on the occupational anamnesis. Details on the acute beryllium lung disease can be found in an English [33] as well as in a Russian [34] monograph.

8.3.2 Chronic Beryllium-Induced Lung Disease: Epidemiological and Clinical Aspects

The chronic beryllium-induced disease of the lung may develop after an acute exposure but the primarily chronic form is more common. It is a chronic inflammatory reaction of the lung accompanied by granuloma formation and development of lung fibrosis. The disease may start about 4 years after first exposure but latency periods longer than 20 years have been reported.

An analysis of the epidemiology of the beryllium lung disease has recently been published for the USA by Eisenbud and Lisson [8]. It concerned a cohort of 622 cases, which became known to the Beryllium Case Registry and 45 cases from other sources. Occupational

Table 8/4

Sources of Cases of Occupational Berylliosis [8].

use of phosphors in fluorescent lamps, neon tubes, etc.		319
beryllium extraction		101
research laboratories		29
beryllium copper	foundries	22
	cold working	11
beryllium metal machining (other than research)		47
ceramics		11
unknown		17
total		557

exposure was responsible for 557 cases. Of the 65 cases which had occurred in members of the general public, 42 were due to air pollution and 23 to contacts with contaminated clothes. In Table 8/4 the sources of cases of occupational berylliosis are listed. The majority of cases stem from use of beryllium in fluorescent lamps, and beryllium extraction also caused a significant number of cases. The incidence of chronic beryllium lung-disease was never higher than a few percent of the exposed persons (Table 8/5). No clear dose-response relationship seems to exist. A quite unusual finding was that the incidence among residents near a beryllium manufacture plant was similar to that among workers of that plant, while the levels of exposure differed by about three orders of magnitude. This corresponds to the current view to be discussed in a later paragraph, that immunepathogenetic mechanisms are responsible for the development of a chronic beryllium lung disease.

The beryllium disease has become very rare due to the setting of hygienic standards [8, 9]. In **Fig. 8-2**, p. 306, the number of cases is plotted against time (years of first exposure).

Table 8/5
Incidence of Berylliosis and Estimated Level of Severity of Exposure [8].

	number exposed	number of cases	incidence in %	estimated level of exposure
residents living within 0.25 mile of the Lorain plant	500	5	1.0	1
fluorescent lamp manufacturing				
Massachusetts	15000	175	1.16	100
Ohio	8000	32	0.4	100
machine shop	225	11	4.9	500
beryllium-copper foundry	1000	13	1.3	500
beryllium extraction				
Lorain, Ohio	1700	22	1.3	1000
Painesville, Ohio	200	0	0.0	1000
Reading, Penn.	4000	51	1.3	1000

References on pp. 313/5

Fig. 8-2

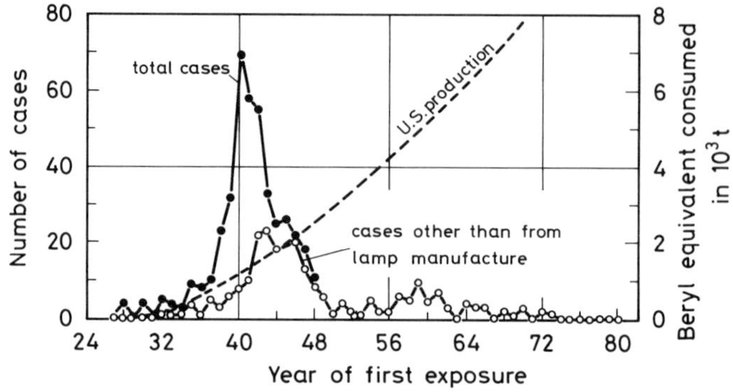

Occupational berylliosis by year of first exposure, 1927 to 1980 (and annual U.S. beryllium consumption in thousands of tons of beryl equivalent) [8].

There is a peak for those first exposed during the years 1939 and 1941 which correlates with the increase in fluorescent lamp manufacturing. No case is reported in this analysis for persons first exposed after 1973. Since no reliable data for incidence were available, an attempt was made to relate the number of beryllium cases to the annual beryllium production, the results are shown in **Fig.** 8-**3**. Though this procedure is certainly not fully satisfactory the sharp drop of the number "cases per ton of beryllium" may be taken as an indicator for an analogous reduction of the incidence. An analysis of the relationship between length of exposure and the cumulative percentage of cases (**Fig.** 8-**4**) showed that roughly two thirds of the cases were exposed for less than 60 months, a significant number of cases even for less than two years.

It is not yet clear whether the high number of cases in the fluorescent lamp manufacturing industry was due to the phosphor ($ZnMnBeSiO_4$) or to the presence of BeO. Cold working with beryllium-copper produced a comparatively low number of cases and is considered as a relatively "safe" form of handling Beryllium [9].

In the following paragraph the characteristic features of the chronic berylliosis are discussed briefly, the readers interested in more medical details may refer to the monographs [33 to 35]. The symptoms of beryllium-induced chronic lung disease develop gradually. Generally they are those of a granulomatous and/or fibrosing chronic lung disease. The first sign may be dyspnea on exertion, which leads the patients to consult a physician. Dry cough, also on exertion or during rest, combined with fatigue, weight loss or anorexia are the additional symptoms of a progressive disease. Upon physical inspection low grade fever, tachycardia, tachypnea, and cyanosis can be noted. The radiographic appearance of the lung corresponds to the type and degree of the interstitial infiltrations. There is, however, no abnormality which is typical or found exclusively in beryllium-induced chronic lung disease. The opacities have been described as "granular", "nodular", and "reticular", depending on their size or arrangement. The hilar lymph nodes may be enlarged, but not to the degree which is found in sarcoidosis, and only in those cases which also show parenchymal infiltrates. The results of the lung function tests will reflect the severity of the changes in the lung, but a detailed description goes far beyond the scope of this article. The microscopic inspection of the lung reveals pathohistological changes, the analysis of which has been important to understand the pathogenetic mechanisms leading to berylliosis. The alveolar walls ar infiltrated by histiocytes, lymphocytes, and plasma cells. Granulomas

Fig. 8-3

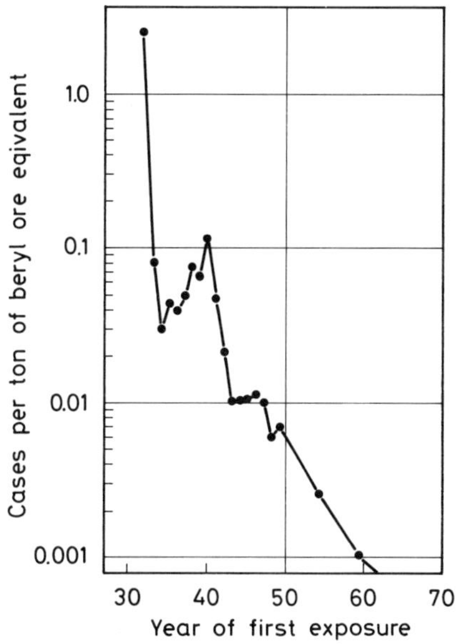

Cases of occupational berylliosis by year of first exposure per ton of beryl ore consumed (incidence rates after 1960 are less than 0.001 cases/ton and are not plotted) [8].

Fig. 8-4

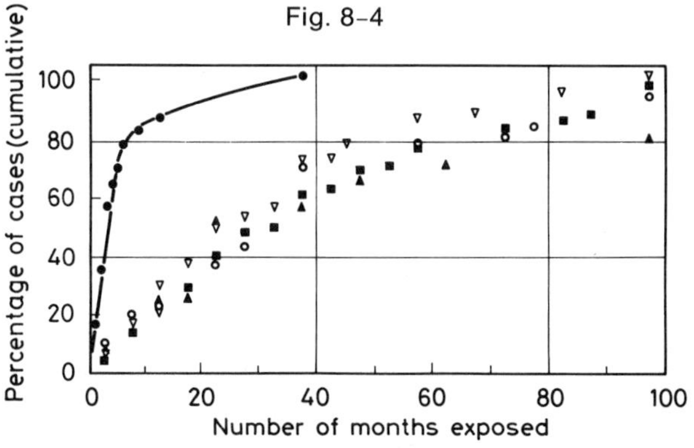

Cumulative chronic (occupational) berylliosis cases versus number of months of exposure for specified cohorts. Numbers in parentheses are numbers of employees in the cohort [8].

are formed in the interstitium, around the bronchi and blood vessels. These granulomas consist of the same classes of cells which also form the granulomas in certain other types of lung diseases to be discussed later. Around the granulomas fibroblasts are activated to produce collagen, a process which gradually leads to the fibrotic induration of the lung. Ultimately the alveolar walls and ducts may be destroyed. The mononuclear cell infiltration of the alveolar walls together with the formation of the non-caseating epitheloid-cell/macrophage granulomas is characteristic not only for the beryllium-induced lung disease but also for other lung diseases like sarcoidosis and hypersensitivity pneumonitis. It is probable

References on pp. 313/5

that these lung diseases are caused by immunepathogenetic mechanisms, which possibly are also important for the development of the berylliosis (see the following sections).

For the diagnosis of the disease, a combination of several criteria has to be fulfilled [33, 35]. Of primary importance for all further diagnostic considerations is an occupational exposure during working life. The roentgenologic signs of a granulomatous-fibrosing chronic lung disease combined with corresponding abnormal lung function tests will be present in various degrees. By lung biopsy or in post-mortem samples the histological changes described above will be found. In cases with chronic granulomatous lung diseases of unknown etiology the presence of beryllium in lung tissue gives very strong arguments to clearly establish the cause of the disease in a given case, but is not fully sufficient for an unequivocal diagnosis [35].

8.3.3 Immunological Aspects

Immunological tests have been used to differentiate between chronic berylliosis and other similar lung diseases, especially sarcoidosis. The first test used was the beryllium patch test [36]. For this test a soluble beryllium compound is applied to the skin, causing an erythematous reaction in sensitized individuals. This test is no longer recommended since an already existing beryllium disease may be aggravated or sensitizing of unaffected persons may occur, and in addition false negative reactions have to be considered.

Two other immunological in vitro tests take advantage of the specific responses of lymphocytes when they are brought in contact with an antigen. Specifically sensitized T-lymphocytes are stimulated to blastogenic transformation when they are incubated in vitro with this antigen. This type of reaction is part of the cell-mediated immune response. The blastogenic transformation can be determined and quantified morphologically or by measuring the incorporation of tritiated thymidin by the cells. It has first been shown by Hanifin et al. [37] that for subjects with hypersensitivity to beryllium the lymphocyte transformation test is positive while there is no reaction with lymphocytes from nonsensitive persons. These results were confirmed in subsequent studies by Deodhar et al. [38] and Preuss et al [39]. The lymphocytes from the majority of patients with chronic berylliosis gave a positive reaction, the degree of the transformation was correlated with the severity of the disease. Positive results in non-diseased workers, healthy controls or patients with other lung diseases were very rare. A persistently positive lymphocyte blast transformation test even under steroid therapy is presently considered as a strong diagnostic criterion for a chronic beryllium disease [33, 40] in cases with beryllium exposure in the past and the respective clinical, radiographic, and histological changes present. It should be kept in mind that a positive lymphocyte transformation test alone is not pathognomonic and that it may be reversible when the exposure is reduced or ceased [41]. Another test is the macrophage inhibition test which is based on the release of factors by sensitized T-lymphocytes in contact with the antigen, which impair the mobility of macrophages. This test is also positive for patients with chronic berylliosis without steroid therapy [42].

As to the prognosis of chronic beryllium-induced lung disease it is very difficult to predict the issue of the disease for a given case. The disease may end by respiratory or heart failure. However, the majority of afflicted persons will be able to lead a relatively normal life though exacerbations always have to be expected. Treatment consists mainly in corticosteroid medication, in addition, of course, to the usual treatment of respiratory diseases. Chelating agents to remove beryllium are of no value.

At present it is widely accepted that the pathogenetic mechanisms which are responsible for the chronic beryllium lung disease are related to the responses of the immune system

Fig. 8-5

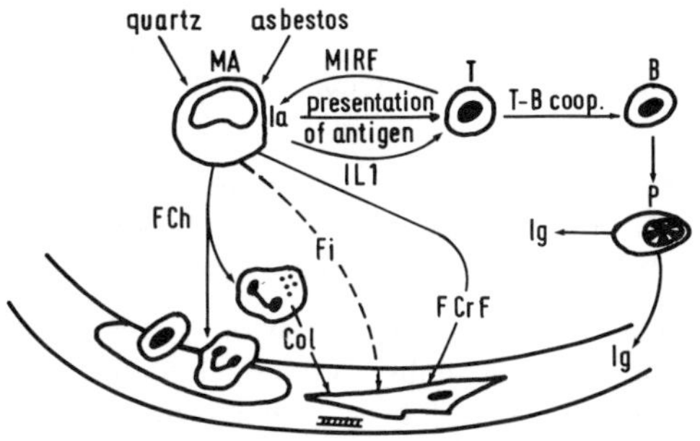

Schematic description of mechanisms possibly responsible for the development of pulmonary fibrosis after inhalation of mineral dusts (MA: macrophage; MIRF: macrophage Ia recruitment factor; FCh: chemotactic factor; IL1: interleukin-1; FCrF: fibroblast growth factor; Col: collagenase; T, B: T- or B-lymphocytes; T-B coop.: T-B lymphocyte cooperation; Ig: immunoglobulin; Fi: fibroblast stimulation factor. (The scheme is modified from (71) with kind permission of the authors.)

to beryllium [33, 35]. Beryllium is incorporated into alveolar macrophages (see Chapter 8.2) and remains present in the lung for a long time. By one of the hypotheses it is assumed that by mechanisms which are unknown so far a beryllium specific antigen is formed. Thereafter T-lymphocytes become sensitized and develop antigen-specific receptors. These cells react with the antigen in the lung, by this reaction so-called lymphokines are released. These mediators cause blood monocytes to invade the lung tissues and activate macrophages. Thus, granulomas are formed. Subsequently other mediators are released by the macrophages which stimulate fibroblasts and ultimately lead to increased extracellular matrix production [35]. A schematic description of this concept is given in **Fig.** 8-**5**. It has recently been possible to induce chronic beryllium disease in guinea pigs [43, 44] and rats [45]. The results of these studies confirmed the view that chronic berylliosis is an immune-system related disease. It is very interesting to note that at least in guinea pigs this immune response to beryllium is genetically controlled. Also in man host factors probably play a decisive role for the onset and issue of chronic beryllium induced lung disease. Studies on this and other [46 to 50] immunepathogenetic mechanisms possibly involved in chronic beryllium disease have gained increased importance since the results obtained may become relevant also for other granulomatous lung diseases, like sarcoidosis.

8.3.4 Carcinogenic Effects

It has been demonstrated already in 1946 by Gardner and Heslington [51] that incorporated beryllium compounds are carcinogenic in rabbits, causing osteogenic sarcomas. This type of bone cancer is frequent neither in animals nor in man. It is well known that it can be induced experimentally by bone-seeking radionuclides but it is rather unique that osteogenic sarcomas are produced by incorporation of a nonradioactive metal. It has meanwhile been confirmed in a series of studies published by different authors (summary in [52 to 54]) that injection of beryllium compounds (intravenously or directly near bone surfaces) or inhalation can cause osteosarcomas in rabbits. Inhalation of various beryllium compounds by rats or their intratracheal instillation led to the occurrence of pulmonary carcinomas. Lung tumors were also found in monkeys that had survived several years after exposures to 35 $\mu g/m^3$ Be (for 1200 to 4000 h). In the same species intrabronchial

References on pp. 313/5

implantation of beryllium oxide also caused lung cancer. From an experimental point of view it is interesting that no tumors were ever observed after beryllium incorporation into guinea pigs or hamsters [52]. The reasons for these species differences in susceptibility are not yet understood, but it has been discussed that differences in immune responses may be involved.

Of course a crucial question is whether carcinogenic effects after exposure to beryllium compounds can be demonstrated also in man. During the past five years several updated reports concerning beryllium carcinogenicity in exposed persons have appeared.

Wagoner et al. [55] studied a cohort of 3055 white male workers from one beryllium extraction, processing, and fabrication facility in Pennsylvania, who were employed between 1942 and 1968. Information about this group regarding date of initial employment and duration of employment is contained in Table 8/6. As an example for detailed epidemiological and statistical analysis performed by the authors, the numbers of observed and expected deaths due to lung cancer according to duration of employment and start of employment are listed in Table 8/7. An excess of lung cancer was noted for the later time intervals

Table 8/6
Distribution of Study Cohort Members According to Calendar Time Period of Initial Employment and Duration of Employment [55].

calendar time period of initial employment	number of employees by duration of employment in months						total No.	medium duration of employment in months
	<1	1 to 5	6 to 11	12 to 35	36 to 59	≧60		
1930 to 1939	0	0	0	7	9	47	63	118.0
1940 to 1949	469	658	266	298	90	224	2005	3.9
1950 to 1959	118	137	64	107	38	221	685	14.5
≧1960	22	53	27	115	52	33	302	20.1
all time periods	609	848	357	527	189	525	3055	7.2

Table 8/7
Observed and Expected Deaths Due to Lung Cancer According to Duration of Employment and Time Since Onset of Employment Among White Males Employed Sometime During January 1942 to September 1968 in a Beryllium Production Facility and Following Through 1975 [55].

interval since onset of employment in years	duration of employment in years a)					
	<5		≧5		total	
	observed	expected	observed	expected	observed	expected
<15	8	7.95	1	1.48	9	9.43
15 to 24	15	11.56	3	2.51	18	14.07
≧25	17	9.07 b)	3	1.72	20	10.79 c)
total	40	28.58 b)	7	5.71	47	34.29 b)

a) Employment histories ascertained only through 1967 to 1968. — b) Significant at $P<0.05$. — c) Significant at $P<0.01$.

since onset of beryllium exposure. Several factors other than beryllium exposure were ruled out as causes of the excess lung cancer cases. The authors concluded that "the findings ... are supportive of the hypothesis that beryllium is carcinogenic in man".

In another study, Infante et al. [56] chose as a cohort all white males who entered in the Beryllium Case Registry while alive during 1952 through 1975. From the total of 421 persons 139 were deceased, of which death certificates could be obtained for 124 cases. The results of the analysis are shown in Table 8/8, which also demonstrates an excess of nonneoplastic lung diseases. The total number of malignant neoplasms was significantly higher than expected. In Table 8/9 the results of an additional analysis are listed with special emphasis on the lung cancer cases, showing that of the 7 lung cancer cases, 6 occurred at an interval greater than 15 years after onset of beryllium exposure where only 2.82 were to be expected. Five of these cases were exposed for less than 1 year.The conclusions of the authors with regard to beryllium carcinogenicity in man were similar to those quoted at the end of the preceeding paragraph.

Table 8/8
Observed and Expected Deaths According to Cause Among White Males Enrolled in the Beryllium Case Registry While Alive July 1, 1952 Through December 31, 1975 [56].

cause of death	observed	expected	SMR [a)]
total malignant	19	12.41	153 [b)]
lung	7	3.30	212
residual	12	9.11	132
heart disease	31	29.88	104
nonneoplastic			
respiratory disease	52	3.17	1640 [c)]
influenza and pneumonia	0	1.55	0
other respiratory disease	52	1.62	3210 [c)]
all other known causes	22	20.43	108
unknown causes	15	—	—
all causes	139	65.89	211 [c)]

a) SMR = standardized mortality ratio = (observed : expected) $\times$ 100.
b) $P < 0.05$.
c) $P < 0.001$.

The third study to be mentioned consists of a survey of 3685 white male employees from beryllium production facilities in Ohio and Pennsylvania, which were compared to workers from the viscose rayon industry [57]. The distribution of the cohorts is shown in Table 8/10. Also in this study a significant excess of lung cancer cases in beryllium exposed workers was seen when compared with those from the rayon industry (Table 8/11).

It is obvious that only a short account of the very detailed and complex epidemiological work on beryllium carcinogenesis in man could be given within the scope of the present review. In spite of some reservations about the results of the epidemiological studies [9, 33] there are strong arguments for the assumption that beryllium compounds are potential carcinogens in man [55 to 59].

References on pp. 313/5

Table 8/9
Observed and Expected Deaths for Respiratory Diseases According to Interval Since Onset of Beryllium Exposure for White Males Entered Into the Beryllium Case Registry While Alive July 1, 1952 Through December 31, 1975 [56].

interval since onset of beryllium exposure in years	cause of death: lung cancer			nonneoplastic respiratory disease [b)]		
	obs.	exp.	SMR [a)]	obs.	exp.	SMR [a)]
less than 15	1	0.49	204	10	0.19	5263 [c)]
greater than or equal to 15	6	2.81	214	42	1.43	2937 [c)]
total	7	3.30	212	52	1.62	3210 [c)]

a) SMR = standardized mortality ratio = (observed : expected) × 100.
b) Excludes influenza and pneumonia.
c) $P < 0.001$.

Table 8/10
Distribution by Age at Hire of the Number Living and Dead for White Males Employed (1937 to 1948) for the Beryllium Manufacturing Plants and for Those Employed (1938 to 1948) in the Viscose Rayon Plant all Followed to End 1976 [57].

age at hire	beryllium industry			rayon industry		
	living	dead	% dead	living	dead	% dead
<25	1241	174	12.3	2459	338	12.1
25 to 34	736	284	27.8	1141	433	27.5
35 to 44	262	326	55.4	366	509	58.2
45 to 54	64	338	84.1	116	378	76.5
55 to 64	16	196	92.5	21	136	86.6
>65	10	38	79.2	2	30	93.8
total	2329	1356	36.8	4105	1824	30.8

8.3.5 Beryllium and Chelating Agents

One of the most thoroughly studied potential antidotes for beryllium intoxication is aurintricarboxylic acid (ATA). In a series of experimental studies Schubert, White, Lindenbaum and coworkers [60 to 64] have shown that ATA injection reduces the acute mortality after injection of lethal doses of beryllium compounds. The protective effect was explained by "local inactivation" of the beryllium due to the formation of inert complexes in the tissues. A protective action of ATA was also reported by Sterner et al. [65]. The same authors have also tested various polyaminopolycarboxylic acids which were all found to be ineffective. The studies on ATA have been reviewed by Tepper et al. [66].

A promising approach to enhance beryllium excretion and to reduce its toxicity, at least from an experimental point of view, seems to be the use of polyaminopolyphosphonic acids.

Table 8/11
Lung Cancer Mortality among Beryllium-Exposed Workers Ages 35 to 74 Years as Contrasted with that Expected on the Basis of Two Cohorts of Workers in the Viscose Rayon Industry Employed for Similar Durations of Time and Followed Over the Same Period of Time [57].

duration of employment in months	lung cancer mortality obs.	exp. [b]	exp. [c]	SMR [a]	SMR [a]
≦12	52	37.60	31.67	138 [d]	164 [b]
13 to 48	14	13.26	10.82	106	129
≧49	14	6.32	8.14	222 [c]	172 [d]
total	80	57.06	50.63	140 [e]	158 [e]

[a] SMR = Standardized mortality ratio = (observed : expected) × 100.
[b] Total viscose rayon employees.
[c] Viscose rayon employees never having transferred from department of initial employment.
[d] $P < 0.05$.
[e] $P < 0.01$.

These chelating agents have also been successfully tested for removing uranium from the body of experimental animals. A detailed description of the theoretical and experimental background for the use of these substances is contained in the Russian monograph edited by Balabucha [67]. Parts of this monograph have already been reviewed in English in the volume of this handbook [68] concerning uranium. Of the various compounds tested, diethylenetriaminepentakis(methyl-phosphonic acid) (DTPP) was the most effective in removing beryllium from the body of mice. It was also superior to other polyaminopolyphosphonic acids in preventing mortality after injection of lethal doses of $BeCl_2$ or $BeSO_4$ into rats, while aurintricarboxylic acid, diethylene-triaminepentaacetate, sodium citrate, and sodium salycilate were completely ineffective at the beryllium dose levels chosen. Chronic toxicity studies with polyamino-polyphosphonic acids [67, 68] indicated that their toxicity may be as low as that of the analogous polyaminopolycarboxylic acids, which have already been successfully used in man. However, no clinical experience with polyaminopolyphosphonic acids in humans with metal intoxications exists. It should finally be mentioned that Basinger et al. [69] recently published results according to which 4,5-dihydroxy-1,3-benzenedisulfonate (Tiron) was a very effective antidote for acute beryllium intoxications in mice.

References:

[1] J. Blake (Ber. Deut. Chem. Ges. **14** [1881] 394/8).
[2] T. L. Brunton, J. T. Cash (Phil. Trans. Roy. Soc. [London] **175** [1884] 197/244).
[3] A. J. Vorwald, A. L. Reeves, E. C. J. Urban (Beryllium Its Ind. Hyg. Aspects **1966** 201/34).
[4] G. Kimmerle (Handb. Exp. Pharmakol. **21** [1966] 1/87).
[5] International Agency for Research on Cancer (IARC Monogr. Eval. Carcinogen. Risk Chem. Human. **23** [1980] 143/204).
[6] H. H. Weber, W. E. Engelhardt (Zentr. Gewerbehyg. Unfallverhüt. **10** [1933] 41/7).
[7] T. L. Shipman, A. Vorwald (Beryllium Its Ind. Hyg. Aspects **1966** 9/17).
[8] M. Eisenbud, J. Lisson (J. Occup. Med. **25** [1983] 196/202).
[9] O. Preuss, H. Oster (Arbeitsmed. Sozialmed. Präventivmed. **15** [1980] 270/5).
[10] J. F. Crowley, J. G. Hamilton, K. G. Scott (J. Biol. Chem. **177** [1949] 975/84).

[11] C. D. Van Cleave, C. T. Kaylor (AMA Arch. Ind. Hyg. Occup. Med. **11** [1955] 375/92).
[12] C. L. Sanders, W. C. Cannon, G. J. Powers, R. R. Adee, D. M. Meier (Arch. Environ. Health **30** [1975] 546/51).
[13] F. W. Klemperer, A. P. Martin, R. E. Liddy (Arch. Biochem. Biophys. **41** [1952] 148/52).
[14] C. D. Van Cleave, C. T. Kaylor (AMA Arch. Ind. Hyg. Occup. Med. **7** [1953] 367/75).
[15] C. T. Kaylor, C. D. Van Cleave (Anat. Record **117** [1953] 467/81).
[16] J. Schubert, M. R. White (J. Lab. Clin. Med. **35** [1950] 854/64).
[17] D. N. Skilleter, R. J. Price (Chem. Biol. Interact. **20** [1978] 383/96).
[18] H. P. Witschi, W. N. Aldridge (Biochem. J. **106** [1968] 811/20).
[19] M. Wiener, A. Seidel (unpublished data).
[20] D. J. Price, J. G. Joshi (Toxicology **31** [1984] 151/63).

[21] A. L. Reeves, A. J. Vorwald (Cancer Res. **27** [1967] 446/51).
[22] A. Taya, G. Hotz, R. Mauser, A. Seidel (Proc. 6th Intern. Symp. Inhaled Part., Cambridge 1985, to be published).
[23] H. Firket (Compt. Rend. Seances Soc. Biol. Ses Fil. **147** [1953] 167/8).
[24] D. H. Groth, C. Kommineni, G. R. Mackay (Environ. Res. **21** [1980] 63/84).
[25] O. G. Archipova (Gig. Tr. Prof. Zabol. **11** [1967] 19/23).
[26] A. J. Vorwald, A. L. Reeves (AMA Arch. Ind. Hyg. Occup. Med. **19** [1959] 190/8).
[27] B. E. Williams, D. N. Skilleter (Biosci. Rept. **3** [1983] 955/62).
[28] V. Bencko, E. V.Vasil'eva (J. Hyg. Epidemiol. Microbiol. Immunol. **27** [1983] 403/17).
[29] J. Vacher, H. B. Stoner (Biochem. Pharmacol. **17** [1968] 93/107).
[30] I. Feldman, J. R. Havill, W. F. Neuman (Arch. Biochem. Biophys. **46** [1953] 443/53).

[31] A. L. Reeves, A. J. Vorwald (J. Occup. Med. **3** [1961] 567/74).
[32] J. A. Hurlbut (RFP-2152 [1974] 1/24; N.S.A. **31** [1975] No. 13610; C.A. **83** [1975] No. 38846; TID-4500-R62 [1974]).
[33] W. R. Parkes (Occupational Lung Disorders, Butterworths, London 1982, pp. 333/58).
[34] A. I. Burnazyana, S. A. Keizer (Berillii: Toksikologiya Gigiena Profilaktika Diagnostika i Lechenie Berillievykh Porazhenii, Energoatomizd., Moscow 1985, pp. 104/10).
[35] R. P. Daniele (in: N. S. Cherniack, N. H. Edelman, Contemporary Issues in Pulmonary Disease, Vol. 2, Churchill Livingstone, New York 1985, pp. 183/92).
[36] G. H. Curtis (AMA Arch. Ind. Health **19** [1959] 150/3).
[37] J. M. Hanifin, W. L. Epstein, M. J. Cline (J. Invest. Dermatol. **55** [1970] 284/8).
[38] S. D. Deodhar, B. Barna, H. S. Van Ordstrand (Chest **63** [1973] 309/13).
[39] O. P. Preuss, S. D. Deodhar, H. S. Van Ordstrand (8th Intern. Conf. Sarcoidosis Other Granulomatous Disord., Cardiff, Wales, 1978 [1980], pp. 711/4).
[40] J. Bargon, H. Kronenberger, L. Bergmann, R. Buhl, J. Meier-Sydow, P. Mitrou (Proc. 4th Congr. Eur. Soc. Pneumology, Milano, Italy, 1985, p. A 14).

[41] W. N. Rom, K. M. Bang, C. Dewitt, R. E. Johns (Arch. Environ. Health **38** [1983] 302/7).
[42] C. D. Price, W. Jones Williams, A. Pugh, D. H. Joynson (J. Clin. Pathol. **30** [1977] 24/8).
[43] B. P. Barna, S. D. Deodhar, T. Chiang, S. Gautam, M. Edinger (Intern. Arch. Allergy Appl. Immunol. **73** [1984] 42/8).
[44] B. P. Barna, S. D. Deodhar, S. Gautam, M. Edinger, T. Chiang, J. T. McMahon (Intern. Arch. Allergy Appl. Immunol. **73** [1984] 49/55).
[45] J.J. Votto, R. W. Barton, J. R. McCormick (Am. Rev. Respir. Disease 124 Suppl. 4 [1984] A 13).
[46] J. G. Hall (Immunology **53** [1984] 105/13).
[47] J. G. Hall, J. O. Spencer (Immunology **53** [1984] 115/20).
[48] J. M. P. Maceira, K. Fukuyama, W. L. Epstein (J. Invest. Dermatol. **83** [1984] 314/6).
[49] K. Behbehani, D. I. Beller, E.R. Unanue (J. Immunol. **134** [1985] 2047/9).

[50] R. J. Price, D. N. Skilleter (Arch. Toxicol. **56** [1985] 207/11).

[51] L. U. Gardner, H. F. Heslington (Fed. Proc., Fed. Am. Soc. Exptl. Biol. **5** [1946] Abstr., p. 221).
[52] A. L. Reeves (Advan. Exptl. Med. Biol. **91** [1977] 13/27).
[53] D. H. Groth (Environ. Res. **21** [1980] 56/62).
[54] M. Kuschner (Environ. Health Perspect. **40** [1981] 101/5).
[55] J. K. Wagoner, P. F. Infante, D. L. Bayliss (Environ. Res. **21** [1980] 15/34).
[56] P. F. Infante, J. K. Wagoner, N. L. Sprince (Environ. Res. **21** [1980] 35/43).
[57] Th. F. Mancuso (Environ. Res. **21** [1980] 48/55).
[58] R. Doll, L. Fishbein, P. Infante, P. Landrigan, J. W. Lloyd, Th. J. Mason, E. Mastromatteo, T. Norseth, G. Pershagen, U. Saffiotti, R. Saracci (Environ. Health Perspect. **40** [1981] 11/20).
[59] D. N. Skilleter (Toxicol. Environ. Chem. **7** [1984] 213/28).
[60] M. R. White, A. J. Finkel, J. Schubert (J. Pharmacol. Exptl. Therap. **102** [1951] 88/93).

[61] J. Schubert, A. Lindenbaum (J. Biol. Chem. **208** [1954] 359/68).
[62] A. Lindenbaum, M. R. White, J. Schubert (Arch. Biochem. Biophys. **52** [1954] 110/32).
[63] M. R. White, J. Schubert (Arch. Biochem. Biophys. **52** [1954] 133/42).
[64] H. Lisco, M. R. White (Brit. J. Exptl. Pathol. **36** [1955] 27/34).
[65] W. Sterner, L. E. Loveless (AMRL-TR-65-135 [1965] 1/29; N66-13576 [1965] 1/29; C.A. **66** [1967] No. 27593).
[66] L. B. Tepper, H. L. Hardy, R. I. Chamberlin (Toxicity of Beryllium Compounds, Elsevier, Amsterdam 1961).
[67] V. S. Balabucha (Uran i Berilli: Problema Vyvedeniya iz Organizma, Atomizdat., Moscow 1976).
[68] A. Seidel (Gmelin Handbook of Inorganic Chemistry "Uranium" A 7, 1982, pp. 300/41).
[69] M. A. Basinger, J. E. Johnson, L. T. Burka, M. M. Jones (Res. Commun. Chem. Pathol. Pharmacol. **36** [1982] 519/22).
[70] H. E. Stockinger (Beryllium Its Industrial Hygiene Aspects, Academic, New York 1966).

[71] J. Bignon, P. Brochard (Rev. Franc. Mal. Respir. **11** [1983] 371/82).

Table of Conversion Factors

Following the notation in Landolt-Börnstein [7], values which have been fixed by convention are indicated by a bold-face last digit. The conversion factor between calorie and Joule that is given here is based on the thermochemical calorie, cal_{thch}, and is defined as 4.1840 J/cal. However, for the conversion of the "Internationale Tafelkalorie", cal_{IT}, into Joule, the factor 4.1868 J/cal is to be used [1, p. 147]. For the conversion factor for the British thermal unit, the Steam Table Btu, Btu_{ST}, is used [1, p. 95].

Force	N	dyn	kp
1 N (Newton)	1	10^5	0.1019716
1 dyn	10^{-5}	1	1.019716×10^{-6}
1 kp	9.8066**5**	9.8066**5** $\times 10^5$	1

Pressure	Pa	bar	kp/m^2	at	atm	Torr	lb/in^2
1 Pa (Pascal) = 1 N/m^2	1	10^{-5}	1.019716×10^{-1}	1.019716×10^{-5}	0.986923×10^{-5}	0.750062×10^{-2}	145.0378×10^{-6}
1 bar = 10^6 dyn/cm^2	10^5	1	10.19716×10^3	1.019716	0.986923	750.062	14.50378
1 kp/m^2 = 1 mm H_2O	9.8066**5**	0.98066**5** $\times 10^{-4}$	1	10^{-4}	0.967841×10^{-4}	0.735559×10^{-1}	1.422335×10^{-3}
1 at = 1 kp/cm^2	0.98066**5** $\times 10^5$	0.98066**5**	10^4	1	0.967841	735.559	14.22335
1 atm = 76**0** Torr	1.0132**5** $\times 10^5$	1.0132**5**	1.033227×10^4	1.033227	1	760	14.69595
1 Torr = 1 mm Hg	133.3224	1.333224×10^{-3}	13.59510	1.359510×10^{-3}	1.315789×10^{-3}	1	19.33678×10^{-3}
1 lb/in^2 = 1 psi	6.89476×10^3	68.9476×10^{-3}	703.069	70.3069×10^{-3}	68.0460×10^{-3}	51.7149	1

Work, Energy, Heat	J	kWh	kcal	Btu	MeV
1 J (Joule) = 1 Ws = 1 Nm = 10^7 erg	1	2.778×10^{-7}	2.39006×10^{-4}	9.4781×10^{-4}	6.242×10^{12}
1 kWh	$3.\mathbf{6} \times 10^{6}$	1	860.4	3412.14	2.247×10^{19}
1 kcal	4184.**0**	1.1622×10^{-3}	1	3.96566	2.6117×10^{16}
1 Btu (British thermal unit)	1055.06	2.93071×10^{-4}	0.25164	1	6.5858×10^{15}
1 MeV	1.602×10^{-13}	4.450×10^{-20}	3.8289×10^{-17}	1.51840×10^{-16}	1

1 eV/mol $\triangleq$ 23.0578 kcal/mol = 96.473 kJ/mol

Power	kW	PS	kp m/s	kcal/s
1 kW = 10^{10} erg/s	1	1.35962	101.972	0.239006
1 PS	0.73550	1	**75**	0.17579
1 kp m/s	$9.8066\mathbf{5} \times 10^{-3}$	0.01333	1	2.34384×10^{-3}
1 kcal/s	4.184**0**	5.6886	426.650	1

References:

[1] A. Sacklowski, Die neuen SI-Einheiten, Goldmann, München 1979. (Conversion tables in an appendix.)
[2] International Union of Pure and Applied Chemistry, Manual of Symbols and Terminology for Physicochemical Quantities and Units, Pergamon, London 1979; Pure Appl. Chem. **51** [1979] 1/41.
[3] The International System of Units (SI), National Bureau of Standards Spec. Publ. No. 330 [1972].
[4] H. Ebert, Physikalisches Taschenbuch, 5th Ed., Vieweg, Wiesbaden 1976.
[5] Kraftwerk Union Information, Technical and Economic Data on Power Engineering, Mülheim/Ruhr 1978.
[6] E. Padelt, H. Laporte, Einheiten und Größenarten der Naturwissenschaften, 3rd Ed., VEB Fachbuchverlag, Leipzig 1976.
[7] Landolt-Börnstein, 6th Ed., Vol. II, Pt. 1, 1971, pp. 1/14.
[8] ISO Standards Handbook 2, Units of Measurement, 2nd Ed., Geneva 1982.